Marvin L. Bittinger • Judith A. Beecher • Barbara L. Johnson

Algebra Foundations
Basic Math, Introductory and Intermediate Algebra

Custom Edition for Lakeland Community College

Taken from:
Algebra Foundations: Basic Math, Introductory and Intermediate Algebra
by Marvin L. Bittinger, Judith A. Beecher, and Barbara L. Johnson

Cover Art: Courtesy of Pearson Learning Solutions.

Taken from:

Algebra Foundations: Basic Math, Introductory and Intermediate Algebra
by Marvin L. Bittinger, Judith A. Beecher, and Barbara L. Johnson
Copyright © 2015 by Pearson Education, Inc.
New York, New York 10013

This special edition published in cooperation with Pearson Learning Solutions.

Pearson Learning Solutions, 330 Hudson Street, New York, New York 10013
A Pearson Education Company
www.pearsoned.com

Printed in the United States of America

000200010272038864

AJ

PEARSON   ISBN 10: 1-323-41812-1
             ISBN 13: 978-1-323-41812-3

Contents

CREDITS

CHAPTER
10

Introduction to Real Numbers and Algebraic Expressions

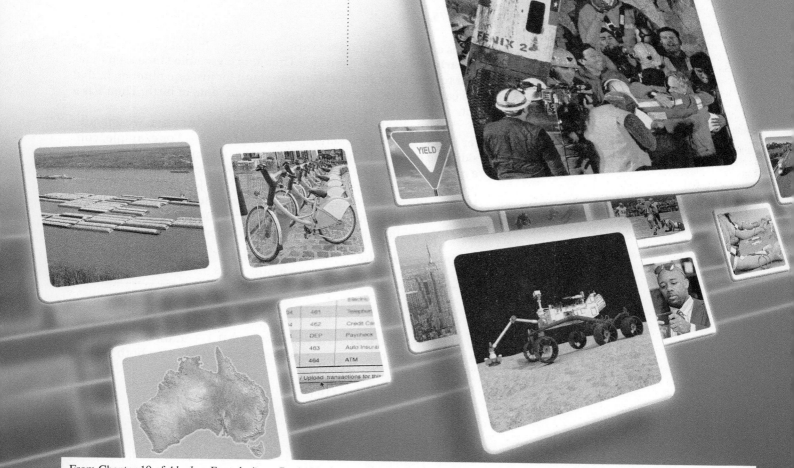

From Chapter 10 of *Algebra Foundations: Basic Mathematics, Introductory Algebra, and Intermediate Algebra*, First Edition.
Marvin J. Bittinger, Judith A. Beecher, and Barbara L. Johnson. Copyright © 2015 by Pearson Education, Inc. All rights reserved.

10.1   Introduction to Algebra

OBJECTIVES

a Evaluate algebraic expressions by substitution.

b Translate phrases to algebraic expressions.

The study of algebra involves the use of equations to solve problems. Equations are constructed from algebraic expressions.

a   EVALUATING ALGEBRAIC EXPRESSIONS

In arithmetic, you have worked with expressions such as

$$49 + 75, \quad 8 \times 6.07, \quad 29 - 14, \quad \text{and} \quad \frac{5}{6}.$$

In algebra, we can use letters to represent numbers and work with *algebraic expressions* such as

$$x + 75, \quad 8 \times y, \quad 29 - t, \quad \text{and} \quad \frac{a}{b}.$$

Sometimes a letter can represent various numbers. In that case, we call the letter a **variable**. Let a = your age. Then a is a variable since a changes from year to year. Sometimes a letter can stand for just one number. In that case, we call the letter a **constant**. Let b = your date of birth. Then b is a constant.

Where do algebraic expressions occur? Most often we encounter them when we are solving applied problems. For example, consider the bar graph shown at left, one that we might find in a book or a magazine. Suppose we want to know how much greater the average population density per square mile is in New Jersey than in Illinois. Using arithmetic, we might simply subtract. But let's see how we can determine this using algebra. We translate the problem into a statement of equality, an equation. It could be done as follows:

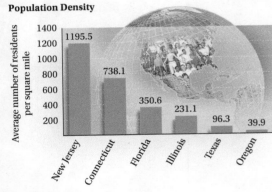

Population Density

Average number of residents per square mile

1400
1200 — 1195.5
1000
800 — 738.1
600
400 — 350.6   231.1
200 — 96.3   39.9

New Jersey   Connecticut   Florida   Illinois   Texas   Oregon

SOURCE: 2010 U.S. Census

Population density in Illinois	plus	How much more	is	Population density in New Jersey
231.1	+	x	=	1195.5.

Note that we have an algebraic expression, $231.1 + x$, on the left of the equals sign. To find the number x, we can subtract 231.1 on both sides of the equation:

$$231.1 + x = 1195.5$$
$$231.1 + x - 231.1 = 1195.5 - 231.1$$
$$x = 964.4.$$

This value of x gives the answer, 964.4 residents per square mile.

We call $231.1 + x$ an *algebraic expression* and $231.1 + x = 1195.5$ an *algebraic equation*. Note that there is no equals sign, $=$, in an algebraic expression.

Do Margin Exercise 1. ▶

An **algebraic expression** consists of variables, constants, numerals, operation signs, and/or grouping symbols. When we replace a variable with a number, we say that we are **substituting** for the variable. When we replace all of the variables in an expression with numbers and carry out the operations in the expression, we are **evaluating the expression**.

EXAMPLE 1 Evaluate $x + y$ when $x = 37$ and $y = 29$.

We substitute 37 for x and 29 for y and carry out the addition:

$$x + y = 37 + 29 = 66.$$

The number 66 is called the **value** of the expression when $x = 37$ and $y = 29$.

Algebraic expressions involving multiplication can be written in several ways. For example, "8 times a" can be written as

$$8 \times a, \quad 8 \cdot a, \quad 8(a), \quad \text{or simply} \quad 8a.$$

Two letters written together without an operation symbol, such as ab, also indicate a multiplication.

EXAMPLE 2 Evaluate $3y$ when $y = 14$.

$$3y = 3(14) = 42$$

Do Exercises 2–4. ▶

EXAMPLE 3 *Area of a Rectangle.* The area A of a rectangle of length l and width w is given by the formula $A = lw$. Find the area when l is 24.5 in. and w is 16 in.

We substitute 24.5 in. for l and 16 in. for w and carry out the multiplication:

$$\begin{aligned} A = lw &= (24.5\text{ in.})(16\text{ in.}) \\ &= (24.5)(16)(\text{in.})(\text{in.}) \\ &= 392\text{ in}^2, \text{ or } 392 \text{ square inches.} \end{aligned}$$

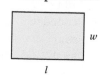

Do Exercise 5. ▶

Algebraic expressions involving division can also be written in several ways. For example, "8 divided by t" can be written as

$$8 \div t, \quad \frac{8}{t}, \quad 8/t, \quad \text{or} \quad 8 \cdot \frac{1}{t},$$

where the fraction bar is a division symbol.

EXAMPLE 4 Evaluate $\dfrac{a}{b}$ when $a = 63$ and $b = 9$.

We substitute 63 for a and 9 for b and carry out the division:

$$\frac{a}{b} = \frac{63}{9} = 7.$$

1. Translate this problem to an equation. Then solve the equation.

   Population Density. The average number of residents per square mile in six U.S. states is shown in the bar graph on the preceding page. How much greater is the population density in Connecticut than in Oregon?

2. Evaluate $a + b$ when $a = 38$ and $b = 26$.

3. Evaluate $x - y$ when $x = 57$ and $y = 29$.

4. Evaluate $4t$ when $t = 15$.

 5. Find the area of a rectangle when l is 24 ft and w is 8 ft.

$$A = lw$$
$$A = (24\text{ ft})(\quad)$$
$$= (24)(\quad)(\text{ft})(\text{ft})$$
$$= 192 \quad, \text{ or}$$
$$192 \text{ square feet}$$

6. Evaluate a/b when $a = 200$ and $b = 8$.

7. Evaluate $10p/q$ when $p = 40$ and $q = 25$.

8. *Commuting via Bicycle.* Find the time it takes to bike 22 mi if the speed is 16 mph.

EXAMPLE 5 Evaluate $\dfrac{12m}{n}$ when $m = 8$ and $n = 16$.

$$\frac{12m}{n} = \frac{12 \cdot 8}{16} = \frac{96}{16} = 6$$

◀ Do Exercises 6 and 7.

EXAMPLE 6 *Commuting Via Bicycle.* Commuting to work via bicycle has increased in popularity with the emerging concept of sharing bicycles. Bikes are picked up and returned at docking stations. The payment is approximately $1.50 per 30 min. Richard bicycles 18 mi to work. The time t, in hours, that it takes to bike 18 mi is given by

$$t = \frac{18}{r},$$

where r is the speed. Find the time for Richard to commute to work if his speed is 15 mph.

We substitute 15 for r and carry out the division:

$$t = \frac{18}{r} = \frac{18}{15} = 1.2 \text{ hr.}$$

◀ Do Exercise 8.

b TRANSLATING TO ALGEBRAIC EXPRESSIONS

We translate problems to equations. The different parts of an equation are translations of word phrases to algebraic expressions. It is easier to translate if we know that certain words often translate to certain operation symbols.

Key Words, Phrases, and Concepts

ADDITION (+)	SUBTRACTION (−)	MULTIPLICATION (·)	DIVISION (÷)
add	subtract	multiply	divide
added to	subtracted from	multiplied by	divided by
sum	difference	product	quotient
total	minus	times	
plus	less than	of	
more than	decreased by		
increased by	take away		

EXAMPLE 7 Translate to an algebraic expression:

    Twice (or two times) some number.

    Think of some number, say, 8. We can write 2 times 8 as 2×8, or $2 \cdot 8$. We multiplied by 2. Do the same thing using a variable. We can use any variable we wish, such as x, y, m, or n. Let's use y to represent some number. If we multiply by 2, we get an expression

    $y \times 2$, $2 \times y$, $2 \cdot y$, or $2y$.

Answers

6. 25 **7.** 16 **8.** 1.375 hr

EXAMPLE 8   Translate to an algebraic expression:

Thirty-eight percent of some number.

Let n = the number. The word "of" translates to a multiplication symbol, so we could write any of the following expressions as a translation:

$$38\% \cdot n, \quad 0.38 \times n, \quad \text{or} \quad 0.38n.$$

EXAMPLE 9   Translate to an algebraic expression:

Seven less than some number.

We let x represent the number. If the number were 10, then 7 less than 10 is $10 - 7$, or 3. If we knew the number to be 34, then 7 less than the number would be $34 - 7$. Thus if the number is x, then the translation is

$$x - 7.$$

EXAMPLE 10   Translate to an algebraic expression:

Eighteen more than a number.

We let t = the number. If the number were 6, then the translation would be $6 + 18$, or $18 + 6$. If we knew the number to be 17, then the translation would be $17 + 18$, or $18 + 17$. Thus if the number is t, then the translation is

$$t + 18, \quad \text{or} \quad 18 + t.$$

EXAMPLE 11   Translate to an algebraic expression:

A number divided by 5.

We let m = the number. If the number were 7, then the translation would be $7 \div 5$, or $7/5$, or $\frac{7}{5}$. If the number were 21, then the translation would be $21 \div 5$, or $21/5$, or $\frac{21}{5}$. If the number is m, then the translation is

$$m \div 5, \quad m/5, \quad \text{or} \quad \frac{m}{5}.$$

EXAMPLE 12   Translate each phrase to an algebraic expression.

PHRASE	ALGEBRAIC EXPRESSION
Five more than some number	$n + 5$, or $5 + n$
Half of a number	$\frac{1}{2}t, \frac{t}{2}$, or $t/2$
Five more than three times some number	$3p + 5$, or $5 + 3p$
The difference of two numbers	$x - y$
Six less than the product of two numbers	$mn - 6$
Seventy-six percent of some number	$76\%z$, or $0.76z$
Four less than twice some number	$2x - 4$

Do Exercises 9–17.

Translate each phrase to an algebraic expression.

9. Eight less than some number

10. Eight more than some number

11. Four less than some number

12. One-third of some number

13. Six more than eight times some number

14. The difference of two numbers

15. Fifty-nine percent of some number

16. Two hundred less than the product of two numbers

17. The sum of two numbers

Answers

9. $x - 8$ 10. $y + 8$, or $8 + y$ 11. $m - 4$
12. $\frac{1}{3} \cdot p$, or $\frac{p}{3}$ 13. $8x + 6$, or $6 + 8x$
14. $a - b$ 15. $59\%x$, or $0.59x$ 16. $xy - 200$
17. $p + q$

☑ Reading Check

Classify each expression as an algebraic expression involving either multiplication or division.

RC1. $3/q$

RC2. $3q$

RC3. $3 \cdot q$

RC4. $\dfrac{3}{q}$

a   Substitute to find values of the expressions in each of the following applied problems.

1. *Commuting Time.* It takes Abigail 24 min less time to commute to work than it does Jayden. Suppose that the variable x stands for the time it takes Jayden to get to work. Then $x - 24$ stands for the time it takes Abigail to get to work. How long does it take Abigail to get to work if it takes Jayden 56 min? 93 min? 105 min?

2. *Enrollment Costs.* At Mountain View Community College, it costs $600 to enroll in the 8 A.M. section of Elementary Algebra. Suppose that the variable n stands for the number of students who enroll. Then $600n$ stands for the total amount of tuition collected for this course. How much is collected if 34 students enroll? 78 students? 250 students?

3. *Distance Traveled.* A driver who drives at a constant speed of r miles per hour for t hours will travel a distance of d miles given by $d = rt$ miles. How far will a driver travel at a speed of 65 mph for 4 hr?

4. *Simple Interest.* The simple interest I on a principal of P dollars at interest rate r for time t, in years, is given by $I = Prt$. Find the simple interest on a principal of $4800 at 3% for 2 years.

5. *Wireless Internet Sign.* The U.S. Department of Transportation has designed a new sign that indicates the availability of wireless internet. The square sign measures 24 in. on each side. Find its area.

Source: *Manual of Uniform Traffic Control Devices*, U.S. Department of Transportation, 2009

6. *Yield Sign.* The U.S. Department of Transportation has designed a new yield sign. Each side of the triangular sign measures 30 in., and the height of the triangle is 26 in. Find its area. The area of a triangle with base b and height h is given by $A = \frac{1}{2}bh$.

Source: *Manual of Uniform Traffic Control Devices*, U.S. Department of Transportation, 2009

7. *Area of a Triangle.* The area A of a triangle with base b and height h is given by $A = \frac{1}{2}bh$. Find the area when $b = 45$ m (meters) and $h = 86$ m.

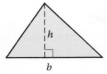

8. *Area of a Parallelogram.* The area A of a parallelogram with base b and height h is given by $A = bh$. Find the area of the parallelogram when the height is 15.4 cm (centimeters) and the base is 6.5 cm.

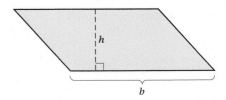

Evaluate.

9. $8x$, when $x = 7$

10. $6y$, when $y = 7$

11. $\dfrac{c}{d}$, when $c = 24$ and $d = 3$

12. $\dfrac{p}{q}$, when $p = 16$ and $q = 2$

13. $\dfrac{3p}{q}$, when $p = 2$ and $q = 6$

14. $\dfrac{5y}{z}$, when $y = 15$ and $z = 25$

15. $\dfrac{x + y}{5}$, when $x = 10$ and $y = 20$

16. $\dfrac{p + q}{2}$, when $p = 2$ and $q = 16$

17. $\dfrac{x - y}{8}$, when $x = 20$ and $y = 4$

18. $\dfrac{m - n}{5}$, when $m = 16$ and $n = 6$

b   Translate each phrase to an algebraic expression. Use any letter for the variable(s) unless directed otherwise.

19. Seven more than some number

20. Some number increased by thirteen

21. Twelve less than some number

22. Fourteen less than some number

23. b more than a

24. c more than d

25. x divided by y

26. c divided by h

27. x plus w

28. s added to t

29. m subtracted from n

30. p subtracted from q

31. Twice some number

32. Three times some number

33. Three multiplied by some number

34. The product of eight and some number

35. Six more than four times some number

36. Two more than six times some number

37. Eight less than the product of two numbers

38. The product of two numbers minus seven

39. Five less than twice some number

40. Six less than seven times some number

41. Three times some number plus eleven

42. Some number times 8 plus 5

43. The sum of four times a number plus three times another number

44. Five times a number minus eight times another number

45. Your salary after a 5% salary increase if your salary before the increase was s

46. The price of a chain saw after a 30% reduction if the price before the reduction was P

47. Aubrey drove at a speed of 65 mph for t hours. How far did she travel? (See Exercise 3.)

48. Liam drove his pickup truck at 55 mph for t hours. How far did he travel? (See Exercise 3.)

49. Lisa had $50 before spending x dollars on pizza. How much money remains?

50. Juan has d dollars before spending $820 on four new tires for his truck. How much did Juan have after the purchase?

51. Sid's part-time job pays $12.50 per hour. How much does he earn for working n hours?

52. Meredith pays her babysitter $10 per hour. What does it cost her to hire the sitter for m hours?

Synthesis

To the student and the instructor: The Synthesis exercises found at the end of most exercise sets challenge students to combine concepts or skills studied in that section or in preceding parts of the text.

Evaluate.

53. $\dfrac{a - 2b + c}{4b - a}$, when $a = 20$, $b = 10$, and $c = 5$

54. $\dfrac{x}{y} - \dfrac{5}{x} + \dfrac{2}{y}$, when $x = 30$ and $y = 6$

55. $\dfrac{12 - c}{c + 12b}$, when $b = 1$ and $c = 12$

56. $\dfrac{2w - 3z}{7y}$, when $w = 5$, $y = 6$, and $z = 1$

The Real Numbers

A **set** is a collection of objects. For our purposes, we will most often be considering sets of numbers. One way to name a set uses what is called **roster notation**. For example, roster notation for the set containing the numbers 0, 2, and 5 is $\{0, 2, 5\}$.

Sets that are part of other sets are called **subsets**. In this section, we become acquainted with the set of *real numbers* and its various subsets.

Two important subsets of the real numbers are listed below using roster notation.

NATURAL NUMBERS

The set of **natural numbers** $= \{1, 2, 3, \dots\}$. These are the numbers used for counting.

WHOLE NUMBERS

The set of **whole numbers** $= \{0, 1, 2, 3, \dots\}$. This is the set of natural numbers and 0.

We can represent these sets on the number line. The natural numbers are to the right of zero. The whole numbers are the natural numbers and zero.

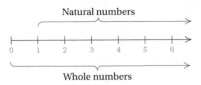

We create a new set, called the *integers*, by starting with the whole numbers, 0, 1, 2, 3, and so on. For each natural number 1, 2, 3, and so on, we obtain a new number to the left of zero on the number line:

For the number 1, there will be an *opposite* number -1 (negative 1).

For the number 2, there will be an *opposite* number -2 (negative 2).

For the number 3, there will be an *opposite* number -3 (negative 3), and so on.

The **integers** consist of the whole numbers and these new numbers.

INTEGERS

The set of **integers** $= \{\dots, -5, -4, -3, -2, -1, 0, 1, 2, 3, 4, 5, \dots\}$.

We picture the integers on the number line as follows.

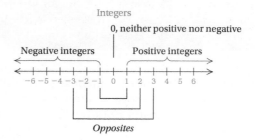

Integers

0, neither positive nor negative

Negative integers   Positive integers

Opposites

We call the integers to the left of zero **negative integers**. The natural numbers are also called **positive integers**. Zero is neither positive nor negative. We call −1 and 1 **opposites** of each other. Similarly, −2 and 2 are opposites, −3 and 3 are opposites, −100 and 100 are opposites, and 0 is its own opposite. Pairs of opposite numbers like −3 and 3 are the same distance from zero. The integers extend infinitely on the number line to the left and right of zero.

a INTEGERS AND THE REAL WORLD

Integers correspond to many real-world problems and situations. The following examples will help you get ready to translate problem situations that involve integers to mathematical language.

EXAMPLE 1 Tell which integer corresponds to this situation: The temperature is 4 degrees below zero.

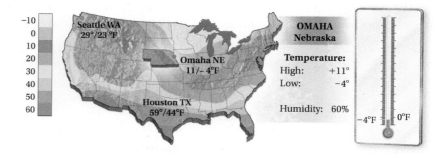

The integer −4 corresponds to the situation. The temperature is −4°.

EXAMPLE 2 *Water Level.* Tell which integer corresponds to this situation: As the water level of the Mississippi River fell during the drought of 2012, barge traffic was restricted, causing a severe decline in shipping volumes. On August 24, the river level at Greenville, Mississippi, was 10 ft below normal.

Source: Rick Jervis, *USA TODAY*, August 24, 2012

The integer −10 corresponds to the drop in water level.

EXAMPLE 3 *Stock Price Change.* Tell which integers correspond to this situation: Hal owns a stock whose price decreased $16 per share over a recent period. He owns another stock whose price increased $2 per share over the same period.

The integer -16 corresponds to the decrease in the value of the first stock. The integer 2 represents the increase in the value of the second stock.

Do Exercises 1–5. ▶

b THE RATIONAL NUMBERS

We created the set of integers by obtaining a negative number for each natural number and also including 0. To create a larger number system, called the set of **rational numbers**, we consider quotients of integers with nonzero divisors. The following are some examples of rational numbers:

$$\frac{2}{3}, \quad -\frac{2}{3}, \quad \frac{7}{1}, \quad 4, \quad -3, \quad 0, \quad \frac{23}{-8}, \quad 2.4, \quad -0.17, \quad 10\frac{1}{2}.$$

The number $-\frac{2}{3}$ (read "negative two-thirds") can also be named $\frac{-2}{3}$ or $\frac{2}{-3}$; that is,

$$-\frac{a}{b} = \frac{-a}{b} = \frac{a}{-b}.$$

The number 2.4 can be named $\frac{24}{10}$ or $\frac{12}{5}$, and -0.17 can be named $-\frac{17}{100}$. We can describe the set of rational numbers as follows.

RATIONAL NUMBERS

The set of **rational numbers** = the set of numbers $\frac{a}{b}$,

where a and b are integers and b is not equal to 0 ($b \neq 0$).

Note that this new set of numbers, the rational numbers, contains the whole numbers, the integers, the arithmetic numbers (also called the nonnegative rational numbers), and the negative rational numbers.

We picture the rational numbers on the number line as follows.

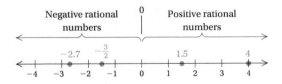

To **graph** a number means to find and mark its point on the number line. Some rational numbers are graphed in the preceding figure.

Tell which integers correspond to each situation.

1. *Temperature High and Low.* The highest recorded temperature in Illinois is 117°F on July 14, 1954, in East St. Louis. The lowest recorded temperature in Illinois is 36°F below zero on January 5, 1999, in Congerville.

   Source: National Climate Data Center, NESDIS, NOAA, U.S. Department of Commerce (through 2010)

2. *Stock Decrease.* The price of a stock decreased $3 per share over a recent period.

3. At 10 sec before liftoff, ignition occurs. At 148 sec after liftoff, the first stage is detached from the rocket.

4. The halfback gained 8 yd on first down. The quarterback was sacked for a 5-yd loss on second down.

5. A submarine dove 120 ft, rose 50 ft, and then dove 80 ft.

Answers

1. 117; −36   2. −3   3. −10; 148
4. 8; −5   5. −120; 50; −80

Graph each number on the number line.

6. $-\dfrac{7}{2}$

<---+----+----+----+----+----+----+----+----+----+----+----+--->
  -6 -5 -4 -3 -2 -1 0 1 2 3 4 5 6

7. 1.4

<---+----+----+----+----+----+----+----+----+----+----+----+--->
  -6 -5 -4 -3 -2 -1 0 1 2 3 4 5 6

8. $-\dfrac{11}{4}$

<---+----+----+----+----+----+----+----+----+----+----+----+--->
  -6 -5 -4 -3 -2 -1 0 1 2 3 4 5 6

EXAMPLES Graph each number on the number line.

4. -3.2 The graph of -3.2 is $\frac{2}{10}$ of the way from -3 to -4.

5. $\dfrac{13}{8}$ The number $\frac{13}{8}$ can also be named $1\frac{5}{8}$, or 1.625. The graph is $\frac{5}{8}$ of the way from 1 to 2.

<---+----+----+----+----+----+----+--->
  -3 -2 -1   0   1   2   3

(mark at $\frac{13}{8}$)

◀ Do Exercises 6–8.

C NOTATION FOR RATIONAL NUMBERS

Each rational number can be named using fraction notation or decimal notation.

EXAMPLE 6 Convert to decimal notation: $-\frac{5}{8}$.

We first find decimal notation for $\frac{5}{8}$. Since $\frac{5}{8}$ means $5 \div 8$, we divide.

```
      0.6 2 5
  8 ) 5.0 0 0
      4 8
        2 0
        1 6
          4 0
          4 0
            0
```

Thus, $\frac{5}{8} = 0.625$, so $-\frac{5}{8} = -0.625$.

Decimal notation for $-\frac{5}{8}$ is -0.625. We consider -0.625 to be a **terminating decimal**. Decimal notation for some numbers repeats.

EXAMPLE 7 Convert to decimal notation: $\frac{7}{11}$.

```
       0.6 3 6 3 ...        Dividing
  1 1 ) 7.0 0 0 0
        6 6
          4 0
          3 3
            7 0
            6 6
              4 0
              3 3
                7
```

We can abbreviate **repeating decimal** notation by writing a bar over the repeating part—in this case, we write $0.\overline{63}$. Thus, $\frac{7}{11} = 0.\overline{63}$.

Answers

6. $-\frac{7}{2}$

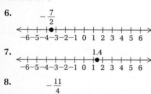

7. 1.4

<---+----+----+----+----+----+----+----+----+----+----+--->
  -6 -5 -4 -3 -2 -1 0 1 2 3 4 5 6

8. $-\frac{11}{4}$

<---+----+----+----+----+----+----+----+----+----+----+--->
  -6 -5 -4 -3 -2 -1 0 1 2 3 4 5 6

Each rational number can be expressed in either terminating decimal notation or repeating decimal notation.

The following are other examples showing how rational numbers can be named using fraction notation or decimal notation:

$$0 = \frac{0}{8}, \quad \frac{27}{100} = 0.27, \quad -8\frac{3}{4} = -8.75, \quad -\frac{13}{6} = -2.1\overline{6}.$$

Do Exercises 9–11. ▶

d THE REAL NUMBERS AND ORDER

Every rational number has a point on the number line. However, there are some points on the line for which there is no rational number. These points correspond to what are called **irrational numbers**.

What kinds of numbers are irrational? One example is the number π, which is used in finding the area and the circumference of a circle: $A = \pi r^2$ and $C = 2\pi r$.

Another example of an irrational number is the square root of 2, named $\sqrt{2}$. It is the length of the diagonal of a square with sides of length 1. It is also the number that when multiplied by itself gives 2—that is, $\sqrt{2} \cdot \sqrt{2} = 2$. There is no rational number that can be multiplied by itself to get 2. But the following are rational *approximations*:

1.4 is an approximation of $\sqrt{2}$ because $(1.4)^2 = 1.96$;

1.41 is a better approximation because $(1.41)^2 = 1.9881$;

1.4142 is an even better approximation because $(1.4142)^2 = 1.99996164$.

We can find rational approximations for square roots using a calculator.

Decimal notation for rational numbers *either* terminates *or* repeats.
Decimal notation for irrational numbers *neither* terminates *nor* repeats.

Some other examples of irrational numbers are $\sqrt{3}$, $-\sqrt{8}$, $\sqrt{11}$, and 0.121221222122221. . . . Whenever we take the square root of a number that is not a perfect square, we will get an irrational number.

The rational numbers and the irrational numbers together correspond to all the points on the number line and make up what is called the **real-number system**.

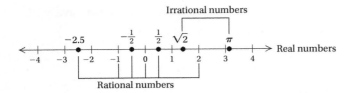

Find decimal notation.

9. $-\frac{3}{8}$

10. $-\frac{6}{11}$

11. $\frac{4}{3}$

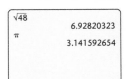
Answers

9. -0.375 10. $-0.\overline{54}$ 11. $1.\overline{3}$

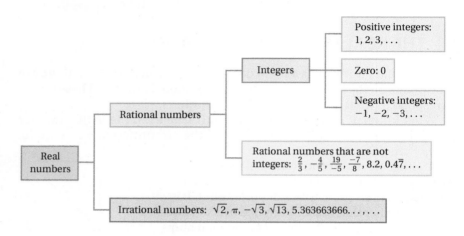

REAL NUMBERS

The set of **real numbers** = The set of all numbers corresponding to points on the number line.

The real numbers consist of the rational numbers and the irrational numbers. The following figure shows the relationships among various kinds of numbers.

CALCULATOR CORNER

Negative Numbers on a Calculator; Converting to Decimal Notation We use the opposite key (-) to enter negative numbers on a graphing calculator. Note that this is different from the subtraction key, (−).

To convert $-\frac{5}{8}$ to decimal notation, we press (-) (5) (÷) (8) **ENTER**. The result is −0.625.

```
-5/8
                -.625
```

EXERCISES: Convert to decimal notation.

1. $-\dfrac{3}{4}$

2. $-\dfrac{9}{20}$

3. $-\dfrac{1}{8}$

4. $-\dfrac{9}{5}$

5. $-\dfrac{27}{40}$

6. $-\dfrac{11}{16}$

7. $-\dfrac{7}{2}$

8. $-\dfrac{19}{25}$

Order

Real numbers are named in order on the number line, increasing as we move from left to right. For any two numbers on the line, the one on the left is less than the one on the right.

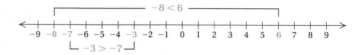

We use the symbol **<** to mean "**is less than**." The sentence $-8 < 6$ means "-8 is less than 6." The symbol **>** means "**is greater than**." The sentence $-3 > -7$ means "-3 is greater than -7." The sentences $-8 < 6$ and $-3 > -7$ are **inequalities**.

EXAMPLES Use either < or > for ☐ to write a true sentence.

8. 2 ☐ 9   Since 2 is to the left of 9, 2 is less than 9, so $2 < 9$.

9. −7 ☐ 3   Since −7 is to the left of 3, we have $-7 < 3$.

10. 6 ☐ −12   Since 6 is to the right of −12, then $6 > -12$.

11. −18 ☐ −5   Since −18 is to the left of −5, we have $-18 < -5$.

12. −2.7 ☐ $-\frac{3}{2}$ The answer is $-2.7 < -\frac{3}{2}$.

13. 1.5 ☐ −2.7   The answer is $1.5 > -2.7$.

14. 1.38 ☐ 1.83   The answer is $1.38 < 1.83$.

15. $-3.45 \,\square\, 1.32$ The answer is $-3.45 < 1.32$.

16. $-4 \,\square\, 0$ The answer is $-4 < 0$.

17. $5.8 \,\square\, 0$ The answer is $5.8 > 0$.

18. $\frac{5}{8} \,\square\, \frac{7}{11}$ We convert to decimal notation: $\frac{5}{8} = 0.625$ and $\frac{7}{11} = 0.6363 \ldots$. Thus, $\frac{5}{8} < \frac{7}{11}$.

19. $-\frac{1}{2} \,\square\, -\frac{1}{3}$ The answer is $-\frac{1}{2} < -\frac{1}{3}$.

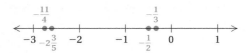

20. $-2\frac{3}{5} \,\square\, -\frac{11}{4}$ The answer is $-2\frac{3}{5} > -\frac{11}{4}$.

Do Exercises 12–19. ▶

Note that both $-8 < 6$ and $6 > -8$ are true. Every true inequality yields another true inequality when we interchange the numbers or the variables and reverse the direction of the inequality sign.

ORDER; $>$, $<$

$a < b$ also has the meaning $b > a$.

EXAMPLES Write another inequality with the same meaning.

21. $-3 > -8$ The inequality $-8 < -3$ has the same meaning.

22. $a < -5$ The inequality $-5 > a$ has the same meaning.

A helpful mental device is to think of an inequality sign as an "arrow" with the arrowhead pointing to the smaller number.

Do Exercises 20 and 21. ▶

Note that all positive real numbers are greater than zero and all negative real numbers are less than zero.

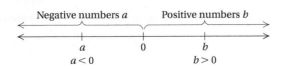

If b is a positive real number, then $b > 0$.

If a is a negative real number, then $a < 0$.

Use either $<$ or $>$ for $\square$ to write a true sentence.

12. $-3 \,\square\, 7$

13. $-8 \,\square\, -5$

14. $7 \,\square\, -10$

15. $3.1 \,\square\, -9.5$

16. $-4.78 \,\square\, -5.01$

17. $-\dfrac{2}{3} \,\square\, -\dfrac{5}{9}$

18. $-\dfrac{11}{8} \,\square\, \dfrac{23}{15}$

19. $0 \,\square\, -9.9$

Write another inequality with the same meaning.

20. $-5 < 7$

21. $x > 4$

Answers

12. $<$ **13.** $<$ **14.** $>$ **15.** $>$ **16.** $>$
17. $<$ **18.** $<$ **19.** $>$ **20.** $7 > -5$
21. $4 < x$

Write true or false for each statement.

22. $-4 \le -6$

23. $7.8 \ge 7.8$

24. $-2 \le \dfrac{3}{8}$

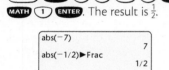

CALCULATOR CORNER

Absolute Value Finding absolute value is the first item in the Catalog on the T1-84 Plus graphing calculator. To find $|-7|$, we first press **2ND** **CATALOG** **ENTER** to copy "abs(" to the home screen. (**CATALOG** is the second operation associated with the **0** numeric key.) Then we press **(-)** **7** **)** **ENTER**. The result is 7.

To find $\left|-\frac{1}{2}\right|$ and express the result as a fraction, we press **2ND** **CATALOG** **ENTER** **(-)** **1** **÷** **2** **)** **MATH** **1** **ENTER**. The result is $\frac{1}{2}$.

```
abs(-7)
                    7
abs(-1/2)▶Frac
                  1/2
```

EXERCISES: Find the absolute value.

1. $|-5|$ **2.** $|17|$

3. $|0|$ **4.** $|6.48|$

5. $|-12.7|$ **6.** $|-0.9|$

7. $\left|-\dfrac{5}{7}\right|$ **8.** $\left|\dfrac{4}{3}\right|$

Find the absolute value.

25. $|8|$ **26.** $|-9|$

27. $\left|-\dfrac{2}{3}\right|$ **28.** $|5.6|$

Answers

22. False   **23.** True   **24.** True   **25.** 8
26. 9   **27.** $\dfrac{2}{3}$ **28.** 5.6

Expressions like $a \le b$ and $b \ge a$ are also inequalities. We read $\boldsymbol{a \le b}$ as "***a* is less than or equal to *b*.**" We read $\boldsymbol{a \ge b}$ as "***a* is greater than or equal to *b*.**"

EXAMPLES Write true or false for each statement.

23. $-3 \le 5.4$ True since $-3 < 5.4$ is true

24. $-3 \le -3$ True since $-3 = -3$ is true

25. $-5 \ge 1\frac{2}{3}$ False since neither $-5 > 1\frac{2}{3}$ nor $-5 = 1\frac{2}{3}$ is true

◀ Do Exercises 22–24.

e ABSOLUTE VALUE

From the number line, we see that numbers like 4 and -4 are the same distance from zero. Distance is always a nonnegative number. We call the distance of a number from zero on the number line the **absolute value** of the number.

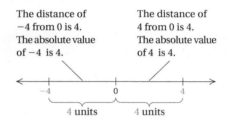

The distance of -4 from 0 is 4. The absolute value of -4 is 4.

The distance of 4 from 0 is 4. The absolute value of 4 is 4.

4 units   4 units

ABSOLUTE VALUE

The **absolute value** of a number is its distance from zero on the number line. We use the symbol $|x|$ to represent the absolute value of a number x.

FINDING ABSOLUTE VALUE

a) If a number is negative, its absolute value is its opposite.

b) If a number is positive or zero, its absolute value is the same as the number.

EXAMPLES Find the absolute value.

26. $|-7|$ The distance of -7 from 0 is 7, so $|-7| = 7$.

27. $|12|$ The distance of 12 from 0 is 12, so $|12| = 12$.

28. $|0|$ The distance of 0 from 0 is 0, so $|0| = 0$.

29. $\left|\frac{3}{2}\right| = \frac{3}{2}$

30. $|-2.73| = 2.73$

◀ Do Exercises 25–28.

✓ Reading Check

Use the number line below for Exercises RC1–RC10.

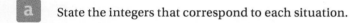

Match each number with its graph.

RC1. $-2\dfrac{5}{7}$ **RC2.** $\left|\dfrac{0}{-8}\right|$ **RC3.** -2.25 **RC4.** $\dfrac{17}{3}$ **RC5.** $|-4|$ **RC6.** $3.\overline{4}$

Write true or false. The letters name numbers on the number line shown above.

RC7. $K < B$ **RC8.** $H < B$ **RC9.** $E < C$ **RC10.** $J > D$

a   State the integers that correspond to each situation.

1. On Wednesday, the temperature was 24° above zero. On Thursday, it was 2° below zero.

2. A student deposited her tax refund of $750 in a savings account. Two weeks later, she withdrew $125 to pay technology fees.

3. *Temperature Extremes.*   The highest temperature ever created in a lab is 7,200,000,000,000°F. The lowest temperature ever created is approximately 460°F below zero.

   Sources: *Live Science; Guinness Book of World Records*

4. *Extreme Climate.*   Verkhoyansk, a river port in northeast Siberia, has the most extreme climate on the planet. Its average monthly winter temperature is 58.5°F below zero, and its average monthly summer temperature is 56.5°F.

   Source: *Guinness Book of World Records*

5. *Empire State Building.*   The Empire State Building has a total height, including the lightning rod at the top, of 1454 ft. The foundation depth is 55 ft below ground level.

   Source: www.empirestatebuildingfacts.com

6. *Shipwreck.*   There are numerous shipwrecks to explore near Bermuda. One of the most frequently visited sites is L'Herminie, a French warship that sank in 1838. This ship is 35 ft below the surface.

   Source: www./10best.com/interests/adventure/ scuba-diving-in-pirate-territory/

b Graph the number on the number line.

7. $\dfrac{10}{3}$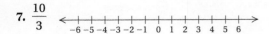

8. $-\dfrac{17}{4}$

9. -5.2

10. 4.78

11. $-4\dfrac{2}{5}$

12. $2\dfrac{6}{11}$

c Convert to decimal notation.

13. $-\dfrac{7}{8}$

14. $-\dfrac{3}{16}$

15. $\dfrac{5}{6}$

16. $\dfrac{5}{3}$

17. $-\dfrac{7}{6}$

18. $-\dfrac{5}{12}$

19. $\dfrac{2}{3}$

20. $-\dfrac{11}{9}$

21. $\dfrac{1}{10}$

22. $\dfrac{1}{4}$

23. $-\dfrac{1}{2}$

24. $\dfrac{9}{8}$

25. $\dfrac{4}{25}$

26. $-\dfrac{7}{20}$

d Use either $<$ or $>$ for $\square$ to write a true sentence.

27. $8 \ \square \ 0$

28. $3 \ \square \ 0$

29. $-8 \ \square \ 3$

30. $6 \ \square \ -6$

31. $-8 \ \square \ 8$

32. $0 \ \square \ -9$

33. $-8 \ \square \ -5$

34. $-4 \ \square \ -3$

35. $-5 \ \square \ -11$

36. $-3 \ \square \ -4$

37. $2.14 \ \square \ 1.24$

38. $-3.3 \ \square \ -2.2$

39. -12.88 □ -6.45

40. 17.2 □ -1.67

41. $-\dfrac{1}{2}$ □ $-\dfrac{2}{3}$

42. $-\dfrac{5}{4}$ □ $-\dfrac{3}{4}$

43. $-\dfrac{2}{3}$ □ $\dfrac{1}{3}$

44. $\dfrac{3}{4}$ □ $-\dfrac{5}{4}$

45. $\dfrac{5}{12}$ □ $\dfrac{11}{25}$

46. $-\dfrac{13}{16}$ □ $-\dfrac{5}{9}$

Write an inequality with the same meaning.

47. $-6 > x$

48. $x < 8$

49. $-10 \leq y$

50. $12 \geq t$

Write true or false.

51. $-5 \leq -6$

52. $-7 \geq -10$

53. $4 \geq 4$

54. $7 \leq 7$

55. $-3 \geq -11$

56. $-1 \leq -5$

57. $0 \geq 8$

58. $-5 \leq 7$

e Find the absolute value.

59. $|-3|$

60. $|-6|$

61. $|11|$

62. $|0|$

63. $\left|-\dfrac{2}{3}\right|$

64. $|325|$

65. $\left|\dfrac{0}{4}\right|$

66. $|14.8|$

67. $|-2.65|$

68. $\left|-3\dfrac{5}{8}\right|$

Skill Maintenance

This heading indicates that the exercises that follow are Skill Maintenance exercises, which review any skill previously studied in the text. You can expect such exercises in every exercise set. Answers to *all* skill maintenance exercises are found at the back of the book. If you miss an exercise, restudy the objective shown in red.

Evaluate.　[10.1a]

69. $\dfrac{5c}{d}$, when $c = 15$ and $d = 25$

70. $\dfrac{2x + y}{3}$, when $x = 12$ and $y = 9$

71. $\dfrac{q - r}{8}$, when $q = 30$ and $r = 6$

72. $\dfrac{w}{4y}$, when $w = 52$ and $y = 13$

Synthesis

List in order from the least to the greatest.

73. $\dfrac{2}{3}, -\dfrac{1}{7}, \dfrac{1}{3}, -\dfrac{2}{7}, -\dfrac{2}{3}, \dfrac{2}{5}, -\dfrac{1}{3}, -\dfrac{2}{5}, \dfrac{9}{8}$

74. $-8\dfrac{7}{8}, 7^1, -5, |-6|, 4, |3|, -8\dfrac{5}{8}, -100, 0, 1^7, \dfrac{7}{2}, -\dfrac{67}{8}$

Given that $0.\overline{3} = \frac{1}{3}$ and $0.\overline{6} = \frac{2}{3}$, express each of the following as a quotient or a ratio of two integers.

75. $0.\overline{9}$

76. $0.\overline{1}$

77. $5.\overline{5}$

10.3

Addition of Real Numbers

OBJECTIVES

a Add real numbers without using the number line.

b Find the opposite, or additive inverse, of a real number.

c Solve applied problems involving addition of real numbers.

SKILL TO REVIEW

Objective 10.1a: Evaluate algebraic expressions by substitution.

1. Evaluate $t - h$ when $t = 1$ and $h = 0.05$.

2. Evaluate $44 - 9q$ when $q = 3$.

In this section, we consider addition of real numbers. First, to gain an understanding, we add using the number line. Then we consider rules for addition.

ADDITION ON THE NUMBER LINE

To do the addition $a + b$ on the number line, start at 0, move to a, and then move according to b.

a) If b is positive, move from a to the right.

b) If b is negative, move from a to the left.

c) If b is 0, stay at a.

EXAMPLE 1 Add: $3 + (-5)$.

We start at 0 and move to 3. Then we move 5 units left since -5 is negative.

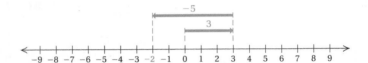

$$3 + (-5) = -2$$

EXAMPLE 2 Add: $-4 + (-3)$.

We start at 0 and move to -4. Then we move 3 units left since -3 is negative.

$$-4 + (-3) = -7$$

EXAMPLE 3 Add: $-4 + 9$.

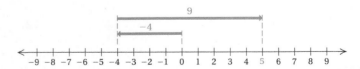

$$-4 + 9 = 5$$

Answers

Skill to Review:
1. 0.95 **2.** 17

EXAMPLE 4 Add: $-5.2 + 0$.

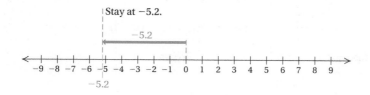

$-5.2 + 0 = -5.2$

Do Exercises 1–6. ▶

a ADDING WITHOUT THE NUMBER LINE

You may have noticed some patterns in the preceding examples. These lead us to rules for adding without using the number line that are more efficient for adding larger numbers.

RULES FOR ADDITION OF REAL NUMBERS
....

1. *Positive numbers*: Add the same as arithmetic numbers. The answer is positive.
2. *Negative numbers*: Add absolute values. The answer is negative.
3. *A positive number and a negative number*:
   - If the numbers have the same absolute value, the answer is 0.
   - If the numbers have different absolute values, subtract the smaller absolute value from the larger. Then:
     a) If the positive number has the greater absolute value, the answer is positive.
     b) If the negative number has the greater absolute value, the answer is negative.
4. *One number is zero*: The sum is the other number.

Rule 4 is known as the **identity property of 0.** It says that for any real number a, $a + 0 = a$.

EXAMPLES Add without using the number line.

5. $-12 + (-7) = -19$ Two negatives. Add the absolute values: $|-12| + |-7| = 12 + 7 = 19$. Make the answer *negative*: -19.

6. $-1.4 + 8.5 = 7.1$ One negative, one positive. Find the absolute values: $|-1.4| = 1.4$; $|8.5| = 8.5$. Subtract the smaller absolute value from the larger: $8.5 - 1.4 = 7.1$. The *positive* number, 8.5, has the larger absolute value, so the answer is *positive*: 7.1.

7. $-36 + 21 = -15$ One negative, one positive. Find the absolute values: $|-36| = 36$; $|21| = 21$. Subtract the smaller absolute value from the larger: $36 - 21 = 15$. The *negative* number, -36, has the larger absolute value, so the answer is *negative*: -15.

Add using the number line.

1. $0 + (-3)$

2. $1 + (-4)$

3. $-3 + (-2)$

4. $-3 + 7$

5. $-2.4 + 2.4$

6. $-\dfrac{5}{2} + \dfrac{1}{2}$

Answers

1. -3 **2.** -3 **3.** -5
4. 4 **5.** 0 **6.** -2

Add without using the number line.

7. $-5 + (-6)$ 8. $-9 + (-3)$

9. $-4 + 6$ 10. $-7 + 3$

11. $5 + (-7)$ 12. $-20 + 20$

13. $-11 + (-11)$ 14. $10 + (-7)$

15. $-0.17 + 0.7$ 16. $-6.4 + 8.7$

17. $-4.5 + (-3.2)$

18. $-8.6 + 2.4$

19. $\dfrac{5}{9} + \left(-\dfrac{7}{9}\right)$

20. $-\dfrac{1}{5} + \left(-\dfrac{3}{4}\right)$

$= -\dfrac{4}{20} + \left(-\dfrac{\boxed{}}{20}\right)$

$= -\dfrac{19}{\boxed{}}$

Add.

21. $(-15) + (-37) + 25 + 42 + (-59) + (-14)$

22. $42 + (-81) + (-28) + 24 + 18 + (-31)$

23. $-2.5 + (-10) + 6 + (-7.5)$

24. $-35 + 17 + 14 + (-27) + 31 + (-12)$

8. $1.5 + (-1.5) = 0$ The numbers have the same absolute value. The sum is 0.

9. $-\dfrac{7}{8} + 0 = -\dfrac{7}{8}$ One number is zero. The sum is $-\dfrac{7}{8}$.

10. $-9.2 + 3.1 = -6.1$

11. $-\dfrac{3}{2} + \dfrac{9}{2} = \dfrac{6}{2} = 3$

12. $-\dfrac{2}{3} + \dfrac{5}{8} = -\dfrac{16}{24} + \dfrac{15}{24} = -\dfrac{1}{24}$

◀ Do Exercises 7–20.

Suppose we want to add several numbers, some positive and some negative, as follows. How can we proceed?

$$15 + (-2) + 7 + 14 + (-5) + (-12)$$

We can change grouping and order as we please when adding. For instance, we can group the positive numbers together and the negative numbers together and add them separately. Then we add the two results.

EXAMPLE 13   Add: $15 + (-2) + 7 + 14 + (-5) + (-12)$.

a) $15 + 7 + 14 = 36$ Adding the positive numbers

b) $-2 + (-5) + (-12) = -19$ Adding the negative numbers

$36 + (-19) = 17$ Adding the results in (a) and (b)

We can also add the numbers in any other order we wish—say, from left to right—as follows:

$$15 + (-2) + 7 + 14 + (-5) + (-12) = 13 + 7 + 14 + (-5) + (-12)$$
$$= 20 + 14 + (-5) + (-12)$$
$$= 34 + (-5) + (-12)$$
$$= 29 + (-12)$$
$$= 17$$

◀ Do Exercises 21–24.

b OPPOSITES, OR ADDITIVE INVERSES

Suppose we add two numbers that are **opposites**, such as 6 and -6. The result is 0. When opposites are added, the result is always 0. Opposites are also called **additive inverses**. Every real number has an opposite, or additive inverse.

> **OPPOSITES, OR ADDITIVE INVERSES**
>
> Two numbers whose sum is 0 are called **opposites**, or **additive inverses**, of each other.

Answers

7. -11 8. -12 9. 2 10. -4
11. -2 12. 0 13. -22 14. 3
15. 0.53 16. 2.3 17. -7.7 18. -6.2
19. $-\dfrac{2}{9}$ 20. $-\dfrac{19}{20}$ 21. -58 22. -56
23. -14 24. -12

Guided Solution:
20. $15, 20$

EXAMPLES Find the opposite, or additive inverse, of each number.

14. 34   The opposite of 34 is -34 because $34 + (-34) = 0$.

15. -8 The opposite of -8 is 8 because $-8 + 8 = 0$.

16. 0   The opposite of 0 is 0 because $0 + 0 = 0$.

17. $-\dfrac{7}{8}$ The opposite of $-\dfrac{7}{8}$ is $\dfrac{7}{8}$ because $-\dfrac{7}{8} + \dfrac{7}{8} = 0$.

Do Exercises 25–30. ▶

To name the opposite, we use the symbol $-$, as follows.

> ### SYMBOLIZING OPPOSITES
>
> The opposite, or additive inverse, of a number a can be named $-a$
> (read "the opposite of a," or "the additive inverse of a").

Note that if we take a number, say, 8, and find its opposite, -8, and then find the opposite of the result, we will have the original number, 8, again.

> ### THE OPPOSITE OF AN OPPOSITE
>
> The **opposite of the opposite** of a number is the number itself.
> (The additive inverse of the additive inverse of a number is the
> number itself.) That is, for any number a,
>
> $$-(-a) = a.$$

EXAMPLE 18   Evaluate $-x$ and $-(-x)$ when $x = 16$.

If $x = 16$, then $-x = -16$.   The opposite of 16 is -16.

If $x = 16$, then $-(-x) = -(-16) = 16$.   The opposite of the opposite of 16 is 16.

EXAMPLE 19   Evaluate $-x$ and $-(-x)$ when $x = -3$.

If $x = -3$, then $-x = -(-3) = 3$.

If $x = -3$, then $-(-x) = -(-(-3)) = -(3) = -3$.

Note that in Example 19 we used a second set of parentheses to show that we are substituting the negative number -3 for x. Symbolism like $--x$ is not considered meaningful.

Do Exercises 31–34. ▶

A symbol such as -8 is usually read "negative 8." It could be read "the additive inverse of 8," because the additive inverse of 8 is negative 8. It could also be read "the opposite of 8," because the opposite of 8 is -8. Thus a symbol like -8 can be read in more than one way. It is never correct to read -8 as "minus 8."

·· **Caution!** ··

A symbol like $-x$, which has a variable, should be read "the opposite of x" or "the additive inverse of x" and *not* "negative x," because we do not know whether x represents a positive number, a negative number, or 0. You can check this in Examples 18 and 19.

··

Find the opposite, or additive inverse, of each number.

25. -4 **26.** 8.7

27. -7.74 **28.** $-\dfrac{8}{9}$

29. 0   **30.** 12

Evaluate $-x$ and $-(-x)$ when:

31. $x = 14$.

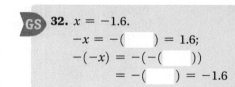

GS **32.** $x = -1.6$.
$-x = -(\;\;\;\;) = 1.6;$
$-(-x) = -(-(\;\;\;\;))$
$= -(\;\;\;\;) = -1.6$

33. $x = \dfrac{2}{3}$.   **34.** $x = -\dfrac{9}{8}$.

Answers

25. 4   **26.** -8.7 **27.** 7.74   **28.** $\dfrac{8}{9}$

29. 0   **30.** -12 **31.** $-14; 14$

32. 1.6; -1.6 **33.** $-\dfrac{2}{3}; \dfrac{2}{3}$ **34.** $\dfrac{9}{8}; -\dfrac{9}{8}$

Guided Solution:
32. $-1.6; -1.6, 1.6$

We can use the symbolism $-a$ to restate the definition of opposite, or additive inverse.

OPPOSITES, OR ADDITIVE INVERSES

For any real number a, the **opposite**, or **additive inverse**, of a, denoted $-a$, is such that

$$a + (-a) = (-a) + a = 0.$$

Signs of Numbers

A negative number is sometimes said to have a "negative sign." A positive number is said to have a "positive sign." When we replace a number with its opposite, we can say that we have "changed its sign."

EXAMPLES Find the opposite. (Change the sign.)

20. -3 $-(-3) = 3$

21. $-\dfrac{2}{13}$ $-\left(-\dfrac{2}{13}\right) = \dfrac{2}{13}$

22. 0 $-(0) = 0$

23. 14 $-(14) = -14$

◀ Do Exercises 35–38.

C APPLICATIONS AND PROBLEM SOLVING

Addition of real numbers occurs in many real-world situations.

EXAMPLE 24 *Banking Transactions.* On August 1st, Martias checks his bank account balance on his phone and sees that it is $54. During the next week, the following transactions were recorded: a debit-card purchase of $71, an overdraft fee of $29, a direct deposit of $160, and an ATM withdrawal of $80. What is Martias's balance at the end of the week?

We let $B =$ the ending balance of the bank account. Then the problem translates to the following:

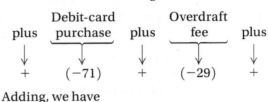

Ending balance	is	Beginning balance	plus	Debit-card purchase	plus	Overdraft fee	plus	Direct deposit	plus	ATM withdrawal
B	$=$	54	$+$	(-71)	$+$	(-29)	$+$	160	$+$	$(-80).$

Adding, we have

$B = 54 + (-71) + (-29) + 160 + (-80)$

$= 214 + (-180)$ Adding the positive numbers and adding the negative numbers

$= 34.$

Martias's balance at the end of the week was $34.

◀ Do Exercise 39.

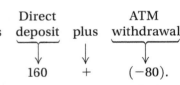
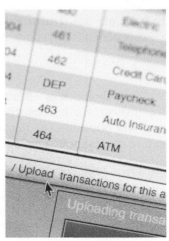

Find the opposite. (Change the sign.)

35. -4

36. -13.4

37. 0

38. $\dfrac{1}{4}$

39. *Change in Class Size.* During the first two weeks of the semester in Jim's algebra class, 4 students withdrew, 8 students enrolled late, and 6 students were dropped as "no shows." By how many students had the class size changed at the end of the first two weeks?

Answers

35. 4   **36.** 13.4   **37.** 0

38. $-\dfrac{1}{4}$ **39.** -2 students

☑ Reading Check

Fill in each blank with either "left" or "right" so that the statements describe the steps when adding numbers with the number line.

RC1. To add 7 + 2, start at 0, move _____ to 7, and then move 2 units _____. The sum is 9.

RC2. To add −3 + (−5), start at 0, move _____ to −3, and then move 5 units _____. The sum is −8.

RC3. To add 4 + (−6), start at 0, move _____ to 4, and then move 6 units _____. The sum is −2.

RC4. To add −8 + 3, start at 0, move _____ to −8, and then move 3 units _____. The sum is −5.

a Add. Do not use the number line except as a check.

1. $2 + (-9)$ **2.** $-5 + 2$ **3.** $-11 + 5$ **4.** $4 + (-3)$ **5.** $-6 + 6$

6. $8 + (-8)$ **7.** $-3 + (-5)$ **8.** $-4 + (-6)$ **9.** $-7 + 0$ **10.** $-13 + 0$

11. $0 + (-27)$ **12.** $0 + (-35)$ **13.** $17 + (-17)$ **14.** $-15 + 15$ **15.** $-17 + (-25)$

16. $-24 + (-17)$ **17.** $18 + (-18)$ **18.** $-13 + 13$ **19.** $-28 + 28$ **20.** $11 + (-11)$

21. $8 + (-5)$ **22.** $-7 + 8$ **23.** $-4 + (-5)$ **24.** $10 + (-12)$ **25.** $13 + (-6)$

26. $-3 + 14$ **27.** $-25 + 25$ **28.** $50 + (-50)$ **29.** $53 + (-18)$ **30.** $75 + (-45)$

31. $-8.5 + 4.7$ **32.** $-4.6 + 1.9$ **33.** $-2.8 + (-5.3)$ **34.** $-7.9 + (-6.5)$ **35.** $-\dfrac{3}{5} + \dfrac{2}{5}$

36. $-\dfrac{4}{3} + \dfrac{2}{3}$ **37.** $-\dfrac{2}{9} + \left(-\dfrac{5}{9}\right)$ **38.** $-\dfrac{4}{7} + \left(-\dfrac{6}{7}\right)$ **39.** $-\dfrac{5}{8} + \dfrac{1}{4}$ **40.** $-\dfrac{5}{6} + \dfrac{2}{3}$

41. $-\dfrac{5}{8} + \left(-\dfrac{1}{6}\right)$ **42.** $-\dfrac{5}{6} + \left(-\dfrac{2}{9}\right)$ **43.** $-\dfrac{3}{8} + \dfrac{5}{12}$ **44.** $-\dfrac{7}{16} + \dfrac{7}{8}$

45. $-\dfrac{1}{6} + \dfrac{7}{10}$ **46.** $-\dfrac{11}{18} + \left(-\dfrac{3}{4}\right)$ **47.** $\dfrac{7}{15} + \left(-\dfrac{1}{9}\right)$ **48.** $-\dfrac{4}{21} + \dfrac{3}{14}$

49. $76 + (-15) + (-18) + (-6)$ **50.** $29 + (-45) + 18 + 32 + (-96)$

51. $-44 + \left(-\dfrac{3}{8}\right) + 95 + \left(-\dfrac{5}{8}\right)$ **52.** $24 + 3.1 + (-44) + (-8.2) + 63$

b   Find the opposite, or additive inverse.

53. 24 **54.** -64 **55.** -26.9 **56.** 48.2

Evaluate $-x$ when:

57. $x = 8.$ **58.** $x = -27.$ **59.** $x = -\dfrac{13}{8}.$ **60.** $x = \dfrac{1}{236}.$

Evaluate $-(-x)$ when:

61. $x = -43.$ **62.** $x = 39.$ **63.** $x = \dfrac{4}{3}.$ **64.** $x = -7.1.$

Find the opposite. (Change the sign.)

65. -24 **66.** -12.3 **67.** $-\dfrac{3}{8}$ **68.** 10

c   Solve.

69. *Tallest Mountain.*   The tallest mountain in the world, when measured from base to peak, is Mauna Kea (White Mountain) in Hawaii. From its base 19,684 ft below sea level in the Hawaiian Trough, it rises 33,480 ft. What is the elevation of the peak above sea level?

Source: *The Guinness Book of Records*

70. *Copy Center Account.*   Rachel's copy-center bill for July was $327. She made a payment of $200 and then made $48 worth of copies in August. How much did she then owe on her account?

71. _Temperature Changes._ One day, the temperature in Lawrence, Kansas, is 32°F at 6:00 A.M. It rises 15° by noon, but falls 50° by midnight when a cold front moves in. What is the final temperature?

72. _Stock Changes._ On a recent day, the price of a stock opened at a value of $61.38. During the day, it rose $4.75, dropped $7.38, and rose $5.13. Find the value of the stock at the end of the day.

73. _"Flipping" Houses._ Buying run-down houses, fixing them up, and reselling them is referred to as "flipping" houses. Charlie and Sophia bought and sold four houses in a recent year. The profits and losses are shown in the following bar graph. Find the sum of the profits and losses.

Flipping Houses

74. _Football Yardage._ In a college football game, the quarterback attempted passes with the following results. Find the total gain or loss.

ATTEMPT	GAIN OR LOSS
1st	13-yd gain
2nd	12-yd loss
3rd	21-yd gain

75. _Credit-Card Bills._ On August 1, Lyle's credit-card bill shows that he owes $470. During the month of August, Lyle makes a payment of $45 to the credit-card company, charges another $160 in merchandise, and then pays off another $500 of his bill. What is the new amount that Lyle owes at the end of August?

76. _Account Balance._ Emma has $460 in a checking account. She uses her debit card for a purchase of $530, makes a deposit of $75, and then writes a check for $90. What is the balance in her account?

Skill Maintenance

77. Evaluate $5a - 2b$ when $a = 9$ and $b = 3$. [10.1a]

78. Write an inequality with the same meaning as the inequality $-3 < y$. [10.2d]

Convert to decimal notation. [10.2c]

79. $-\dfrac{1}{12}$

80. $\dfrac{5}{8}$

Find the absolute value. [10.2e]

81. $|0|$

82. $|-21.4|$

Synthesis

83. For what numbers x is $-x$ negative?

84. For what numbers x is $-x$ positive?

85. If a is positive and b is negative, then $-a + b$ is which of the following?

  A. Positive   **B.** Negative

  C. 0   **D.** Cannot be determined without more information

86. If $a = b$ and a and b are negative, then $-a + (-b)$ is which of the following?

  A. Positive   **B.** Negative

  C. 0   **D.** Cannot be determined without more information

10.4   Subtraction of Real Numbers

OBJECTIVES

a   Subtract real numbers and simplify combinations of additions and subtractions.

b   Solve applied problems involving subtraction of real numbers.

a   SUBTRACTION

We now consider subtraction of real numbers.

SUBTRACTION

The difference $a - b$ is the number c for which $a = b + c$.

Consider, for example, $45 - 17$. *Think*: What number can we add to 17 to get 45? Since $45 = 17 + 28$, we know that $45 - 17 = 28$. Let's consider an example whose answer is a negative number.

EXAMPLE 1   Subtract: $3 - 7$.

Think: What number can we add to 7 to get 3? The number must be negative. Since $7 + (-4) = 3$, we know the number is -4: $3 - 7 = -4$. That is, $3 - 7 = -4$ because $7 + (-4) = 3$.

◀ Do Exercises 1–3.

The definition above does not provide the most efficient way to do subtraction. We can develop a faster way to subtract. As a rationale for the faster way, let's compare $3 + 7$ and $3 - 7$ on the number line.

To find $3 + 7$ on the number line, we start at 0, move to 3, and then move 7 units farther to the right since 7 is positive.

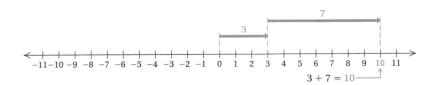

To find $3 - 7$, we do the "opposite" of adding 7: We move 7 units to the *left* to do the subtracting. This is the same as *adding* the opposite of 7, -7, to 3.

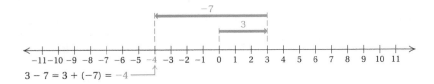

◀ Do Exercises 4–6.

Look for a pattern in the examples shown at right.

SUBTRACTING	ADDING AN OPPOSITE
$5 - 8 = -3$	$5 + (-8) = -3$
$-6 - 4 = -10$	$-6 + (-4) = -10$
$-7 - (-2) = -5$	$-7 + 2 = -5$

Subtract.

1. $-6 - 4$

   Think: What number can be added to 4 to get -6:
   $$\square + 4 = -6?$$

2. $-7 - (-10)$

   Think: What number can be added to -10 to get -7:
   $$\square + (-10) = -7?$$

3. $-7 - (-2)$

   Think: What number can be added to -2 to get -7:
   $$\square + (-2) = -7?$$

Subtract. Use the number line, doing the "opposite" of addition.

4. $5 - 9$

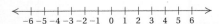

5. $-3 - 2$

6. $-4 - (-3)$

Answers

1. -10 2. 3   3. -5 4. -4
5. -5 6. -1

Do Exercises 7–10. ▶

Perhaps you have noticed that we can subtract by adding the opposite of the number being subtracted. This can always be done.

SUBTRACTING BY ADDING THE OPPOSITE

For any real numbers a and b,

$$a - b = a + (-b).$$

(To subtract, add the opposite, or additive inverse, of the number being subtracted.)

Complete the addition and compare with the subtraction.

7. $4 - 6 = -2$;
$4 + (-6) = $ _____

8. $-3 - 8 = -11$;
$-3 + (-8) = $ _____

9. $-5 - (-9) = 4$;
$-5 + 9 = $ _____

10. $-5 - (-3) = -2$;
$-5 + 3 = $ _____

This is the method generally used for quick subtraction of real numbers.

EXAMPLES Subtract.

2. $2 - 6 = 2 + (-6) = -4$ The opposite of 6 is -6. We change the subtraction to addition and add the opposite. *Check*: $-4 + 6 = 2$.

3. $4 - (-9) = 4 + 9 = 13$ The opposite of -9 is 9. We change the subtraction to addition and add the opposite. *Check*: $13 + (-9) = 4$.

4. $-4.2 - (-3.6) = -4.2 + 3.6 = -0.6$ Adding the opposite. *Check*: $-0.6 + (-3.6) = -4.2$.

5. $-\dfrac{1}{2} - \left(-\dfrac{3}{4}\right) = -\dfrac{1}{2} + \dfrac{3}{4}$ Adding the opposite.
$$= -\dfrac{2}{4} + \dfrac{3}{4} = \dfrac{1}{4}$$
Check: $\dfrac{1}{4} + \left(-\dfrac{3}{4}\right) = -\dfrac{1}{2}$.

Subtract.

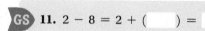 **11.** $2 - 8 = 2 + ($ ☐ $) = $ ☐

12. $-6 - 10$

13. $12.4 - 5.3$

14. $-8 - (-11)$

15. $-8 - (-8)$

16. $\dfrac{2}{3} - \left(-\dfrac{5}{6}\right)$

Do Exercises 11–16. ▶

EXAMPLES Subtract by adding the opposite of the number being subtracted.

6. $3 - 5$ *Think*: "Three minus five is three plus the opposite of five"
$$3 - 5 = 3 + (-5) = -2$$

7. $\dfrac{1}{8} - \dfrac{7}{8}$ *Think*: "One-eighth minus seven-eighths is one-eighth plus the opposite of seven-eighths"
$$\dfrac{1}{8} - \dfrac{7}{8} = \dfrac{1}{8} + \left(-\dfrac{7}{8}\right) = -\dfrac{6}{8}, \text{ or } -\dfrac{3}{4}$$

8. $-4.6 - (-9.8)$ *Think*: "Negative four point six minus negative nine point eight is negative four point six plus the opposite of negative nine point eight"
$$-4.6 - (-9.8) = -4.6 + 9.8 = 5.2$$

9. $-\dfrac{3}{4} - \dfrac{7}{5}$ *Think*: "Negative three-fourths minus seven-fifths is negative three-fourths plus the opposite of seven-fifths"
$$-\dfrac{3}{4} - \dfrac{7}{5} = -\dfrac{3}{4} + \left(-\dfrac{7}{5}\right) = -\dfrac{15}{20} + \left(-\dfrac{28}{20}\right) = -\dfrac{43}{20}$$

Subtract by adding the opposite of the number being subtracted.

17. $3 - 11$

18. $12 - 5$

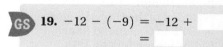 **19.** $-12 - (-9) = -12 + $ ☐
$$= \text{☐}$$

20. $-12.4 - 10.9$

21. $-\dfrac{4}{5} - \left(-\dfrac{4}{5}\right)$

Answers

7. -2 8. -11 9. 4   10. -2 11. -6
12. -16 13. 7.1   14. 3   15. 0   16. $\dfrac{3}{2}$
17. -8 18. 7   19. -3 20. -23.3 21. 0
Guided Solutions:
11. $-8, -6$ 19. $9, -3$

Do Exercises 17–21. ▶

When several additions and subtractions occur together, we can make them all additions.

EXAMPLES Simplify.

10. $8 - (-4) - 2 - (-4) + 2 = 8 + 4 + (-2) + 4 + 2$ Adding the opposite
$$= 16$$

11. $8.2 - (-6.1) + 2.3 - (-4) = 8.2 + 6.1 + 2.3 + 4 = 20.6$

12. $\dfrac{3}{4} - \left(-\dfrac{1}{12}\right) - \dfrac{5}{6} - \dfrac{2}{3} = \dfrac{9}{12} + \dfrac{1}{12} + \left(-\dfrac{10}{12}\right) + \left(-\dfrac{8}{12}\right)$

$$= \dfrac{9 + 1 + (-10) + (-8)}{12}$$

$$= \dfrac{-8}{12} = -\dfrac{8}{12} = -\dfrac{2}{3}$$

Simplify.

22. $-6 - (-2) - (-4) - 12 + 3$

23. $\dfrac{2}{3} - \dfrac{4}{5} - \left(-\dfrac{11}{15}\right) + \dfrac{7}{10} - \dfrac{5}{2}$

24. $-9.6 + 7.4 - (-3.9) - (-11)$

◀ Do Exercises 22–24.

b APPLICATIONS AND PROBLEM SOLVING

Let's now see how we can use subtraction of real numbers to solve applied problems.

EXAMPLE 13 *Surface Temperatures on Mars.* Surface temperatures on Mars vary from −128°C during polar night to 27°C at the equator during midday at the closest point in orbit to the sun. Find the difference between the highest value and the lowest value in this temperature range.
Source: Mars Institute

We let D = the difference in the temperatures. Then the problem translates to the following subtraction:

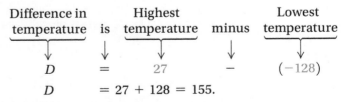

Difference in temperature	is	Highest temperature	minus	Lowest temperature
D	$=$	27	$-$	(-128)

$D = 27 + 128 = 155.$

The difference in the temperatures is 155°C.

◀ Do Exercise 25.

25. *Temperature Extremes.* The highest temperature ever recorded in the United States is 134°F in Greenland Ranch, California, on July 10, 1913. The lowest temperature ever recorded is −80°F in Prospect Creek, Alaska, on January 23, 1971. How much higher was the temperature in Greenland Ranch than the temperature in Prospect Creek?

Source: National Oceanographic and Atmospheric Administration

Answers

22. −9 **23.** $-\dfrac{6}{5}$ **24.** 12.7 **25.** 214°F

✓ **Reading Check**

Match the expression with an expression from the column on the right that names the same number.

RC1. $18 - 6$

RC2. $-18 - (-6)$

RC3. $-18 - 6$

RC4. $18 - (-6)$

a) $18 + 6$
b) $-18 + 6$
c) $18 + (-6)$
d) $-18 + (-6)$

a Subtract.

1. $2 - 9$

2. $3 - 8$

3. $-8 - (-2)$

4. $-6 - (-8)$

5. $-11 - (-11)$

6. $-6 - (-6)$

7. $12 - 16$

8. $14 - 19$

9. $20 - 27$

10. $30 - 4$

11. $-9 - (-3)$

12. $-7 - (-9)$

13. $-40 - (-40)$

14. $-9 - (-9)$

15. $7 - (-7)$

16. $4 - (-4)$

17. $8 - (-3)$

18. $-7 - 4$

19. $-6 - 8$

20. $6 - (-10)$

21. $-4 - (-9)$

22. $-14 - 2$

23. $-6 - (-5)$

24. $-4 - (-3)$

25. $8 - (-10)$

26. $5 - (-6)$

27. $-5 - (-2)$

28. $-3 - (-1)$

29. $-7 - 14$

30. $-9 - 16$

31. $0 - (-5)$

32. $0 - (-1)$

33. $-8 - 0$

34. $-9 - 0$

35. $7 - (-5)$

36. $7 - (-4)$

37. $2 - 25$

38. $18 - 63$

39. $-42 - 26$

40. $-18 - 63$

41. $-71 - 2$

42. $-49 - 3$

43. $24 - (-92)$

44. $48 - (-73)$

45. $-50 - (-50)$

46. $-70 - (-70)$

47. $-\dfrac{3}{8} - \dfrac{5}{8}$

48. $\dfrac{3}{9} - \dfrac{9}{9}$

49. $\dfrac{3}{4} - \dfrac{2}{3}$

50. $\dfrac{5}{8} - \dfrac{3}{4}$

51. $-\dfrac{3}{4} - \dfrac{2}{3}$

52. $-\dfrac{5}{8} - \dfrac{3}{4}$

53. $-\dfrac{5}{8} - \left(-\dfrac{3}{4}\right)$

54. $-\dfrac{3}{4} - \left(-\dfrac{2}{3}\right)$

55. $6.1 - (-13.8)$

56. $1.5 - (-3.5)$

57. $-2.7 - 5.9$

58. $-3.2 - 5.8$

59. $0.99 - 1$

60. $0.87 - 1$

61. $-79 - 114$

62. $-197 - 216$

63. $0 - (-500)$

64. $500 - (-1000)$

65. $-2.8 - 0$

66. $6.04 - 1.1$

67. $7 - 10.53$

68. $8 - (-9.3)$

69. $\dfrac{1}{6} - \dfrac{2}{3}$

70. $-\dfrac{3}{8} - \left(-\dfrac{1}{2}\right)$

71. $-\dfrac{4}{7} - \left(-\dfrac{10}{7}\right)$

72. $\dfrac{12}{5} - \dfrac{12}{5}$

73. $-\dfrac{7}{10} - \dfrac{10}{15}$

74. $-\dfrac{4}{18} - \left(-\dfrac{2}{9}\right)$

75. $\dfrac{1}{5} - \dfrac{1}{3}$

76. $-\dfrac{1}{7} - \left(-\dfrac{1}{6}\right)$

77. $\dfrac{5}{12} - \dfrac{7}{16}$

78. $-\dfrac{1}{35} - \left(-\dfrac{9}{40}\right)$

79. $-\dfrac{2}{15} - \dfrac{7}{12}$

80. $\dfrac{2}{21} - \dfrac{9}{14}$

Simplify.

81. $18 - (-15) - 3 - (-5) + 2$

82. $22 - (-18) + 7 + (-42) - 27$

83. $-31 + (-28) - (-14) - 17$

84. $-43 - (-19) - (-21) + 25$

85. $-34 - 28 + (-33) - 44$

86. $39 + (-88) - 29 - (-83)$

87. $-93 - (-84) - 41 - (-56)$

88. $84 + (-99) + 44 - (-18) - 43$

89. $-5.4 - (-30.9) + 30.8 + 40.2 - (-12)$

90. $14.9 - (-50.7) + 20 - (-32.8)$

91. $-\dfrac{7}{12} + \dfrac{3}{4} - \left(-\dfrac{5}{8}\right) - \dfrac{13}{24}$

92. $-\dfrac{11}{16} + \dfrac{5}{32} - \left(-\dfrac{1}{4}\right) + \dfrac{7}{8}$

b Solve.

93. *Elevations in Asia.* The elevation of the highest point in Asia, Mt. Everest, Nepal–Tibet, is 29,035 ft. The lowest elevation, at the Dead Sea, Israel–Jordan, is −1348 ft. What is the difference in the elevations of the two locations?

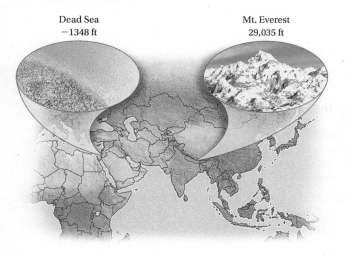

Dead Sea
−1348 ft

Mt. Everest
29,035 ft

94. *Ocean Depth.* The deepest point in the Pacific Ocean is the Marianas Trench, with a depth of 10,924 m. The deepest point in the Atlantic Ocean is the Puerto Rico Trench, with a depth of 8605 m. What is the difference in the elevation of the two trenches?

Source: *The World Almanac and Book of Facts*

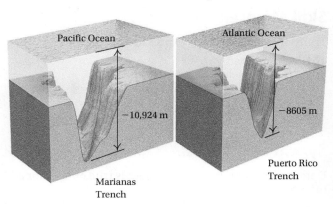

95. Francisca has a charge of $476.89 on her credit card, but she then returns a sweater that cost $128.95. How much does she now owe on her credit card?

96. Jacob has $825 in a checking account. What is the balance in his account after he has written a check for $920 to pay for a laptop?

97. *Difference in Elevation.* The highest elevation in Japan is 3776 m above sea level at Fujiyama. The lowest elevation in Japan is 4 m below sea level at Hachirogata. Find the difference in the elevations.

Source: *The CIA World Factbook* 2012

98. *Difference in Elevation.* The lowest elevation in North America—Death Valley, California—is 282 ft below sea level. The highest elevation in North America—Mount McKinley, Alaska—is 20,320 ft above sea level. Find the difference in elevation between the highest point and the lowest point.

Source: National Geographic Society

99. *Low Points on Continents.* The lowest point in Africa is Lake Assal, which is 512 ft below sea level. The lowest point in South America is the Valdes Peninsula, which is 131 ft below sea level. How much lower is Lake Assal than the Valdes Peninsula?

Source: National Geographic Society

100. *Temperature Records.* The greatest recorded temperature change in one 24-hr period occurred between January 14 and January 15, 1972, in Loma, Montana, where the temperature rose from $-54°$F to $49°$F. By how much did the temperature rise?

Source: National Weather Service

101. *Surface Temperature on Mercury.* Surface temperatures on Mercury vary from $840°$F on the equator when the planet is closest to the sun to $-290°$F at night. Find the difference between these two temperatures.

102. *Run Differential.* In baseball, the difference between the number of runs that a team scores and the number of runs that it allows its opponents to score is called the *run differential*. That is,

$$\text{Run differential} = \frac{\text{Number of}}{\text{runs scored}} - \frac{\text{Number of}}{\text{runs allowed}}.$$

Teams strive for a positive run differential.

Source: Major League Baseball

a) In a recent season, the Kansas City Royals scored 676 runs and allowed 746 runs to be scored on them. Find the run differential.

b) In a recent season, the Atlanta Braves scored 700 runs and allowed 600 runs to be scored on them. Find the run differential.

Skill Maintenance

Translate to an algebraic expression.   [10.1b]

103. 7 more than y

104. 41 less than t

105. h subtracted from a

106. The product of 6 and c

107. r more than s

108. x less than y

Synthesis

Determine whether each statement is true or false for all integers a and b. If false, give an example to show why. Examples may vary.

109. $a - 0 = 0 - a$

110. $0 - a = a$

111. If $a \neq b$, then $a - b \neq 0$.

112. If $a = -b$, then $a + b = 0$.

113. If $a + b = 0$, then a and b are opposites.

114. If $a - b = 0$, then $a = -b$.

Mid-Chapter Review

Concept Reinforcement

Determine whether each statement is true or false.

_____ **1.** All rational numbers can be named using fraction notation. [10.2c]

_____ **2.** If $a > b$, then a lies to the left of b on the number line. [10.2d]

_____ **3.** The absolute value of a number is always nonnegative. [10.2e]

_____ **4.** We can translate "7 less than y" as $7 - y$. [10.1b]

Guided Solutions

 Fill in each blank with the number that creates a correct statement or solution.

5. Evaluate $-x$ and $-(-x)$ when $x = -4$. [10.3b]

$$-x = -(\boxed{}) = \boxed{};$$
$$-(-x) = -(-(\boxed{})) = -(\boxed{}) = \boxed{}$$

Subtract. [10.4a]

6. $5 - 13 = 5 + (\boxed{}) = \boxed{}$

7. $-6 - 7 = -6 + (\boxed{}) = \boxed{}$

Mixed Review

Evaluate. [10.1a]

8. $\dfrac{3m}{n}$, when $m = 8$ and $n = 6$

9. $\dfrac{a + b}{2}$, when $a = 5$ and $b = 17$

Translate each phrase to an algebraic expression. Use any letter for the variable. [10.1b]

10. Three times some number

11. Five less than some number

12. State the integers that correspond to this situation: Jerilyn deposited $450 in her checking account. Later that week, she wrote a check for $79. [10.2a]

13. Graph -3.5 on the number line. [10.2b]

$$\longleftarrow \!\!+\!\!+\!\!+\!\!+\!\!+\!\!+\!\!+\!\!+\!\!+\!\!+\!\!+\!\!+\!\!+\!\!\longrightarrow$$
$$-6\ -5\ -4\ -3\ -2\ -1\ \ 0\ \ 1\ \ 2\ \ 3\ \ 4\ \ 5\ \ 6$$

Convert to decimal notation. [10.2c]

14. $-\dfrac{4}{5}$

15. $\dfrac{7}{3}$

Use either $<$ or $>$ for $\square$ to write a true sentence. [10.2d]

16. $-5\ \square\ -3$

17. $-9.9\ \square\ -10.1$

Write true or false. [10.2d]

18. $-8 \geq -5$ **19.** $-4 \leq -4$

Find the absolute value. [10.2e]

22. $|15.6|$ **23.** $|-18|$

Find the opposite, or additive inverse, of each number. [10.3b]

26. -5.6 **27.** $\dfrac{7}{4}$

30. Evaluate $-x$ when x is -19.   [10.3b]

Compute and simplify. [10.3a], [10.4a]

32. $7 + (-9)$ **33.** $-\dfrac{3}{8} + \dfrac{1}{4}$

36. $\dfrac{2}{3} + \left(-\dfrac{9}{8}\right)$ **37.** $-4.2 + (-3.9)$

40. $-4.1 - 6.3$ **41.** $5 - (-11)$

44. $-8 - (-4)$ **45.** $-\dfrac{1}{2} - \dfrac{5}{6}$

48. $16 - (-9) - 20 - (-4)$

50. $17 - (-25) + 15 - (-18)$

Write an inequality with the same meaning. [10.2d]

20. $y < 5$ **21.** $-3 \geq t$

24. $|0|$ **25.** $\left|-\dfrac{12}{5}\right|$

28. 0 **29.** -49

31. Evaluate $-(-x)$ when x is 2.3.   [10.3b]

34. $3.6 + (-3.6)$ **35.** $-8 + (-9)$

38. $-14 + 5$ **39.** $19 + (-21)$

42. $-\dfrac{1}{4} - \left(-\dfrac{3}{5}\right)$ **43.** $12 - 24$

46. $12.3 - 14.1$ **47.** $6 - (-7)$

49. $-4 + (-10) - (-3) - 12$

51. $-9 + (-3) + 16 - (-10)$

Solve.   [10.3c], [10.4b]

52. *Temperature Change.*   In chemistry lab, Ben works with a substance whose initial temperature is 25°C. During an experiment, the temperature falls to -8°C. Find the difference between the two temperatures.

53. *Stock Price Change.*   The price of a stock opened at $56.12. During the day, it dropped $1.18, then rose $1.22, and then dropped $1.36. Find the value of the stock at the end of the day.

Understanding Through Discussion and Writing

54. Give three examples of rational numbers that are not integers. Explain.   [10.2b]

55. Give three examples of irrational numbers. Explain the difference between an irrational number and a rational number.   [10.2b, d]

56. Explain in your own words why the sum of two negative numbers is always negative.   [10.3a]

57. If a negative number is subtracted from a positive number, will the result always be positive? Why or why not?   [10.4a]

Multiplication of Real Numbers

10.5

a   MULTIPLICATION

Multiplication of real numbers is very much like multiplication of arithmetic numbers. The only difference is that we must determine whether the product is positive or negative.

OBJECTIVES

a   Multiply real numbers.

b   Solve applied problems involving multiplication of real numbers.

Multiplication of a Positive Number and a Negative Number

To see how to multiply a positive number and a negative number, consider the pattern of the following.

This number decreases by 1 each time.

$$
\begin{aligned}
4 \cdot 5 &= 20 \\
3 \cdot 5 &= 15 \\
2 \cdot 5 &= 10 \\
1 \cdot 5 &= 5 \\
0 \cdot 5 &= 0 \\
-1 \cdot 5 &= -5 \\
-2 \cdot 5 &= -10 \\
-3 \cdot 5 &= -15
\end{aligned}
$$

This number decreases by 5 each time.

Do Exercise 1. ▶

1. Complete, as in the example.

$$
\begin{aligned}
4 \cdot 10 &= 40 \\
3 \cdot 10 &= 30 \\
2 \cdot 10 &= \\
1 \cdot 10 &= \\
0 \cdot 10 &= \\
-1 \cdot 10 &= \\
-2 \cdot 10 &= \\
-3 \cdot 10 &=
\end{aligned}
$$

According to this pattern, it looks as though the product of a negative number and a positive number is negative. That is the case, and we have the first part of the rule for multiplying real numbers.

THE PRODUCT OF A POSITIVE NUMBER AND A NEGATIVE NUMBER

To multiply a positive number and a negative number, multiply their absolute values. The product is negative.

EXAMPLES   Multiply.

1. $8(-5) = -40$

2. $-\dfrac{1}{3} \cdot \dfrac{5}{7} = -\dfrac{5}{21}$

3. $(-7.2)5 = -36$

Do Exercises 2–5. ▶

Multiply.

2. $-3 \cdot 6$ **3.** $20 \cdot (-5)$

4. $-\dfrac{2}{3} \cdot \dfrac{5}{6}$ **5.** $-4.23(7.1)$

Answers

1. $20; 10; 0; -10; -20; -30$ **2.** -18
3. -100 **4.** $-\dfrac{5}{9}$ **5.** -30.033

Multiplication of Two Negative Numbers

How do we multiply two negative numbers? Again, we look for a pattern.

This number decreases by 1 each time. ⟶

$$4 \cdot (-5) = -20$$
$$3 \cdot (-5) = -15$$
$$2 \cdot (-5) = -10$$
$$1 \cdot (-5) = -5$$
$$0 \cdot (-5) = 0$$
$$-1 \cdot (-5) = 5$$
$$-2 \cdot (-5) = 10$$
$$-3 \cdot (-5) = 15$$

⟵ This number increases by 5 each time.

6. Complete, as in the example.

$$3 \cdot (-10) = -30$$
$$2 \cdot (-10) = -20$$
$$1 \cdot (-10) =$$
$$0 \cdot (-10) =$$
$$-1 \cdot (-10) =$$
$$-2 \cdot (-10) =$$
$$-3 \cdot (-10) =$$

◀ Do Exercise 6.

According to the pattern, it appears that the product of two negative numbers is positive. That is actually so, and we have the second part of the rule for multiplying real numbers.

> ### THE PRODUCT OF TWO NEGATIVE NUMBERS
>
> To multiply two negative numbers, multiply their absolute values.
> The product is positive.

◀ Do Exercises 7–12.

The following is another way to consider the rules that we have for multiplication.

> To multiply two nonzero real numbers:
>
> **a)** Multiply the absolute values.
> **b)** If the signs are the same, the product is positive.
> **c)** If the signs are different, the product is negative.

Multiply.

7. $-9 \cdot (-3)$

8. $-16 \cdot (-2)$

9. $-7 \cdot (-5)$

10. $-\dfrac{4}{7}\left(-\dfrac{5}{9}\right)$

11. $-\dfrac{3}{2}\left(-\dfrac{4}{9}\right)$

12. $-3.25(-4.14)$

Multiplication by Zero

The only case that we have not considered is multiplying by zero. As with nonnegative numbers, the product of any real number and 0 is 0.

> ### THE MULTIPLICATION PROPERTY OF ZERO
>
> For any real number a,
> $$a \cdot 0 = 0 \cdot a = 0.$$
> (The product of 0 and any real number is 0.)

EXAMPLES Multiply.

4. $(-3)(-4) = 12$

5. $-1.6(2) = -3.2$

6. $-19 \cdot 0 = 0$

7. $\left(-\dfrac{5}{6}\right)\left(-\dfrac{1}{9}\right) = \dfrac{5}{54}$

8. $0 \cdot (-452) = 0$

9. $23 \cdot 0 \cdot \left(-8\dfrac{2}{3}\right) = 0$

Answers

6. $-10; 0; 10; 20; 30$ **7.** 27 **8.** 32 **9.** 35
10. $\dfrac{20}{63}$ **11.** $\dfrac{2}{3}$ **12.** 13.455

Do Exercises 13–18. ▶

Multiplying More Than Two Numbers

When multiplying more than two real numbers, we can choose order and grouping as we please.

EXAMPLES Multiply.

10. $-8 \cdot 2(-3) = -16(-3)$ Multiplying the first two numbers
$$= 48$$

11. $-8 \cdot 2(-3) = 24 \cdot 2$ Multiplying the negative numbers. Every pair of negative numbers gives a positive product.
$$= 48$$

12. $-3(-2)(-5)(4) = 6(-5)(4)$ Multiplying the first two numbers
$$= (-30)4$$
$$= -120$$

13. $\left(-\dfrac{1}{2}\right)(8)\left(-\dfrac{2}{3}\right)(-6) = (-4)4$ Multiplying the first two numbers and the last two numbers
$$= -16$$

14. $-5 \cdot (-2) \cdot (-3) \cdot (-6) = 10 \cdot 18 = 180$

15. $(-3)(-5)(-2)(-3)(-6) = (-30)(18) = -540$

Considering that the product of a pair of negative numbers is positive, we see the following pattern.

> The product of an even number of negative numbers is positive.
> The product of an odd number of negative numbers is negative.

Do Exercises 19–24. ▶

EXAMPLE 16 Evaluate $2x^2$ when $x = 3$ and when $x = -3$.
$$2x^2 = 2(3)^2 = 2(9) = 18;$$
$$2x^2 = 2(-3)^2 = 2(9) = 18$$

Let's compare the expressions $(-x)^2$ and $-x^2$.

EXAMPLE 17 Evaluate $(-x)^2$ and $-x^2$ when $x = 5$.
$$(-x)^2 = (-5)^2 = (-5)(-5) = 25;$$ Substitute 5 for x. Then evaluate the power.

$$-x^2 = -(5)^2 = -(25) = -25$$ Substitute 5 for x. Evaluate the power. Then find the opposite.

In Example 17, we see that the expressions $(-x)^2$ and $-x^2$ are *not* equivalent. That is, they do not have the same value for every allowable replacement of the variable by a real number. To find $(-x)^2$, we take the opposite and then square. To find $-x^2$, we find the square and then take the opposite.

Multiply.

13. $5(-6)$

14. $(-5)(-6)$

15. $(-3.2) \cdot 10$

16. $\left(-\dfrac{4}{5}\right)\left(\dfrac{10}{3}\right)$

17. $0 \cdot (-34.2)$

18. $-\dfrac{5}{7} \cdot 0 \cdot \left(-4\dfrac{2}{3}\right)$

Multiply.

19. $5 \cdot (-3) \cdot 2$

20. $-3 \times (-4.1) \times (-2.5)$

21. $-\dfrac{1}{2} \cdot \left(-\dfrac{4}{3}\right) \cdot \left(-\dfrac{5}{2}\right)$

22. $-2 \cdot (-5) \cdot (-4) \cdot (-3)$

23. $(-4)(-5)(-2)(-3)(-1)$

24. $(-1)(-1)(-2)(-3)(-1)(-1)$

Answers

13. -30 **14.** 30 **15.** -32 **16.** $-\dfrac{8}{3}$
17. 0 **18.** 0 **19.** -30 **20.** -30.75
21. $-\dfrac{5}{3}$ **22.** 120 **23.** -120 **24.** 6

25. Evaluate $3x^2$ when $x = 4$ and when $x = -4$.

26. Evaluate $(-x)^2$ and $-x^2$ when $x = 2$.

27. Evaluate $(-x)^2$ and $-x^2$ when $x = -3$.

EXAMPLE 18   Evaluate $(-a)^2$ and $-a^2$ when $a = -4$.

To make sense of the substitutions and computations, we introduce extra grouping symbols into the expressions.

$$(-a)^2 = [-(-4)]^2 = [4]^2 = 16;$$
$$-a^2 = -(-4)^2 = -(16) = -16$$

◀ Do Exercises 25–27.

b   APPLICATIONS AND PROBLEM SOLVING

We now consider multiplication of real numbers in real-world applications.

EXAMPLE 19   *Mine Rescue.*   The San Jose copper and gold mine near Copiapó, Chile, collapsed on August 5, 2010, trapping 33 miners. Each miner was safely brought out of the mine with a specially designed capsule that could be lowered into the mine at −137 feet per minute. It took approximately 15 minutes to lower the capsule to the miners' location. Determine how far below the surface of the earth the miners were trapped.

Source: Reuters News

28. *Chemical Reaction.*   During a chemical reaction, the temperature in a beaker increased by 3°C every minute until 1:34 P.M. If the temperature was −17°C at 1:10 P.M., when the reaction began, what was the temperature at 1:34 P.M.?

Since the capsule moved −137 feet per minute and it took 15 minutes to reach the miners, we have the depth d given by

$$d = 15 \cdot (-137) = -2055.$$

Thus the miners were trapped at −2055 ft.

◀ Do Exercise 28.

Answers

25. 48; 48   **26.** 4; −4   **27.** 9; −9   **28.** 55°C

10.5   Exercise Set

✓ Reading Check

Fill in the blank with either "positive" or "negative."

RC1. To multiply a positive number and a negative number, multiply their absolute values. The answer is _____.

RC2. To multiply two negative numbers, multiply their absolute values. The answer is _____.

RC3. The product of an even number of negative numbers is _____.

RC4. The product of an odd number of negative numbers is _____.

Evaluate.

RC5. -3^2

RC6. $(-3)^2$

RC7. $-\left(\dfrac{1}{2}\right)^2$

RC8. $-\left(-\dfrac{1}{2}\right)^2$

a Multiply.

1. $-4 \cdot 2$ 2. $-3 \cdot 5$ 3. $8 \cdot (-3)$ 4. $9 \cdot (-5)$ 5. $-9 \cdot 8$

6. $-10 \cdot 3$ 7. $-8 \cdot (-2)$ 8. $-2 \cdot (-5)$ 9. $-7 \cdot (-6)$ 10. $-9 \cdot (-2)$

11. $15 \cdot (-8)$ 12. $-12 \cdot (-10)$ 13. $-14 \cdot 17$ 14. $-13 \cdot (-15)$ 15. $-25 \cdot (-48)$

16. $39 \cdot (-43)$ 17. $-3.5 \cdot (-28)$ 18. $97 \cdot (-2.1)$ 19. $9 \cdot (-8)$ 20. $7 \cdot (-9)$

21. $4 \cdot (-3.1)$ 22. $3 \cdot (-2.2)$ 23. $-5 \cdot (-6)$ 24. $-6 \cdot (-4)$ 25. $-7 \cdot (-3.1)$

26. $-4 \cdot (-3.2)$ 27. $\dfrac{2}{3} \cdot \left(-\dfrac{3}{5}\right)$ 28. $\dfrac{5}{7} \cdot \left(-\dfrac{2}{3}\right)$ 29. $-\dfrac{3}{8} \cdot \left(-\dfrac{2}{9}\right)$

30. $-\dfrac{5}{8} \cdot \left(-\dfrac{2}{5}\right)$ 31. -6.3×2.7 32. -4.1×9.5 33. $7 \cdot (-4) \cdot (-3) \cdot 5$

34. $9 \cdot (-2) \cdot (-6) \cdot 7$ 35. $-\dfrac{2}{3} \cdot \dfrac{1}{2} \cdot \left(-\dfrac{6}{7}\right)$ 36. $-\dfrac{1}{8} \cdot \left(-\dfrac{1}{4}\right) \cdot \left(-\dfrac{3}{5}\right)$ 37. $-3 \cdot (-4) \cdot (-5)$

38. $-2 \cdot (-5) \cdot (-7)$ 39. $-2 \cdot (-5) \cdot (-3) \cdot (-5)$ 40. $-3 \cdot (-5) \cdot (-2) \cdot (-1)$

41. $-4 \cdot (-1.8) \cdot 7$ 42. $-8 \cdot (-1.3) \cdot (-5)$ 43. $-\dfrac{1}{9}\left(-\dfrac{2}{3}\right)\left(\dfrac{5}{7}\right)$

44. $-\dfrac{7}{2}\left(-\dfrac{5}{7}\right)\left(-\dfrac{2}{5}\right)$ 45. $4 \cdot (-4) \cdot (-5) \cdot (-12)$ 46. $-2 \cdot (-3) \cdot (-4) \cdot (-5)$

47. $0.07 \cdot (-7) \cdot 6 \cdot (-6)$

48. $80 \cdot (-0.8) \cdot (-90) \cdot (-0.09)$

49. $\left(-\dfrac{5}{6}\right)\left(\dfrac{1}{8}\right)\left(-\dfrac{3}{7}\right)\left(-\dfrac{1}{7}\right)$

50. $\left(\dfrac{4}{5}\right)\left(-\dfrac{2}{3}\right)\left(-\dfrac{15}{7}\right)\left(\dfrac{1}{2}\right)$

51. $(-14) \cdot (-27) \cdot 0$

52. $7 \cdot (-6) \cdot 5 \cdot (-4) \cdot 3 \cdot (-2) \cdot 1 \cdot 0$

53. $(-8)(-9)(-10)$

54. $(-7)(-8)(-9)(-10)$

55. $(-6)(-7)(-8)(-9)(-10)$

56. $(-5)(-6)(-7)(-8)(-9)(-10)$

57. $(-1)^{12}$

58. $(-1)^9$

59. Evaluate $(-x)^2$ and $-x^2$ when $x = 4$ and when $x = -4$.

60. Evaluate $(-x)^2$ and $-x^2$ when $x = 10$ and when $x = -10$.

61. Evaluate $(-y)^2$ and $-y^2$ when $y = \frac{2}{5}$ and when $y = -\frac{2}{5}$.

62. Evaluate $(-w)^2$ and $-w^2$ when $w = \frac{1}{10}$ and when $w = -\frac{1}{10}$.

63. Evaluate $-(-t)^2$ and $-t^2$ when $t = 3$ and when $t = -3$.

64. Evaluate $-(-s)^2$ and $-s^2$ when $s = 1$ and when $s = -1$.

65. Evaluate $(-3x)^2$ and $-3x^2$ when $x = 7$ and when $x = -7$.

66. Evaluate $(-2x)^2$ and $-2x^2$ when $x = 3$ and when $x = -3$.

67. Evaluate $5x^2$ when $x = 2$ and when $x = -2$.

68. Evaluate $2x^2$ when $x = 5$ and when $x = -5$.

69. Evaluate $-2x^3$ when $x = 1$ and when $x = -1$.

70. Evaluate $-3x^3$ when $x = 2$ and when $x = -2$.

b   Solve.

71. *Chemical Reaction.*   The temperature of a chemical compound was 0°C at 11:00 A.M. During a reaction, it dropped 3°C per minute until 11:08 A.M. What was the temperature at 11:08 A.M.?

72. *Chemical Reaction.*   The temperature of a chemical compound was −5°C at 3:20 P.M. During a reaction, it increased 2°C per minute until 3:52 P.M. What was the temperature at 3:52 P.M.?

73. *Weight Loss.* Dave lost 2 lb each week for a period of 10 weeks. Express his total weight change as an integer.

74. *Stock Loss.* Each day for a period of 5 days, the value of a stock that Lily owned dropped $3. Express Lily's total loss as an integer.

75. *Stock Price.* The price of a stock began the day at $23.75 per share and dropped $1.38 per hour for 8 hr. What was the price of the stock after 8 hr?

76. *Population Decrease.* The population of Bloomtown was 12,500. It decreased 380 each year for 4 years. What was the population of the town after 4 years?

77. *Diver's Position.* After diving 95 m below sea level, a diver rises at a rate of 7 m/min for 9 min. Where is the diver in relation to the surface at the end of the 9-min period?

78. *Bank Account Balance.* Karen had $68 in her bank account. After she used her debit card to make seven purchases at $13 each, what was the balance in her bank account?

79. *Drop in Temperature.* The temperature in Osgood was 62°F at 2:00 P.M. It dropped 6°F per hour for the next 4 hr. What was the temperature at the end of the 4-hr period?

80. *Juice Consumption.* Oliver bought a 64-oz container of cranberry juice and drank 8 oz per day for a week. How much juice was left in the container at the end of the week?

Skill Maintenance

81. Evaluate $\dfrac{x - 2y}{3}$ when $x = 20$ and $y = 7$. [10.1a]

82. Evaluate $\dfrac{d - e}{3d}$ when $d = 5$ and $e = 1$. [10.1a]

Subtract. [10.4a]

83. $-\dfrac{1}{2} - \left(-\dfrac{1}{6}\right)$

84. $8 - 12.3$

85. $31 - (-13)$

86. $-\dfrac{5}{12} - \left(-\dfrac{1}{3}\right)$

Write true or false. [10.2d]

87. $-10 > -12$

88. $0 \leq -1$

89. $4 < -8$

90. $-7 \geq -6$

Synthesis

91. If a is positive and b is negative, then $-ab$ is which of the following?

A. Positive
B. Negative
C. 0
D. Cannot be determined without more information

92. If a is positive and b is negative, then $(-a)(-b)$ is which of the following?

A. Positive
B. Negative
C. 0
D. Cannot be determined without more information

93. Of all possible quotients of the numbers 10, $-\frac{1}{2}$, -5, and $\frac{1}{5}$, which two produce the largest quotient? Which two produce the smallest quotient?

OBJECTIVES

a   Divide integers.

b   Find the reciprocal of a real number.

c   Divide real numbers.

d   Solve applied problems involving division of real numbers.

SKILL TO REVIEW

Objective 10.5a: Multiply real numbers.

Multiply.

1. $-\dfrac{2}{9} \cdot \dfrac{4}{11}$ 2. $\dfrac{13}{2} \cdot \left(-\dfrac{2}{25}\right)$

Divide.

1. $6 \div (-3)$

Think: What number multiplied by -3 gives 6?

2. $\dfrac{-15}{-3}$

Think: What number multiplied by -3 gives -15?

3. $-24 \div 8$

Think: What number multiplied by 8 gives -24?

4. $\dfrac{-48}{-6}$ 5. $\dfrac{30}{-5}$

6. $\dfrac{30}{-7}$

Answers

Skill to Review:

1. $-\dfrac{8}{99}$ 2. $-\dfrac{13}{25}$

Margin Exercises:

1. -2 2. 5   3. -3 4. 8

5. -6 6. $-\dfrac{30}{7}$

We now consider division of real numbers. The definition of division results in rules for division that are the same as those for multiplication.

a   DIVISION OF INTEGERS

> **DIVISION**
>
> The quotient $a \div b$, or $\dfrac{a}{b}$, where $b \neq 0$, is that unique real number c for which $a = b \cdot c$.

Let's use the definition to divide integers.

EXAMPLES   Divide, if possible. Check your answer.

1. $14 \div (-7) = -2$ *Think*: What number multiplied by -7 gives 14? That number is -2. *Check*: $(-2)(-7) = 14$.

2. $\dfrac{-32}{-4} = 8$ *Think*: What number multiplied by -4 gives -32? That number is 8. *Check*: $8(-4) = -32$.

3. $\dfrac{-10}{7} = -\dfrac{10}{7}$ *Think*: What number multiplied by 7 gives -10? That number is $-\frac{10}{7}$. *Check*: $-\frac{10}{7} \cdot 7 = -10$.

4. $\dfrac{-17}{0}$ is **not defined**.   *Think*: What number multiplied by 0 gives -17? There is no such number because the product of 0 and *any* number is 0.

The rules for division are the same as those for multiplication.

> To multiply or divide two real numbers (where the divisor is nonzero):
>
> a) Multiply or divide the absolute values.
>
> b) If the signs are the same, the answer is positive.
>
> c) If the signs are different, the answer is negative.

◀ Do Margin Exercises 1–6.

Excluding Division by 0

Example 4 shows why we cannot divide -17 by 0. We can use the same argument to show why we cannot divide any nonzero number b by 0. Consider $b \div 0$. We look for a number that when multiplied by 0 gives b. There is no such number because the product of 0 and any number is 0. Thus we cannot divide a nonzero number b by 0.

On the other hand, if we divide 0 by 0, we look for a number c such that $0 \cdot c = 0$. But $0 \cdot c = 0$ for any number c. Thus it appears that $0 \div 0$ could be any number we choose. Getting any answer we want when we divide 0 by 0 would be very confusing. Thus we agree that division by 0 is not defined.

EXCLUDING DIVISION BY 0

Division by 0 is not defined.

$$a \div 0, \text{ or } \frac{a}{0}, \text{ is not defined for all real numbers } a.$$

Dividing 0 by Other Numbers

Note that

$$0 \div 8 = 0 \text{ because } 0 = 0 \cdot 8; \qquad \frac{0}{-5} = 0 \text{ because } 0 = 0 \cdot (-5).$$

DIVIDENDS OF 0

Zero divided by any nonzero real number is 0:

$$\frac{0}{a} = 0; \quad a \neq 0.$$

EXAMPLES Divide.

5. $0 \div (-6) = 0$ **6.** $\dfrac{0}{12} = 0$ **7.** $\dfrac{-3}{0}$ is not defined.

Divide, if possible.

7. $\dfrac{-5}{0}$ **8.** $\dfrac{0}{-3}$

Do Exercises 7 and 8. ▶

b   RECIPROCALS

When two numbers like $\frac{1}{2}$ and 2 are multiplied, the result is 1. Such numbers are called **reciprocals** of each other. Every nonzero real number has a reciprocal, also called a **multiplicative inverse**.

RECIPROCALS

Two numbers whose product is 1 are called **reciprocals**, or **multiplicative inverses**, of each other.

EXAMPLES Find the reciprocal.

8. $\dfrac{7}{8}$ The reciprocal of $\dfrac{7}{8}$ is $\dfrac{8}{7}$ because $\dfrac{7}{8} \cdot \dfrac{8}{7} = 1$.

9. -5 The reciprocal of -5 is $-\dfrac{1}{5}$ because $-5\left(-\dfrac{1}{5}\right) = 1$.

10. 3.9 The reciprocal of 3.9 is $\dfrac{1}{3.9}$ because $3.9\left(\dfrac{1}{3.9}\right) = 1$.

11. $-\dfrac{1}{2}$ The reciprocal of $-\dfrac{1}{2}$ is -2 because $\left(-\dfrac{1}{2}\right)(-2) = 1$.

12. $-\dfrac{2}{3}$ The reciprocal of $-\dfrac{2}{3}$ is $-\dfrac{3}{2}$ because $\left(-\dfrac{2}{3}\right)\left(-\dfrac{3}{2}\right) = 1$.

13. $\dfrac{3y}{8x}$ The reciprocal of $\dfrac{3y}{8x}$ is $\dfrac{8x}{3y}$ because $\left(\dfrac{3y}{8x}\right)\left(\dfrac{8x}{3y}\right) = 1$.

Answers

7. Not defined   **8.** 0

Find the reciprocal.

9. $\dfrac{2}{3}$ 10. $-\dfrac{5}{4}$

11. -3 12. $-\dfrac{1}{5}$

13. 1.3 14. $\dfrac{a}{6b}$

RECIPROCAL PROPERTIES

For $a \ne 0$, the reciprocal of a can be named $\dfrac{1}{a}$ and the reciprocal of $\dfrac{1}{a}$ is a.

The reciprocal of a nonzero number $\dfrac{a}{b}$ can be named $\dfrac{b}{a}$.

The number 0 has no reciprocal.

◀ **Do Exercises 9–14.**

The reciprocal of a positive number is also a positive number, because the product of the two numbers must be the positive number 1. The reciprocal of a negative number is also a negative number, because the product of the two numbers must be the positive number 1.

THE SIGN OF A RECIPROCAL

The reciprocal of a number has the same sign as the number itself.

⋯⋯⋯⋯⋯⋯⋯⋯⋯⋯⋯⋯⋯⋯⋯⋯⋯⋯⋯ Caution! ⋯⋯⋯⋯⋯⋯⋯⋯⋯⋯⋯⋯

It is important *not* to confuse *opposite* with *reciprocal.* Keep in mind that the opposite, or additive inverse, of a number is what we add to the number to get 0. The reciprocal, or multiplicative inverse, is what we multiply the number by to get 1.

⋯⋯⋯⋯⋯⋯⋯⋯⋯⋯⋯⋯⋯⋯⋯⋯⋯⋯⋯⋯⋯⋯⋯⋯⋯⋯⋯⋯⋯⋯⋯⋯⋯⋯⋯⋯⋯⋯

15. Complete the following table.

NUMBER	OPPOSITE	RECIPROCAL
$\dfrac{2}{9}$		
$-\dfrac{7}{4}$		
0		
1		
-8		
-4.7		

Compare the following.

NUMBER	OPPOSITE (Change the sign.)	RECIPROCAL (Invert but do not change the sign.)
$-\dfrac{3}{8}$	$\dfrac{3}{8}$	$-\dfrac{8}{3}$
19	-19	$\dfrac{1}{19}$
$\dfrac{18}{7}$	$-\dfrac{18}{7}$	$\dfrac{7}{18}$
-7.9	7.9	$-\dfrac{1}{7.9}$, or $-\dfrac{10}{79}$
0	0	None

$\left(-\dfrac{3}{8}\right)\left(-\dfrac{8}{3}\right) = 1$

$-\dfrac{3}{8} + \dfrac{3}{8} = 0$

◀ **Do Exercise 15.**

c DIVISION OF REAL NUMBERS

We know that we can subtract by adding an opposite. Similarly, we can divide by multiplying by a reciprocal.

RECIPROCALS AND DIVISION

For any real numbers a and b, $b \neq 0$,

$$a \div b = \frac{a}{b} = a \cdot \frac{1}{b}.$$

(To divide, multiply by the reciprocal of the divisor.)

EXAMPLES Rewrite each division as a multiplication.

14. $-4 \div 3$ $-4 \div 3$ is the same as $-4 \cdot \frac{1}{3}$

15. $\dfrac{6}{-7}$ $\dfrac{6}{-7} = 6\left(-\dfrac{1}{7}\right)$

16. $\dfrac{3}{5} \div \left(-\dfrac{9}{7}\right)$ $\dfrac{3}{5} \div \left(-\dfrac{9}{7}\right) = \dfrac{3}{5}\left(-\dfrac{7}{9}\right)$

17. $\dfrac{x+2}{5}$ $\dfrac{x+2}{5} = (x+2)\dfrac{1}{5}$ Parentheses are necessary here.

18. $\dfrac{-17}{1/b}$ $\dfrac{-17}{1/b} = -17 \cdot b$

Do Exercises 16–20. ▶

When actually doing division calculations, we sometimes multiply by a reciprocal and we sometimes divide directly. With fraction notation, it is usually better to multiply by a reciprocal. With decimal notation, it is usually better to divide directly.

EXAMPLES Divide by multiplying by the reciprocal of the divisor.

19. $\dfrac{2}{3} \div \left(-\dfrac{5}{4}\right) = \dfrac{2}{3} \cdot \left(-\dfrac{4}{5}\right) = -\dfrac{8}{15}$

20. $-\dfrac{5}{6} \div \left(-\dfrac{3}{4}\right) = -\dfrac{5}{6} \cdot \left(-\dfrac{4}{3}\right) = \dfrac{20}{18} = \dfrac{10 \cdot 2}{9 \cdot 2} = \dfrac{10}{9} \cdot \dfrac{2}{2} = \dfrac{10}{9}$

·········· **Caution!** ··········

Be careful *not* to change the sign when taking a reciprocal!

···

21. $-\dfrac{3}{4} \div \dfrac{3}{10} = -\dfrac{3}{4} \cdot \left(\dfrac{10}{3}\right) = -\dfrac{30}{12} = -\dfrac{5 \cdot 6}{2 \cdot 6} = -\dfrac{5}{2} \cdot \dfrac{6}{6} = -\dfrac{5}{2}$

Do Exercises 21 and 22. ▶

Rewrite each division as a multiplication.

16. $\dfrac{4}{7} \div \left(-\dfrac{3}{5}\right)$

17. $\dfrac{5}{-8}$

18. $\dfrac{a-b}{7}$

19. $\dfrac{-23}{1/a}$

20. $-5 \div 7$

Divide by multiplying by the reciprocal of the divisor.

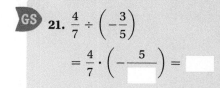

GS 21. $\dfrac{4}{7} \div \left(-\dfrac{3}{5}\right)$

$= \dfrac{4}{7} \cdot \left(-\dfrac{5}{\boxed{}}\right) = \boxed{}$

22. $-\dfrac{12}{7} \div \left(-\dfrac{3}{4}\right)$

Answers

16. $\dfrac{4}{7} \cdot \left(-\dfrac{5}{3}\right)$ **17.** $5 \cdot \left(-\dfrac{1}{8}\right)$ **18.** $(a-b) \cdot \dfrac{1}{7}$

19. $-23 \cdot a$ **20.** $-5 \cdot \left(\dfrac{1}{7}\right)$ **21.** $-\dfrac{20}{21}$

22. $\dfrac{16}{7}$

Guided Solution:
21. $3, -\dfrac{20}{21}$

With decimal notation, it is easier to carry out long division than to multiply by the reciprocal.

EXAMPLES Divide.

22. $-27.9 \div (-3) = \dfrac{-27.9}{-3} = 9.3$ Do the long division $3)\overline{27.9}$. The answer is positive.

$$\begin{array}{r} 9.3 \\ 3)\overline{27.9} \end{array}$$

Divide.

23. $21.7 \div (-3.1)$

24. $-20.4 \div (-4)$

23. $-6.3 \div 2.1 = -3$ Do the long division $2.1_\wedge)\overline{6.3_\wedge}$. The answer is negative.

$$\begin{array}{r} 3. \\ 2.1)\overline{6.3_\wedge} \end{array}$$

◀ **Do Exercises 23 and 24.**

Consider the following:

1. $\dfrac{2}{3} = \dfrac{2}{3} \cdot 1 = \dfrac{2}{3} \cdot \dfrac{-1}{-1} = \dfrac{2(-1)}{3(-1)} = \dfrac{-2}{-3}$.   Thus, $\dfrac{2}{3} = \dfrac{-2}{-3}$.

(A negative number divided by a negative number is positive.)

2. $-\dfrac{2}{3} = -1 \cdot \dfrac{2}{3} = \dfrac{-1}{1} \cdot \dfrac{2}{3} = \dfrac{-1 \cdot 2}{1 \cdot 3} = \dfrac{-2}{3}$.   Thus, $-\dfrac{2}{3} = \dfrac{-2}{3}$.

(A negative number divided by a positive number is negative.)

3. $\dfrac{-2}{3} = \dfrac{-2}{3} \cdot 1 = \dfrac{-2}{3} \cdot \dfrac{-1}{-1} = \dfrac{-2(-1)}{3(-1)} = \dfrac{2}{-3}$.   Thus, $-\dfrac{2}{3} = \dfrac{2}{-3}$.

(A positive number divided by a negative number is negative.)

We can use the following properties to make sign changes in fraction notation.

Find two equal expressions for each number with negative signs in different places.

> ### SIGN CHANGES IN FRACTION NOTATION
>
> For any numbers a and b, $b \neq 0$:
>
> **1.** $\dfrac{-a}{-b} = \dfrac{a}{b}$
>
> (The opposite of a number a divided by the opposite of another number b is the same as the quotient of the two numbers a and b.)
>
> **2.** $\dfrac{-a}{b} = \dfrac{a}{-b} = -\dfrac{a}{b}$
>
> (The opposite of a number a divided by another number b is the same as the number a divided by the opposite of the number b, and both are the same as the opposite of a *divided by* b.)

25. $\dfrac{-5}{6} = \dfrac{5}{\boxed{}} = -\dfrac{\boxed{}}{6}$ **GS**

26. $-\dfrac{8}{7}$

27. $\dfrac{10}{-3}$

◀ **Do Exercises 25–27.**

d APPLICATIONS AND PROBLEM SOLVING

EXAMPLE 24 *Chemical Reaction.* During a chemical reaction, the temperature in a beaker decreased every minute by the same number of degrees. The temperature was 56°F at 10:10 A.M. By 10:42 A.M., the temperature had dropped to -12°F. By how many degrees did it change each minute?

Answers

23. -7 **24.** 5.1 **25.** $\dfrac{5}{-6}; -\dfrac{5}{6}$ **26.** $\dfrac{8}{-7}; \dfrac{-8}{7}$

27. $\dfrac{-10}{3}; -\dfrac{10}{3}$

Guided Solution:
25. $-6, 5$

We first determine by how many degrees d the temperature changed altogether. We subtract -12 from 56:

$$d = 56 - (-12) = 56 + 12 = 68.$$

The temperature changed a total of 68°. We can express this as $-68°$ since the temperature dropped.

The amount of time t that passed was $42 - 10$, or 32 min. Thus the number of degrees T that the temperature dropped each minute is given by

$$T = \frac{d}{t} = \frac{-68}{32} = -2.125.$$

The change was $-2.125°$F per minute.

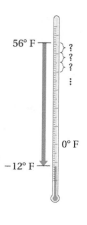

Do Exercise 28. ▷

28. *Chemical Reaction.* During a chemical reaction, the temperature in a beaker decreased every minute by the same number of degrees. The temperature was 71°F at 2:12 P.M. By 2:37 P.M., the temperature had changed to $-14°$F. By how many degrees did it change each minute?

Answer

28. $-3.4°$F per minute

CALCULATOR CORNER

Operations on the Real Numbers To perform operations on the real numbers on a graphing calculator, recall that negative numbers are entered using the opposite key, (-), rather than the subtraction operation key, ⊖. Consider the sum $-5 + (-3.8)$. On a graphing calculator, the parentheses are not necessary. The result is -8.8. Note that it is not incorrect to enter the parentheses. The result will be the same if this is done. We can also subtract, multiply, and divide real numbers. At right, we see $10 - (-17)$, $-5 \cdot (-7)$, and $45 \div (-9)$. Again, it is not necessary to use parentheses.

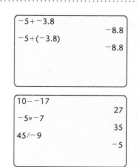

EXERCISES: Use a calculator to perform each operation.

1. $-8 + 4$
2. $-7 - (-5)$
3. $-8 \cdot 4$
4. $-7 \div (-5)$
5. $1.2 - (-1.5)$
6. $-7.6 + (-1.9)$
7. $1.2 \div (-1.5)$
8. $-7.6 \cdot (-1.9)$

10.6 Exercise Set

✓ Reading Check

Choose the word or the number below the blank that will make the sentence true.

RC1. The numbers 4 and -4 are called _____ of each other.
   <u>opposites/reciprocals</u>

RC2. The multiplicative inverse, or reciprocal, of a number is what we multiply the number by to get ____.
   <u>0/1</u>

RC3. The additive inverse, or opposite, of a number is what we add to the number to get ____.
   <u>0/1</u>

RC4. The numbers $-\frac{9}{4}$ and $-\frac{4}{9}$ are called _____ of each other.
   <u>opposites/reciprocals</u>

a   Divide, if possible. Check each answer.

1. $48 \div (-6)$

2. $\dfrac{42}{-7}$

3. $\dfrac{28}{-2}$

4. $24 \div (-12)$

5. $\dfrac{-24}{8}$

6. $-18 \div (-2)$

7. $\dfrac{-36}{-12}$

8. $-72 \div (-9)$

9. $\dfrac{-72}{9}$

10. $\dfrac{-50}{25}$

11. $-100 \div (-50)$

12. $\dfrac{-200}{8}$

13. $-108 \div 9$

14. $\dfrac{-63}{-7}$

15. $\dfrac{200}{-25}$

16. $-300 \div (-16)$

17. $\dfrac{75}{0}$

18. $\dfrac{0}{-5}$

19. $\dfrac{0}{-2.6}$

20. $\dfrac{-23}{0}$

b   Find the reciprocal.

21. $\dfrac{15}{7}$

22. $\dfrac{3}{8}$

23. $-\dfrac{47}{13}$

24. $-\dfrac{31}{12}$

25. 13

26. -10

27. -32

28. 15

29. $\dfrac{1}{-7.1}$

30. $\dfrac{1}{-4.9}$

31. $\dfrac{1}{9}$

32. $\dfrac{1}{16}$

33. $\dfrac{1}{4y}$

34. $\dfrac{-1}{8a}$

35. $\dfrac{2a}{3b}$

36. $\dfrac{-4y}{3x}$

C Rewrite each division as a multiplication.

37. $4 \div 17$

38. $5 \div (-8)$

39. $\dfrac{8}{-13}$

40. $-\dfrac{13}{47}$

41. $\dfrac{13.9}{-1.5}$

42. $-\dfrac{47.3}{21.4}$

43. $\dfrac{2}{3} \div \left(-\dfrac{4}{5}\right)$

44. $\dfrac{3}{4} \div \left(-\dfrac{7}{10}\right)$

45. $\dfrac{\dfrac{x}{1}}{y}$

46. $\dfrac{13}{\dfrac{1}{x}}$

47. $\dfrac{3x + 4}{5}$

48. $\dfrac{4y - 8}{-7}$

Divide.

49. $\dfrac{3}{4} \div \left(-\dfrac{2}{3}\right)$

50. $\dfrac{7}{8} \div \left(-\dfrac{1}{2}\right)$

51. $-\dfrac{5}{4} \div \left(-\dfrac{3}{4}\right)$

52. $-\dfrac{5}{9} \div \left(-\dfrac{5}{6}\right)$

53. $-\dfrac{2}{7} \div \left(-\dfrac{4}{9}\right)$

54. $-\dfrac{3}{5} \div \left(-\dfrac{5}{8}\right)$

55. $-\dfrac{3}{8} \div \left(-\dfrac{8}{3}\right)$

56. $-\dfrac{5}{8} \div \left(-\dfrac{6}{5}\right)$

57. $-\dfrac{5}{6} \div \dfrac{2}{3}$

58. $-\dfrac{7}{16} \div \dfrac{3}{8}$

59. $-\dfrac{9}{4} \div \dfrac{5}{12}$

60. $-\dfrac{3}{5} \div \dfrac{7}{10}$

61. $\dfrac{-11}{-13}$

62. $\dfrac{-21}{-25}$

63. $-6.6 \div 3.3$

64. $-44.1 \div (-6.3)$

65. $\dfrac{48.6}{-3}$ **66.** $\dfrac{-1.9}{20}$ **67.** $\dfrac{-12.5}{5}$ **68.** $\dfrac{-17.8}{3.2}$

69. $11.25 \div (-9)$ **70.** $-9.6 \div (-6.4)$ **71.** $\dfrac{-9}{17-17}$ **72.** $\dfrac{-8}{-5+5}$

d To determine percent increase or decrease, we first determine the change by subtracting the new amount from the original amount. Then we divide the change, which is *positive* if an increase, or *negative* if a decrease, by the original amount. Finally, we convert the decimal answer to percent notation.

73. *Passports.* In 2006, approximately 71 million valid passports were in circulation in the United States. This number increased to approximately 102 million in 2011. What is the percent increase?

Source: U.S. Department of State

74. *Super Bowl Spending.* The average amount spent during the Super Bowl per television viewer increased from $59.33 in 2011 to $63.87 in 2012. What is the percent increase?

Source: Retail Advertising and Marketing Association, Super Bowl Consumer Intentions and Actions Survey

75. *Pieces of Mail.* The number of pieces of mail handled by the U.S. Postal Service decreased from 212 billion in 2007 to 168 billion in 2011. What is the percent decrease?

Source: U.S. Postal Service

76. *Beef Consumption.* Beef consumption per capita in the United States decreased from 67.5 lb in 1990 to 57.4 in 2011. What is the percent decrease?

Source: U.S. Department of Agriculture

Skill Maintenance

Simplify.

77. $\dfrac{1}{4} - \dfrac{1}{2}$ [10.4a]

78. $-9 - 3 + 17$ [10.4a]

79. $35 \cdot (-1.2)$ [10.5a]

80. $4 \cdot (-6) \cdot (-2) \cdot (-1)$ [10.5a]

81. $13.4 + (-4.9)$ [10.3a]

82. $-\dfrac{3}{8} - \left(-\dfrac{1}{4}\right)$ [10.4a]

Convert to decimal notation. [10.2c]

83. $-\dfrac{1}{11}$ **84.** $\dfrac{11}{12}$ **85.** $\dfrac{15}{4}$ **86.** $-\dfrac{10}{3}$

Synthesis

87. Find the reciprocal of -10.5. What happens if you take the reciprocal of the result?

88. Determine those real numbers a for which the opposite of a is the same as the reciprocal of a.

Determine whether each expression represents a positive number or a negative number when a and b are negative.

89. $\dfrac{-a}{b}$ **90.** $\dfrac{-a}{-b}$ **91.** $-\left(\dfrac{a}{-b}\right)$ **92.** $-\left(\dfrac{-a}{b}\right)$ **93.** $-\left(\dfrac{-a}{-b}\right)$

Properties of Real Numbers

a EQUIVALENT EXPRESSIONS

In solving equations and doing other kinds of work in algebra, we manipulate expressions in various ways. For example, instead of $x + x$, we might write $2x$, knowing that the two expressions represent the same number for any allowable replacement of x. In that sense, the expressions $x + x$ and $2x$ are **equivalent**, as are $\dfrac{3}{x}$ and $\dfrac{3x}{x^2}$, even though 0 is not an allowable replacement because division by 0 is not defined.

EQUIVALENT EXPRESSIONS

Two expressions that have the same value for all allowable replacements are called **equivalent**.

The expressions $x + 3x$ and $5x$ are *not* equivalent, as we see in Margin Exercise 2.

Do Margin Exercises 1 and 2. ▶

In this section, we will consider several laws of real numbers that will allow us to find equivalent expressions. The first two laws are the *identity properties of 0 and 1*.

THE IDENTITY PROPERTY OF 0

For any real number a,

$$a + 0 = 0 + a = a.$$

(The number 0 is the *additive identity*.)

THE IDENTITY PROPERTY OF 1

For any real number a,

$$a \cdot 1 = 1 \cdot a = a.$$

(The number 1 is the *multiplicative identity*.)

We often refer to the use of the identity property of 1 as "multiplying by 1." We can use this method to find equivalent fraction expressions. Recall from arithmetic that to multiply with fraction notation, we multiply the numerators and multiply the denominators.

EXAMPLE 1 Write a fraction expression equivalent to $\frac{2}{3}$ with a denominator of $3x$:

$$\frac{2}{3} = \frac{\square}{3x}.$$

OBJECTIVES

a Find equivalent fraction expressions and simplify fraction expressions.

b Use the commutative laws and the associative laws to find equivalent expressions.

c Use the distributive laws to multiply expressions like 8 and $x - y$.

d Use the distributive laws to factor expressions like $4x - 12 + 24y$.

e Collect like terms.

SKILL TO REVIEW

Objective 10.3a: Add real numbers.

Add.

1. $-16 + 5$

2. $29 + (-23)$

Complete the table by evaluating each expression for the given values.

1.

VALUE	x + x	2x
$x = 3$		
$x = -6$		
$x = 4.8$		

2.

VALUE	x + 3x	5x
$x = 2$		
$x = -6$		
$x = 4.8$		

Answers

Skill to Review:
1. −11   2. 6

Margin Exercises:
1. 6, 6; −12, −12; 9.6, 9.6   2. 8, 10; −24, −30; 19.2, 24

3. Write a fraction expression equivalent to $\frac{3}{4}$ with a denominator of 8: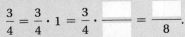

$$\frac{3}{4} = \frac{3}{4} \cdot 1 = \frac{3}{4} \cdot \frac{}{} = \frac{}{8}.$$

4. Write a fraction expression equivalent to $\frac{3}{4}$ with a denominator of $4t$:

$$\frac{3}{4} = \frac{3}{4} \cdot 1 = \frac{3}{4} \cdot \frac{}{} = \frac{}{4t}.$$

Simplify.

5. $\dfrac{3y}{4y}$

6. $-\dfrac{16m}{12m}$

7. $\dfrac{5xy}{40y}$

8. $\dfrac{18p}{24pq} = \dfrac{6p \cdot 3}{6p \cdot \boxed{}}$

$\qquad = \dfrac{6p}{6p} \cdot \dfrac{\boxed{}}{4q}$

$\qquad = 1 \cdot \dfrac{3}{4q} = \dfrac{3}{4q}$

9. Evaluate $x + y$ and $y + x$ when $x = -2$ and $y = 3$.

10. Evaluate xy and yx when $x = -2$ and $y = 5$.

Answers

3. $\frac{6}{8}$ **4.** $\frac{3t}{4t}$ **5.** $\frac{3}{4}$ **6.** $-\frac{4}{3}$ **7.** $\frac{x}{8}$

8. $\frac{3}{4q}$ **9.** $1; 1$ **10.** $-10; -10$

Guided Solutions:

3. $\frac{2}{2}, 6$ **4.** $\frac{t}{t}, 3t$ **8.** $4q, 3$

Note that $3x = 3 \cdot x$. We want fraction notation for $\frac{2}{3}$ that has a denominator of $3x$, but the denominator 3 is missing a factor of x. Thus we multiply by 1, using x/x as an equivalent expression for 1:

$$\frac{2}{3} = \frac{2}{3} \cdot 1 = \frac{2}{3} \cdot \frac{x}{x} = \frac{2x}{3x}.$$

The expressions $2/3$ and $2x/(3x)$ are equivalent. They have the same value for any allowable replacement. Note that $2x/3x$ is not defined for a replacement of 0, but for all nonzero real numbers, the expressions $2/3$ and $2x/(3x)$ have the same value.

◀ **Do Exercises 3 and 4.**

In algebra, we consider an expression like $2/3$ to be "simplified" from $2x/(3x)$. To find such simplified expressions, we use the identity property of 1 to remove a factor of 1.

EXAMPLE 2 Simplify: $-\dfrac{20x}{12x}$.

$$-\frac{20x}{12x} = -\frac{5 \cdot 4x}{3 \cdot 4x} \qquad \text{We look for the largest factor common to both the numerator and the denominator and factor each.}$$

$$= -\frac{5}{3} \cdot \frac{4x}{4x} \qquad \text{Factoring the fraction expression}$$

$$= -\frac{5}{3} \cdot 1 \qquad \frac{4x}{4x} = 1$$

$$= -\frac{5}{3} \qquad \text{Removing a factor of 1 using the identity property of 1}$$

EXAMPLE 3 Simplify: $\dfrac{14ab}{56a}$.

$$\frac{14ab}{56a} = \frac{14a \cdot b}{14a \cdot 4} = \frac{14a}{14a} \cdot \frac{b}{4} = 1 \cdot \frac{b}{4} = \frac{b}{4}$$

◀ **Do Exercises 5–8.**

b THE COMMUTATIVE LAWS AND THE ASSOCIATIVE LAWS

The Commutative Laws

Let's examine the expressions $x + y$ and $y + x$, as well as xy and yx.

EXAMPLE 4 Evaluate $x + y$ and $y + x$ when $x = 4$ and $y = 3$.

We substitute 4 for x and 3 for y in both expressions:

$$x + y = 4 + 3 = 7; \qquad y + x = 3 + 4 = 7.$$

EXAMPLE 5 Evaluate xy and yx when $x = 3$ and $y = -12$.

We substitute 3 for x and -12 for y in both expressions:

$$xy = 3 \cdot (-12) = -36; \qquad yx = (-12) \cdot 3 = -36.$$

◀ **Do Exercises 9 and 10.**

The expressions $x + y$ and $y + x$ have the same values no matter what the variables stand for. Thus they are equivalent. Therefore, when we add two numbers, the order in which we add does not matter. Similarly, the expressions xy and yx are equivalent. They also have the same values, no matter what the variables stand for. Therefore, when we multiply two numbers, the order in which we multiply does not matter.

The following are examples of general patterns or laws.

THE COMMUTATIVE LAWS

Addition.   For any numbers a and b,

$$a + b = b + a.$$

(We can change the order when adding without affecting the answer.)

Multiplication.   For any numbers a and b,

$$ab = ba.$$

(We can change the order when multiplying without affecting the answer.)

Using a commutative law, we know that $x + 2$ and $2 + x$ are equivalent. Similarly, $3x$ and $x(3)$ are equivalent. Thus, in an algebraic expression, we can replace one with the other and the result will be equivalent to the original expression.

EXAMPLE 6   Use the commutative laws to write an equivalent expression:
(a) $y + 5$; **(b)** mn; **(c)** $7 + xy$.

a) An expression equivalent to $y + 5$ is $5 + y$ by the commutative law of addition.

b) An expression equivalent to mn is nm by the commutative law of multiplication.

c) An expression equivalent to $7 + xy$ is $xy + 7$ by the commutative law of addition. Another expression equivalent to $7 + xy$ is $7 + yx$ by the commutative law of multiplication. Another equivalent expression is $yx + 7$.

Do Exercises 11–13. ▶

The Associative Laws

Now let's examine the expressions $a + (b + c)$ and $(a + b) + c$. Note that these expressions involve the use of parentheses as *grouping* symbols, and they also involve three numbers. Calculations within parentheses are to be done first.

EXAMPLE 7   Calculate and compare: $3 + (8 + 5)$ and $(3 + 8) + 5$.

$$3 + (8 + 5) = 3 + 13 \qquad \text{Calculating within parentheses first;} $$
$$\text{adding 8 and 5}$$
$$= 16;$$
$$(3 + 8) + 5 = 11 + 5 \qquad \text{Calculating within parentheses first;}$$
$$\text{adding 3 and 8}$$
$$= 16$$

Use a commutative law to write an equivalent expression.

11. $x + 9$

12. pq

13. $xy + t$

Answers

11. $9 + x$ **12.** qp
13. $t + xy$, or $yx + t$, or $t + yx$

The two expressions in Example 7 name the same number. Moving the parentheses to group the additions differently does not affect the value of the expression.

14. Calculate and compare:
$8 + (9 + 2)$ and $(8 + 9) + 2$.

EXAMPLE 8 Calculate and compare: $3 \cdot (4 \cdot 2)$ and $(3 \cdot 4) \cdot 2$.

$$3 \cdot (4 \cdot 2) = 3 \cdot 8 = 24; \qquad (3 \cdot 4) \cdot 2 = 12 \cdot 2 = 24$$

15. Calculate and compare:
$10 \cdot (5 \cdot 3)$ and $(10 \cdot 5) \cdot 3$.

◀ Do Exercises 14 and 15.

You may have noted that when only addition is involved, numbers can be grouped in any way we please without affecting the answer. When only multiplication is involved, numbers can also be grouped in any way we please without affecting the answer.

THE ASSOCIATIVE LAWS

Addition. For any numbers a, b, and c,
$$a + (b + c) = (a + b) + c.$$
(Numbers can be grouped in any manner for addition.)

Multiplication. For any numbers a, b, and c,
$$a \cdot (b \cdot c) = (a \cdot b) \cdot c.$$
(Numbers can be grouped in any manner for multiplication.)

EXAMPLE 9 Use an associative law to write an equivalent expression: **(a)** $(y + z) + 3$; **(b)** $8(xy)$.

Use an associative law to write an equivalent expression.

16. $r + (s + 7)$

a) An expression equivalent to $(y + z) + 3$ is $y + (z + 3)$ by the associative law of addition.

b) An expression equivalent to $8(xy)$ is $(8x)y$ by the associative law of multiplication.

17. $9(ab)$

◀ Do Exercises 16 and 17.

The associative laws say that numbers can be grouped in any way we please when only additions or only multiplications are involved. Thus we often omit the parentheses. For example,

$$x + (y + 2) \quad \text{means} \quad x + y + 2, \quad \text{and} \quad (lw)h \quad \text{means} \quad lwh.$$

Using the Commutative Laws and the Associative Laws Together

EXAMPLE 10 Use the commutative laws and the associative laws to write at least three expressions equivalent to $(x + 5) + y$.

a) $(x + 5) + y = x + (5 + y)$ Using the associative law first and then
$= x + (y + 5)$ using the commutative law

b) $(x + 5) + y = y + (x + 5)$ Using the commutative law twice
$= y + (5 + x)$

c) $(x + 5) + y = (5 + x) + y$ Using the commutative law first and
$= 5 + (x + y)$ then the associative law

Answers
14. 19; 19   **15.** 150; 150   **16.** $(r + s) + 7$
17. $(9a)b$

EXAMPLE 11 Use the commutative laws and the associative laws to write at least three expressions equivalent to $(3x)y$.

a) $(3x)y = 3(xy)$ Using the associative law first and then using the commutative law
$\quad\quad\quad = 3(yx)$

b) $(3x)y = y(3x)$ Using the commutative law twice
$\quad\quad\quad = y(x \cdot 3)$

c) $(3x)y = (x \cdot 3)y$ Using the commutative law, and then the associative law, and then the commutative law again
$\quad\quad\quad = x(3y)$
$\quad\quad\quad = x(y \cdot 3)$

Do Exercises 18 and 19. ▶

Use the commutative laws and the associative laws to write at least three equivalent expressions.

18. $4(tu)$

19. $r + (2 + s)$

C   THE DISTRIBUTIVE LAWS

The *distributive laws* are the basis of many procedures in both arithmetic and algebra. They are probably the most important laws that we use to manipulate algebraic expressions. The distributive law of multiplication over addition involves two operations: addition and multiplication.

Let's begin by considering a multiplication problem from arithmetic:

$$
\begin{array}{r}
4\ 5 \\
\times\ \ 7 \\
\hline
3\ 5 \\
2\ 8\ 0 \\
\hline
3\ 1\ 5
\end{array}
$$

← This is $7 \cdot 5$.
← This is $7 \cdot 40$.
← This is the sum $7 \cdot 5 + 7 \cdot 40$.

To carry out the multiplication, we actually added two products. That is,

$$7 \cdot 45 = 7(5 + 40) = 7 \cdot 5 + 7 \cdot 40.$$

Let's examine this further. If we wish to multiply a sum of several numbers by a factor, we can either add and then multiply, or multiply and then add.

EXAMPLE 12 Compute in two ways: $5 \cdot (4 + 8)$.

a) $\quad 5 \cdot (\underbrace{4 + 8})$ Adding within parentheses first, and then multiplying

$= 5 \cdot \quad 12$
$= 60$

b) $5 \cdot (4 + 8) = \underbrace{(5 \cdot 4)} + \underbrace{(5 \cdot 8)}$ Distributing the multiplication to terms within parentheses first and then adding

$\quad\quad\quad\quad = \quad 20 \ \ + \ \ 40$
$\quad\quad\quad\quad = \quad 60$

Do Exercises 20–22. ▶

Compute.

20. a) $7 \cdot (3 + 6)$
  b) $(7 \cdot 3) + (7 \cdot 6)$

21. a) $2 \cdot (10 + 30)$
  b) $(2 \cdot 10) + (2 \cdot 30)$

22. a) $(2 + 5) \cdot 4$
  b) $(2 \cdot 4) + (5 \cdot 4)$

THE DISTRIBUTIVE LAW OF MULTIPLICATION OVER ADDITION

For any numbers a, b, and c,

$$a(b + c) = ab + ac.$$

Answers

18. $(4t)u, (tu)4, t(4u)$; answers may vary
19. $(2 + r) + s, (r + s) + 2, s + (r + 2)$; answers may vary   **20. (a)** $7 \cdot 9 = 63$;
(b) $21 + 42 = 63$ **21. (a)** $2 \cdot 40 = 80$;
(b) $20 + 60 = 80$ **22. (a)** $7 \cdot 4 = 28$;
(b) $8 + 20 = 28$

In the statement of the distributive law, we know that in an expression such as $ab + ac$, the multiplications are to be done first according to the rules for order of operations. So, instead of writing $(4 \cdot 5) + (4 \cdot 7)$, we can write $4 \cdot 5 + 4 \cdot 7$. However, in $a(b + c)$, we cannot omit the parentheses. If we did, we would have $ab + c$, which means $(ab) + c$. For example, $3(4 + 2) = 3(6) = 18$, but $3 \cdot 4 + 2 = 12 + 2 = 14$.

Another distributive law relates multiplication and subtraction. This law says that to multiply by a difference, we can either subtract and then multiply, or multiply and then subtract.

THE DISTRIBUTIVE LAW OF MULTIPLICATION OVER SUBTRACTION

For any numbers a, b, and c,
$$a(b - c) = ab - ac.$$

Calculate.

23. a) $4(5 - 3)$
   b) $4 \cdot 5 - 4 \cdot 3$

24. a) $-2 \cdot (5 - 3)$
   b) $-2 \cdot 5 - (-2) \cdot 3$

25. a) $5 \cdot (2 - 7)$
   b) $5 \cdot 2 - 5 \cdot 7$

We often refer to "*the* distributive law" when we mean *either* or *both* of these laws.

◀ **Do Exercises 23–25.**

What do we mean by the *terms* of an expression? **Terms** are separated by addition signs. If there are subtraction signs, we can find an equivalent expression that uses addition signs.

EXAMPLE 13 What are the terms of $3x - 4y + 2z$?

We have
$$3x - 4y + 2z = 3x + (-4y) + 2z.$$ Separating parts with $+$ signs

The terms are $3x$, $-4y$, and $2z$.

◀ **Do Exercises 26 and 27.**

What are the terms of each expression?

26. $5x - 8y + 3$

27. $-4y - 2x + 3z$

The distributive laws are a basis for **multiplying** algebraic expressions. In an expression like $8(a + 2b - 7)$, we multiply each term inside the parentheses by 8:

$$8(a + 2b - 7) = 8 \cdot a + 8 \cdot 2b - 8 \cdot 7 = 8a + 16b - 56.$$

EXAMPLES Multiply.

14. $9(x - 5) = 9 \cdot x - 9 \cdot 5$ Using the distributive law of multiplication over subtraction

    $= 9x - 45$

15. $\frac{2}{3}(w + 1) = \frac{2}{3} \cdot w + \frac{2}{3} \cdot 1$ Using the distributive law of multiplication over addition

    $= \frac{2}{3}w + \frac{2}{3}$

16. $\frac{4}{3}(s - t + w) = \frac{4}{3}s - \frac{4}{3}t + \frac{4}{3}w$ Using both distributive laws

◀ **Do Exercises 28–30.**

Multiply.

28. $3(x - 5)$

29. $5(x + 1)$

30. $\frac{3}{5}(p + q - t)$

Answers

23. (a) $4 \cdot 2 = 8$; (b) $20 - 12 = 8$
24. (a) $-2 \cdot 2 = -4$; (b) $-10 + 6 = -4$
25. (a) $5(-5) = -25$; (b) $10 - 35 = -25$
26. $5x, -8y, 3$ **27.** $-4y, -2x, 3z$ **28.** $3x - 15$
29. $5x + 5$ **30.** $\frac{3}{5}p + \frac{3}{5}q - \frac{3}{5}t$

EXAMPLE 17 Multiply: $-4(x - 2y + 3z)$.

$$-4(x - 2y + 3z) = -4 \cdot x - (-4)(2y) + (-4)(3z)$$

Using both distributive laws

$$= -4x - (-8y) + (-12z)$$

Multiplying

$$= -4x + 8y - 12z$$

We can also do this problem by first finding an equivalent expression with all plus signs and then multiplying:

$$-4(x - 2y + 3z) = -4[x + (-2y) + 3z]$$
$$= -4 \cdot x + (-4)(-2y) + (-4)(3z)$$
$$= -4x + 8y - 12z.$$

Do Exercises 31–33. ▶

EXAMPLES Name the property or the law illustrated by each equation.

Equation	*Property*
18. $5x = x(5)$	Commutative law of multiplication
19. $a + (8.5 + b) = (a + 8.5) + b$	Associative law of addition
20. $0 + 11 = 11$	Identity property of 0
21. $(-5s)t = -5(st)$	Associative law of multiplication
22. $\frac{3}{4} \cdot 1 = \frac{3}{4}$	Identity property of 1
23. $12.5(w - 3) = 12.5w - 12.5(3)$	Distributive law of multiplication over subtraction
24. $y + \frac{1}{2} = \frac{1}{2} + y$	Commutative law of addition

Do Exercises 34–40. ▶

Multiply.

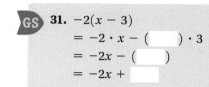

GS **31.** $-2(x - 3)$
$$= -2 \cdot x - (\quad) \cdot 3$$
$$= -2x - (\quad)$$
$$= -2x + \boxed{}$$

32. $5(x - 2y + 4z)$

33. $-5(x - 2y + 4z)$

Name the property or the law illustrated by each equation.

34. $(-8a)b = -8(ab)$

35. $p \cdot 1 = p$

36. $m + 34 = 34 + m$

37. $2(t + 5) = 2t + 2(5)$

38. $0 + k = k$

39. $-8x = x(-8)$

40. $x + (4.3 + b) = (x + 4.3) + b$

---

d FACTORING

Factoring is the reverse of multiplying. To factor, we can use the distributive laws in reverse:

$$ab + ac = a(b + c) \quad \text{and} \quad ab - ac = a(b - c).$$

> **FACTORING**
>
> To **factor** an expression is to find an equivalent expression that is a product.

To factor $9x - 45$, for example, we find an equivalent expression that is a product: $9(x - 5)$. This reverses the multiplication that we did in Example 14. When all the terms of an expression have a factor in common, we can "factor it out" using the distributive laws. Note the following.

$9x$ has the factors $9, -9, 3, -3, 1, -1, x, -x, 3x, -3x, 9x, -9x$;

-45 has the factors $1, -1, 3, -3, 5, -5, 9, -9, 15, -15, 45, -45$

We generally remove the largest common factor. In this case, that factor is 9. Thus,

$$9x - 45 = 9 \cdot x - 9 \cdot 5$$
$$= 9(x - 5).$$

Remember that an expression has been factored when we have found an equivalent expression that is a product. Above, we note that $9x - 45$ and $9(x - 5)$ are equivalent expressions. The expression $9x - 45$ is the difference of $9x$ and 45; the expression $9(x - 5)$ is the product of 9 and $(x - 5)$.

EXAMPLES Factor.

25. $5x - 10 = 5 \cdot x - 5 \cdot 2$ Try to do this step mentally.

$ = 5(x - 2)$ You can check by multiplying.

26. $ax - ay + az = a(x - y + z)$

27. $9x + 27y - 9 = 9 \cdot x + 9 \cdot 3y - 9 \cdot 1 = 9(x + 3y - 1)$

Note in Example 27 that you might, at first, just factor out a 3, as follows:

$$9x + 27y - 9 = 3 \cdot 3x + 3 \cdot 9y - 3 \cdot 3$$
$$= 3(3x + 9y - 3).$$

At this point, the mathematics is correct, but the answer is not because there is another factor of 3 that can be factored out, as follows:

$$3 \cdot 3x + 3 \cdot 9y - 3 \cdot 3 = 3(3x + 9y - 3)$$
$$= 3(3 \cdot x + 3 \cdot 3y - 3 \cdot 1)$$
$$= 3 \cdot 3(x + 3y - 1)$$
$$= 9(x + 3y - 1).$$

We now have a correct answer, but it took more work than we did in Example 27. Thus it is better to look for the *greatest common factor* at the outset.

EXAMPLES Factor. Try to write just the answer, if you can.

28. $5x - 5y = 5(x - y)$

29. $-3x + 6y - 9z = -3(x - 2y + 3z)$

We generally factor out a negative factor when the first term is negative. The way we factor can depend on the situation in which we are working. We might also factor the expression in Example 29 as follows:

$$-3x + 6y - 9z = 3(-x + 2y - 3z).$$

30. $18z - 12x - 24 = 6(3z - 2x - 4)$

31. $\frac{1}{2}x + \frac{3}{2}y - \frac{1}{2} = \frac{1}{2}(x + 3y - 1)$

Remember that you can always check factoring by multiplying. Keep in mind that an expression is factored when it is written as a product.

◀ **Do Exercises 41–46.**

Factor.

41. $6x - 12$

42. $3x - 6y + 9$

43. $bx + by - bz$

44. $16a - 36b + 42$ **GS**
$= 2 \cdot 8a - \boxed{} \cdot 18b + 2 \cdot 21$
$= \boxed{} (8a - 18b + 21)$

45. $\dfrac{3}{8}x - \dfrac{5}{8}y + \dfrac{7}{8}$

46. $-12x + 32y - 16z$

e COLLECTING LIKE TERMS

Terms such as $5x$ and $-4x$, whose variable factors are exactly the same, are called **like terms**. Similarly, numbers, such as -7 and 13, are like terms. Also, $3y^2$ and $9y^2$ are like terms because the variables are raised to the same power. Terms such as $4y$ and $5y^2$ are not like terms, and $7x$ and $2y$ are not like terms.

The process of **collecting like terms** is also based on the distributive laws. We can apply a distributive law when a factor is on the right-hand side because of the commutative law of multiplication.

Later in this text, terminology like "collecting like terms" and "combining like terms" will also be referred to as "simplifying."

EXAMPLES Collect like terms. Try to write just the answer, if you can.

32. $4x + 2x = (4 + 2)x = 6x$ Factoring out x using a distributive law

33. $2x + 3y - 5x - 2y = 2x - 5x + 3y - 2y$
$$= (2 - 5)x + (3 - 2)y = -3x + 1y = -3x + y$$

34. $3x - x = 3x - 1x = (3 - 1)x = 2x$

35. $x - 0.24x = 1 \cdot x - 0.24x = (1 - 0.24)x = 0.76x$

36. $x - 6x = 1 \cdot x - 6 \cdot x = (1 - 6)x = -5x$

37. $4x - 7y + 9x - 5 + 3y - 8 = 13x - 4y - 13$

38. $\frac{2}{3}a - b + \frac{4}{5}a + \frac{1}{4}b - 10 = \frac{2}{3}a - 1 \cdot b + \frac{4}{5}a + \frac{1}{4}b - 10$
$$= (\tfrac{2}{3} + \tfrac{4}{5})a + (-1 + \tfrac{1}{4})b - 10$$
$$= (\tfrac{10}{15} + \tfrac{12}{15})a + (-\tfrac{4}{4} + \tfrac{1}{4})b - 10$$
$$= \tfrac{22}{15}a - \tfrac{3}{4}b - 10$$

Do Exercises 47–53.

Collect like terms.

47. $6x - 3x$ **48.** $7x - x$

49. $x - 9x$ **50.** $x - 0.41x$

51. $5x + 4y - 2x - y$

GS 52. $3x - 7x - 11 + 8y + 4 - 13y$
$$= (3 - \boxed{})x + (8 - 13)y + (\boxed{} + 4)$$
$$= \boxed{}\ x + (\boxed{})y + (\boxed{})$$
$$= -4x - 5y - 7$$

53. $-\dfrac{2}{3} - \dfrac{3}{5}x + y + \dfrac{7}{10}x - \dfrac{2}{9}y$

Answers

47. $3x$ **48.** $6x$ **49.** $-8x$ **50.** $0.59x$
51. $3x + 3y$ **52.** $-4x - 5y - 7$
53. $\dfrac{1}{10}x + \dfrac{7}{9}y - \dfrac{2}{3}$

Guided Solution:
52. $7, -11, -4, -5, -7$

10.7 Exercise Set

For Extra Help

MyMathLab®   MathXL°   PRACTICE   WATCH   READ   REVIEW

✓ Reading Check

Choose from the column on the right an equation that illustrates the property or law.

RC1. Associative law of multiplication

RC2. Identity property of 1

RC3. Distributive law of multiplication over subtraction

RC4. Commutative law of addition

RC5. Identity property of 0

RC6. Commutative law of multiplication

RC7. Associative law of addition

a) $3 \cdot 5 = 5 \cdot 3$

b) $8 + (\tfrac{1}{2} + 9) = (8 + \tfrac{1}{2}) + 9$

c) $5(6 + 3) = 5 \cdot 6 + 5 \cdot 3$

d) $3 + 0 = 3$

e) $3 + 5 = 5 + 3$

f) $5(6 - 3) = 5 \cdot 6 - 5 \cdot 3$

g) $8 \cdot \left(\dfrac{1}{2} \cdot 9\right) = \left(8 \cdot \dfrac{1}{2}\right) \cdot 9$

h) $\dfrac{6}{5} \cdot 1 = \dfrac{6}{5}$

a Find an equivalent expression with the given denominator.

1. $\dfrac{3}{5} = \dfrac{\square}{5y}$

2. $\dfrac{5}{8} = \dfrac{\square}{8t}$

3. $\dfrac{2}{3} = \dfrac{\square}{15x}$

4. $\dfrac{6}{7} = \dfrac{\square}{14y}$

5. $\dfrac{2}{x} = \dfrac{\square}{x^2}$

6. $\dfrac{4}{9x} = \dfrac{\square}{9xy}$

Simplify.

7. $-\dfrac{24a}{16a}$

8. $-\dfrac{42t}{18t}$

9. $-\dfrac{42ab}{36ab}$

10. $-\dfrac{64pq}{48pq}$

11. $\dfrac{20st}{15t}$

12. $\dfrac{21w}{7wz}$

b Write an equivalent expression. Use a commutative law.

13. $y + 8$

14. $x + 3$

15. mn

16. yz

17. $9 + xy$

18. $11 + ab$

19. $ab + c$

20. $rs + t$

Write an equivalent expression. Use an associative law.

21. $a + (b + 2)$

22. $3(vw)$

23. $(8x)y$

24. $(y + z) + 7$

25. $(a + b) + 3$

26. $(5 + x) + y$

27. $3(ab)$

28. $(6x)y$

Use the commutative laws and the associative laws to write three equivalent expressions.

29. $(a + b) + 2$

30. $(3 + x) + y$

31. $5 + (v + w)$

32. $6 + (x + y)$

33. $(xy)3$

34. $(ab)5$

35. $7(ab)$

36. $5(xy)$

c Multiply.

37. $2(b + 5)$

38. $4(x + 3)$

39. $7(1 + t)$

40. $4(1 + y)$

41. $6(5x + 2)$

42. $9(6m + 7)$

43. $7(x + 4 + 6y)$

44. $4(5x + 8 + 3p)$

45. $7(x - 3)$

46. $15(y - 6)$

47. $-3(x - 7)$

48. $1.2(x - 2.1)$

49. $\dfrac{2}{3}(b - 6)$

50. $\dfrac{5}{8}(y + 16)$

51. $7.3(x - 2)$

52. $5.6(x - 8)$

53. $-\dfrac{3}{5}(x - y + 10)$

54. $-\dfrac{2}{3}(a + b - 12)$

55. $-9(-5x - 6y + 8)$

56. $-7(-2x - 5y + 9)$

57. $-4(x - 3y - 2z)$

58. $8(2x - 5y - 8z)$

59. $3.1(-1.2x + 3.2y - 1.1)$

60. $-2.1(-4.2x - 4.3y - 2.2)$

List the terms of each expression.

61. $4x + 3z$

62. $8x - 1.4y$

63. $7x + 8y - 9z$

64. $8a + 10b - 18c$

d   Factor. Check by multiplying.

65. $2x + 4$

66. $5y + 20$

67. $30 + 5y$

68. $7x + 28$

69. $14x + 21y$

70. $18a + 24b$

71. $14t - 7$

72. $25m - 5$

73. $8x - 24$

74. $10x - 50$

75. $18a - 24b$

76. $32x - 20y$

77. $-4y + 32$ **78.** $-6m + 24$ **79.** $5x + 10 + 15y$ **80.** $9a + 27b + 81$

81. $16m - 32n + 8$ **82.** $6x + 10y - 2$ **83.** $12a + 4b - 24$ **84.** $8m - 4n + 12$

85. $8x + 10y - 22$ **86.** $9a + 6b - 15$ **87.** $ax - a$ **88.** $by - 9b$

89. $ax - ay - az$ **90.** $cx + cy - cz$ **91.** $-18x + 12y + 6$ **92.** $-14x + 21y + 7$

93. $\dfrac{2}{3}x - \dfrac{5}{3}y + \dfrac{1}{3}$ **94.** $\dfrac{3}{5}a + \dfrac{4}{5}b - \dfrac{1}{5}$ **95.** $36x - 6y + 18z$ **96.** $8a - 4b + 20c$

e   Collect like terms.

97. $9a + 10a$ **98.** $12x + 2x$ **99.** $10a - a$

100. $-16x + x$ **101.** $2x + 9z + 6x$ **102.** $3a - 5b + 7a$

103. $7x + 6y^2 + 9y^2$ **104.** $12m^2 + 6q + 9m^2$ **105.** $41a + 90 - 60a - 2$

106. $42x - 6 - 4x + 2$ **107.** $23 + 5t + 7y - t - y - 27$ **108.** $45 - 90d - 87 - 9d + 3 + 7d$

109. $\dfrac{1}{2}b + \dfrac{1}{2}b$ **110.** $\dfrac{2}{3}x + \dfrac{1}{3}x$ **111.** $2y + \dfrac{1}{4}y + y$

112. $\dfrac{1}{2}a + a + 5a$ **113.** $11x - 3x$ **114.** $9t - 17t$

115. $6n - n$

116. $100t - t$

117. $y - 17y$

118. $3m - 9m + 4$

119. $-8 + 11a - 5b + 6a - 7b + 7$

120. $8x - 5x + 6 + 3y - 2y - 4$

121. $9x + 2y - 5x$

122. $8y - 3z + 4y$

123. $11x + 2y - 4x - y$

124. $13a + 9b - 2a - 4b$

125. $2.7x + 2.3y - 1.9x - 1.8y$

126. $6.7a + 4.3b - 4.1a - 2.9b$

127. $\dfrac{13}{2}a + \dfrac{9}{5}b - \dfrac{2}{3}a - \dfrac{3}{10}b - 42$

128. $\dfrac{11}{4}x + \dfrac{2}{3}y - \dfrac{4}{5}x - \dfrac{1}{6}y + 12$

Skill Maintenance

Compute and simplify. [10.3a], [10.4a], [10.5a], [10.6a, c]

129. $18 - (-20)$

130. $-3.8 + (-1.1)$

131. $-\dfrac{4}{15} \cdot (-15)$

132. $-500 \div (-50)$

133. $\dfrac{2}{7} \div \left(-\dfrac{7}{2}\right)$

134. $2 \cdot (-53)$

135. $-\dfrac{1}{2} + \dfrac{3}{2}$

136. $-6 - 28$

137. Evaluate $9w$ when $w = 20$. [10.1a]

138. Find the absolute value: $\left| -\dfrac{4}{13} \right|$. [10.2e]

Write true or false. [10.2d]

139. $-43 < -40$

140. $-3 \geq 0$

141. $-6 \leq -6$

142. $0 > -4$

Synthesis

Determine whether the expressions are equivalent. Explain why if they are. Give an example if they are not. Examples may vary.

143. $3t + 5$ and $3 \cdot 5 + t$

144. $4x$ and $x + 4$

145. $5m + 6$ and $6 + 5m$

146. $(x + y) + z$ and $z + (x + y)$

147. Factor: $q + qr + qrs + qrst$.

148. Collect like terms:
$21x + 44xy + 15y - 16x - 8y - 38xy + 2y + xy.$

Simplifying Expressions; Order of Operations

OBJECTIVES

a Find an equivalent expression for an opposite without parentheses, where an expression has several terms.

b Simplify expressions by removing parentheses and collecting like terms.

c Simplify expressions with parentheses inside parentheses.

d Simplify expressions using the rules for order of operations.

SKILL TO REVIEW

Objective 10.7c: Use the distributive laws to multiply expressions like 8 and $x - y$.

Multiply.

1. $4(x + 5)$

2. $-7(a + b)$

We now expand our ability to manipulate expressions by first considering opposites of sums and differences. Then we simplify expressions involving parentheses.

a OPPOSITES OF SUMS

What happens when we multiply a real number by -1? Consider the following products:

$$-1(7) = -7, \quad -1(-5) = 5, \quad -1(0) = 0.$$

From these examples, it appears that when we multiply a number by -1, we get the opposite, or additive inverse, of that number.

THE PROPERTY OF -1

For any real number a,

$$-1 \cdot a = -a.$$

(Negative one times a is the opposite, or additive inverse, of a.)

The property of -1 enables us to find expressions equivalent to opposites of sums.

EXAMPLES Find an equivalent expression without parentheses.

1. $-(3 + x) = -1(3 + x)$ Using the property of -1

$\qquad\qquad\quad = -1 \cdot 3 + (-1)x$ Using a distributive law, multiplying each term by -1

$\qquad\qquad\quad = -3 + (-x)$ Using the property of -1

$\qquad\qquad\quad = -3 - x$

2. $-(3x + 2y + 4) = -1(3x + 2y + 4)$ Using the property of -1

$\qquad\qquad\qquad\quad = -1(3x) + (-1)(2y) + (-1)4$ Using a distributive law

$\qquad\qquad\qquad\quad = -3x - 2y - 4$ Using the property of -1

◀ Do Margin Exercises 1 and 2.

Find an equivalent expression without parentheses.

1. $-(x + 2)$

2. $-(5x + 2y + 8)$

Suppose we want to remove parentheses in an expression like

$$-(x - 2y + 5).$$

We can first rewrite any subtractions inside the parentheses as additions. Then we take the opposite of each term:

$$-(x - 2y + 5) = -[x + (-2y) + 5]$$
$$= -x + 2y + (-5) = -x + 2y - 5.$$

The most efficient method for removing parentheses is to replace each term in the parentheses with its opposite ("change the sign of every term"). Doing so for $-(x - 2y + 5)$, we obtain $-x + 2y - 5$ as an equivalent expression.

Answers

Skill to Review:
1. $4x + 20$ 2. $-7a - 7b$

Margin Exercises:
1. $-x - 2$ 2. $-5x - 2y - 8$

EXAMPLES Find an equivalent expression without parentheses.

3. $-(5 - y) = -5 + y$ Changing the sign of each term

4. $-(2a - 7b - 6) = -2a + 7b + 6$

5. $-(-3x + 4y + z - 7w - 23) = 3x - 4y - z + 7w + 23$

Do Exercises 3–6. ▶

b   REMOVING PARENTHESES AND SIMPLIFYING

When a sum is added to another expression, as in $5x + (2x + 3)$, we can simply remove, or drop, the parentheses and collect like terms because of the associative law of addition:

$$5x + (2x + 3) = 5x + 2x + 3 = 7x + 3.$$

On the other hand, when a sum is subtracted from another expression, as in $3x - (4x + 2)$, we cannot simply drop the parentheses. However, we can subtract by adding an opposite. We then remove parentheses by changing the sign of each term inside the parentheses and collecting like terms.

EXAMPLE 6 Remove parentheses and simplify.

$$3x - (4x + 2) = 3x + [-(4x + 2)] \quad \text{Adding the opposite of } (4x + 2)$$

$$= 3x + (-4x - 2) \quad \text{Changing the sign of each term inside the parentheses}$$

$$= 3x - 4x - 2$$

$$= -x - 2 \quad \text{Collecting like terms}$$

···················· **Caution!** ····················

Note that $3x - (4x + 2) \neq 3x - 4x + 2$. You cannot simply drop the parentheses.

Do Exercises 7 and 8. ▶

    In practice, the first three steps of Example 6 are usually combined by changing the sign of each term in parentheses and then collecting like terms.

EXAMPLES Remove parentheses and simplify.

7. $5y - (3y + 4) = 5y - 3y - 4$ Removing parentheses by changing the sign of every term inside the parentheses

$$= 2y - 4 \quad \text{Collecting like terms}$$

8. $3x - 2 - (5x - 8) = 3x - 2 - 5x + 8$

$$= -2x + 6$$

9. $(3a + 4b - 5) - (2a - 7b + 4c - 8)$

$$= 3a + 4b - 5 - 2a + 7b - 4c + 8$$

$$= a + 11b - 4c + 3$$

Do Exercises 9–11. ▶

Find an equivalent expression without parentheses. Try to do this in one step.

3. $-(6 - t)$

4. $-(x - y)$

5. $-(-4a + 3t - 10)$

6. $-(18 - m - 2n + 4z)$

Remove parentheses and simplify.

7. $5x - (3x + 9)$

8. $5y - 2 - (2y - 4)$

Remove parentheses and simplify.

9. $6x - (4x + 7)$

10. $8y - 3 - (5y - 6)$

11. $(2a + 3b - c) - (4a - 5b + 2c)$

Answers

3. $-6 + t$ **4.** $-x + y$ **5.** $4a - 3t + 10$
6. $-18 + m + 2n - 4z$ **7.** $2x - 9$
8. $3y + 2$ **9.** $2x - 7$ **10.** $3y + 3$
11. $-2a + 8b - 3c$

Next, consider subtracting an expression consisting of several terms multiplied by a number other than 1 or -1.

Remove parentheses and simplify.

12. $y - 9(x + y)$

13. $5a - 3(7a - 6)$
$= 5a - \boxed{} + \boxed{}$
$= \boxed{} + 18$

14. $4a - b - 6(5a - 7b + 8c)$

15. $5x - \dfrac{1}{4}(8x + 28)$

16. $4.6(5x - 3y) - 5.2(8x + y)$

EXAMPLE 10 Remove parentheses and simplify.

$$\begin{aligned} x - 3(x + y) &= x + [-3(x + y)] &&\text{Adding the opposite of } 3(x + y) \\ &= x + [-3x - 3y] &&\text{Multiplying } x + y \text{ by } -3 \\ &= x - 3x - 3y \\ &= -2x - 3y &&\text{Collecting like terms} \end{aligned}$$

EXAMPLES Remove parentheses and simplify.

11. $3y - 2(4y - 5) = 3y - 8y + 10$ Multiplying each term in the parentheses by -2

$$= -5y + 10$$

12. $(2a + 3b - 7) - 4(-5a - 6b + 12)$
$$= 2a + 3b - 7 + 20a + 24b - 48 = 22a + 27b - 55$$

13. $2y - \frac{1}{3}(9y - 12) = 2y - 3y + 4 = -y + 4$

14. $6(5x - 3y) - 2(8x + y) = 30x - 18y - 16x - 2y = 14x - 20y$

◀ Do Exercises 12–16.

c PARENTHESES WITHIN PARENTHESES

In addition to parentheses, some expressions contain other grouping symbols such as brackets [ ] and braces { }.

> When more than one kind of grouping symbol occurs, do the computations in the innermost ones first. Then work from the inside out.

Simplify.

17. $12 - (8 + 2)$

18. $9 - [10 - (13 + 6)]$
$= 9 - [10 - (\boxed{})]$
$= 9 - [\boxed{}]$
$= 9 + \boxed{}$
$= 18$

19. $[24 \div (-2)] \div (-2)$

20. $5(3 + 4) -$
$\{8 - [5 - (9 + 6)]\}$

EXAMPLES Simplify.

15. $2[3 - (7 + 3)] = 2[3 - 10] = 2[-7] = -14$

16. $8 - [9 - (12 + 5)] = 8 - [9 - 17]$ Computing $12 + 5$
$$\begin{aligned} &= 8 - [-8] &&\text{Computing } 9 - 17 \\ &= 8 + 8 = 16 \end{aligned}$$

17. $\left[-4 - 2\left(-\frac{1}{2}\right)\right] \div \frac{1}{4} = [-4 + 1] \div \frac{1}{4}$ Working within parentheses
$$\begin{aligned} &= -3 \div \frac{1}{4} &&\text{Computing } -4 + 1 \\ &= -3 \cdot 4 = -12 \end{aligned}$$

18. $4(2 + 3) - \{7 - [4 - (8 + 5)]\}$
$$\begin{aligned} &= 4(5) - \{7 - [4 - 13]\} &&\text{Working with the innermost parentheses first} \\ &= 4(5) - \{7 - [-9]\} &&\text{Computing } 4 - 13 \\ &= 4(5) - 16 &&\text{Computing } 7 - [-9] \\ &= 20 - 16 = 4 \end{aligned}$$

◀ Do Exercises 17–20.

Answers
12. $-9x - 8y$ 13. $-16a + 18$
14. $-26a + 41b - 48c$ 15. $3x - 7$
16. $-18.6x - 19y$ 17. 2   18. 18   19. 6
20. 17

Guided Solutions:
13. $21a, 18, -16a$ 18. $19, -9, 9$

EXAMPLE 19 Simplify.

$$[5(x + 2) - 3x] - [3(y + 2) - 7(y - 3)]$$
$$= [5x + 10 - 3x] - [3y + 6 - 7y + 21] \quad \text{Working with the innermost parentheses first}$$

$$= [2x + 10] - [-4y + 27] \quad \text{Collecting like terms within brackets}$$

$$= 2x + 10 + 4y - 27 \quad \text{Removing brackets}$$
$$= 2x + 4y - 17 \quad \text{Collecting like terms}$$

Do Exercise 21. ▶

21. Simplify:
$$[3(x + 2) + 2x] - [4(y + 2) - 3(y - 2)].$$

d ORDER OF OPERATIONS

When several operations are to be done in a calculation or a problem, we apply the following.

RULES FOR ORDER OF OPERATIONS

1. Do all calculations within grouping symbols before operations outside.
2. Evaluate all exponential expressions.
3. Do all multiplications and divisions in order from left to right.
4. Do all additions and subtractions in order from left to right.

These rules are consistent with the way in which most computers and scientific calculators perform calculations.

EXAMPLE 20 Simplify: $-34 \cdot 56 - 17$.

There are no parentheses or powers, so we start with the third step.

$$-34 \cdot 56 - 17 = -1904 - 17 \quad \text{Doing all multiplications and divisions in order from left to right}$$

$$= -1921 \quad \text{Doing all additions and subtractions in order from left to right} \quad ▦$$

EXAMPLE 21 Simplify: $25 \div (-5) + 50 \div (-2)$.

There are no calculations inside parentheses and no powers. The parentheses with (-5) and (-2) are used only to represent the negative numbers. We begin by doing all multiplications and divisions.

$$\underbrace{25 \div (-5)} + \underbrace{50 \div (-2)}$$

$$= -5 + (-25) \quad \text{Doing all multiplications and divisions in order from left to right}$$

$$= -30 \quad \text{Doing all additions and subtractions in order from left to right}$$

Do Exercises 22–24. ▶

Simplify.

22. $23 - 42 \cdot 30$

23. $32 \div 8 \cdot 2$

24. $-24 \div 3 - 48 \div (-4)$

Answers

21. $5x - y - 8$ **22.** -1237 **23.** 8   **24.** 4

EXAMPLE 22 Simplify: $-2^4 + 51 \cdot 4 - (37 + 23 \cdot 2)$.

$$-2^4 + 51 \cdot 4 - (37 + 23 \cdot 2)$$

$= -2^4 + 51 \cdot 4 - (37 + 46)$ Following the rules for order of operations within the parentheses first

$= -2^4 + 51 \cdot 4 - 83$ Completing the addition inside parentheses

$= -16 + 51 \cdot 4 - 83$ Evaluating exponential expressions. Note that $-2^4 \neq (-2)^4$.

$= -16 + 204 - 83$ Doing all multiplications

$= 188 - 83$ Doing all additions and subtractions in order from left to right

$= 105$

Simplify.

25. $-4^3 + 52 \cdot 5 + 5^3 - (4^2 - 48 \div 4)$ **(GS)**

$= \boxed{} + 52 \cdot 5 + 125 - (\boxed{} - 48 \div 4)$

$= -64 + 52 \cdot 5 + 125 - (16 - \boxed{})$

$= -64 + 52 \cdot 5 + 125 - 4$

$= -64 + \boxed{} + 125 - 4$

$= \boxed{} + 125 - 4$

$= 321 - 4$

$= 317$

A fraction bar can play the role of a grouping symbol.

EXAMPLE 23 Simplify: $\dfrac{-64 \div (-16) \div (-2)}{2^3 - 3^2}$.

An equivalent expression with brackets as grouping symbols is

$$[-64 \div (-16) \div (-2)] \div [2^3 - 3^2].$$

This shows, in effect, that we do the calculations in the numerator and then in the denominator, and divide the results:

$$\frac{-64 \div (-16) \div (-2)}{2^3 - 3^2} = \frac{4 \div (-2)}{8 - 9} = \frac{-2}{-1} = 2.$$

26. $\dfrac{5 - 10 - 5 \cdot 23}{2^3 + 3^2 - 7}$

◀ **Do Exercises 25 and 26.**

Answers

25. 317   **26.** −12

Guided Solution:
25. −64, 16, 12, 260, 196

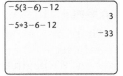

CALCULATOR CORNER

Order of Operations and Grouping Symbols   Parentheses are necessary in some calculations. To simplify $-5(3 - 6) - 12$, we must use parentheses. The result is 3. Without parentheses, the computation is $-5 \cdot 3 - 6 - 12$, and the result is −33.

When a negative number is raised to an even power, parentheses also must be used. To find −3 raised to the fourth power, for example, we must use parentheses. The result is 81. Without parentheses, the computation is $-3^4 = -1 \cdot 3^4 = -1 \cdot 81 = -81$.

To simplify an expression like $\dfrac{49 - 104}{7 + 4}$, we must enter it as $(49 - 104) \div (7 + 4)$.

The result is −5.

EXERCISES: Calculate.

1. $-8 + 4(7 - 9)$ **2.** $-3[2 + (-5)]$ **3.** $(-7)^6$ **4.** $(-17)^5$

5. -7^6 **6.** -17^5 **7.** $\dfrac{38 - 178}{5 + 30}$ **8.** $\dfrac{311 - 17^2}{2 - 13}$

☑ Reading Check

In each of Exercises 1–6, name the operation that should be performed first. Do not calculate.

RC1. $10 - 4 \cdot 2 + 5$

RC2. $10 - 4(2 + 5)$

RC3. $(10 - 4) \cdot 2 + 5$

RC4. $5[2(10 \div 5) - 3]$

RC5. $5(10 \div 2 + 5 - 3)$

RC6. $5 \cdot 2 - 4 \cdot 8 \div 2$

a   Find an equivalent expression without parentheses.

1. $-(2x + 7)$

2. $-(8x + 4)$

3. $-(8 - x)$

4. $-(a - b)$

5. $-(4a - 3b + 7c)$

6. $-(x - 4y - 3z)$

7. $-(6x - 8y + 5)$

8. $-(4x + 9y + 7)$

9. $-(3x - 5y - 6)$

10. $-(6a - 4b - 7)$

11. $-(-8x - 6y - 43)$

12. $-(-2a + 9b - 5c)$

b   Remove parentheses and simplify.

13. $9x - (4x + 3)$

14. $4y - (2y + 5)$

15. $2a - (5a - 9)$

16. $12m - (4m - 6)$

17. $2x + 7x - (4x + 6)$

18. $3a + 2a - (4a + 7)$

19. $2x - 4y - 3(7x - 2y)$

20. $3a - 9b - 1(4a - 8b)$

21. $15x - y - 5(3x - 2y + 5z)$

22. $4a - b - 4(5a - 7b + 8c)$

23. $(3x + 2y) - 2(5x - 4y)$

24. $(-6a - b) - 5(2b + a)$

25. $(12a - 3b + 5c) - 5(-5a + 4b - 6c)$

26. $(-8x + 5y - 12) - 6(2x - 4y - 10)$

c Simplify.

27. $9 - 2(5 - 4)$

28. $6 - 5(8 - 4)$

29. $8[7 - 6(4 - 2)]$

30. $10[7 - 4(7 - 5)]$

31. $[4(9 - 6) + 11] - [14 - (6 + 4)]$

32. $[7(8 - 4) + 16] - [15 - (7 + 8)]$

33. $[10(x + 3) - 4] + [2(x - 1) + 6]$

34. $[9(x + 5) - 7] + [4(x - 12) + 9]$

35. $[7(x + 5) - 19] - [4(x - 6) + 10]$

36. $[6(x + 4) - 12] - [5(x - 8) + 14]$

37. $3\{[7(x - 2) + 4] - [2(2x - 5) + 6]\}$

38. $4\{[8(x - 3) + 9] - [4(3x - 2) + 6]\}$

39. $4\{[5(x - 3) + 2] - 3[2(x + 5) - 9]\}$

40. $3\{[6(x - 4) + 5] - 2[5(x + 8) - 3]\}$

d Simplify.

41. $8 - 2 \cdot 3 - 9$

42. $8 - (2 \cdot 3 - 9)$

43. $(8 - 2) \div (3 - 9)$

44. $(8 - 2) \div 3 - 9$

45. $[(-24) \div (-3)] \div \left(-\frac{1}{2}\right)$

46. $[32 \div (-2)] \div \left(-\frac{1}{4}\right)$

47. $16 \cdot (-24) + 50$

48. $10 \cdot 20 - 15 \cdot 24$

49. $2^4 + 2^3 - 10$

50. $40 - 3^2 - 2^3$

51. $5^3 + 26 \cdot 71 - (16 + 25 \cdot 3)$

52. $4^3 + 10 \cdot 20 + 8^2 - 23$

53. $4 \cdot 5 - 2 \cdot 6 + 4$

54. $4 \cdot (6 + 8)/(4 + 3)$

55. $4^3/8$

56. $5^3 - 7^2$

57. $8(-7) + 6(-5)$

58. $10(-5) + 1(-1)$

59. $19 - 5(-3) + 3$

60. $14 - 2(-6) + 7$

61. $9 \div (-3) + 16 \div 8$

62. $-32 - 8 \div 4 - (-2)$

63. $-4^2 + 6$

64. $-5^2 + 7$

65. $-8^2 - 3$

66. $-9^2 - 11$

67. $12 - 20^3$

68. $20 + 4^3 \div (-8)$

69. $2 \cdot 10^3 - 5000$

70. $-7(3^4) + 18$

71. $6[9 - (3 - 4)]$

72. $8[3(6 - 13) - 11]$

73. $-1000 \div (-100) \div 10$

74. $256 \div (-32) \div (-4)$

75. $8 - (7 - 9)$

76. $(16 - 6) \cdot \dfrac{1}{2} + 9$

77. $\dfrac{10 - 6^2}{9^2 + 3^2}$

78. $\dfrac{5^2 - 4^3 - 3}{9^2 - 2^2 - 1^5}$

79. $\dfrac{3(6 - 7) - 5 \cdot 4}{6 \cdot 7 - 8(4 - 1)}$

80. $\dfrac{20(8 - 3) - 4(10 - 3)}{10(2 - 6) - 2(5 + 2)}$

81. $\dfrac{|2^3 - 3^2| + |12 \cdot 5|}{-32 \div (-16) \div (-4)}$

82. $\dfrac{|3 - 5|^2 - |7 - 13|}{|12 - 9| + |11 - 14|}$

Skill Maintenance

Evaluate.   [10.1a]

83. $\dfrac{x - y}{y}$, when $x = 38$ and $y = 2$

84. $a - 3b$, when $a = 50$ and $b = 5$

Find the absolute value.   [10.2e]

85. $|-0.4|$

86. $\left|\dfrac{15}{2}\right|$

Find the reciprocal.   [10.6b]

87. -9

88. $\dfrac{7}{3}$

Subtract.   [10.4a]

89. $5 - 30$

90. $-5 - 30$

91. $-5 - (-30)$

92. $5 - (-30)$

Synthesis

Simplify.

93. $x - [f - (f - x)] + [x - f] - 3x$

94. $x - \{x - 1 - [x - 2 - (x - 3 - \{x - 4 - [x - 5 - (x - 6)]\})]\}$

95. ▦ Use your calculator to do the following.
 a) Evaluate $x^2 + 3$ when $x = 7$, when $x = -7$, and when $x = -5.013$.
 b) Evaluate $1 - x^2$ when $x = 5$, when $x = -5$, and when $x = -10.455$.

96. Express $3^3 + 3^3 + 3^3$ as a power of 3.

Find the average.

97. $-15, 20, 50, -82, -7, -2$

98. $-1, 1, 2, -2, 3, -8, -10$

Vocabulary Reinforcement

Complete each statement with the correct term from the column on the right. Some of the choices may not be used.

1. The set of _____ is
   $\{ \dots, -5, -4, -3, -2, -1, 0, 1, 2, 3, 4, 5, \dots \}$. [10.2a]

2. Two numbers whose sum is 0 are called _____ of each other. [10.3b]

3. The _____ of addition says that $a + b = b + a$ for any real numbers a and b. [10.7b]

4. The _____ states that for any real number a, $a \cdot 1 = 1 \cdot a = a$. [10.7a]

5. The _____ of multiplication says that $a(bc) = (ab)c$ for any real numbers a, b, and c. [10.7b]

6. Two numbers whose product is 1 are called _____ of each other. [10.6b]

7. The equation $y + 0 = y$ illustrates the _____. [10.7a]

natural numbers

whole numbers

integers

real numbers

multiplicative inverses

additive inverses

commutative law

associative law

distributive law

identity property of 0

identity property of 1

property of -1

Concept Reinforcement

Determine whether each statement is true or false.

_____ 1. Every whole number is also an integer. [10.2a]

_____ 2. The product of an even number of negative numbers is positive. [10.5a]

_____ 3. The product of a number and its multiplicative inverse is -1. [10.6b]

_____ 4. $a < b$ also has the meaning $b \geq a$. [10.2d]

Study Guide

Objective 10.1a   Evaluate algebraic expressions by substitution.

Example   Evaluate $y - z$ when $y = 5$ and $z = -7$.
   $y - z = 5 - (-7) = 5 + 7 = 12$

Practice Exercise

1. Evaluate $2a + b$ when $a = -1$ and $b = 16$.

Objective 10.2d   Determine which of two real numbers is greater and indicate which, using $<$ or $>$.

Example   Use $<$ or $>$ for ☐ to write a true sentence:
   -5 ☐ -12.
   Since -5 is to the right of -12 on the number line, we have $-5 > -12$.

Practice Exercise

2. Use $<$ or $>$ for ☐ to write a true sentence:
   -6 ☐ -3.

Objective 10.2e Find the absolute value of a real number.

Example Find the absolute value: **(a)** $|21|$;
(b) $|-3.2|$; **(c)** $|0|$.

a) The number is positive, so the absolute value is the same as the number.
$$|21| = 21$$

b) The number is negative, so we make it positive.
$$|-3.2| = 3.2$$

c) The number is 0, so the absolute value is the same as the number.
$$|0| = 0$$

Practice Exercise

3. Find the absolute value: $\left|-\dfrac{5}{4}\right|$.

Objective 10.3a Add real numbers without using the number line.

Example Add without using the number line:
(a) $-13 + 4$; **(b)** $-2 + (-3)$.

a) We have a negative number and a positive number. The absolute values are 13 and 4. The difference is 9. The negative number has the larger absolute value, so the answer is negative.
$$-13 + 4 = -9$$

b) We have two negative numbers. The sum of the absolute values is $2 + 3$, or 5. The answer is negative.
$$-2 + (-3) = -5$$

Practice Exercise

4. Add without using the number line:
$$-5.6 + (-2.9).$$

Objective 10.4a Subtract real numbers.

Example Subtract: $-4 - (-6)$.
$$-4 - (-6) = -4 + 6 = 2$$

Practice Exercise

5. Subtract: $7 - 9$.

Objective 10.5a Multiply real numbers.

Example Multiply: **(a)** $-1.9(4)$; **(b)** $-7(-6)$.

a) The signs are different, so the answer is negative.
$$-1.9(4) = -7.6$$

b) The signs are the same, so the answer is positive.
$$-7(-6) = 42$$

Practice Exercise

6. Multiply: $-8(-7)$.

Objective 10.6a Divide integers.

Example Divide: **(a)** $15 \div (-3)$; **(b)** $-72 \div (-9)$.

a) The signs are different, so the answer is negative.
$$15 \div (-3) = -5$$

b) The signs are the same, so the answer is positive.
$$-72 \div (-9) = 8$$

Practice Exercise

7. Divide: $-48 \div 6$.

Objective 10.6c   Divide real numbers.

Example   Divide: **(a)** $-\dfrac{1}{4} \div \dfrac{3}{5}$; **(b)** $-22.4 \div (-4)$.

a) We multiply by the reciprocal of the divisor:

$$-\frac{1}{4} \div \frac{3}{5} = -\frac{1}{4} \cdot \frac{5}{3} = -\frac{5}{12}.$$

b) We carry out the long division. The answer is positive.

$$
\begin{array}{r}
5.6 \\
4\overline{)22.4} \\
\underline{20} \\
2\,4 \\
\underline{2\,4} \\
0
\end{array}
$$

Practice Exercise

8. Divide: $-\dfrac{3}{4} \div \left(-\dfrac{5}{3}\right)$.

Objective 10.7a   Simplify fraction expressions.

Example   Simplify: $-\dfrac{18x}{15x}$.

$$-\frac{18x}{15x} = -\frac{6 \cdot 3x}{5 \cdot 3x}$$ 　Factoring the numerator and the denominator

$$= -\frac{6}{5} \cdot \frac{3x}{3x}$$ 　Factoring the fraction expression

$$= -\frac{6}{5} \cdot 1$$ 　$\dfrac{3x}{3x} = 1$

$$= -\frac{6}{5}$$ 　Removing a factor of 1

Practice Exercise

9. Simplify: $\dfrac{45y}{27y}$.

Objective 10.7c   Use the distributive laws to multiply expressions like 8 and $x - y$.

Example   Multiply: $3(4x - y + 2z)$.

$$3(4x - y + 2z) = 3 \cdot 4x - 3 \cdot y + 3 \cdot 2z$$
$$= 12x - 3y + 6z$$

Practice Exercise

10. Multiply: $5(x + 3y - 4z)$.

Objective 10.7d   Use the distributive laws to factor expressions like $4x - 12 + 24y$.

Example   Factor: $12a - 8b + 4c$.

$$12a - 8b + 4c = 4 \cdot 3a - 4 \cdot 2b + 4 \cdot c$$
$$= 4(3a - 2b + c)$$

Practice Exercise

11. Factor: $27x + 9y - 36z$.

Objective 10.7e   Collect like terms.

Example   Collect like terms: $3x - 5y + 8x + y$.

$$3x - 5y + 8x + y = 3x + 8x - 5y + y$$
$$= 3x + 8x - 5y + 1 \cdot y$$
$$= (3 + 8)x + (-5 + 1)y$$
$$= 11x - 4y$$

Practice Exercise

12. Collect like terms: $6a - 4b - a + 2b$.

Objective 10.8b Simplify expressions by removing parentheses and collecting like terms.

Example Remove parentheses and simplify: $5x - 2(3x - y)$. $5x - 2(3x - y) = 5x - 6x + 2y = -x + 2y$	**Practice Exercise** **13.** Remove parentheses and simplify: $8a - b - (4a + 3b)$.

Objective 10.8d Simplify expressions using the rules for order of operations.

Example Simplify: $12 - (7 - 3 \cdot 6)$. $\begin{aligned} 12 - (7 - 3 \cdot 6) &= 12 - (7 - 18) \\ &= 12 - (-11) \\ &= 12 + 11 \\ &= 23 \end{aligned}$	**Practice Exercise** **14.** Simplify: $75 \div (-15) + 24 \div 8$.

Review Exercises

The review exercises that follow are for practice. Answers are at the back of the book. If you miss an exercise, restudy the objective indicated in red after the exercise or the direction line that precedes it.

1. Evaluate $\dfrac{x - y}{3}$ when $x = 17$ and $y = 5$. [10.1a]

2. Translate to an algebraic expression: [10.1b]
Nineteen percent of some number.

3. Tell which integers correspond to this situation: [10.2a]

Josh earned $620 for one week's work. While driving to work one day, he received a speeding ticket for $125.

Find the absolute value. [10.2e]

4. $|-38|$

5. $|126|$

Graph the number on the number line. [10.2b]

6. -2.5

7. $\dfrac{8}{9}$

Use either $<$ or $>$ for $\square$ to write a true sentence. [10.2d]

8. $-3 \ \square \ 10$

9. $-1 \ \square \ -6$

10. $0.126 \ \square \ -12.6$

11. $-\dfrac{2}{3} \ \square \ -\dfrac{1}{10}$

12. Write another inequality with the same meaning as $-3 < x$. [10.2d]

Write true or false. [10.2d]

13. $-9 \leq 9$

14. $-11 \geq -3$

Find the opposite. [10.3b]

15. 3.8

16. $-\dfrac{3}{4}$

Find the reciprocal. [10.6b]

17. $\dfrac{3}{8}$

18. -7

19. Evaluate $-x$ when $x = -34$. [10.3b]

20. Evaluate $-(-x)$ when $x = 5$. [10.3b]

Compute and simplify.

21. $4 + (-7)$ [10.3a]

22. $6 + (-9) + (-8) + 7$ [10.3a]

23. $-3.8 + 5.1 + (-12) + (-4.3) + 10$ [10.3a]

24. $-3 - (-7) + 7 - 10$ [10.4a]

25. $-\dfrac{9}{10} - \dfrac{1}{2}$ [10.4a]

26. $-3.8 - 4.1$ [10.4a]

27. $-9 \cdot (-6)$ [10.5a]

28. $-2.7(3.4)$ [10.5a]

29. $\dfrac{2}{3} \cdot \left(-\dfrac{3}{7}\right)$ [10.5a]

30. $3 \cdot (-7) \cdot (-2) \cdot (-5)$ [10.5a]

31. $35 \div (-5)$ [10.6a]

32. $-5.1 \div 1.7$ [10.6c]

33. $-\dfrac{3}{11} \div \left(-\dfrac{4}{11}\right)$ [10.6c]

Simplify. [10.8d]

34. $2(-3.4 - 12.2) - 8(-7)$

35. $\dfrac{-12(-3) - 2^3 - (-9)(-10)}{3 \cdot 10 + 1}$

36. $-16 \div 4 - 30 \div (-5)$

37. $\dfrac{-4[7 - (10 - 13)]}{|-2(8) - 4|}$

Solve.

38. On the first, second, and third downs, a football team had these gains and losses: 5-yd gain, 12-yd loss, and 15-yd gain, respectively. Find the total gain (or loss). [10.3c]

39. Chang's total assets are \$2140. He borrows \$2500. What are his total assets now? [10.4b]

40. *Stock Price.* The value of EFX Corp. stock began the day at \$17.68 per share and dropped \$1.63 per hour for 8 hr. What was the price of the stock after 8 hr? [10.5b]

41. *Bank Account Balance.* Yuri had \$68 in his bank account. After using his debit card to buy seven equally priced tee shirts, the balance in his account was $-\$64.65$. What was the price of each shirt? [10.6d]

Multiply. [10.7c]

42. $5(3x - 7)$ **43.** $-2(4x - 5)$

44. $10(0.4x + 1.5)$ **45.** $-8(3 - 6x)$

Factor. [10.7d]

46. $2x - 14$ **47.** $-6x + 6$

48. $5x + 10$ **49.** $-3x + 12y - 12$

Collect like terms.   [10.7e]

50. $11a + 2b - 4a - 5b$

51. $7x - 3y - 9x + 8y$

52. $6x + 3y - x - 4y$

53. $-3a + 9b + 2a - b$

Remove parentheses and simplify.

54. $2a - (5a - 9)$ [10.8b]

55. $3(b + 7) - 5b$ [10.8b]

56. $3[11 - 3(4 - 1)]$ [10.8c]

57. $2[6(y - 4) + 7]$ [10.8c]

58. $[8(x + 4) - 10] - [3(x - 2) + 4]$ [10.8c]

59. $5\{[6(x - 1) + 7] - [3(3x - 4) + 8]\}$ [10.8c]

60. Factor out the greatest common factor:
$18x - 6y + 30$.   [10.7d]
 A. $2(9x - 2y + 15)$ **B.** $3(6x - 2y + 10)$
 C. $6(3x + 5)$ **D.** $6(3x - y + 5)$

61. Which expression is *not* equivalent to $mn + 5$?
[10.7b]
 A. $nm + 5$ **B.** $5n + m$
 C. $5 + mn$ **D.** $5 + nm$

Synthesis

Simplify.   [10.2e], [10.4a], [10.6a], [10.8d]

62. $-\left| \dfrac{7}{8} - \left(-\dfrac{1}{2}\right) - \dfrac{3}{4} \right|$

63. $(|2.7 - 3| + 3^2 - |-3|) \div (-3)$

64. $2000 - 1990 + 1980 - 1970 + \cdots + 20 - 10$

65. Find a formula for the perimeter of the figure
below.   [10.7e]

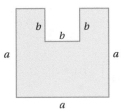

Understanding Through Discussion and Writing

1. Without actually performing the addition, explain
why the sum of all integers from -50 to 50 is 0.
[10.3b]

2. What rule have we developed that would tell you
the sign of $(-7)^8$ and of $(-7)^{11}$ without doing the
computations? Explain.   [10.5a]

3. Explain how multiplication can be used to justify
why a negative number divided by a negative
number is positive.   [10.6c]

4. Explain how multiplication can be used to justify
why a negative number divided by a positive number is negative.   [10.6c]

5. The distributive law was introduced before the dis-
cussion on collecting like terms. Why do you think
this was done?   [10.7c, e]

6. ▦ Jake keys in $18/2 \cdot 3$ on his calculator and
expects the result to be 3. What mistake is he
making?   [10.8d]

For Extra Help   For step-by-step test solutions, access the Chapter Test Prep Videos in MyMathLab® or on YouTube (search "BittingerAlgebraFoundations" and click on "Channels").

1. Evaluate $\dfrac{3x}{y}$ when $x = 10$ and $y = 5$.

2. Translate to an algebraic expression: Nine less than some number.

Use either $<$ or $>$ for $\square$ to write a true sentence.

3. $-3 \ \square \ -8$

4. $-\dfrac{1}{2} \ \square \ -\dfrac{1}{8}$

5. $-0.78 \ \square \ -0.87$

6. Write an inequality with the same meaning as $x < -2$.

7. Write true or false: $-13 \leq -3$.

Simplify.

8. $|-7|$

9. $\left|\dfrac{9}{4}\right|$

10. $|-2.7|$

Find the opposite.

11. $\dfrac{2}{3}$

12. -1.4

Find the reciprocal.

13. -2

14. $\dfrac{4}{7}$

15. Evaluate $-x$ when $x = -8$.

Compute and simplify.

16. $3.1 - (-4.7)$

17. $-8 + 4 + (-7) + 3$

18. $-\dfrac{1}{5} + \dfrac{3}{8}$

19. $2 - (-8)$

20. $3.2 - 5.7$

21. $\dfrac{1}{8} - \left(-\dfrac{3}{4}\right)$

22. $4 \cdot (-12)$

23. $-\dfrac{1}{2} \cdot \left(-\dfrac{3}{8}\right)$

24. $-45 \div 5$

25. $-\dfrac{3}{5} \div \left(-\dfrac{4}{5}\right)$

26. $4.864 \div (-0.5)$

27. $-2(16) - |2(-8) - 5^3|$

28. $-20 \div (-5) + 36 \div (-4)$

29. Isabella kept track of the changes in the stock market over a period of 5 weeks. By how many points had the market risen or fallen over this time?

WEEK 1	WEEK 2	WEEK 3	WEEK 4	WEEK 5
Down 13 pts	Down 16 pts	Up 36 pts	Down 11 pts	Up 19 pts

30. *Difference in Elevation.* The lowest elevation in Australia, Lake Eyre, is 15 m below sea level. The highest elevation in Australia, Mount Kosciuszko, is 2229 m. Find the difference in elevation between the highest point and the lowest point.

Source: *The CIA World Factbook,* 2012

Lake Eyre
−15 m

Sydney

Mt. Kosciuszko
2229 m

31. *Population Decrease.* The population of Stone City was 18,600. It dropped 420 each year for 6 years. What was the population of the city after 6 years?

32. *Chemical Experiment.* During a chemical reaction, the temperature in a beaker decreased every minute by the same number of degrees. The temperature was 16°C at 11:08 A.M. By 11:52 A.M., the temperature had dropped to −17°C. By how many degrees did it change each minute?

Multiply.

33. $3(6 - x)$

34. $-5(y - 1)$

Factor.

35. $12 - 22x$

36. $7x + 21 + 14y$

Simplify.

37. $6 + 7 - 4 - (-3)$

38. $5x - (3x - 7)$

39. $4(2a - 3b) + a - 7$

40. $4\{3[5(y - 3) + 9] + 2(y + 8)\}$

41. $256 \div (-16) \div 4$

42. $2^3 - 10[4 - 3(-2 + 18)]$

43. Which of the following is *not* a true statement?

    A. $-5 \leq -5$ **B.** $-5 < -5$

    C. $-5 \geq -5$ **D.** $-5 = -5$

Synthesis

Simplify.

44. $|-27 - 3(4)| - |-36| + |-12|$

45. $a - \{3a - [4a - (2a - 4a)]\}$

46. Find a formula for the perimeter of the figure shown here.

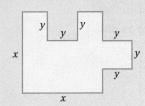

CHAPTER
11

Solving Equations and Inequalities

From Chapter 11 of *Algebra Foundations: Basic Mathematics, Introductory Algebra, and Intermediate Algebra,* First Edition.
Marvin J. Bittinger, Judith A. Beecher, and Barbara L. Johnson. Copyright © 2015 by Pearson Education, Inc. All rights reserved.

11.1 Solving Equations: The Addition Principle

OBJECTIVES

a Determine whether a given number is a solution of a given equation.

b Solve equations using the addition principle.

SKILL TO REVIEW

Objective 10.1a: Evaluate algebraic expressions by substitution.

1. Evaluate $x - 7$ when $x = 5$.
2. Evaluate $2x + 3$ when $x = -1$.

Determine whether each equation is true, false, or neither.

1. $5 - 8 = -4$

2. $12 + 6 = 18$

3. $x + 6 = 7 - x$

a EQUATIONS AND SOLUTIONS

In order to solve problems, we must learn to solve equations.

> **EQUATION**
>
> An **equation** is a number sentence that says that the expressions on either side of the equals sign, $=$, represent the same number.

Here are some examples of equations:

$$3 + 2 = 5, \quad 14 - 10 = 1 + 3, \quad x + 6 = 13, \quad 3x - 2 = 7 - x.$$

Equations have expressions on each side of the equals sign. The sentence "$14 - 10 = 1 + 3$" asserts that the expressions $14 - 10$ and $1 + 3$ name the same number.

Some equations are true. Some are false. Some are neither true nor false.

EXAMPLES Determine whether each equation is true, false, or neither.

1. $3 + 2 = 5$ The equation is *true*.
2. $7 - 2 = 4$ The equation is *false*.
3. $x + 6 = 13$ The equation is *neither* true nor false, because we do not know what number x represents.

◀ Do Margin Exercises 1–3.

> **SOLUTION OF AN EQUATION**
>
> Any replacement for the variable that makes an equation true is called a **solution** of the equation. To solve an equation means to find *all* of its solutions.

One way to determine whether a number is a solution of an equation is to evaluate the expression on each side of the equals sign by substitution. If the values are the same, then the number is a solution.

Answers

Skill to Review:
1. -2 2. 1

Margin Exercises:
1. False   2. True   3. Neither

EXAMPLE 4 Determine whether 7 is a solution of $x + 6 = 13$.

We have

$$\begin{array}{c|c} x + 6 = 13 & \text{Writing the equation} \\ \hline 7 + 6 \ ? \ 13 & \text{Substituting 7 for } x \\ 13 & \text{TRUE} \end{array}$$

Since the left-hand side and the right-hand side are the same, 7 is a solution. No other number makes the equation true, so the only solution is the number 7.

EXAMPLE 5 Determine whether 19 is a solution of $7x = 141$.

We have

$$\begin{array}{c|c} 7x = 141 & \text{Writing the equation} \\ \hline 7(19) \ ? \ 141 & \text{Substituting 19 for } x \\ 133 & \text{FALSE} \end{array}$$

Since the left-hand side and the right-hand side are not the same, 19 is not a solution of the equation.

Do Exercises 4–7. ▷

Determine whether the given number is a solution of the given equation.

4. 8; $x + 4 = 12$

5. 0; $x + 4 = 12$

6. −3; $7 + x = -4$

7. $-\dfrac{3}{5}$; $-5x = 3$

b USING THE ADDITION PRINCIPLE

Consider the equation

$$x = 7.$$

We can easily see that the solution of this equation is 7. If we replace x with 7, we get

$$7 = 7, \quad \text{which is true.}$$

Now consider the equation of Example 4: $x + 6 = 13$. In Example 4, we discovered that the solution of this equation is also 7, but the fact that 7 is the solution is not as obvious. We now begin to consider principles that allow us to start with an equation like $x + 6 = 13$ and end up with an *equivalent equation,* like $x = 7$, in which the variable is alone on one side and for which the solution is easier to find.

EQUIVALENT EQUATIONS

Equations with the same solutions are called **equivalent equations**.

One of the principles that we use in solving equations involves addition. An equation $a = b$ says that a and b stand for the same number. Suppose this is true, and we add a number c to the number a. We get the same answer if we add c to b, because a and b are the same number.

THE ADDITION PRINCIPLE FOR EQUATIONS

For any real numbers a, b, and c,

$$a = b \quad \text{is equivalent to} \quad a + c = b + c.$$

Let's solve the equation $x + 6 = 13$ using the addition principle. We want to get x alone on one side. To do so, we use the addition principle, choosing to add -6 because $6 + (-6) = 0$:

$$x + 6 = 13$$
$$x + 6 + (-6) = 13 + (-6) \qquad \text{Using the addition principle:}$$
$$\text{adding } -6 \text{ on both sides}$$
$$x + 0 = 7 \qquad \text{Simplifying}$$
$$x = 7. \qquad \text{Identity property of } 0: \ x + 0 = x$$

The solution of $x + 6 = 13$ is 7.

◀ **Do Exercise 8.**

8. Solve $x + 2 = 11$ using the addition principle.
$$x + 2 = 11$$
$$x + 2 + (-2) = 11 + (\quad)$$
$$x + \boxed{} = 9$$
$$x = \boxed{}$$

When we use the addition principle, we sometimes say that we "add the same number on both sides of the equation." This is also true for subtraction, since we can express every subtraction as an addition. That is, since

$$a - c = b - c \quad \text{is equivalent to} \quad a + (-c) = b + (-c),$$

the addition principle tells us that we can "subtract the same number on both sides of the equation."

EXAMPLE 6 Solve: $x + 5 = -7$.

We have

$$x + 5 = -7$$
$$x + 5 - 5 = -7 - 5 \qquad \text{Using the addition principle: adding } -5 \text{ on}$$
$$\text{both sides or subtracting 5 on both sides}$$
$$x + 0 = -12 \qquad \text{Simplifying}$$
$$x = -12. \qquad \text{Identity property of } 0$$

The solution of the original equation is -12. The equations $x + 5 = -7$ and $x = -12$ are *equivalent*.

◀ **Do Exercise 9.**

9. Solve using the addition principle, subtracting 5 on both sides:
$$x + 5 = -8.$$

Now we use the addition principle to solve an equation that involves a subtraction.

EXAMPLE 7 Solve: $a - 4 = 10$.

We have

$$a - 4 = 10$$
$$a - 4 + 4 = 10 + 4 \qquad \text{Using the addition principle:}$$
$$\text{adding 4 on both sides}$$
$$a + 0 = 14 \qquad \text{Simplifying}$$
$$a = 14. \qquad \text{Identity property of } 0$$

Check:
$$\frac{a - 4 = 10}{14 - 4 \ ? \ 10}$$
$$10 \ \bigg| \qquad \text{TRUE}$$

The solution is 14.

10. Solve: $t - 3 = 19$.

◀ **Do Exercise 10.**

Answers

8. 9   **9.** -13 **10.** 22

Guided Solution:

8. $-2, 0, 9$

EXAMPLE 8 Solve: $-6.5 = y - 8.4$.

We have

$$-6.5 = y - 8.4$$

$$-6.5 + 8.4 = y - 8.4 + 8.4$$ Using the addition principle: adding 8.4 on both sides to eliminate -8.4 on the right

$$1.9 = y.$$

Check:

$$\begin{array}{c|c} \hline -6.5 = y - 8.4 \\ \hline -6.5 \ ? \ 1.9 - 8.4 \\ \ \ \ \ \ \ \mid -6.5 \quad \text{TRUE} \end{array}$$

The solution is 1.9.

Note that equations are reversible. That is, if $a = b$ is true, then $b = a$ is true. Thus when we solve $-6.5 = y - 8.4$, we can reverse it and solve $y - 8.4 = -6.5$ if we wish.

Do Exercises 11 and 12. ▶

Solve.

11. $8.7 = n - 4.5$

12. $y + 17.4 = 10.9$

EXAMPLE 9 Solve: $-\dfrac{2}{3} + x = \dfrac{5}{2}$.

We have

$$-\frac{2}{3} + x = \frac{5}{2}$$

$$\frac{2}{3} - \frac{2}{3} + x = \frac{2}{3} + \frac{5}{2}$$ Adding $\frac{2}{3}$ on both sides

$$x = \frac{2}{3} + \frac{5}{2}$$

$$x = \frac{2}{3} \cdot \frac{2}{2} + \frac{5}{2} \cdot \frac{3}{3}$$ Multiplying by 1 to obtain equivalent fraction expressions with the least common denominator 6

$$x = \frac{4}{6} + \frac{15}{6}$$

$$x = \frac{19}{6}.$$

Check:

$$\begin{array}{c|c} \hline -\dfrac{2}{3} + x = \dfrac{5}{2} \\ \hline -\dfrac{2}{3} + \dfrac{19}{6} \ ? \ \dfrac{5}{2} \\ -\dfrac{4}{6} + \dfrac{19}{6} \\ \dfrac{15}{6} \\ \dfrac{5}{2} \quad \text{TRUE} \end{array}$$

The solution is $\dfrac{19}{6}$.

Do Exercises 13 and 14. ▶

Solve.

13. $x + \dfrac{1}{2} = -\dfrac{3}{2}$

14. $t - \dfrac{13}{4} = \dfrac{5}{8}$

Answers

11. 13.2 **12.** -6.5 **13.** -2 **14.** $\dfrac{31}{8}$

✓ Reading Check

Choose from the column on the right the most appropriate first step in solving each equation.

RC1. $9 = x - 4$

RC2. $3 + x = -15$

RC3. $x - 3 = 9$

RC4. $x + 4 = 3$

a) Add -4 on both sides.
b) Add 15 on both sides.
c) Subtract 3 on both sides.
d) Subtract 9 on both sides.
e) Add 3 on both sides.
f) Add 4 on both sides.

a   Determine whether the given number is a solution of the given equation.

1. 15; $x + 17 = 32$ **2.** 35; $t + 17 = 53$ **3.** 21; $x - 7 = 12$ **4.** 36; $a - 19 = 17$

5. -7; $6x = 54$ **6.** -9; $8y = -72$ **7.** 30; $\frac{x}{6} = 5$ **8.** 49; $\frac{y}{8} = 6$

9. 20; $5x + 7 = 107$ **10.** 9; $9x + 5 = 86$ **11.** -10; $7(y - 1) = 63$ **12.** -5; $6(y - 2) = 18$

b   Solve using the addition principle. Don't forget to check!

13. $x + 2 = 6$
Check: $x + 2 = 6$?

14. $y + 4 = 11$
Check: $y + 4 = 11$?

15. $x + 15 = -5$
Check: $x + 15 = -5$?

16. $t + 10 = 44$
Check: $t + 10 = 44$?

17. $x + 6 = -8$
Check: $x + 6 = -8$?

18. $z + 9 = -14$ **19.** $x + 16 = -2$ **20.** $m + 18 = -13$ **21.** $x - 9 = 6$ **22.** $x - 11 = 12$

23. $x - 7 = -21$ **24.** $x - 3 = -14$ **25.** $5 + t = 7$ **26.** $8 + y = 12$ **27.** $-7 + y = 13$

28. $-8 + y = 17$ **29.** $-3 + t = -9$ **30.** $-8 + t = -24$ **31.** $x + \dfrac{1}{2} = 7$ **32.** $24 = -\dfrac{7}{10} + r$

33. $12 = a - 7.9$ **34.** $2.8 + y = 11$ **35.** $r + \dfrac{1}{3} = \dfrac{8}{3}$ **36.** $t + \dfrac{3}{8} = \dfrac{5}{8}$

37. $m + \dfrac{5}{6} = -\dfrac{11}{12}$ **38.** $x + \dfrac{2}{3} = -\dfrac{5}{6}$ **39.** $x - \dfrac{5}{6} = \dfrac{7}{8}$ **40.** $y - \dfrac{3}{4} = \dfrac{5}{6}$

41. $-\dfrac{1}{5} + z = -\dfrac{1}{4}$ **42.** $-\dfrac{1}{8} + y = -\dfrac{3}{4}$ **43.** $7.4 = x + 2.3$ **44.** $8.4 = 5.7 + y$

45. $7.6 = x - 4.8$ **46.** $8.6 = x - 7.4$ **47.** $-9.7 = -4.7 + y$ **48.** $-7.8 = 2.8 + x$

49. $5\dfrac{1}{6} + x = 7$ **50.** $5\dfrac{1}{4} = 4\dfrac{2}{3} + x$ **51.** $q + \dfrac{1}{3} = -\dfrac{1}{7}$ **52.** $52\dfrac{3}{8} = -84 + x$

Skill Maintenance

53. Divide: $\dfrac{2}{3} \div \left(-\dfrac{4}{9}\right)$.   [10.6c]

54. Add: $-8.6 + 3.4$.   [10.3a]

55. Subtract: $-\dfrac{2}{3} - \left(-\dfrac{5}{8}\right)$.   [10.4a]

56. Multiply: $(-25.4)(-6.8)$.   [10.5a]

Translate to an algebraic expression.   [10.1b]

57. Jane had $83 before paying x dollars for a pair of tennis shoes. How much does she have left?

58. Justin drove his S-10 pickup truck 65 mph for t hours. How far did he drive?

Synthesis

Solve.

59. $x + \dfrac{4}{5} = -\dfrac{2}{3} - \dfrac{4}{15}$ **60.** $x + x = x$ **61.** $16 + x - 22 = -16$

62. $x + 4 = 5 + x$ **63.** $x + 3 = 3 + x$ **64.** $|x| + 6 = 19$

OBJECTIVE

a | Solve equations using the multiplication principle.

SKILL TO REVIEW

Objective 10.6b: Find the reciprocal of a real number.

Find the reciprocal.

1. 5

2. $-\dfrac{5}{4}$

a USING THE MULTIPLICATION PRINCIPLE

Suppose that $a = b$ is true, and we multiply a by some number c. We get the same number if we multiply b by c, because a and b are the same number.

THE MULTIPLICATION PRINCIPLE FOR EQUATIONS

For any real numbers a, b, and c, $c \neq 0$,

$$a = b \quad \text{is equivalent to} \quad a \cdot c = b \cdot c.$$

When using the multiplication principle, we sometimes say that we "multiply on both sides of the equation by the same number."

EXAMPLE 1   Solve: $5x = 70$.

To get x alone on one side, we multiply by the *multiplicative inverse,* or *reciprocal*, of 5. Then we get the *multiplicative identity* 1 times x, or $1 \cdot x$, which simplifies to x. This allows us to eliminate 5 on the left.

$$5x = 70 \qquad \text{The reciprocal of 5 is } \tfrac{1}{5}.$$

$$\frac{1}{5} \cdot 5x = \frac{1}{5} \cdot 70 \qquad \begin{array}{l}\text{Multiplying by } \tfrac{1}{5} \text{ to get } 1 \cdot x \text{ and} \\ \text{eliminate 5 on the left}\end{array}$$

$$1 \cdot x = 14 \qquad \text{Simplifying}$$

$$x = 14 \qquad \text{Identity property of 1: } 1 \cdot x = x$$

Check:
$$\begin{array}{c} 5x = 70 \\ \hline 5 \cdot 14 \ ? \ 70 \\ 70 \ \Big| \qquad \text{TRUE} \end{array}$$

The solution is 14.

The multiplication principle also tells us that we can "divide on both sides of the equation by the same nonzero number." This is because dividing is the same as multiplying by a reciprocal. That is,

$$\frac{a}{c} = \frac{b}{c} \quad \text{is equivalent to} \quad a \cdot \frac{1}{c} = b \cdot \frac{1}{c}, \quad \text{when } c \neq 0.$$

In an expression like $5x$ in Example 1, the number 5 is called the **coefficient**. Example 1 could be done as follows, dividing by 5 on both sides, the coefficient of x.

EXAMPLE 2   Solve: $5x = 70$.

$$5x = 70$$

$$\frac{5x}{5} = \frac{70}{5} \qquad \text{Dividing by 5 on both sides}$$

$$1 \cdot x = 14 \qquad \text{Simplifying}$$

$$x = 14 \qquad \text{Identity property of 1. The solution is 14.}$$

Answers

Skill to Review:

1. $\dfrac{1}{5}$ 2. $-\dfrac{4}{5}$

Do Exercises 1 and 2. ▶

EXAMPLE 3   Solve: $-4x = 92$.

We have

$$-4x = 92$$

$$\frac{-4x}{-4} = \frac{92}{-4}$$ Using the multiplication principle. Dividing by -4 on both sides is the same as multiplying by $-\frac{1}{4}$.

$$1 \cdot x = -23$$ Simplifying

$$x = -23.$$ Identity property of 1

Check: $$\frac{-4x = 92}{-4(-23) \overset{?}{|} 92}$$
$$92 \overset{|}{} \quad \text{TRUE}$$

The solution is -23.

Do Exercise 3. ▶

EXAMPLE 4   Solve: $-x = 9$.

We have

$$-x = 9$$

$$-1 \cdot x = 9$$ Using the property of -1: $-x = -1 \cdot x$

$$\frac{-1 \cdot x}{-1} = \frac{9}{-1}$$ Dividing by -1 on both sides: $-1/(-1) = 1$

$$1 \cdot x = -9$$

$$x = -9.$$

Check: $$\frac{-x = 9}{-(-9) \overset{?}{|} 9}$$
$$9 \overset{|}{} \quad \text{TRUE}$$

The solution is -9.

Do Exercise 4. ▶

We can also solve the equation $-x = 9$ by multiplying as follows.

EXAMPLE 5   Solve: $-x = 9$.

We have

$$-x = 9$$

$$-1 \cdot (-x) = -1 \cdot 9$$ Multiplying by -1 on both sides

$$-1 \cdot (-1) \cdot x = -9$$ $-x = (-1) \cdot x$

$$1 \cdot x = -9$$ $-1 \cdot (-1) = 1$

$$x = -9.$$

The solution is -9.

Do Exercise 5. ▶

GS **1.** Solve $6x = 90$ by multiplying on both sides.

$$6x = 90$$

$$\frac{1}{6} \cdot 6x = \boxed{} \cdot 90$$

$$1 \cdot x = 15$$

$$\boxed{} = 15$$

Check: $$\frac{6x = 90}{6 \cdot \boxed{} \overset{?}{|} 90}$$
$$90 \overset{|}{} \quad \text{TRUE}$$

2. Solve $4x = -7$ by dividing on both sides.

$$4x = -7$$

$$\frac{4x}{4} = \frac{-7}{\boxed{}}$$

$$1 \cdot x = -\frac{7}{4}$$

$$\boxed{} = -\frac{7}{4}$$

Don't forget to check.

3. Solve: $-6x = 108$.

4. Solve. Divide on both sides.
$$-x = -10$$

5. Solve. Multiply on both sides.
$$-x = -10$$

Answers

1. 15   **2.** $-\dfrac{7}{4}$ **3.** -18 **4.** 10   **5.** 10

Guided Solutions:

1. $\dfrac{1}{6}, x, 15$ **2.** $4, x$

In practice, it is generally more convenient to divide on both sides of the equation if the coefficient of the variable is in decimal notation or is an integer. If the coefficient is in fraction notation, it is usually more convenient to multiply by a reciprocal.

EXAMPLE 6 Solve: $\frac{3}{8} = -\frac{5}{4}x.$

$$\frac{3}{8} = -\frac{5}{4}x$$

The reciprocal of $-\frac{5}{4}$ is $-\frac{4}{5}$. There is no sign change.

$$-\frac{4}{5} \cdot \frac{3}{8} = -\frac{4}{5} \cdot \left(-\frac{5}{4}x\right)$$

Multiplying by $-\frac{4}{5}$ to get $1 \cdot x$ and eliminate $-\frac{5}{4}$ on the right

$$-\frac{12}{40} = 1 \cdot x$$

$$-\frac{3}{10} = 1 \cdot x$$ Simplifying

$$-\frac{3}{10} = x$$ Identity property of 1

Check:

$$\frac{3}{8} = -\frac{5}{4}x$$

$$\frac{3}{8} \quad ? \quad -\frac{5}{4}\left(-\frac{3}{10}\right)$$

$$\frac{3}{8} \qquad \text{TRUE}$$

The solution is $-\frac{3}{10}$.

As noted in Section 2.1, if $a = b$ is true, then $b = a$ is true. Thus we can reverse the equation $\frac{3}{8} = -\frac{5}{4}x$ and solve $-\frac{5}{4}x = \frac{3}{8}$ if we wish.

◀ Do Exercise 6.

EXAMPLE 7 Solve: $1.16y = 9744.$

$$1.16y = 9744$$

$$\frac{1.16y}{1.16} = \frac{9744}{1.16}$$ Dividing by 1.16 on both sides

$$y = \frac{9744}{1.16}$$

$$y = 8400$$ Simplifying

Check:

$$1.16y = 9744$$

$$1.16(8400) \quad ? \quad 9744$$

$$9744 \qquad \text{TRUE}$$

The solution is 8400.

◀ Do Exercises 7 and 8.

6. Solve: $\frac{2}{3} = -\frac{5}{6}y.$

$$\frac{2}{3} = -\frac{5}{6}y$$

$$\boxed{} \cdot \frac{2}{3} = -\frac{6}{5} \cdot \left(-\frac{5}{6}y\right)$$

$$-\frac{\boxed{}}{15} = 1 \cdot y$$

$$-\frac{\boxed{}}{5} = y$$

Solve.

7. $1.12x = 8736$

8. $6.3 = -2.1y$

Answers

6. $-\frac{4}{5}$ **7.** 7800 **8.** −3

Guided Solution:

6. $-\frac{6}{5}$, 12, 4

Now we use the multiplication principle to solve an equation that in-volves division.

EXAMPLE 8 Solve: $\dfrac{-y}{9} = 14$.

$$\frac{-y}{9} = 14$$

$$9 \cdot \frac{-y}{9} = 9 \cdot 14 \qquad \text{Multiplying by 9 on both sides}$$

$$-y = 126$$

$$-1 \cdot (-y) = -1 \cdot 126 \qquad \text{Multiplying by } -1 \text{ on both sides}$$

$$y = -126$$

Check:

$$\frac{-y}{9} = 14$$

$$\frac{-(-126)}{9} \;\overset{?}{\vert}\; 14$$

$$\frac{126}{9}$$

$$14 \;\Big\vert\; \qquad \text{TRUE}$$

The solution is -126.

There are other ways to solve the equation in Example 8. One is by multiplying by -9 on both sides as follows:

$$-9 \cdot \frac{-y}{9} = -9 \cdot 14$$

$$\frac{9y}{9} = -126$$

$$y = -126.$$

9. Solve: $-14 = \dfrac{-y}{2}$.

Do Exercise 9. ▶

Answer
9. 28

For Extra Help

MyMathLab® MathXL®
          PRACTICE   WATCH   READ   REVIEW

☑ **Reading Check**

Choose from the column on the right the most appropriate first step in solving each equation.

RC1. $3 = -\dfrac{1}{12}x$

RC2. $-6x = 12$

RC3. $12x = -6$

RC4. $\dfrac{1}{6}x = 12$

a) Divide by 12 on both sides.
b) Multiply by 6 on both sides.
c) Multiply by 12 on both sides.
d) Divide by -6 on both sides.
e) Divide by 6 on both sides.
f) Multiply by -12 on both sides.

a Solve using the multiplication principle. Don't forget to check!

1. $6x = 36$

Check: $\dfrac{6x = 36}{?}$

2. $3x = 51$

Check: $\dfrac{3x = 51}{?}$

3. $5y = 45$

Check: $\dfrac{5y = 45}{?}$

4. $8y = 72$

Check: $\dfrac{8y = 72}{?}$

5. $84 = 7x$

6. $63 = 9x$

7. $-x = 40$

8. $-x = 53$

9. $-1 = -z$

10. $-47 = -t$

11. $7x = -49$

12. $8x = -56$

13. $-12x = 72$

14. $-15x = 105$

15. $-21w = -126$

16. $-13w = -104$

17. $\dfrac{t}{7} = -9$

18. $\dfrac{y}{5} = -6$

19. $\dfrac{n}{-6} = 8$

20. $\dfrac{y}{-8} = 11$

21. $\dfrac{3}{4}x = 27$

22. $\dfrac{4}{5}x = 16$

23. $-\dfrac{2}{3}x = 6$

24. $-\dfrac{3}{8}x = 12$

25. $\dfrac{-t}{3} = 7$

26. $\dfrac{-x}{6} = 9$

27. $-\dfrac{m}{3} = \dfrac{1}{5}$

28. $\dfrac{1}{8} = -\dfrac{y}{5}$

29. $-\dfrac{3}{5}r = \dfrac{9}{10}$

30. $-\dfrac{2}{5}y = \dfrac{4}{15}$

31. $-\dfrac{3}{2}r = -\dfrac{27}{4}$

32. $-\dfrac{3}{8}x = -\dfrac{15}{16}$

33. $6.3x = 44.1$

34. $2.7y = 54$

35. $-3.1y = 21.7$

36. $-3.3y = 6.6$

37. $38.7m = 309.6$

38. $29.4m = 235.2$

39. $-\dfrac{2}{3}y = -10.6$

40. $-\dfrac{9}{7}y = 12.06$

41. $\dfrac{-x}{5} = 10$

42. $\dfrac{-x}{8} = -16$

43. $-\dfrac{t}{2} = 7$

44. $\dfrac{m}{-3} = 10$

Skill Maintenance

Collect like terms. [10.7e]

45. $3x + 4x$

46. $6x + 5 - 7x$

47. $-4x + 11 - 6x + 18x$

48. $8y - 16y - 24y$

Remove parentheses and simplify. [10.8b]

49. $3x - (4 + 2x)$

50. $2 - 5(x + 5)$

51. $8y - 6(3y + 7)$

52. $-2a - 4(5a - 1)$

Translate to an algebraic expression. [10.1b]

53. Patty drives her van for 8 hr at a speed of r miles per hour. How far does she drive?

54. A triangle has a height of 10 meters and a base of b meters. What is the area of the triangle?

Synthesis

Solve.

55. $-0.2344m = 2028.732$

56. $0 \cdot x = 0$

57. $0 \cdot x = 9$

58. $4|x| = 48$

59. $2|x| = -12$

Solve for x.

60. $ax = 5a$

61. $3x = \dfrac{b}{a}$

62. $cx = a^2 + 1$

63. $\dfrac{a}{b}x = 4$

64. A student makes a calculation and gets an answer of 22.5. On the last step, she multiplies by 0.3 when she should have divided by 0.3. What is the correct answer?

OBJECTIVES

a   Solve equations using both the addition principle and the multiplication principle.

b   Solve equations in which like terms may need to be collected.

c   Solve equations by first removing parentheses and collecting like terms; solve equations with an infinite number of solutions and equations with no solutions.

SKILL TO REVIEW

Objective 10.7e: Collect like terms.

Collect like terms.

1. $q + 5t - 1 + 5q - t$

2. $7d + 16 - 11w - 2 - 10d$

---

a   APPLYING BOTH PRINCIPLES

Consider the equation $3x + 4 = 13$. It is more complicated than those we discussed in the preceding two sections. In order to solve such an equation, we first isolate the x-term, $3x$, using the addition principle. Then we apply the multiplication principle to get x by itself.

EXAMPLE 1   Solve: $3x + 4 = 13$.

$$3x + 4 = 13$$
$$3x + 4 - 4 = 13 - 4 \qquad \text{Using the addition principle: subtracting 4 on both sides}$$

First isolate the x-term. → $3x = 9$ Simplifying

$$\frac{3x}{3} = \frac{9}{3} \qquad \text{Using the multiplication principle: dividing by 3 on both sides}$$

Then isolate x. → $x = 3$ Simplifying

Check:
$$\begin{array}{c|c} 3x + 4 = 13 \\ \hline 3 \cdot 3 + 4 \ ? \ 13 \\ 9 + 4 \\ 13 \ \bigg| \ \text{TRUE} \end{array}$$

We use the rules for order of operations to carry out the check. We find the product $3 \cdot 3$. Then we add 4.

The solution is 3.

◀ Do Margin Exercise 1.

1. Solve: $9x + 6 = 51$.

EXAMPLE 2   Solve: $-5x - 6 = 16$.

$$-5x - 6 = 16$$
$$-5x - 6 + 6 = 16 + 6 \qquad \text{Adding 6 on both sides}$$
$$-5x = 22$$
$$\frac{-5x}{-5} = \frac{22}{-5} \qquad \text{Dividing by } -5 \text{ on both sides}$$
$$x = -\frac{22}{5}, \text{ or } -4\frac{2}{5} \qquad \text{Simplifying}$$

Check:
$$\begin{array}{c|c} -5x - 6 = 16 \\ \hline -5\left(-\dfrac{22}{5}\right) - 6 \ ? \ 16 \\ 22 - 6 \\ 16 \ \bigg| \ \text{TRUE} \end{array}$$

The solution is $-\frac{22}{5}$.

◀ Do Margin Exercises 2 and 3.

Solve.

2. $8x - 4 = 28$

3. $-\dfrac{1}{2}x + 3 = 1$

Answers

Skill to Review:
1. $6q + 4t - 1$
2. $-3d + 14 - 11w$

Margin Exercises:
1. 5   **2.** 4   **3.** 4

EXAMPLE 3 Solve: $45 - t = 13$.

$$45 - t = 13$$
$$-45 + 45 - t = -45 + 13 \qquad \text{Adding } -45 \text{ on both sides}$$
$$-t = -32$$
$$-1(-t) = -1(-32) \qquad \text{Multiplying by } -1 \text{ on both sides}$$
$$t = 32$$

The number 32 checks and is the solution.

Do Exercise 4. ▶

GS **4.** Solve: $-18 - m = -57$.

$$-18 - m = -57$$
$$18 - 18 - m = \boxed{} - 57$$
$$\boxed{} = -39$$
$$\boxed{} (-m) = -1(-39)$$
$$\boxed{} = 39$$

EXAMPLE 4 Solve: $16.3 - 7.2y = -8.18$.

$$16.3 - 7.2y = -8.18$$
$$-16.3 + 16.3 - 7.2y = -16.3 + (-8.18) \qquad \text{Adding } -16.3 \text{ on both sides}$$
$$-7.2y = -24.48$$
$$\frac{-7.2y}{-7.2} = \frac{-24.48}{-7.2} \qquad \text{Dividing by } -7.2 \text{ on both sides}$$
$$y = 3.4$$

Check:

$$\frac{16.3 - 7.2y = -8.18}{16.3 - 7.2(3.4) \; ? \; -8.18}$$
$$16.3 - 24.48$$
$$-8.18 \quad | \qquad \text{TRUE}$$

The solution is 3.4.

Do Exercises 5 and 6. ▶

Solve.

5. $-4 - 8x = 8$

6. $41.68 = 4.7 - 8.6y$

b   COLLECTING LIKE TERMS

If there are like terms on one side of the equation, we collect them before using the addition principle or the multiplication principle.

EXAMPLE 5 Solve: $3x + 4x = -14$.

$$3x + 4x = -14$$
$$7x = -14 \qquad \text{Collecting like terms}$$
$$\frac{7x}{7} = \frac{-14}{7} \qquad \text{Dividing by 7 on both sides}$$
$$x = -2$$

The number -2 checks, so the solution is -2.

Do Exercises 7 and 8. ▶

    If there are like terms on opposite sides of the equation, we get them on the same side by using the addition principle. Then we collect them. In other words, we get all the terms with a variable on one side of the equation and all the terms without a variable on the other side.

Solve.

7. $4x + 3x = -21$

8. $x - 0.09x = 728$

Answers

4. 39 **5.** $-\dfrac{3}{2}$ **6.** -4.3 **7.** -3 **8.** 800

Guided Solution:
4. $18, -m, -1, m$

EXAMPLE 6 Solve: $2x - 2 = -3x + 3$.

$$2x - 2 = -3x + 3$$

$2x - 2 + 2 = -3x + 3 + 2$ Adding 2

$2x = -3x + 5$ Collecting like terms

$2x + 3x = -3x + 3x + 5$ Adding $3x$

$5x = 5$ Simplifying

$\dfrac{5x}{5} = \dfrac{5}{5}$ Dividing by 5

$x = 1$ Simplifying

Check:

$$\begin{array}{c|c} 2x - 2 = -3x + 3 \\ \hline 2 \cdot 1 - 2 \; ? \; -3 \cdot 1 + 3 \\ 2 - 2 \;\;\Big|\;\; -3 + 3 \\ 0 \;\;\Big|\;\; 0 \end{array}$$ TRUE

Substituting in the original equation

The solution is 1.

◀ **Do Exercises 9 and 10.**

In Example 6, we used the addition principle to get all the terms with an x on one side of the equation and all the terms without an x on the other side. Then we collected like terms and proceeded as before. If there are like terms on one side at the outset, they should be collected first.

EXAMPLE 7 Solve: $6x + 5 - 7x = 10 - 4x + 3$.

$$6x + 5 - 7x = 10 - 4x + 3$$

$-x + 5 = 13 - 4x$ Collecting like terms

$4x - x + 5 = 13 - 4x + 4x$ Adding $4x$ to get all terms with a variable on one side

$3x + 5 = 13$ Simplifying; that is, collecting like terms

$3x + 5 - 5 = 13 - 5$ Subtracting 5

$3x = 8$ Simplifying

$\dfrac{3x}{3} = \dfrac{8}{3}$ Dividing by 3

$x = \dfrac{8}{3}$ Simplifying

The number $\frac{8}{3}$ checks, so it is the solution.

◀ **Do Exercises 11 and 12.**

Clearing Fractions and Decimals

In general, equations are easier to solve if they do not contain fractions or decimals. Consider, for example, the equations

$$\frac{1}{2}x + 5 = \frac{3}{4} \quad \text{and} \quad 2.3x + 7 = 5.4.$$

Solve.

9. $7y + 5 = 2y + 10$

10. $5 - 2y = 3y - 5$

Solve.

11. $7x - 17 + 2x = 2 - 8x + 15$ **GS**

$\boxed{} \cdot x - 17 = 17 - 8x$

$8x + 9x - 17 = 17 - 8x + \boxed{}$

$\boxed{} \cdot x - 17 = 17$

$17x - 17 + 17 = 17 + \boxed{}$

$17x = 34$

$\dfrac{17x}{17} = \dfrac{34}{\boxed{}}$

$\boxed{} = 2$

12. $3x - 15 = 5x + 2 - 4x$

If we multiply by 4 on both sides of the first equation and by 10 on both sides of the second equation, we have

$$4\left(\frac{1}{2}x + 5\right) = 4 \cdot \frac{3}{4} \quad \text{and} \quad 10(2.3x + 7) = 10 \cdot 5.4$$

$$4 \cdot \frac{1}{2}x + 4 \cdot 5 = 4 \cdot \frac{3}{4} \quad \text{and} \quad 10 \cdot 2.3x + 10 \cdot 7 = 10 \cdot 5.4$$

$$2x + 20 = 3 \quad \text{and} \quad 23x + 70 = 54.$$

The first equation has been "cleared of fractions" and the second equation has been "cleared of decimals." Both resulting equations are equivalent to the original equations and are easier to solve. *It is your choice* whether to clear fractions or decimals, but doing so often eases computations.

The easiest way to clear an equation of fractions is to multiply *every term on both sides* by the **least common multiple of all the denominators**.

EXAMPLE 8 Solve: $\frac{2}{3}x - \frac{1}{6} + \frac{1}{2}x = \frac{7}{6} + 2x$.

The denominators are 3, 6, and 2. The number 6 is the least common multiple of all the denominators. We multiply by 6 on both sides of the equation.

$$6\left(\frac{2}{3}x - \frac{1}{6} + \frac{1}{2}x\right) = 6\left(\frac{7}{6} + 2x\right) \qquad \text{Multiplying by 6 on both sides}$$

$$6 \cdot \frac{2}{3}x - 6 \cdot \frac{1}{6} + 6 \cdot \frac{1}{2}x = 6 \cdot \frac{7}{6} + 6 \cdot 2x \qquad \text{Using the distributive law (\textit{Caution!} Be sure to multiply \textit{all} the terms by 6.)}$$

$$4x - 1 + 3x = 7 + 12x \qquad \text{Simplifying. Note that the fractions are cleared.}$$

$$7x - 1 = 7 + 12x \qquad \text{Collecting like terms}$$

$$7x - 1 - 12x = 7 + 12x - 12x \qquad \text{Subtracting } 12x$$

$$-5x - 1 = 7 \qquad \text{Collecting like terms}$$

$$-5x - 1 + 1 = 7 + 1 \qquad \text{Adding 1}$$

$$-5x = 8 \qquad \text{Collecting like terms}$$

$$\frac{-5x}{-5} = \frac{8}{-5} \qquad \text{Dividing by } -5$$

$$x = -\frac{8}{5}$$

Check:

$$\frac{2}{3}x - \frac{1}{6} + \frac{1}{2}x = \frac{7}{6} + 2x$$

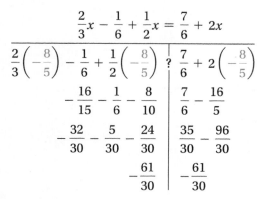

············ **Caution!** ············

Check the possible solution in the *original* equation rather than in the equation that has been cleared of fractions.

The solution is $-\frac{8}{5}$.

CALCULATOR CORNER

Checking Possible Solutions   There are several ways to check the possible solutions of an equation on a calculator. One of the most straightforward methods is to substitute and carry out the calculations on each side of the equation just as we do when we check by hand. To check the possible solution, 1, in Example 6, for instance, we first substitute 1 for *x* in the expression on the left side of the equation. We get 0. Next, we substitute 1 for *x* in the expression on the right side of the equation. Again we get 0. Since the two sides of the equation have the same value when *x* is 1, we know that 1 is the solution of the equation.

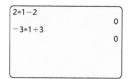

EXERCISE:

1. Use substitution to check the solutions found in Examples 1–5.

13. Solve: $\dfrac{7}{8}x - \dfrac{1}{4} + \dfrac{1}{2}x = \dfrac{3}{4} + x.$ **GS**

$$\frac{7}{8}x - \frac{1}{4} + \frac{1}{2}x = \frac{3}{4} + x$$

$$8 \cdot \left(\frac{7}{8}x - \frac{1}{4} + \frac{1}{2}x\right) = \boxed{} \cdot \left(\frac{3}{4} + x\right)$$

$$8 \cdot \frac{7}{8}x - \boxed{} \cdot \frac{1}{4} + 8 \cdot \frac{1}{2}x$$

$$= 8 \cdot \frac{3}{4} + \boxed{} \cdot x$$

$$\boxed{}\,x - \boxed{} + 4x = 6 + 8x$$

$$\boxed{}\,x - 2 = 6 + 8x$$

$$11x - 2 - 8x = 6 + 8x - \boxed{}$$

$$3x - 2 = \boxed{}$$

$$3x - 2 + \boxed{} = 6 + 2$$

$$3x = \boxed{}$$

$$\frac{3x}{3} = \frac{8}{\boxed{}}$$

$$x = \frac{8}{3}$$

14. Solve: $41.68 = 4.7 - 8.6y.$

Solve.

15. $2(2y + 3) = 14$

16. $5(3x - 2) = 35$

◀ **Do Exercise 13.**

To illustrate clearing decimals, we repeat Example 4, but this time we clear the equation of decimals first. Compare the methods.

To clear an equation of decimals, we count the greatest number of decimal places in any one number. If the greatest number of decimal places is 1, we multiply every term on both sides by 10; if it is 2, we multiply by 100; and so on.

EXAMPLE 9 Solve: $16.3 - 7.2y = -8.18$.

The greatest number of decimal places in any one number is *two*. Multiplying by 100, which has *two* 0's, will clear all decimals.

$$100(16.3 - 7.2y) = 100(-8.18) \qquad \text{Multiplying by 100 on both sides}$$

$$100(16.3) - 100(7.2y) = 100(-8.18) \qquad \text{Using the distributive law}$$

$$1630 - 720y = -818 \qquad \text{Simplifying}$$

$$1630 - 720y - 1630 = -818 - 1630 \qquad \text{Subtracting 1630}$$

$$-720y = -2448 \qquad \text{Collecting like terms}$$

$$\frac{-720y}{-720} = \frac{-2448}{-720} \qquad \text{Dividing by } -720$$

$$y = \frac{17}{5}, \text{ or } 3.4$$

The number $\frac{17}{5}$, or 3.4, checks, as shown in Example 4, so it is the solution.

◀ **Do Exercise 14.**

C EQUATIONS CONTAINING PARENTHESES

To solve certain kinds of equations that contain parentheses, we first use the distributive laws to remove the parentheses. Then we proceed as before.

EXAMPLE 10 Solve: $8x = 2(12 - 2x)$.

$$8x = 2(12 - 2x)$$

$$8x = 24 - 4x \qquad \text{Using the distributive laws to multiply and remove parentheses}$$

$$8x + 4x = 24 - 4x + 4x \qquad \text{Adding } 4x \text{ to get all the } x\text{-terms on one side}$$

$$12x = 24 \qquad \text{Collecting like terms}$$

$$\frac{12x}{12} = \frac{24}{12} \qquad \text{Dividing by 12}$$

$$x = 2$$

The number 2 checks, so the solution is 2.

◀ **Do Exercises 15 and 16.**

Answers

13. $\dfrac{8}{3}$ **14.** $-\dfrac{43}{10}$, or -4.3 **15.** 2   **16.** 3

Guided Solution:
13. 8, 8, 8, 7, 2, 11, 8x, 6, 2, 8, 3

Here is a procedure for solving the types of equation discussed in this section.

AN EQUATION-SOLVING PROCEDURE

....

1. Multiply on both sides to clear the equation of fractions or decimals. (This is optional, but it can ease computations.)
2. If parentheses occur, multiply to remove them using the *distributive laws.*
3. Collect like terms on each side, if necessary.
4. Get all terms with variables on one side and all numbers (constant terms) on the other side, using the *addition principle.*
5. Collect like terms again, if necessary.
6. Multiply or divide to solve for the variable, using the *multiplication principle.*
7. Check all possible solutions in the original equation.

EXAMPLE 11 Solve: $2 - 5(x + 5) = 3(x - 2) - 1$.

$$2 - 5(x + 5) = 3(x - 2) - 1$$

$$2 - 5x - 25 = 3x - 6 - 1 \qquad \text{Using the distributive laws to multiply and remove parentheses}$$

$$-5x - 23 = 3x - 7 \qquad \text{Collecting like terms}$$

$$-5x - 23 + 5x = 3x - 7 + 5x \qquad \text{Adding } 5x$$

$$-23 = 8x - 7 \qquad \text{Collecting like terms}$$

$$-23 + 7 = 8x - 7 + 7 \qquad \text{Adding } 7$$

$$-16 = 8x \qquad \text{Collecting like terms}$$

$$\frac{-16}{8} = \frac{8x}{8} \qquad \text{Dividing by 8}$$

$$-2 = x$$

Check:

$$\begin{array}{c|c} \multicolumn{2}{c}{2 - 5(x + 5) = 3(x - 2) - 1} \\ \hline 2 - 5(-2 + 5) \;?\; 3(-2 - 2) - 1 \\ 2 - 5(3) \;\big|\; 3(-4) - 1 \\ 2 - 15 \;\big|\; -12 - 1 \\ -13 \;\big|\; -13 \qquad \text{TRUE} \end{array}$$

The solution is -2.

Do Exercises 17 and 18. ▷

Equations with Infinitely Many Solutions

The types of equations we have considered thus far in Sections 2.1–2.3 have all had exactly one solution. We now look at two other possibilities.
   Consider

$$3 + x = x + 3.$$

Let's explore the equation and possible solutions in Margin Exercises 19–22.

Do Exercises 19–22. ▷

Solve.

17. $3(7 + 2x) = 30 + 7(x - 1)$

18. $4(3 + 5x) - 4 = 3 + 2(x - 2)$

Determine whether the given number is a solution of the given equation.

19. $10; \; 3 + x = x + 3$

20. $-7; \; 3 + x = x + 3$

21. $\frac{1}{2}; \; 3 + x = x + 3$

22. $0; \; 3 + x = x + 3$

Answers

17. -2 18. $-\frac{1}{2}$ 19. Yes 20. Yes

21. Yes 22. Yes

We know by the commutative law of addition that the equation $3 + x = x + 3$ holds for any replacement of x with a real number. (See Section 1.7.) We have confirmed some of these solutions in Margin Exercises 19–22. Suppose we try to solve this equation using the addition principle:

$$3 + x = x + 3$$
$$-x + 3 + x = -x + x + 3 \qquad \text{Adding } -x$$
$$3 = 3. \qquad \text{True}$$

We end with a true equation. The original equation holds for all real-number replacements. Every real number is a solution. Thus the number of solutions is **infinite**.

EXAMPLE 12 Solve: $7x - 17 = 4 + 7(x - 3)$.

$$7x - 17 = 4 + 7(x - 3)$$
$$7x - 17 = 4 + 7x - 21 \qquad \text{Using the distributive law to multiply and remove parentheses}$$
$$7x - 17 = 7x - 17 \qquad \text{Collecting like terms}$$
$$-7x + 7x - 17 = -7x + 7x - 17 \qquad \text{Adding } -7x$$
$$-17 = -17 \qquad \text{True for all real numbers}$$

Every real number is a solution. There are infinitely many solutions.

Equations with No Solution

Now consider

$$3 + x = x + 8.$$

Let's explore the equation and possible solutions in Margin Exercises 23–26.

◀ **Do Exercises 23–26.**

None of the replacements in Margin Exercises 23–26 is a solution of the given equation. In fact, there are no solutions. Let's try to solve this equation using the addition principle:

$$3 + x = x + 8$$
$$-x + 3 + x = -x + x + 8 \qquad \text{Adding } -x$$
$$3 = 8. \qquad \text{False}$$

We end with a false equation. The original equation is false for all real-number replacements. Thus it has **no** solution.

EXAMPLE 13 Solve: $3x + 4(x + 2) = 11 + 7x$.

$$3x + 4(x + 2) = 11 + 7x$$
$$3x + 4x + 8 = 11 + 7x \qquad \text{Using the distributive law to multiply and remove parentheses}$$
$$7x + 8 = 11 + 7x \qquad \text{Collecting like terms}$$
$$7x + 8 - 7x = 11 + 7x - 7x \qquad \text{Subtracting } 7x$$
$$8 = 11 \qquad \text{False}$$

There are no solutions.

◀ **Do Exercises 27 and 28.**

Determine whether the given number is a solution of the given equation.

23. 10; $3 + x = x + 8$

24. -7; $3 + x = x + 8$

25. $\dfrac{1}{2}$; $3 + x = x + 8$

26. 0; $3 + x = x + 8$

Solve.

27. $30 + 5(x + 3) = -3 + 5x + 48$

28. $2x + 7(x - 4) = 13 + 9x$

When solving an equation, if the result is

- an equation of the form $x = a$, where a is a real number, then there is one solution, the number a;
- a true equation like $3 = 3$ or $-1 = -1$, then every real number is a solution;
- a false equation like $3 = 8$ or $-4 = 5$, then there is no solution.

Answers

23. No   24. No   25. No   26. No
27. All real numbers   28. No solution

✓ Reading Check

Choose from the column on the right the operation that will clear each equation of fractions or decimals.

RC1. $\frac{2}{5}x - 5 + \frac{1}{2}x = \frac{3}{10} + x$

RC2. $0.003y - 0.1 = 0.03 + y$

RC3. $\frac{1}{4} - 8t + \frac{5}{6} = t - \frac{1}{12}$

RC4. $0.5 + 2.15y = 1.5y - 10$

RC5. $\frac{3}{5} - x = \frac{2}{7}x + 4$

a) Multiply by 1000 on both sides.
b) Multiply by 35 on both sides.
c) Multiply by 12 on both sides.
d) Multiply by 10 on both sides.
e) Multiply by 100 on both sides.

a   Solve. Don't forget to check!

1. $5x + 6 = 31$
Check: $5x + 6 = 31$
?

2. $7x + 6 = 13$
Check: $7x + 6 = 13$
?

3. $8x + 4 = 68$
Check: $8x + 4 = 68$
?

4. $4y + 10 = 46$
Check: $4y + 10 = 46$
?

5. $4x - 6 = 34$

6. $5y - 2 = 53$

7. $3x - 9 = 33$

8. $4x - 19 = 5$

9. $7x + 2 = -54$

10. $5x + 4 = -41$

11. $-45 = 3 + 6y$

12. $-91 = 9t + 8$

13. $-4x + 7 = 35$

14. $-5x - 7 = 108$

15. $\frac{5}{4}x - 18 = -3$

16. $\frac{3}{2}x - 24 = -36$

b   Solve.

17. $5x + 7x = 72$
Check: $5x + 7x = 72$
?

18. $8x + 3x = 55$
Check: $8x + 3x = 55$
?

19. $8x + 7x = 60$
Check: $8x + 7x = 60$
?

20. $8x + 5x = 104$
Check: $8x + 5x = 104$
?

21. $4x + 3x = 42$

22. $7x + 18x = 125$

23. $-6y - 3y = 27$

24. $-5y - 7y = 144$

25. $-7y - 8y = -15$

26. $-10y - 3y = -39$

27. $x + \dfrac{1}{3}x = 8$

28. $x + \dfrac{1}{4}x = 10$

29. $10.2y - 7.3y = -58$

30. $6.8y - 2.4y = -88$

31. $8y - 35 = 3y$

32. $4x - 6 = 6x$

33. $8x - 1 = 23 - 4x$

34. $5y - 2 = 28 - y$

35. $2x - 1 = 4 + x$

36. $4 - 3x = 6 - 7x$

37. $6x + 3 = 2x + 11$

38. $14 - 6a = -2a + 3$

39. $5 - 2x = 3x - 7x + 25$

40. $-7z + 2z - 3z - 7 = 17$

41. $4 + 3x - 6 = 3x + 2 - x$

42. $5 + 4x - 7 = 4x - 2 - x$

43. $4y - 4 + y + 24 = 6y + 20 - 4y$

44. $5y - 7 + y = 7y + 21 - 5y$

Solve. Clear fractions or decimals first.

45. $\dfrac{7}{2}x + \dfrac{1}{2}x = 3x + \dfrac{3}{2} + \dfrac{5}{2}x$

46. $\dfrac{7}{8}x - \dfrac{1}{4} + \dfrac{3}{4}x = \dfrac{1}{16} + x$

47. $\dfrac{2}{3} + \dfrac{1}{4}t = \dfrac{1}{3}$

48. $-\dfrac{3}{2} + x = -\dfrac{5}{6} - \dfrac{4}{3}$

49. $\dfrac{2}{3} + 3y = 5y - \dfrac{2}{15}$

50. $\dfrac{1}{2} + 4m = 3m - \dfrac{5}{2}$

51. $\dfrac{5}{3} + \dfrac{2}{3}x = \dfrac{25}{12} + \dfrac{5}{4}x + \dfrac{3}{4}$

52. $1 - \dfrac{2}{3}y = \dfrac{9}{5} - \dfrac{y}{5} + \dfrac{3}{5}$

53. $2.1x + 45.2 = 3.2 - 8.4x$

54. $0.96y - 0.79 = 0.21y + 0.46$

55. $1.03 - 0.62x = 0.71 - 0.22x$

56. $1.7t + 8 - 1.62t = 0.4t - 0.32 + 8$

57. $\dfrac{2}{7}x - \dfrac{1}{2}x = \dfrac{3}{4}x + 1$

58. $\dfrac{5}{16}y + \dfrac{3}{8}y = 2 + \dfrac{1}{4}y$

C   Solve.

59. $3(2y - 3) = 27$

60. $8(3x + 2) = 30$

61. $40 = 5(3x + 2)$

62. $9 = 3(5x - 2)$

63. $-23 + y = y + 25$

64. $17 - t = -t + 68$

65. $-23 + x = x - 23$

66. $y - \dfrac{2}{3} = -\dfrac{2}{3} + y$

67. $2(3 + 4m) - 9 = 45$

68. $5x + 5(4x - 1) = 20$

69. $5r - (2r + 8) = 16$

70. $6b - (3b + 8) = 16$

71. $6 - 2(3x - 1) = 2$

72. $10 - 3(2x - 1) = 1$

73. $5(d + 4) = 7(d - 2)$

74. $3(t - 2) = 9(t + 2)$

75. $8(2t + 1) = 4(7t + 7)$

76. $7(5x - 2) = 6(6x - 1)$

77. $5x + 5 - 7x = 15 - 12x + 10x - 10$

78. $3 - 7x + 10x - 14 = 9 - 6x + 9x - 20$

79. $22x - 5 - 15x + 3 = 10x - 4 - 3x + 11$

80. $11x - 6 - 4x + 1 = 9x - 8 - 2x + 12$

81. $3(r - 6) + 2 = 4(r + 2) - 21$

82. $5(t + 3) + 9 = 3(t - 2) + 6$

83. $19 - (2x + 3) = 2(x + 3) + x$

84. $13 - (2c + 2) = 2(c + 2) + 3c$

85. $2[4 - 2(3 - x)] - 1 = 4[2(4x - 3) + 7] - 25$

86. $5[3(7 - t) - 4(8 + 2t)] - 20 = -6[2(6 + 3t) - 4]$

87. $11 - 4(x + 1) - 3 = 11 + 2(4 - 2x) - 16$

88. $6(2x - 1) - 12 = 7 + 12(x - 1)$

89. $22x - 1 - 12x = 5(2x - 1) + 4$

90. $2 + 14x - 9 = 7(2x + 1) - 14$

91. $0.7(3x + 6) = 1.1 - (x + 2)$

92. $0.9(2x + 8) = 20 - (x + 5)$

Skill Maintenance

93. Divide: $-22.1 \div 3.4$. [10.6c]

94. Multiply: $-22.1(3.4)$. [10.5a]

95. Factor: $7x - 21 - 14y$. [10.7d]

96. Factor: $8y - 88x + 8$. [10.7d]

Simplify.

97. $-3 + 2(-5)^2(-3) - 7$ [10.8d]

98. $3x + 2[4 - 5(2x - 1)]$ [10.8c]

99. $23(2x - 4) - 15(10 - 3x)$ [10.8b]

100. $256 \div 64 \div 4^2$ [10.8d]

Synthesis

Solve.

101. $\frac{2}{3}\left(\frac{7}{8} - 4x\right) - \frac{5}{8} = \frac{3}{8}$

102. $\frac{1}{4}(8y + 4) - 17 = -\frac{1}{2}(4y - 8)$

103. $\frac{4 - 3x}{7} = \frac{2 + 5x}{49} - \frac{x}{14}$

104. The width of a rectangle is 5 ft, its length is $(3x + 2)$ ft, and its area is 75 ft^2. Find x.

Formulas

a   EVALUATING FORMULAS

A **formula** is a "recipe" for doing a certain type of calculation. Formulas are often given as equations. When we replace the variables in an equation with numbers and calculate the result, we are **evaluating** the formula. Evaluating was introduced in Section 10.1.

Let's consider a formula that has to do with weather. Suppose you see a flash of lightning during a storm. Then a few seconds later, you hear thunder. Your distance from the place where the lightning struck is given by the formula $M = \frac{1}{5}t$, where t is the number of seconds from the lightning flash to the sound of the thunder and M is in miles.

EXAMPLE 1   *Distance from Lightning.*   Consider the formula $M = \frac{1}{5}t$. Suppose it takes 10 sec for the sound of thunder to reach you after you have seen a flash of lightning. How far away did the lightning strike?

We substitute 10 for t and calculate M:

$$M = \frac{1}{5}t = \frac{1}{5}(10) = 2.$$

The lightning struck 2 mi away.

Do Margin Exercise 1. ▷

EXAMPLE 2   *Cost of Operating a Microwave Oven.*   The cost C of operating a microwave oven for 1 year is given by the formula

$$C = \frac{W \times h \times 365}{1000} \cdot e,$$

where $W =$ the wattage, $h =$ the number of hours used per day, and $e =$ the energy cost per kilowatt-hour. Find the cost of operating a 1500-W microwave oven for 0.25 hr per day if the energy cost is $0.13 per kilowatt-hour.

Substituting, we have

$$C = \frac{W \times h \times 365}{1000} \cdot e = \frac{1500 \times 0.25 \times 365}{1000} \cdot \$0.13 \approx \$17.79.$$

The cost for operating a 1500-W microwave oven for 0.25 hr per day for 1 year is about $17.79.

Do Margin Exercise 2. ▷

OBJECTIVES

a Evaluate a formula.

b Solve a formula for a specified letter.

SKILL TO REVIEW

Objective 11.3a: Solve equations using both the addition principle and the multiplication principle.

Solve.

1. $28 = 7 - 3a$

2. $\frac{1}{2}x - 22 = -20$

1. *Storm Distance.*   Refer to Example 1. Suppose that it takes the sound of thunder 14 sec to reach you. How far away is the storm?

2. *Microwave Oven.*   Refer to Example 2. Determine the cost of operating an 1100-W microwave oven for 0.5 hr per day for 1 year if the energy cost is $0.16 per kilowatt-hour.

Answers

Skill to Review:
1. −7   **2.** 4

Margin Exercises:
1. 2.8 mi   **2.** $32.12

3. *Socks from Cotton.* Refer to Example 3. Determine the number of socks that can be made from 65 bales of cotton.

EXAMPLE 3 *Socks from Cotton.* Consider the formula $S = 4321x$, where S is the number of socks of average size that can be produced from x bales of cotton. You see a shipment of 300 bales of cotton taken off a ship. How many socks can be made from the cotton?

Source: *Country Woman Magazine*

We substitute 300 for x and calculate S:

$$S = 4321x = 4321(300) = 1,296,300.$$

Thus, 1,296,300 socks can be made from 300 bales of cotton.

◀ **Do Exercise 3.**

b SOLVING FORMULAS

Refer to Example 3. Suppose a clothing company wants to produce S socks and needs to know how many bales of cotton to order. If this calculation is to be repeated many times, it might be helpful to first solve the formula for x:

$$S = 4321x$$

$$\frac{S}{4321} = x. \qquad \text{Dividing by 4321}$$

Then we can substitute a number for S and calculate x. For example, if the number of socks S to be produced is 432,100, then

$$x = \frac{S}{4321} = \frac{432,100}{4321} = 100.$$

The company would need to order 100 bales of cotton.

EXAMPLE 4 Solve for z: $H = \frac{1}{4}z$.

$$H = \tfrac{1}{4}z \qquad \text{We want this letter alone.}$$
$$4 \cdot H = 4 \cdot \tfrac{1}{4}z \qquad \text{Multiplying by 4 on both sides}$$
$$4H = z$$

For $H = 2$ in Example 4, $z = 4H = 4(2)$, or 8.

EXAMPLE 5 *Distance, Rate, and Time.* Solve for t: $d = rt$.

$$d = rt \qquad \text{We want this letter alone.}$$
$$\frac{d}{r} = \frac{rt}{r} \qquad \text{Dividing by } r$$
$$\frac{d}{r} = \frac{r}{r} \cdot t$$
$$\frac{d}{r} = t \qquad \text{Simplifying}$$

◀ **Do Exercises 4–6.**

4. Solve for q: $B = \dfrac{1}{3}q$.

5. Solve for m: $n = mz$.

6. *Electricity.* Solve for I: $V = IR$. (This formula relates voltage V, current I, and resistance R.)

Answers

3. 280,865 socks   **4.** $q = 3B$
5. $m = \dfrac{n}{z}$ **6.** $I = \dfrac{V}{R}$

EXAMPLE 6 Solve for x: $y = x + 3$.

$$y = x + 3 \qquad \text{We want this letter alone.}$$
$$y - 3 = x + 3 - 3 \qquad \text{Subtracting 3}$$
$$y - 3 = x \qquad \text{Simplifying}$$

EXAMPLE 7 Solve for x: $y = x - a$.

$$y = x - a \qquad \text{We want this letter alone.}$$
$$y + a = x - a + a \qquad \text{Adding } a$$
$$y + a = x \qquad \text{Simplifying}$$

Do Exercises 7–9. ▷

Solve for x.

7. $y = x + 5$

8. $y = x - 7$

9. $y = x - b$

EXAMPLE 8 Solve for y: $6y = 3x$.

$$6y = 3x \qquad \text{We want this letter alone.}$$
$$\frac{6y}{6} = \frac{3x}{6} \qquad \text{Dividing by 6}$$
$$y = \frac{x}{2}, \text{ or } \frac{1}{2}x \qquad \text{Simplifying}$$

EXAMPLE 9 Solve for y: $by = ax$.

$$by = ax \qquad \text{We want this letter alone.}$$
$$\frac{by}{b} = \frac{ax}{b} \qquad \text{Dividing by } b$$
$$y = \frac{ax}{b} \qquad \text{Simplifying}$$

10. Solve for y: $9y = 5x$.

11. Solve for p: $ap = bt$.

Do Exercises 10 and 11. ▷

EXAMPLE 10 Solve for x: $ax + b = c$.

$$ax + b = c \qquad \text{We want this letter alone.}$$
$$ax + b - b = c - b \qquad \text{Subtracting } b$$
$$ax = c - b \qquad \text{Simplifying}$$
$$\frac{ax}{a} = \frac{c - b}{a} \qquad \text{Dividing by } a$$
$$x = \frac{c - b}{a} \qquad \text{Simplifying}$$

GS 12. Solve for x: $y = mx + b$.
$$y = mx + b$$
$$y - \boxed{} = mx + b - b$$
$$y - b = \boxed{}$$
$$\frac{y - b}{m} = \frac{mx}{\boxed{}}$$
$$\frac{y - b}{m} = \boxed{}$$

13. Solve for Q: $tQ - p = a$.

Do Exercises 12 and 13. ▷

Answers

7. $x = y - 5$ 8. $x = y + 7$

9. $x = y + b$ 10. $y = \frac{5x}{9}$, or $\frac{5}{9}x$

11. $p = \frac{bt}{a}$ 12. $x = \frac{y - b}{m}$

13. $Q = \frac{a + p}{t}$

Guided Solution:
12. b, mx, m, x

To solve a formula for a given letter, identify the letter and:

1. Multiply on both sides to clear fractions or decimals, if that is needed.
2. Collect like terms on each side, if necessary.
3. Get all terms with the letter to be solved for on one side of the equation and all other terms on the other side.
4. Collect like terms again, if necessary.
5. Solve for the letter in question.

EXAMPLE 11 *Circumference.* Solve for r: $C = 2\pi r$. This is a formula for the circumference C of a circle of radius r.

$$C = 2\pi r \qquad \text{We want this letter alone.}$$

$$\frac{C}{2\pi} = \frac{2\pi r}{2\pi} \qquad \text{Dividing by } 2\pi$$

$$\frac{C}{2\pi} = r$$

EXAMPLE 12 *Averages.* Solve for a: $A = \dfrac{a + b + c}{3}$. This is a formula for the average A of three numbers a, b, and c.

$$A = \frac{a + b + c}{3} \qquad \text{We want the letter } a \text{ alone.}$$

$$3 \cdot A = 3 \cdot \frac{a + b + c}{3} \qquad \text{Multiplying by 3 on both sides}$$

$$3A = a + b + c \qquad \text{Simplifying}$$

$$3A - b - c = a \qquad \text{Subtracting } b \text{ and } c$$

◀ **Do Exercises 14 and 15.**

14. *Circumference.* Solve for D:
$$C = \pi D.$$
This is a formula for the circumference C of a circle of diameter D.

15. *Averages.* Solve for c:
$$A = \frac{a + b + c + d}{4}.$$

Answers

14. $D = \dfrac{C}{\pi}$ 15. $c = 4A - a - b - d$

✓ Reading Check

Solve each equation for the indicated letter and choose the correct solution from the column on the right.

RC1. Solve $4x = 7y$ for x.

RC2. Solve $y = \dfrac{1}{4}x - w$ for w.

RC3. Solve $xz + 4y = w$ for x.

RC4. Solve $z = w + 4$ for w.

a) $w = z - 4$

b) $x = \dfrac{7y}{4}$, or $\dfrac{7}{4}y$

c) $w = \dfrac{1}{4}x - y$

d) $x = \dfrac{w - 4y}{z}$

  a Solve.

1. *Wavelength of a Musical Note.*   The wavelength w, in meters per cycle, of a musical note is given by

$$w = \frac{r}{f},$$

where r is the speed of the sound, in meters per second, and f is the frequency, in cycles per second. The speed of sound in air is 344 m/sec. What is the wavelength of a note whose frequency in air is 24 cycles per second?

2. *Furnace Output.*   Contractors in the Northeast use the formula $B = 30a$ to determine the minimum furnace output B, in British thermal units (Btu's), for a well-insulated house with a square feet of flooring. Determine the minimum furnace output for an 1800-ft² house that is well insulated.

Source: U.S. Department of Energy

3. *Calorie Density.*   The calorie density D, in calories per ounce, of a food that contains c calories and weighs w ounces is given by

$$D = \frac{c}{w}.$$

Eight ounces of fat-free milk contains 84 calories. Find the calorie density of fat-free milk.

Source: *Nutrition Action Healthletter,* March 2000, p. 9. Center for Science in the Public Interest, Suite 300; 1875 Connecticut Ave NW, Washington, D.C. 20008.

4. *Size of a League Schedule.*   When all n teams in a league play every other team twice, a total of N games are played, where

$$N = n^2 - n.$$

A soccer league has 7 teams and all teams play each other twice. How many games are played?

5. *Distance, Rate, and Time.*   The distance d that a car will travel at a rate, or speed, r in time t is given by

$$d = rt.$$

a) A car travels at 75 miles per hour (mph) for 4.5 hr. How far will it travel?

b) Solve the formula for t.

6. *Surface Area of a Cube.*   The surface area A of a cube with side s is given by

$$A = 6s^2.$$

a) Find the surface area of a cube with sides of 3 in.

b) Solve the formula for s^2.

7. College Enrollment. At many colleges, the number of "full-time-equivalent" students f is given by

$$f = \frac{n}{15},$$

where n is the total number of credits for which students have enrolled in a given semester.

a) Determine the number of full-time-equivalent students on a campus in which students registered for a total of 21,345 credits.

b) Solve the formula for n.

8. Electrical Power. The power rating P, in watts, of an electrical appliance is determined by

$$P = I \cdot V,$$

where I is the current, in amperes, and V is measured in volts.

a) A microwave oven requires 12 amps of current and the voltage in the house is 115 volts. What is the wattage of the microwave?

b) Solve the formula for I; for V.

b Solve for the indicated letter.

9. $y = 5x$, for x

10. $d = 55t$, for t

11. $a = bc$, for c

12. $y = mx$, for x

13. $n = m + 11$, for m

14. $z = t + 21$, for t

15. $y = x - \frac{3}{5}$, for x

16. $y = x - \frac{2}{3}$, for x

17. $y = 13 + x$, for x

18. $t = 6 + s$, for s

19. $y = x + b$, for x

20. $y = x + A$, for x

21. $y = 5 - x$, for x

22. $y = 10 - x$, for x

23. $y = a - x$, for x

24. $y = q - x$, for x

25. $8y = 5x$, for y

26. $10y = -5x$, for y

27. $By = Ax$, for x

28. $By = Ax$, for y

29. $W = mt + b$, for t

30. $W = mt - b$, for t

31. $y = bx + c$, for x

32. $y = bx - c$, for x

33. *Area of a Parallelogram:*
$A = bh$, for h
(Area A, base b, height h)

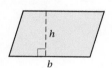

34. *Distance, Rate, Time:*
$d = rt$, for r
(Distance d, speed r, time t)

35. *Perimeter of a Rectangle:*
$P = 2l + 2w$, for w
(Perimeter P, length l, width w)

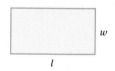

36. *Area of a Circle:*
$A = \pi r^2$, for r^2
(Area A, radius r)

37. *Average of Two Numbers:*
$A = \dfrac{a + b}{2}$, for a

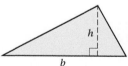

38. *Area of a Triangle:*
$A = \dfrac{1}{2}bh$, for b

39. $A = \dfrac{a + b + c}{3}$, for b

40. $A = \dfrac{a + b + c}{3}$, for c

41. $A = at + b$, for t

42. $S = rx + s$, for x

43. $Ax + By = c$, for x

44. $Q = \dfrac{p - q}{2}$, for p

45. *Force:*
$$F = ma, \text{ for } a$$
(Force F, mass m, acceleration a)

46. *Simple Interest:*
$$I = Prt, \text{ for } P$$
(Interest I, principal P, interest rate r, time t)

47. *Relativity:*
$$E = mc^2, \text{ for } c^2$$
(Energy E, mass m, speed of light c)

48. $Ax + By = c$, for y

49. $v = \dfrac{3k}{t}$, for t

50. $P = \dfrac{ab}{c}$, for c

Skill Maintenance

51. Evaluate $\dfrac{3x - 2y}{y}$ when $x = 6$ and $y = 2$.   [10.1a]

52. Remove parentheses and simplify:
$$4a - 8b - 5(5a - 4b). \quad [10.8b]$$

Subtract.   [10.4a]

53. $-45.8 - (-32.6)$

54. $-\dfrac{2}{3} - \dfrac{5}{6}$

55. $87\dfrac{1}{2} - 123$

Add.   [10.3a]

56. $-\dfrac{5}{12} + \dfrac{1}{4}$

57. $0.082 + (-9.407)$

58. $-2\dfrac{1}{2} + 6\dfrac{1}{4}$

Solve.

59. $2y - 3 + y = 8 - 5y$ [11.3b]

60. $10x + 4 = 3x - 2 + x$ [11.3b]

61. $2(5x + 6) = x - 15$ [11.3c]

62. $5a = 3(6 - 3a)$ [11.3c]

Synthesis

Solve.

63. $H = \dfrac{2}{a - b}$, for b; for a

64. $P = 4m + 7mn$, for m

65. In $A = lw$, if l and w both double, what is the effect on A?

66. In $P = 2a + 2b$, if P doubles, do a and b necessarily both double?

67. In $A = \frac{1}{2}bh$, if b increases by 4 units and h does not change, what happens to A?

68. Solve for F: $D = \dfrac{1}{E + F}$.

Mid-Chapter Review

Concept Reinforcement

Determine whether each statement is true or false.

_____ **1.** $3 - x = 4x$ and $5x = -3$ are equivalent equations. [11.1b]

_____ **2.** For any real numbers a, b, and c, $a = b$ is equivalent to $a + c = b + c$. [11.1b]

_____ **3.** We can use the multiplication principle to divide on both sides of an equation by the same nonzero number. [11.2a]

_____ **4.** Every equation has at least one solution. [11.3c]

Guided Solutions

 Fill in each blank with the number, variable, or expression that creates a correct statement or solution.

Solve. [11.1b], [11.2a]

5.
$$x + 5 = -3$$
$$x + 5 - 5 = -3 - \boxed{}$$
$$x + \boxed{} = -8$$
$$x = \boxed{}$$

6.
$$-6x = 42$$
$$\frac{-6x}{-6} = \frac{42}{\boxed{}}$$
$$\boxed{} \cdot x = -7$$
$$x = \boxed{}$$

7. Solve for y: $5y + z = t$. [11.4b]
$$5y + z = t$$
$$5y + z - z = t - \boxed{}$$
$$5y = \boxed{}$$
$$\frac{5y}{5} = \frac{t - z}{\boxed{}}$$
$$y = \frac{\boxed{}}{5}$$

Mixed Review

Solve. [11.1b], [11.2a], [11.3a, b, c]

8. $x + 5 = 11$

9. $x + 9 = -3$

10. $8 = t + 1$

11. $-7 = y + 3$

12. $x - 6 = 14$

13. $y - 7 = -2$

14. $-\dfrac{3}{2} + z = -\dfrac{3}{4}$

15. $-3.3 = -1.9 + t$

16. $7x = 42$

17. $17 = -t$

18. $6x = -54$

19. $-5y = -85$

20. $\dfrac{x}{7} = 3$

21. $\dfrac{2}{3}x = 12$

22. $-\dfrac{t}{5} = 3$

23. $\dfrac{3}{4}x = -\dfrac{9}{8}$

24. $3x + 2 = 5$

25. $5x + 4 = -11$

26. $6x - 7 = 2$

27. $-4x - 9 = -5$

28. $6x + 5x = 33$

29. $-3y - 4y = 49$

30. $3x - 4 = 12 - x$

31. $5 - 6x = 9 - 8x$

32. $4y - \dfrac{3}{2} = \dfrac{3}{4} + 2y$

33. $\dfrac{4}{5} + \dfrac{1}{6}t = \dfrac{1}{10}$

34. $0.21n - 1.05 = 2.1 - 0.14n$

35. $5(3y - 1) = -35$

36. $7 - 2(5x + 3) = 1$

37. $-8 + t = t - 8$

38. $z + 12 = -12 + z$

39. $4(3x + 2) = 5(2x - 1)$

40. $8x - 6 - 2x = 3(2x - 4) + 6$

Solve for the indicated letter.   [11.4b]

41. $A = 4b$, for b

42. $y = x - 1.5$, for x

43. $n = s - m$, for m

44. $4t = 9w$, for t

45. $B = at - c$, for t

46. $M = \dfrac{x + y + z}{2}$, for y

Understanding Through Discussion and Writing

47. Explain the difference between equivalent expressions and equivalent equations.   [10.7a], [11.1b]

48. Are the equations $x = 5$ and $x^2 = 25$ equivalent? Why or why not?   [11.1b]

49. When solving an equation using the addition principle, how do you determine which number to add or subtract on both sides of the equation?   [11.1b]

50. Explain the following mistake made by a fellow student.   [11.1b]

$$x + \frac{1}{3} = -\frac{5}{3}$$
$$x = -\frac{4}{3}$$

51. When solving an equation using the multiplication principle, how do you determine by what number to multiply or divide on both sides of the equation?   [11.2a]

52. Devise an application in which it would be useful to solve the equation $d = rt$ for r.   [11.4b]

Applications of Percent

11.5

a   TRANSLATING AND SOLVING

Many applied problems involve percent. Here we begin to see how equation solving can enhance our problem-solving skills.

In solving percent problems, we first *translate* the problem to an equation. Then we *solve* the equation using the techniques discussed in Sections 11.1–11.3. The key words in the translation are as follows.

> ### KEY WORDS IN PERCENT TRANSLATIONS
>
> **"Of"** translates to "·" or "×".
>
> **"Is"** translates to "=".
>
> **"What number"** or **"what percent"** translates to any letter.
>
> **"%"** translates to "$\times \frac{1}{100}$" or "$\times 0.01$".

OBJECTIVE

a   Solve applied problems involving percent.

SKILL TO REVIEW

Objective 11.2a: Solve equations using the multiplication principle.

Solve.

1. $20 = 0.05a$
2. $0.3z = 327$

EXAMPLE 1   Translate:

28%   of   5   is   what number?
↓   ↓   ↓   ↓   ↓
28%   ·   5   =   a   This is a percent equation.

EXAMPLE 2   Translate:

45%   of   what number   is   28?
↓   ↓   ↓   ↓   ↓
45%   ×   b   =   28

EXAMPLE 3   Translate:

What percent   of   90   is   7?
     ↓   ↓   ↓   ↓   ↓
     n   ·   90   =   7

Do Margin Exercises 1–6. ▶

Percent problems are actually of three different types. Although the method we present does *not* require that you be able to identify which type we are studying, it is helpful to know them. Let's begin by using a specific example to find a standard form for a percent problem.

Translate to an equation. Do not solve.

1. 13% of 80 is what number?

2. What number is 60% of 70?

3. 43 is 20% of what number?

GS 4. 110% of what number is 30?

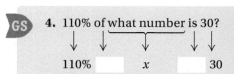

110% [ ] x [ ] 30

5. 16 is what percent of 80?

6. What percent of 94 is 10.5?

Answers

Skill to Review:
1. 400   **2.** 1090

Answers to Margin Exercises 1–6 and Guided Solution are on p. 118.

We know that

15 is 25% of 60, or 15 = 25% × 60.

We can think of this as:

> Amount = Percent number × Base.

Each of the three types of percent problem depends on which of the three pieces of information is missing in the statement

Amount = Percent number × Base.

1. Finding the *amount* (the result of taking the percent)

Example: What number is 25% of 60?

Translation: y = 25% · 60

2. Finding the *base* (the number you are taking the percent of)

Example: 15 is 25% of what number?

Translation: 15 = 25% · y

3. Finding the *percent number* (the percent itself)

Example: 15 is what percent of 60?

Translation: 15 = y · 60

Finding the Amount

EXAMPLE 4 What number is 11% of 49?

What number is 11% of 49?

Translate: a = 11% × 49

Solve: The letter is by itself. To solve the equation, we need only convert 11% to decimal notation and multiply:

$a = 11\% \times 49 = 0.11 \times 49 = 5.39.$

Thus, 5.39 is 11% of 49. The answer is 5.39.

◀ Do Exercise 7.

7. What number is 2.4% of 80?

Finding the Base

EXAMPLE 5 3 is 16% of what number?

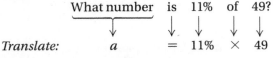

3 is 16% of what number?

Translate: 3 = 16% × b

$3 = 0.16 \times b$ Converting 16% to decimal notation

Solve: In this case, the letter is not by itself. To solve the equation, we divide by 0.16 on both sides:

$$3 = 0.16 \times b$$

$$\frac{3}{0.16} = \frac{0.16 \times b}{0.16} \qquad \text{Dividing by 0.16}$$

$$18.75 = b. \qquad \text{Simplifying}$$

The answer is 18.75.

Do Exercise 8. ▶

Finding the Percent Number

In solving these problems, you *must* remember to convert to percent notation after you have solved the equation.

EXAMPLE 6 $32 is what percent of $50?

$$
\begin{array}{ccccc}
\$32 & \text{is} & \text{what percent} & \text{of} & \$50? \\
\downarrow & \downarrow & \downarrow & \downarrow & \downarrow \\
\end{array}
$$

Translate: $32 = p \times 50$

Solve: To solve the equation, we divide by 50 on both sides and convert the answer to percent notation:

$$32 = p \times 50$$

$$\frac{32}{50} = \frac{p \times 50}{50} \qquad \text{Dividing by 50}$$

$$0.64 = p$$

$$64\% = p. \qquad \text{Converting to percent notation}$$

Thus, $32 is 64% of $50. The answer is 64%.

Do Exercise 9. ▶

EXAMPLE 7   *Donated Girl Scout Cookies.*   Through Operation Cookie Drop, Girl Scout cookies can be donated to all branches of the U.S. military. In 2012, the Girl Scouts of Western Washington sold 3,093,834 boxes of cookies. Of this number, 4.11% were donated to military personnel. How many boxes of cookies did the Girl Scouts of Western Washington donate to Operation Cookie Drop?

Source: Girl Scouts of the USA

To solve this problem, we first reword and then translate. We let $c =$ the number of boxes of cookies donated.

Rewording: What number   is   4.11%   of   3,093,834?

Translating: $c \quad = \quad 4.11\% \quad \times \quad 3{,}093{,}834$

Solve: The letter is by itself. To solve the equation, we need only convert 4.11% to decimal notation and multiply:

$$c = 4.11\% \times 3{,}093{,}834 = 0.0411 \times 3{,}093{,}834 \approx 127{,}157.$$

Thus, 127,157 is about 4.11% of 3,093,834, so 127,157 boxes of Girl Scout cookies were donated to Operation Cookie Drop in 2012.

Do Exercise 10. ▶

GS 8. 25.3 is 22% of what number?

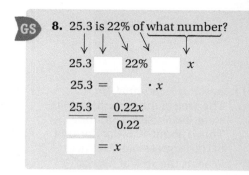

$$25.3 = \boxed{} \cdot x$$

$$\frac{25.3}{\boxed{}} = \frac{0.22x}{0.22}$$

$$\boxed{} = x$$

9. What percent of $50 is $18?

10. *Haitian Population Ages 0–14.* The population of Haiti is approximately 9,720,000. Of this number, 35.9% are ages 0–14. How many Haitians are ages 14 and younger? Round to the nearest 1000.

Source: Central Intelligence Agency

MCT KidNews 05/05

Answers

8. 115   **9.** 36%
10. About 3,489,000

Guided Solution:
8. =, · , 0.22, 0.22, 115

EXAMPLE 8 *Motor Vehicle Production.* In 2010, 7.632 million motor vehicles were produced in the United States. This was 10.4% of the world production of motor vehicles. How many motor vehicles were produced worldwide in 2010?

Sources: Automotive News Data Center; R. L. Polk

To solve this problem, we first reword and then translate. We let P = the total worldwide production, in millions, of motor vehicles in 2010.

$$
\begin{array}{ccccc}
\textit{Rewording:} & 7.632 & \text{is} & 10.4\% & \text{of} & \text{what number?} \\
& \downarrow & \downarrow & \downarrow & \downarrow & \downarrow \\
\textit{Translating:} & 7.632 & = & 10.4\% & \times & P
\end{array}
$$

Solve: To solve the equation, we convert 10.4% to decimal notation and divide by 0.104 on both sides:

$$7.632 = 10.4\% \times P$$
$$7.632 = 0.104 \times P \qquad \text{Converting to decimal notation}$$
$$\frac{7.632}{0.104} = \frac{0.104 \times P}{0.104} \qquad \text{Dividing by 0.104}$$
$$73.4 \approx P. \qquad \text{Simplifying and rounding to the nearest tenth}$$

About 73.4 million motor vehicles were produced worldwide in 2010.

◀ **Do Exercise 11.**

EXAMPLE 9 *Employment Outlook.* Jobs at United States auto plants and parts factories (including domestic and foreign-owned) totaled approximately 650,000 in 2012. This number is expected to grow to 756,800 in 2015. What is the percent increase?

Sources: Center for Automotive Research; IHS Global Insight

To solve the problem, we must first determine the amount of the increase:

$$
\begin{array}{ccccc}
\text{Jobs in 2015} & \text{minus} & \text{Jobs in 2012} & = & \text{Increase} \\
\downarrow & \downarrow & \downarrow & & \downarrow \\
756{,}800 & - & 650{,}000 & = & 106{,}800.
\end{array}
$$

Using the job increase of 106,800, we reword and then translate. We let p = the percent increase. We want to know, "what percent of the number of jobs in 2012 is 106,800?"

$$
\begin{array}{ccccc}
\textit{Rewording:} & 106{,}800 & \text{is} & \text{what percent} & \text{of} & 650{,}000 \\
& \downarrow & \downarrow & \downarrow & \downarrow & \downarrow \\
\textit{Translating:} & 106{,}800 & = & p & \times & 650{,}000
\end{array}
$$

Solve: To solve the equation, we divide by 650,000 on both sides and convert the answer to percent notation:

$$106{,}800 = p \times 650{,}000$$
$$\frac{106{,}800}{650{,}000} = \frac{p \times 650{,}000}{650{,}000} \qquad \text{Dividing by 650,000}$$
$$0.164 \approx p \qquad \text{Simplifying}$$
$$16.4\% \approx p. \qquad \text{Converting to percent notation}$$

The percent increase is about 16.4%.

◀ **Do Exercise 12.**

11. *Areas of Texas and Alaska.* The area of the second largest state, Texas, is 268,581 mi². This is about 40.5% of the area of the largest state, Alaska. What is the area of Alaska?

12. *Median Income.* The U.S. median family income in 2004 was $49,800. This number decreased to $45,800 in 2010. What is the percent decrease?

Source: Federal Reserve's Survey of Consumer Finances, June 2012

Answers

11. About 663,163 mi²
12. About 8.0% decrease

11.5 Exercise Set

For Extra Help

MyMathLab

 MathXL®

 PRACTICE

 WATCH

 READ   REVIEW

✓ **Reading Check**

Match each question with the most appropriate translation from the column on the right.

RC1. 13 is 82% of what number?

RC2. What number is 13% of 82?

RC3. 82 is what percent of 13?

RC4. 82 is 13% of what number?

RC5. 13 is what percent of 82?

RC6. What number is 82% of 13?

a) $82 = 13\% \cdot b$

b) $a = 13\% \cdot 82$

c) $a = 82\% \cdot 13$

d) $13 = 82\% \cdot b$

e) $82 = p \cdot 13$

f) $13 = p \cdot 82$

a   Solve.

1. What percent of 180 is 36?

2. What percent of 76 is 19?

3. 45 is 30% of what number?

4. 20.4 is 24% of what number?

5. What number is 65% of 840?

6. What number is 50% of 50?

7. 30 is what percent of 125?

8. 57 is what percent of 300?

9. 12% of what number is 0.3?

10. 7 is 175% of what number?

11. 2 is what percent of 40?

12. 16 is what percent of 40?

13. What percent of 68 is 17?

14. What percent of 150 is 39?

15. What number is 35% of 240?

16. What number is 1% of one million?

17. What percent of 575 is 138?

18. What percent of 60 is 75?

19. What percent of 300 is 48?

20. What percent of 70 is 70?

21. 14 is 30% of what number?

22. 54 is 24% of what number?

23. What number is 2% of 40?

24. What number is 40% of 2?

25. 0.8 is 16% of what number?

26. 40 is 2% of what number?

27. 54 is 135% of what number?

28. 8 is 2% of what number?

World Population by Continent. It has been projected that in 2050, the world population will be 8909 million, or 8.909 billion. The following circle graph shows the breakdown of this total population by continent.

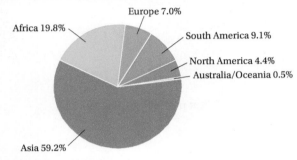

World Population by Continent, 2050

Europe 7.0%
Africa 19.8%
South America 9.1%
North America 4.4%
Australia/Oceania 0.5%
Asia 59.2%

SOURCE: Central Intelligence Agency

Using the data in the figure, complete the following table of projected populations in 2050. Round to the nearest million.

	Continent	Population		Continent	Population
29.	South America		**30.**	Europe	
31.	Asia		**32.**	North America	
33.	Africa		**34.**	Australia/Oceania	

35. *eBook Revenue.* Net revenue from 2011 adult book sales totaled approximately $2360 million. eBook revenue accounted for 41% of this amount. What was the net revenue from adult eBook sales in 2011? Round to the nearest million.

Source: Association of American Publishers

36. *Hardcover Book Revenue.* Net revenue from 2011 adult book sales totaled approximately $2360 million. Revenue from hardcover books accounted for 54.8% of this amount. What was the net revenue from adult hardcover sales in 2011? Round to the nearest million.

Source: Association of American Publishers

37. *Student Loans.* To finance her community college education, Sarah takes out a Stafford loan for $6500. After a year, Sarah decides to pay off the interest, which is 3.4% of $6500. How much will she pay?

38. *Student Loans.* Paul takes out a PLUS loan for $5400. After a year, Paul decides to pay off the interest, which is 7.9% of $5000. How much will he pay?

39. *Tattoos.* Of the 237,400,000 adults ages 18 and older in the United States, approximately 49,854,000 have at least one tattoo. What percent of adults ages 18 and older have at least one tattoo?

Sources: U.S. Census Bureau; Harris Poll of 2016 adults; UPI.com

40. *Boston Marathon.* The 2012 Boston Marathon was the 116th running of the race. Since its first race, the United States has won the men's open division 43 times. What percent of the years did the United States win the men's open?

Source: Boston Athletic Association

41. *Tipping.* William left a $1.50 tip for a meal that cost $12.
   a) What percent of the cost of the meal was the tip?
   b) What was the total cost of the meal including the tip?

42. *Tipping.* Sam, Selena, Rachel, and Clement left a 20% tip for a meal that cost $75.
   a) How much was the tip?
   b) What was the total cost of the meal including the tip?

43. *Tipping.* David left a 15% tip of $4.65 for a meal.
   a) What was the cost of the meal before the tip?
   b) What was the total cost of the meal including the tip?

44. *Tipping.* Addison left an 18% tip of $6.75 for a meal.
   a) What was the cost of the meal before the tip?
   b) What was the total cost of the meal including the tip?

45. *City Park Space.* Portland, Oregon, has 12,959 acres of park space. This is 15.1% of the acreage of the entire city. What is the total acreage of Portland?

Source: Indy Parks and Recreation master plan

46. *Junk Mail.* About 46.2 billion pieces of unopened junk mail end up in landfills each year. This is about 44% of all the junk mail that is sent annually. How many pieces of junk mail are sent annually?

Source: Globaljunkmailcrisis.org

47. *Employment Growth.* In 1980, there were 1.7 million licensed registered nurses in the United States. This number increased to 3.1 million in 2012. What is the percent increase?

Source: Maria Sonnenberg, *Florida Today*, May 14, 2012

48. *Artificial-Tree Sales.* From 2008 to 2012, sales of artificial Christmas trees grew from $950 million to approximately $1070 million. What is the percent increase?

Sources: BalsamHill.com

49. *Newspaper Advertiser Spending.* In 2000, advertisers spent $48.6 billion in newspapers. This number dropped to $22.8 billion in 2010. What is the percent decrease?

Source: "Black and White, Read No More" by Paul Glader, *The American Legion Magazine*, August 2012

50. *New Magazines.* In 2010, 301 magazines were launched in North America. In 2011, only 273 were launched. What is the percent decrease?

Source: MediaFinder

51. *Dog Bites.* Over one-third of all homeowner insurance liability claims paid in 2011 were for dog bites. The average cost of dog-bite claims in the United States increased from $19,162 in 2004 to $29,396 in 2011. What is the percent increase?

Sources: Insurance Information Institute; Adam Belz, *USA TODAY*

52. *International Students.* In 2001–2002, 582,996 international students were enrolled in U.S. colleges. By 2011–2012, this number had increased to 764,495. What is the percent increase?

Source: Institute of International Education, Open Doors 2012

53. *Little League Baseball.* The number of Little League baseball players worldwide declined from 2.6 million in 1997 to only 2.1 million in 2011. What is the percent decrease?

Source: Little League International

54. *Adoptions from Russia.* The number of Russian child adoptions by Americans has declined from 4381 in 1999 to only 962 in 2011. What is the percent decrease?

Source: U.S. Department of State

Skill Maintenance

Multiply. [10.7c]

55. $3(4 + q)$

56. $-\dfrac{1}{2}(-10x + 42)$

Simplify. [10.7a]

57. $\dfrac{75yw}{40y}$

58. $-\dfrac{18b}{12b}$

Simplify. [10.8c]

59. $-2[3 - 5(7 - 2)]$

60. $[3(x + 4) - 6] - [8 + 2(x - 5)]$

Synthesis

61. It has been determined that at the age of 15, a boy has reached 96.1% of his final adult height. Jaraan is 6 ft 4 in. at the age of 15. What will his final adult height be?

62. It has been determined that at the age of 10, a girl has reached 84.4% of her final adult height. Dana is 4 ft 8 in. at the age of 10. What will her final adult height be?

a FIVE STEPS FOR SOLVING PROBLEMS

We have discussed many new equation-solving tools in this chapter and used them for applications and problem solving. Here we consider a five-step strategy that can be very helpful in solving problems.

FIVE STEPS FOR PROBLEM SOLVING IN ALGEBRA

1. *Familiarize* yourself with the problem situation.
2. *Translate* the problem to an equation.
3. *Solve* the equation.
4. *Check* the answer in the original problem.
5. *State* the answer to the problem clearly.

Of the five steps, the most important is probably the first one: becoming familiar with the problem situation. The box below lists some hints for familiarization.

FAMILIARIZING YOURSELF WITH A PROBLEM

• If a problem is given in words, read it carefully. Reread the problem, perhaps aloud. Try to verbalize the problem as though you were explaining it to someone else.

• Choose a variable (or variables) to represent the unknown and clearly state what the variable represents. Be descriptive! For example, let L = the length, d = the distance, and so on.

• Make a drawing and label it with known information, using specific units if given. Also, indicate unknown information.

• Find further information. Look up formulas or definitions with which you are not familiar. (Geometric formulas appear in chapter 9, page 552, of this text.) Consult the Internet or a reference librarian.

• Create a table that lists all the information you have available. Look for patterns that may help in the translation to an equation.

• Think of a possible answer and check the guess. Note the manner in which the guess is checked.

EXAMPLE 1 *Cycling in Vietnam.* National Highway 1, which runs along the coast of Vietnam, is considered one of the top routes for avid bicyclists. While on sabbatical, a history professor spent six weeks biking 1720 km on National Highway 1 from Hanoi through Ha Tinh to Ho Chi Minh City (commonly known as Saigon). At Ha Tinh, he was four times as far from Ho Chi Minh City as he was from Hanoi. How far had he biked and how far did he still need to bike in order to reach the end?

Sources: www.smh.com; *Lonely Planet's Best in 2010*

1. Familiarize. Let's look at a map.

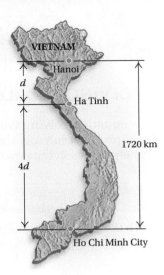

To become familiar with the problem, let's guess a possible distance that the professor is from Hanoi—say, 400 km. Four times 400 km is 1600 km. Since 400 km + 1600 km = 2000 km and 2000 km is greater than 1720 km, we see that our guess is too large. Rather than guess again, let's use the equation-solving skills that we learned in this chapter. We let

$d =$ the distance, in kilometers, to Hanoi, and

$4d =$ the distance, in kilometers, to Ho Chi Minh City.

(We also could let $d =$ the distance to Ho Chi Minh City and $\frac{1}{4}d =$ the distance to Hanoi.)

2. Translate. From the map, we see that the lengths of the two parts of the trip must add up to 1720 km. This leads to our translation.

$$
\underbrace{\text{Distance to Hanoi}}_{d} \quad \underset{+}{\text{plus}} \quad \underbrace{\text{Distance to Ho Chi Minh}}_{4d} \quad \underset{=}{\text{is}} \quad \underset{1720}{1720\text{ km.}}
$$

3. Solve. We solve the equation:

$$d + 4d = 1720$$
$$5d = 1720 \qquad \text{Collecting like terms}$$
$$\frac{5d}{5} = \frac{1720}{5} \qquad \text{Dividing by 5}$$
$$d = 344.$$

4. Check. As we expected, d is less than 400 km. If $d = 344$ km, then $4d = 1376$ km. Since 344 km + 1376 km = 1720 km, the answer checks.

5. State. At Ha Tinh, the professor had biked 344 km from Hanoi and had 1376 km to go to reach Ho Chi Minh City.

◀ Do Exercise 1.

---

1. *Running.* Yiannis Kouros of Australia holds the record for the greatest distance run in 24 hr by running 188 mi. After 8 hr, he was approximately twice as far from the finish line as he was from the start line. How far had he run?

Source: Australian Ultra Runners Association

Answer

1. $62\frac{2}{3}$ mi

EXAMPLE 2 *Knitted Scarf.* Lily knitted a scarf with shades of orange and red yarn, starting with an orange section, then a medium-red section, and finally a dark-red section. The medium-red section is one-half the length of the orange section. The dark-red section is one-fourth the length of the orange section. The scarf is 7 ft long. Find the length of each section of the scarf.

1. **Familiarize.** Because the lengths of the medium-red section and the dark-red section are expressed in terms of the length of the orange section, we let

$$x = \text{the length of the orange section.}$$

Then $\frac{1}{2}x =$ the length of the medium-red section

and $\frac{1}{4}x =$ the length of the dark-red section.

We make a drawing and label it.

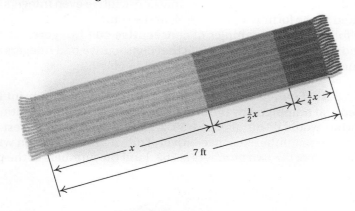

2. **Translate.** From the statement of the problem and the drawing, we know that the lengths add up to 7 ft. This gives us our translation:

Length of orange section	plus	Length of medium-red section	plus	Length of dark-red section	is	Total length
x	$+$	$\frac{1}{2}x$	$+$	$\frac{1}{4}x$	$=$	$7.$

3. **Solve.** First, we clear fractions and then carry out the solution as follows:

$$x + \frac{1}{2}x + \frac{1}{4}x = 7 \qquad \text{The LCM of the denominators is 4.}$$

$$4\left(x + \frac{1}{2}x + \frac{1}{4}x\right) = 4 \cdot 7 \qquad \text{Multiplying by the LCM, 4}$$

$$4 \cdot x + 4 \cdot \frac{1}{2}x + 4 \cdot \frac{1}{4}x = 4 \cdot 7 \qquad \text{Using the distributive law}$$

$$4x + 2x + x = 28 \qquad \text{Simplifying}$$

$$7x = 28 \qquad \text{Collecting like terms}$$

$$\frac{7x}{7} = \frac{28}{7} \qquad \text{Dividing by 7}$$

$$x = 4.$$

2. *Gourmet Sandwiches.* A sandwich shop specializes in sandwiches prepared in buns of length 18 in. Jenny, Emma, and Sarah buy one of these sandwiches and take it back to their apartment. Since they have different appetites, Jenny cuts the sandwich in such a way that Emma gets one-half of what Jenny gets and Sarah gets three-fourths of what Jenny gets. Find the length of each person's sandwich.

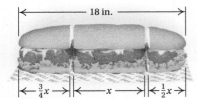

4. Check. Do we have an answer to the *original problem*? If the length of the orange section is 4 ft, then the length of the medium-red section is $\frac{1}{2} \cdot 4$ ft, or 2 ft, and the length of the dark-red section is $\frac{1}{4} \cdot 4$ ft, or 1 ft. The sum of these lengths is 7 ft, so the answer checks.

5. State. The length of the orange section is 4 ft, the length of the medium-red section is 2 ft, and the length of the dark-red section is 1 ft. (Note that we must include the unit, feet, in the answer.)

◀ **Do Exercise 2.**

Recall that the set of integers = $\{\ldots, -5, -4, -3, -2, -1, 0, 1, 2, 3, 4, 5, \ldots\}$. Before we solve the next problem, we need to learn some additional terminology regarding integers.

The following are examples of **consecutive integers:** 16, 17, 18, 19, 20; and $-31, -30, -29, -28$. Note that consecutive integers can be represented in the form $x, x + 1, x + 2$, and so on.

The following are examples of **consecutive even integers:** 16, 18, 20, 22, 24; and $-52, -50, -48, -46$. Note that consecutive even integers can be represented in the form $x, x + 2, x + 4$, and so on.

The following are examples of **consecutive odd integers:** 21, 23, 25, 27, 29; and $-71, -69, -67, -65$. Note that consecutive odd integers can be represented in the form $x, x + 2, x + 4$, and so on.

EXAMPLE 3 *Limited-Edition Prints.* A limited-edition print is usually signed and numbered by the artist. For example, a limited edition with only 50 prints would be numbered 1/50, 2/50, 3/50, and so on. An estate donates two prints numbered consecutively from a limited edition with 150 prints. The sum of the two numbers is 263. Find the numbers of the prints.

1. Familiarize. The numbers of the prints are consecutive integers. If we let $x =$ the smaller number, then $x + 1 =$ the larger number. Since there are 150 prints in the edition, the first number must be 149 or less. If we guess that $x = 138$, then $x + 1 = 139$. The sum of the numbers is 277. We see that the numbers need to be smaller. We could continue guessing and solve the problem this way, but let's work on developing algebra skills.

Answer

2. Jenny: 8 in.; Emma: 4 in.; Sarah: 6 in.

2. Translate. We reword the problem and translate as follows:

Rewording: First integer plus Second integer is 263

Translating: x + $(x + 1)$ = 263.

3. Solve. We solve the equation:

$$x + (x + 1) = 263$$
$$2x + 1 = 263 \qquad \text{Collecting like terms}$$
$$2x + 1 - 1 = 263 - 1 \qquad \text{Subtracting 1}$$
$$2x = 262$$
$$\frac{2x}{2} = \frac{262}{2} \qquad \text{Dividing by 2}$$
$$x = 131.$$

If $x = 131$, then $x + 1 = 132$.

4. Check. Our possible answers are 131 and 132. These are consecutive positive integers and $131 + 132 = 263$, so the answers check.

5. State. The print numbers are 131/150 and 132/150.

Do Exercise 3. ▶

EXAMPLE 4 *Delivery Truck Rental.* An appliance business needs to rent a delivery truck for 6 days while one of its trucks is being repaired. The cost of renting a 16-ft truck is $29.95 per day plus $0.29 per mile. If $550 is budgeted for the rental, how many miles can be driven and stay within budget?

1. Familiarize. Suppose the van is driven 1100 mi. The cost is given by the daily charge plus the mileage charge, so we have

6($29.95) + Cost per mile times Number of miles

$179.70 + $0.29 · 1100,

which is $498.70. We see that the van can be driven more than 1100 mi on the business' budget of $550. This process familiarizes us with the way in which a calculation is made.

 3. *Interstate Mile Markers.* The sum of two consecutive mile markers on I-90 in upstate New York is 627. (On I-90 in New York, the marker numbers increase from east to west.) Find the numbers on the markers.

Source: New York State Department of Transportation

Let $x =$ the first marker and $x + 1 =$ the second marker.

Translate and *Solve*:

First marker + Second marker = 627

☐ + (☐) = 627

☐ + 1 = 627

$$2x + 1 - 1 = 627 - \boxed{}$$
$$2x = \boxed{}$$
$$\frac{2x}{\boxed{}} = \frac{626}{2}$$
$$x = 313.$$

If $x = 313$, then $x + 1 = \boxed{}$. The mile markers are ☐ and 314.

Answer
3. 313 and 314

Guided Solution:
3. $x, x + 1, 2x, 1, 626, 2, 314, 313$

We let m = the number of miles that can be driven on the budget of $550.

2. Translate. We reword the problem and translate as follows:

Daily cost	plus	Cost per mile	times	Number of miles	is	Budget
↓	↓	↓	↓	↓	↓	↓
6($29.95)	+	$0.29	·	m	=	$550.

3. Solve. We solve the equation:

$$6(29.95) + 0.29m = 550$$
$$179.70 + 0.29m = 550$$
$$0.29m = 370.30 \qquad \text{Subtracting 179.70}$$
$$\frac{0.29m}{0.29} = \frac{370.30}{0.29} \qquad \text{Dividing by 0.29}$$
$$m \approx 1277. \qquad \text{Rounding to the nearest one}$$

4. Check. We check our answer in the original problem. The cost for driving 1277 mi is 1277($0.29) = $370.33. The rental for 6 days is 6($29.95) = $179.70. The total cost is then

$$\$370.33 + \$179.70 \approx \$550,$$

which is the $550 budget that was allowed.

5. State. The truck can be driven 1277 mi on the truck-rental allotment.

◀ **Do Exercise 4.**

4. *Delivery Truck Rental.*
Refer to Example 4. The business decides to increase its 6-day rental budget to $625. How many miles can be driven for $625?

EXAMPLE 5 *Perimeter of a Lacrosse Field.* The perimeter of a lacrosse field is 340 yd. The length is 50 yd longer than the width. Find the dimensions of the field.

Source: www.sportsknowhow.com

1. Familiarize. We first make a drawing.

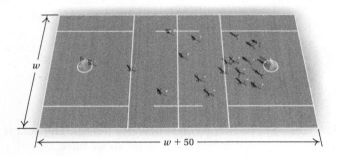

We let w = the width of the rectangle. Then $w + 50$ = the length. The perimeter P of a rectangle is the distance around the rectangle and is given by the formula $2l + 2w = P$, where

l = the length   and w = the width.

Answer

4. 1536 mi

2. Translate. To translate the problem, we substitute $w + 50$ for l and 340 for P:

···· **Caution!** ····

$$2l + 2w = P$$
$$2(w + 50) + 2w = 340.$$

Parentheses are necessary here.

3. Solve. We solve the equation:

$$2(w + 50) + 2w = 340$$
$$2w + 100 + 2w = 340 \quad \text{Using the distributive law}$$
$$4w + 100 = 340 \quad \text{Collecting like terms}$$
$$4w + 100 - 100 = 340 - 100 \quad \text{Subtracting 100}$$
$$4w = 240$$
$$\frac{4w}{4} = \frac{240}{4} \quad \text{Dividing by 4}$$
$$w = 60.$$

Thus the possible dimensions are

$$w = 60 \text{ yd} \quad \text{and} \quad l = w + 50 = 60 + 50, \text{ or } 110 \text{ yd}.$$

4. Check. If the width is 60 yd and the length is 110 yd, then the perimeter is $2(60 \text{ yd}) + 2(110 \text{ yd})$, or 340 yd. This checks.

5. State. The width is 60 ft and the length is 110 yd.

Do Exercise 5. ▶

Do Exercise 5.

5. *Perimeter of High School Basketball Court.* The perimeter of a standard high school basketball court is 268 ft. The length is 34 ft longer than the width. Find the dimensions of the court.

Source: Indiana High School Athletic Association

························· **Caution!** ·························

Always be sure to answer the original problem completely. For instance, in Example 1, we need to find *two* numbers: the distances from *each* city to the biker. Similarly, in Example 3, we need to find two print numbers, and in Example 5, we need to find two dimensions, not just the width.

EXAMPLE 6 *Roof Gable.* In a triangular gable end of a roof, the angle of the peak is twice as large as the angle of the back side of the house. The measure of the angle on the front side is 20° greater than the angle on the back side. How large are the angles?

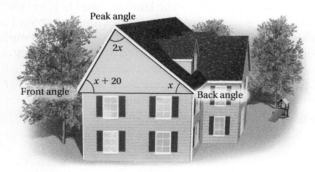

Peak angle

2x

x + 20

x

Front angle   Back angle

1. Familiarize. We first make a drawing as shown above. We let

measure of back angle $= x$.

Then   measure of peak angle $= 2x$

and   measure of front angle $= x + 20$.

Answer

5. Length: 84 ft; width: 50 ft

2. **Translate.** To translate, we need to know that the sum of the measures of the angles of a triangle is 180°. You might recall this fact from geometry or you can look it up in a geometry book or in the list of formulas inside the back cover of this book. We translate as follows:

Measure of back angle plus Measure of peak angle plus Measure of front angle is 180°

$$x + 2x + (x + 20) = 180°.$$

3. **Solve.** We solve the equation:

$$x + 2x + (x + 20) = 180$$
$$4x + 20 = 180$$
$$4x + 20 - 20 = 180 - 20$$
$$4x = 160$$
$$\frac{4x}{4} = \frac{160}{4}$$
$$x = 40.$$

The possible measures for the angles are as follows:

Back angle: $x = 40°$;
Peak angle: $2x = 2(40) = 80°$;
Front angle: $x + 20 = 40 + 20 = 60°$.

4. **Check.** Consider our answers: 40°, 80°, and 60°. The peak is twice the back, and the front is 20° greater than the back. The sum is 180°. The angles check.

5. **State.** The measures of the angles are 40°, 80°, and 60°.

.... **Caution!** ....

Units are important in answers. Remember to include them, where appropriate.

....

6. The second angle of a triangle is three times as large as the first. The third angle measures 30° more than the first angle. Find the measures of the angles.

◀ **Do Exercise 6.**

EXAMPLE 7 *Fastest Roller Coasters.* The average top speed of the three fastest steel roller coasters in the United States is 116 mph. The third-fastest roller coaster, Superman: The Escape (located at Six Flags Magic Mountain, Valencia, California), reaches a top speed of 28 mph less than the fastest roller coaster, Kingda Ka (located at Six Flags Great Adventure, Jackson, New Jersey). The second-fastest roller coaster, Top Thrill Dragster (located at Cedar Point, Sandusky, Ohio), has a top speed of 120 mph. What is the top speed of the fastest steel roller coaster?

Source: Coaster Grotto

Answer

6. First: 30°; second: 90°; third: 60°

694 CHAPTER 11 Solving Equations and Inequalities

1. Familiarize. The **average** of a set of numbers is the sum of the numbers divided by the number of addends.

We are given that the second-fastest speed is 120 mph. Suppose the three top speeds are 131, 120, and 103. The average is then

$$\frac{131 + 120 + 103}{3} = \frac{354}{3} = 118,$$

which is too high. Instead of continuing to guess, let's use the equation-solving skills we have learned in this chapter. We let x = the top speed of the fastest roller coaster. Then $x - 28$ = the top speed of the third-fastest roller coaster.

2. Translate. We reword the problem and translate as follows:

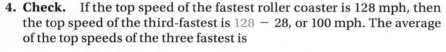

$$\frac{x + 120 + (x - 28)}{3} = 116.$$

3. Solve. We solve as follows:

$$\frac{x + 120 + (x - 28)}{3} = 116$$

$$3 \cdot \frac{x + 120 + (x - 28)}{3} = 3 \cdot 116 \qquad \text{Multiplying by 3 on both sides to clear the fraction}$$

$$x + 120 + (x - 28) = 348$$

$$2x + 92 = 348 \qquad \text{Collecting like terms}$$

$$2x = 256 \qquad \text{Subtracting 92}$$

$$x = 128. \qquad \text{Dividing by 2}$$

4. Check. If the top speed of the fastest roller coaster is 128 mph, then the top speed of the third-fastest is $128 - 28$, or 100 mph. The average of the top speeds of the three fastest is

$$\frac{128 + 120 + 100}{3} = \frac{348}{3} = 116 \text{ mph.}$$

The answer checks.

5. State. The top speed of the fastest steel roller coaster in the United States is 128 mph.

Do Exercise 7. ▶

7. *Average Test Score.* Sam's average score on his first three math tests is 77. He scored 62 on the first test. On the third test, he scored 9 more than he scored on his second test. What did he score on the second and third tests?

Answer

7. Second: 80; third: 89

EXAMPLE 8 *Simple Interest.* An investment is made at 3% simple interest for 1 year. It grows to $746.75. How much was originally invested (the principal)?

1. **Familiarize.** Suppose that $100 was invested. Recalling the formula for simple interest, $I = Prt$, we know that the interest for 1 year on $100 at 3% simple interest is given by $I = \$100 \cdot 0.03 \cdot 1 = \$3$. Then, at the end of the year, the amount in the account is found by adding the principal and the interest:

$$
\begin{array}{ccccc}
\text{Principal} & + & \text{Interest} & = & \text{Amount} \\
\downarrow & & \downarrow & & \downarrow \\
\$100 & + & \$3 & = & \$103.
\end{array}
$$

In this problem, we are working backward. We are trying to find the principal, which is the original investment. We let $x =$ the principal. Then the interest earned is 3%x.

2. **Translate.** We reword the problem and then translate:

$$
\begin{array}{ccccc}
\text{Principal} & + & \text{Interest} & = & \text{Amount} \\
\downarrow & & \downarrow & & \downarrow \\
x & + & 3\%x & = & 746.75.
\end{array}
$$

Interest is 3% of the principal.

3. **Solve.** We solve the equation:

$$x + 3\%x = 746.75$$
$$x + 0.03x = 746.75 \qquad \text{Converting to decimal notation}$$
$$1x + 0.03x = 746.75 \qquad \text{Identity property of 1}$$
$$(1 + 0.03)x = 746.75$$
$$1.03x = 746.75 \qquad \text{Collecting like terms}$$
$$\frac{1.03x}{1.03} = \frac{746.75}{1.03} \qquad \text{Dividing by 1.03}$$
$$x = 725.$$

4. **Check.** We check by taking 3% of $725 and adding it to $725:

$$3\% \times \$725 = 0.03 \times 725 = \$21.75.$$

Then $725 + $21.75 = $746.75, so $725 checks.

5. **State.** The original investment was $725.

◀ Do Exercise 8.

EXAMPLE 9 *Selling a House.* The Patels are planning to sell their house. If they want to be left with $130,200 after paying 7% of the selling price to a realtor as a commission, for how much must they sell the house?

1. **Familiarize.** Suppose the Patels sell the house for $138,000. A 7% commission can be determined by finding 7% of $138,000:

$$7\% \text{ of } \$138,000 = 0.07(\$138,000) = \$9660.$$

Subtracting this commission from $138,000 would leave the Patels with

$$\$138,000 - \$9660 = \$128,340.$$

This shows that in order for the Patels to clear $130,200, the house must sell for more than $138,000. Our guess shows us how to translate to an equation. We let $x =$ the selling price, in dollars. With a 7% commission, the realtor would receive 0.07x.

---

8. *Simple Interest.* An investment is made at 5% simple interest for 1 year. It grows to $2520. How much was originally invested (the principal)?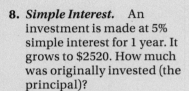

Let $x =$ the principal. Then the interest earned is 5%x.

Translate and *Solve*:

$$
\begin{array}{ccccc}
\text{Principal} & + & \text{Interest} & = & \text{Amount} \\
\downarrow & & \downarrow & & \downarrow \\
x & + & \boxed{} & = & 2520 \\
\end{array}
$$

$$x + 0.05x = 2520$$
$$(1 + \boxed{})x = 2520$$
$$\boxed{}\,x = 2520$$
$$\frac{1.05x}{1.05} = \frac{2520}{\boxed{}}$$
$$x = 2400.$$

Answer

8. $2400

Guided Solution:

8. 5%x, 0.05, 1.05, 1.05

---

2. **Translate.** We reword the problem and translate as follows:

$$\underbrace{\text{Selling price}}_{x} \quad \underbrace{\text{less}}_{-} \quad \underbrace{\text{Commission}}_{0.07x} \quad \underbrace{\text{is}}_{=} \quad \underbrace{\text{Amount remaining}}_{130{,}200.}$$

3. **Solve.** We solve the equation:

$$x - 0.07x = 130{,}200$$
$$1x - 0.07x = 130{,}200$$
$$(1 - 0.07)x = 130{,}200$$
$$0.93x = 130{,}200 \qquad \text{Collecting like terms. Had we noted that after the commission has been paid, 93\% remains, we could have begun with this equation.}$$

$$\frac{0.93x}{0.93} = \frac{130{,}200}{0.93} \qquad \text{Dividing by 0.93}$$

$$x = 140{,}000.$$

4. **Check.** To check, we first find 7% of $140,000:

$$7\% \text{ of } \$140{,}000 = 0.07(\$140{,}000) = \$9800. \qquad \text{This is the commission.}$$

Next, we subtract the commission to find the remaining amount:

$$\$140{,}000 - \$9800 = \$130{,}200.$$

Since, after the commission, the Patels are left with $130,200, our answer checks. Note that the $140,000 selling price is greater than $138,000, as predicted in the *Familiarize* step.

5. **State.** To be left with $130,200, the Patels must sell the house for $140,000.

Do Exercise 9. ▶

······················· Caution! ·······················

The problem in Example 9 is easy to solve with algebra. Without algebra, it is not. A common error in such a problem is to take 7% of the price after commission and then subtract or add. Note that 7% of the selling price ($7\% \cdot \$140{,}000 = \$9800$) is not equal to 7% of the amount that the Patels want to be left with ($7\% \cdot \$130{,}200 = \$9114$).

···

9. *Selling a Condominium.* An investor needs to sell a condominium in New York City. If she wants to be left with $761,400 after paying a 6% commission, for how much must she sell the condominium?

Answer

8. $810,000

Translating for Success

1. **Angle Measures.** The measure of the second angle of a triangle is 51° more than that of the first angle. The measure of the third angle is 3° less than twice the first angle. Find the measures of the angles.

2. **Sales Tax.** Tina paid $3976 for a used car. This amount included 5% for sales tax. How much did the car cost before tax?

3. **Perimeter.** The perimeter of a rectangle is 2347 ft. The length is 28 ft greater than the width. Find the length and the width.

4. **Fraternity or Sorority Membership.** At Arches Tech University, 3976 students belong to a fraternity or a sorority. This is 35% of the total enrollment. What is the total enrollment at Arches Tech?

5. **Fraternity or Sorority Membership.** At Moab Tech University, thirty-five percent of the students belong to a fraternity or a sorority. The total enrollment of the university is 11,360 students. How many students belong to either a fraternity or a sorority?

The goal of these matching questions is to practice step (2), Translate, of the five-step problem-solving process. Translate each word problem to an equation and select a correct translation from equations A–O.

A. $x + (x - 3) + \frac{4}{5}x = 384$

B. $x + (x + 51) + (2x - 3) = 180$

C. $x + (x + 96) = 180$

D. $2 \cdot 96 + 2x = 3976$

E. $x + (x + 1) + (x + 2) = 384$

F. $3976 = x \cdot 11{,}360$

G. $2x + 2(x + 28) = 2347$

H. $3976 = x + 5\%x$

I. $x + (x + 28) = 2347$

J. $x = 35\% \cdot 11{,}360$

K. $x + 96 = 3976$

L. $x + (x + 3) + \frac{4}{5}x = 384$

M. $x + (x + 2) + (x + 4) = 384$

N. $35\% \cdot x = 3976$

O. $2x + (x + 28) = 2347$

Answers on page A-18

---

6. **Island Population.** There are 180 thousand people living on a small Caribbean island. The women outnumber the men by 96 thousand. How many men live on the island?

7. **Wire Cutting.** A 384-m wire is cut into three pieces. The second piece is 3 m longer than the first. The third is four-fifths as long as the first. How long is each piece?

8. **Locker Numbers.** The numbers on three adjoining lockers are consecutive integers whose sum is 384. Find the integers.

9. **Fraternity or Sorority Membership.** The total enrollment at Canyonlands Tech University is 11,360 students. Of these, 3976 students belong to a fraternity or a sorority. What percent of the students belong to a fraternity or a sorority?

10. **Width of a Rectangle.** The length of a rectangle is 96 ft. The perimeter of the rectangle is 3976 ft. Find the width.

✓ Reading Check

Choose from the column on the right the word that completes each step in the five steps for problem solving.

RC1. _____ yourself with the problem situation.

RC2. _____ the problem to an equation.

RC3. _____ the equation.

RC4. _____ the answer in the original problem.

RC5. _____ the answer to the problem clearly.

Solve
Familiarize
State
Translate
Check

a   Solve. *Although you might find the answer quickly in some other way, practice using the five-step problem-solving strategy.*

1. *Medals of Honor.* In 1863, the U.S. Secretary of War presented the first Medals of Honor. The two wars with the most Medals of Honor awarded are the Civil War and World War II. There were 464 recipients of this medal for World War II. This number is 1058 fewer than the number of recipients for the Civil War. How many Medals of Honor were awarded for valor in the Civil War?

Sources: U.S. Army Center of Military History; U.S. Department of Defense

2. *Milk Alternatives.* Milk alternatives such as rice, soy, almond, and flax are becoming more available and increasingly popular. A cup of almond milk contains only 60 calories. This number is 89 calories less than the number of calories in a cup of whole milk. How many calories are in a cup of whole milk?

Source: "Nutrition Udder Chaos," by Janet Kinosian, *AARP Magazine*, August/September, 2012

3. *Pipe Cutting.* A 240-in. pipe is cut into two pieces. One piece is three times the length of the other. Find the lengths of the pieces.

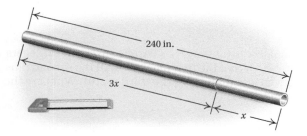

4. *Board Cutting.* A 72-in. board is cut into two pieces. One piece is 2 in. longer than the other. Find the lengths of the pieces.

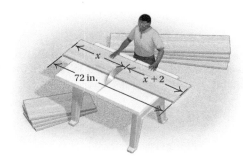

5. *Public Transit Systems.*   In the first quarter of 2012, the ridership on the public transit system in Boston was 99.2 million. This number is 77.4 million more than the ridership in San Diego over the same period of time. What was the ridership in San Diego during the first quarter of 2012?

Source: American Public Transportation Association

6. *Home Listing Price.*   In 2011, the average listing price of a home in Hawaii was $72,000 more than three times the average listing price of a home in Arizona. The average listing price of a home in Hawaii was $876,000. What was the average listing price of a home in Arizona?

Source: Trulia

7. *500 Festival Mini-Marathon.*   On May 4, 2013, 35,000 runners participated in the 13.1-mi One America 500 Festival Mini-Marathon. If a runner stopped at a water station that is twice as far from the start as from the finish, how far is the runner from the finish? Round the answer to the nearest hundredth of a mile.

Source: www.500festival.com

8. *Airport Control Tower.*   At a height of 385 ft, the FAA airport traffic control tower in Atlanta is the tallest traffic control tower in the United States. Its height is 59 ft greater than the height of the tower at the Memphis airport. How tall is the traffic control tower at the Memphis airport?

Source: Federal Aviation Administration

9. *Consecutive Apartment Numbers.*   The apartments in Vincent's apartment house are numbered consecutively on each floor. The sum of his number and his next-door neighbor's number is 2409. What are the two numbers?

10. *Consecutive Post Office Box Numbers.*   The sum of the numbers on two consecutive post office boxes is 547. What are the numbers?

11. *Consecutive Ticket Numbers.*   The numbers on Sam's three raffle tickets are consecutive integers. The sum of the numbers is 126. What are the numbers?

12. *Consecutive Ages.*   The ages of Whitney, Wesley, and Wanda are consecutive integers. The sum of their ages is 108. What are their ages?

13. *Consecutive Odd Integers.*   The sum of three consecutive odd integers is 189. What are the integers?

14. *Consecutive Integers.*   Three consecutive integers are such that the first plus one-half the second plus seven less than twice the third is 2101. What are the integers?

15. Photo Size. A hotel orders a large photo for its newly renovated lobby. The perimeter of the photo is 292 in. The width is 2 in. more than three times the height. Find the dimensions of the photo.

16. Two-by-Four. The perimeter of a cross section or end of a "two-by-four" piece of lumber is 10 in. The length is 2 in. more than the width. Find the actual dimensions of the cross section of a two-by-four.

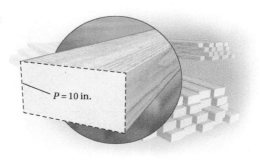

17. Price of Coffee Beans. A student-owned and -operated coffee shop near a campus purchases gourmet coffee beans from Costa Rica. During a recent 30%-off sale, a 3-lb bag could be purchased for $44.10. What is the regular price of a 3-lb bag?

18. Price of an iPad Case. Makayla paid $33.15 for an iPad case during a 15%-off sale. What was the regular price?

19. Price of a Security Wallet. Caleb paid $26.70, including a 7% sales tax, for a security wallet. How much did the wallet itself cost?

20. Price of a Car Battery. Tyler paid $117.15, including a 6.5% sales tax, for a car battery. How much did the battery itself cost?

21. Parking Costs. A hospital parking lot charges $1.50 for the first hour or part thereof, and $1.00 for each additional hour or part thereof. A weekly pass costs $27.00 and allows unlimited parking for 7 days. Suppose that each visit Hailey makes to the hospital lasts $1\frac{1}{2}$ hr. What is the minimum number of times that Hailey would have to visit per week to make it worthwhile for her to buy the pass?

22. Van Rental. Value Rent-A-Car rents vans at a daily rate of $84.45 plus 55¢ per mile. Molly rents a van to deliver electrical parts to her customers. She is allotted a daily budget of $250. How many miles can she drive for $250? (*Hint*: 60¢ = $0.60.)

23. Triangular Field. The second angle of a triangular field is three times as large as the first angle. The third angle is 40° greater than the first angle. How large are the angles?

24. Triangular Parking Lot. The second angle of a triangular parking lot is four times as large as the first angle. The third angle is 45° less than the sum of the other two angles. How large are the angles?

25. *Triangular Backyard.* A home has a triangular backyard. The second angle of the triangle is 5° more than the first angle. The third angle is 10° more than three times the first angle. Find the angles of the triangular yard.

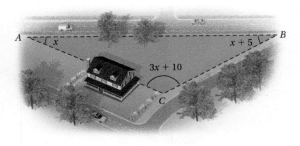

26. *Boarding Stable.* A rancher needs to form a triangular horse pen using ropes next to a stable. The second angle is three times the first angle. The third angle is 15° less than the first angle. Find the angles of the triangular pen.

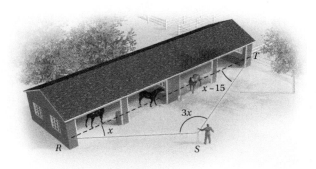

27. *Stock Prices.* Diego's investment in a technology stock grew 28% to $448. How much did he invest?

28. *Savings Interest.* Ella invested money in a savings account at a rate of 6% simple interest. After 1 year, she has $6996 in the account. How much did Ella originally invest?

29. *Credit Cards.* The balance on Will's credit card grew 2%, to $870, in one month. What was his balance at the beginning of the month?

30. *Loan Interest.* Alvin borrowed money from a cousin at a rate of 10% simple interest. After 1 year, $7194 paid off the loan. How much did Alvin borrow?

31. *Taxi Fares.* In New Orleans, Louisiana, taxis charge an initial charge of $3.50 plus $2.00 per mile. How far can one travel for $39.50?

Source: www.taxifarefinders.com

32. *Taxi Fares.* In Baltimore, Maryland, taxis charge an initial charge of $1.80 plus $2.20 per mile. How far can one travel for $26?

Source: www.taxifarefinders.com

33. *Tipping.* Isabella left a 15% tip for a meal. The total cost of the meal, including the tip, was $44.39. What was the cost of the meal before the tip was added?

34. *Tipping.* Nicolas left a 20% tip for a meal. The total cost of the meal, including the tip, was $24.90. What was the cost of the meal before the tip was added?

35. *Average Test Score.* Mariana averaged 84 on her first three history exams. The first score was 67. The second score was 7 less than the third score. What did she score on the second and third exams?

36. *Average Price.* David paid an average of $34 per shirt for a recent purchase of three shirts. The price of one shirt was twice as much as another, and the remaining shirt cost $27. What were the prices of the other two shirts?

37. If you double a number and then add 16, you get $\frac{2}{3}$ of the original number. What is the original number?

38. If you double a number and then add 85, you get $\frac{3}{4}$ of the original number. What is the original number?

Skill Maintenance

Calculate.

39. $-\frac{4}{5} - \frac{3}{8}$ [10.4a]

40. $-\frac{4}{5} + \frac{3}{8}$ [10.3a]

41. $-\frac{4}{5} \cdot \frac{3}{8}$ [10.5a]

42. $-\frac{4}{5} \div \frac{3}{8}$ [10.6c]

43. $\frac{1}{10} \div \left(-\frac{1}{100}\right)$ [10.6c]

44. $-25.6 \div (-16)$ [10.6c]

45. $-25.6\,(-16)$ [10.5a]

46. $-25.6 - (-16)$ [10.4a]

47. $-25.6 + (-16)$ [10.3a]

48. $(-0.02) \div (-0.2)$ [10.6c]

49. Use a commutative law to write an equivalent expression for $12 + yz$. [10.7b]

50. Use an associative law to write an equivalent expression for $(c + 4) + d$. [10.7b]

Synthesis

51. Apples are collected in a basket for six people. One-third, one-fourth, one-eighth, and one-fifth are given to four people, respectively. The fifth person gets ten apples, leaving one apple for the sixth person. Find the original number of apples in the basket.

52. *Test Questions.* A student scored 78 on a test that had 4 seven-point fill-in questions and 24 three-point multiple-choice questions. The student answered one fill-in question incorrectly. How many multiple-choice questions did the student answer correctly?

53. The area of this triangle is 2.9047 in². Find x.

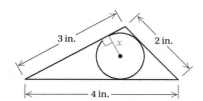

54. Susanne goes to the bank to get $20 in quarters, dimes, and nickels to use to make change at her yard sale. She gets twice as many quarters as dimes and 10 more nickels than dimes. How many of each type of coin does she get?

11.7 Solving Inequalities

OBJECTIVES

a Determine whether a given number is a solution of an inequality.

b Graph an inequality on the number line.

c Solve inequalities using the addition principle.

d Solve inequalities using the multiplication principle.

e Solve inequalities using the addition principle and the multiplication principle together.

SKILL TO REVIEW

Objective 10.2d: Determine whether an inequality like $-3 \leq 5$ is true or false.

Write true or false.

1. $-6 \leq -8$ **2.** $1 \geq 1$

Determine whether each number is a solution of the inequality.

1. $x > 3$
   a) 2   b) 0
   c) -5 d) 15.4
   e) 3   f) $-\dfrac{2}{5}$

2. $x \leq 6$
   a) 6   b) 0
   c) -4.3 d) 25
   e) -6 f) $\dfrac{5}{8}$

We now extend our equation-solving principles to the solving of inequalities.

a SOLUTIONS OF INEQUALITIES

In Section 10.2, we defined the symbols > (is greater than), < (is less than), $\geq$ (is greater than or equal to), and $\leq$ (is less than or equal to).

An **inequality** is a number sentence with >, <, $\geq$, or $\leq$ as its verb—for example,

$$-4 > t, \quad x < 3, \quad 2x + 5 \geq 0, \quad \text{and} \quad -3y + 7 \leq -8.$$

Some replacements for a variable in an inequality make it true and some make it false. (There are some exceptions to this statement, but we will not consider them here.)

> ### SOLUTION OF AN INEQUALITY
>
> A replacement that makes an inequality true is called a **solution**. The set of all solutions is called the **solution set**. When we have found the set of all solutions of an inequality, we say that we have **solved** the inequality.

EXAMPLES   Determine whether each number is a solution of $x < 2$.

1. -2.7 Since $-2.7 < 2$ is true, -2.7 is a solution.
2. 2   Since $2 < 2$ is false, 2 is not a solution.

EXAMPLES   Determine whether each number is a solution of $y \geq 6$.

3. 6   Since $6 \geq 6$ is true, 6 is a solution.
4. $-\dfrac{4}{3}$ Since $-\dfrac{4}{3} \geq 6$ is false, $-\dfrac{4}{3}$ is not a solution.

◀ Do Margin Exercises 1 and 2.

b GRAPHS OF INEQUALITIES

Some solutions of $x < 2$ are $-3, 0, 1, 0.45, -8.9, -\pi, \frac{5}{8}$, and so on. In fact, there are infinitely many real numbers that are solutions. Because we cannot list them all individually, it is helpful to make a drawing that represents all the solutions.

A **graph** of an inequality is a drawing that represents its solutions. An inequality in one variable can be graphed on the number line. An inequality in two variables can be graphed on the coordinate plane. We will study such graphs in Chapter 3.

Answers

Skill to Review:
1. False   **2.** True

Margin Exercises:
1. (a) No; **(b)** no; **(c)** no; **(d)** yes; **(e)** no; **(f)** no
2. (a) Yes; **(b)** yes; **(c)** yes; **(d)** no; **(e)** yes; **(f)** yes

EXAMPLE 5 Graph: $x < 2$.

The solutions of $x < 2$ are all those numbers less than 2. They are shown on the number line by shading all points to the left of 2. The parenthesis at 2 indicates that 2 *is not* part of the graph.

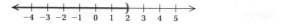

EXAMPLE 6 Graph: $x \geq -3$.

The solutions of $x \geq -3$ are shown on the number line by shading the point for -3 and all points to the right of -3. The bracket at -3 indicates that -3 *is* part of the graph.

EXAMPLE 7 Graph: $-3 \leq x < 2$.

The inequality $-3 \leq x < 2$ is read "-3 is less than or equal to x *and* x is less than 2," or "x is greater than or equal to -3 *and* x is less than 2." In order to be a solution of this inequality, a number must be a solution of both $-3 \leq x$ and $x < 2$. The number 1 is a solution, as are -1.7, 0, 1.5, and $\frac{3}{8}$. We can see from the following graphs that the solution set consists of the numbers that overlap in the two solution sets in Examples 5 and 6.

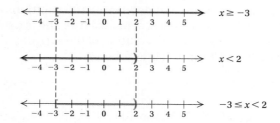

The parenthesis at 2 means that 2 *is not* part of the graph. The bracket at -3 means that -3 *is* part of the graph. The other solutions are shaded.

Do Exercises 3–5. ▶

Graph.

3. $x \leq 4$

4. $x > -2$

5. $-2 < x \leq 4$

C   SOLVING INEQUALITIES USING THE ADDITION PRINCIPLE

Consider the true inequality $3 < 7$. If we add 2 on both sides, we get another true inequality:

$$3 + 2 < 7 + 2, \quad \text{or} \quad 5 < 9.$$

Similarly, if we add -4 on both sides of $x + 4 < 10$, we get an *equivalent* inequality:

$$x + 4 + (-4) < 10 + (-4),$$

or $\qquad\qquad x < 6.$

To say that $x + 4 < 10$ and $x < 6$ are **equivalent** is to say that they have the same solution set. For example, the number 3 is a solution of $x + 4 < 10$. It is also a solution of $x < 6$. The number -2 is a solution of $x < 6$. It is also a solution of $x + 4 < 10$. Any solution of one inequality is a solution of the other—they are equivalent.

Answers

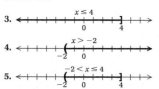

THE ADDITION PRINCIPLE FOR INEQUALITIES

For any real numbers a, b, and c:

$a < b$ is equivalent to $a + c < b + c$;

$a > b$ is equivalent to $a + c > b + c$;

$a \leq b$ is equivalent to $a + c \leq b + c$;

$a \geq b$ is equivalent to $a + c \geq b + c$.

In other words, when we add or subtract the same number on both sides of an inequality, the direction of the inequality symbol is not changed.

As with equation solving, when solving inequalities, our goal is to isolate the variable on one side. Then it is easier to determine the solution set.

EXAMPLE 8 Solve: $x + 2 > 8$. Then graph.

We use the addition principle, subtracting 2 on both sides:

$$x + 2 - 2 > 8 - 2$$
$$x > 6.$$

From the inequality $x > 6$, we can determine the solutions directly. Any number greater than 6 makes the last sentence true and is a solution of that sentence. Any such number is also a solution of the original sentence. Thus the inequality is solved. The graph is as follows:

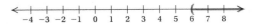

We cannot check all the solutions of an inequality by substitution, as we usually can for an equation, because there are too many of them. A partial check can be done by substituting a number greater than 6—say, 7—into the original inequality:

$$\frac{x + 2 > 8}{7 + 2 \;?\; 8}$$
$$9 \quad\quad \text{TRUE}$$

Since $9 > 8$ is true, 7 is a solution. This is a partial check that any number greater than 6 is a solution.

EXAMPLE 9 Solve: $3x + 1 \leq 2x - 3$. Then graph.

We have

$$3x + 1 \leq 2x - 3$$
$$3x + 1 - 1 \leq 2x - 3 - 1 \qquad \text{Subtracting 1}$$
$$3x \leq 2x - 4 \qquad \text{Simplifying}$$
$$3x - 2x \leq 2x - 4 - 2x \qquad \text{Subtracting } 2x$$
$$x \leq -4. \qquad \text{Simplifying}$$

Any number less than or equal to -4 is a solution. The graph is as follows:

In Example 9, any number less than or equal to −4 is a solution. The following are some solutions:

$$-4, \quad -5, \quad -6, \quad -\frac{13}{3}, \quad -204.5, \quad \text{and} \quad -18\pi.$$

Besides drawing a graph, we can also describe all the solutions of an inequality using **set notation**. We could just begin to list them in a set using roster notation (see p. 9), as follows:

$$\left\{ -4, -5, -6, -\frac{13}{3}, -204.5, -18\pi, \ldots \right\}.$$

We can never list them all this way, however. Seeing this set without knowing the inequality makes it difficult for us to know what real numbers we are considering. There is, however, another kind of notation that we can use. It is

$$\{x | x \leq -4\},$$

which is read

"The set of all x such that x is less than or equal to −4."

This shorter notation for sets is called **set-builder notation**.
From now on, we will use this notation when solving inequalities.

Do Exercises 6–8. ▶

EXAMPLE 10   Solve: $x + \frac{1}{3} > \frac{5}{4}$.

We have

$$x + \frac{1}{3} > \frac{5}{4}$$
$$x + \frac{1}{3} - \frac{1}{3} > \frac{5}{4} - \frac{1}{3} \qquad \text{Subtracting } \frac{1}{3}$$
$$x > \frac{5}{4} \cdot \frac{3}{3} - \frac{1}{3} \cdot \frac{4}{4} \qquad \begin{array}{l}\text{Multiplying by 1 to obtain} \\ \text{a common denominator}\end{array}$$
$$x > \frac{15}{12} - \frac{4}{12}$$
$$x > \frac{11}{12}.$$

Any number greater than $\frac{11}{12}$ is a solution. The solution set is

$$\left\{ x | x > \frac{11}{12} \right\},$$

which is read

"The set of all x such that x is greater than $\frac{11}{12}$."

When solving inequalities, you may obtain an answer like $\frac{11}{12} < x$. Recall from Chapter 1 that this has the same meaning as $x > \frac{11}{12}$. Thus the solution set in Example 10 can be described as $\left\{ x | \frac{11}{12} < x \right\}$ or as $\left\{ x | x > \frac{11}{12} \right\}$. The latter is used most often.

Do Exercises 9 and 10. ▶

d SOLVING INEQUALITIES USING THE MULTIPLICATION PRINCIPLE

There is a multiplication principle for inequalities that is similar to that for equations, but it must be modified. When we are multiplying on both sides by a negative number, the direction of the inequality symbol must be changed.

Solve. Then graph.

6. $x + 3 > 5$

<-----+---+---+---+---+---+---+---+---+---+----->
   −5 −4 −3 −2 −1 0 1 2 3 4 5

7. $x - 1 \leq 2$

<-----+---+---+---+---+---+---+---+---+---+----->
   −5 −4 −3 −2 −1 0 1 2 3 4 5

8. $5x + 1 < 4x - 2$

<-----+---+---+---+---+---+---+---+---+---+----->
   −5 −4 −3 −2 −1 0 1 2 3 4 5

Solve.

9. $x + \dfrac{2}{3} \geq \dfrac{4}{5}$

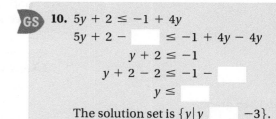

GS **10.** $5y + 2 \leq -1 + 4y$

$5y + 2 - \boxed{} \leq -1 + 4y - 4y$

$y + 2 \leq -1$

$y + 2 - 2 \leq -1 - \boxed{}$

$y \leq \boxed{}$

The solution set is $\{y | y \ \boxed{} \ -3\}$.

Answers

6. $\{x | x > 2\}$;

<-----+---+---+---+---+---+---+----->
          0 2

7. $\{x | x \leq 3\}$;

<-----+---+---+---+---+---+---+----->
          0   3

8. $\{x | x < -3\}$;

<-----+---+---+---+---+---+---+----->
   −3   0

9. $\left\{ x | x \geq \dfrac{2}{15} \right\}$ **10.** $\{y | y \leq -3\}$

Guided Solution:
10. $4y, 2, -3, \leq$

Consider the true inequality $3 < 7$. If we multiply on both sides by a *positive* number, like 2, we get another true inequality:

$$3 \cdot 2 < 7 \cdot 2, \quad \text{or} \quad 6 < 14. \qquad \text{True}$$

If we multiply on both sides by a *negative* number, like -2, and we do not change the direction of the inequality symbol, we get a *false* inequality:

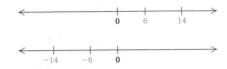

$$3 \cdot (-2) < 7 \cdot (-2), \quad \text{or} \quad -6 < -14. \qquad \text{False}$$

The fact that $6 < 14$ is true but $-6 < -14$ is false stems from the fact that the negative numbers, in a sense, mirror the positive numbers. That is, whereas 14 is to the *right* of 6 on the number line, the number -14 is to the *left* of -6. Thus, if we reverse (change the direction of) the inequality symbol, we get a *true* inequality: $-6 > -14$.

THE MULTIPLICATION PRINCIPLE FOR INEQUALITIES

For any real numbers a and b, and any *positive* number c:

   $a < b$ is equivalent to $ac < bc$;
   $a > b$ is equivalent to $ac > bc$.

For any real numbers a and b, and any *negative* number c:

   $a < b$ is equivalent to $ac > bc$;
   $a > b$ is equivalent to $ac < bc$.

Similar statements hold for $\leq$ and $\geq$.

   In other words, when we multiply or divide by a positive number on both sides of an inequality, the direction of the inequality symbol stays the same. When we multiply or divide by a negative number on both sides of an inequality, the direction of the inequality symbol is reversed.

EXAMPLE 11 Solve: $4x < 28$. Then graph.

   We have

$$4x < 28$$

$$\frac{4x}{4} < \frac{28}{4} \qquad \text{Dividing by 4}$$

$$\text{The symbol stays the same.}$$

$$x < 7. \qquad \text{Simplifying}$$

The solution set is $\{x \mid x < 7\}$. The graph is as follows:

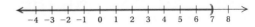

◀ Do Exercises 11 and 12.

Solve. Then graph.

11. $8x < 64$

<-- number line: -12 -8 -4 0 4 8 12 -->

12. $5y \geq 160$

<-- number line: -80 -60 -40 -20 0 20 40 60 80 -->

Answers

11. $\{x \mid x < 8\}$;

<-- number line: 0 8 -->

12. $\{y \mid y \geq 32\}$;

<-- number line: 0 30 32 -->

EXAMPLE 12 Solve: $-2y < 18$. Then graph.

$$-2y < 18$$

$$\frac{-2y}{-2} > \frac{18}{-2} \qquad \text{Dividing by } -2$$

The symbol must be reversed!

$$y > -9. \qquad \text{Simplifying}$$

The solution set is $\{y\,|\,y > -9\}$. The graph is as follows:

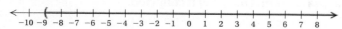

Do Exercises 13 and 14.

Solve.

13. $-4x \le 24$

14. $-5y > 13$

e USING THE PRINCIPLES TOGETHER

All of the equation-solving techniques used in Sections 2.1–2.3 can be used with inequalities, provided we remember to reverse the inequality symbol when multiplying or dividing on both sides by a negative number.

EXAMPLE 13 Solve: $6 - 5x > 7$.

$$6 - 5x > 7$$

$$-6 + 6 - 5x > -6 + 7 \qquad \text{Adding } -6. \text{ The symbol stays the same.}$$

$$-5x > 1 \qquad \text{Simplifying}$$

$$\frac{-5x}{-5} < \frac{1}{-5} \qquad \text{Dividing by } -5$$

The symbol must be reversed because we are dividing by a *negative* number, -5.

$$x < -\frac{1}{5}. \qquad \text{Simplifying}$$

The solution set is $\left\{x\,\middle|\,x < -\frac{1}{5}\right\}$.

Do Exercise 15.

15. Solve: $7 - 4x < 8$.

EXAMPLE 14 Solve: $17 - 5y > 8y - 9$.

$$-17 + 17 - 5y > -17 + 8y - 9 \qquad \text{Adding } -17. \text{ The symbol stays the same.}$$

$$-5y > 8y - 26 \qquad \text{Simplifying}$$

$$-8y - 5y > -8y + 8y - 26 \qquad \text{Adding } -8y$$

$$-13y > -26 \qquad \text{Simplifying}$$

$$\frac{-13y}{-13} < \frac{-26}{-13} \qquad \text{Dividing by } -13$$

The symbol must be reversed because we are dividing by a *negative* number, -13.

$$y < 2$$

The solution set is $\{y\,|\,y < 2\}$.

16. Solve. Begin by subtracting 24 on both sides.

$$24 - 7y \le 11y - 14$$

Do Exercise 16.

Answers

13. $\{x\,|\,x \ge -6\}$ 14. $\left\{y\,\middle|\,y < -\frac{13}{5}\right\}$

15. $\left\{x\,\middle|\,x > -\frac{1}{4}\right\}$ 16. $\left\{y\,\middle|\,y \ge \frac{19}{9}\right\}$

Typically, we solve an equation or an inequality by isolating the variable on the left side. When we are solving an inequality, however, there are situations in which isolating the variable on the right side will eliminate the need to reverse the inequality symbol. Let's solve the inequality in Example 14 again, but this time we will isolate the variable on the right side.

EXAMPLE 15 Solve: $17 - 5y > 8y - 9$.

Note that if we add $5y$ on both sides, the coefficient of the y-term will be positive after like terms have been collected.

$$17 - 5y + 5y > 8y - 9 + 5y \qquad \text{Adding } 5y$$
$$17 > 13y - 9 \qquad \text{Simplifying}$$
$$17 + 9 > 13y - 9 + 9 \qquad \text{Adding } 9$$
$$26 > 13y \qquad \text{Simplifying}$$
$$\frac{26}{13} > \frac{13y}{13} \qquad \begin{array}{l}\text{Dividing by 13. We leave the} \\ \text{inequality symbol the same} \\ \text{because we are dividing by a} \\ \text{positive number.}\end{array}$$
$$2 > y$$

The solution set is $\{y \mid 2 > y\}$, or $\{y \mid y < 2\}$.

◀ **Do Exercise 17.**

EXAMPLE 16 Solve: $3(x - 2) - 1 < 2 - 5(x + 6)$.

First, we use the distributive law to remove parentheses. Next, we collect like terms and then use the addition and multiplication principles for inequalities to get an equivalent inequality with x alone on one side.

$$3(x - 2) - 1 < 2 - 5(x + 6)$$
$$3x - 6 - 1 < 2 - 5x - 30 \qquad \begin{array}{l}\text{Using the distributive law to} \\ \text{multiply and remove parentheses}\end{array}$$
$$3x - 7 < -5x - 28 \qquad \text{Collecting like terms}$$
$$3x + 5x < -28 + 7 \qquad \begin{array}{l}\text{Adding } 5x \text{ and } 7 \text{ to get all } x\text{-terms} \\ \text{on one side and all other terms} \\ \text{on the other side}\end{array}$$
$$8x < -21 \qquad \text{Simplifying}$$
$$x < \frac{-21}{8}, \text{ or } -\frac{21}{8}. \qquad \text{Dividing by 8}$$

The solution set is $\left\{ x \mid x < -\frac{21}{8} \right\}$.

◀ **Do Exercise 18.**

17. Solve. Begin by adding $7y$ on both sides.
$$24 - 7y \le 11y - 14$$

18. Solve:
$$3(7 + 2x) \le 30 + 7(x - 1).$$
$$\boxed{} + 6x \le 30 + 7x - \boxed{}$$
$$21 + 6x \le \boxed{} + 7x$$
$$21 + 6x - 6x \le 23 + 7x - \boxed{}$$
$$21 \le 23 + \boxed{}$$
$$21 - \boxed{} \le 23 + x - 23$$
$$-2 \le x, \text{ or}$$
$$x \boxed{} -2$$
The solution set is $\{x \mid x \ge \boxed{}\}$.

EXAMPLE 17 Solve: $16.3 - 7.2p \le -8.18$.

The greatest number of decimal places in any one number is *two*. Multiplying by 100, which has two 0's, will clear decimals. Then we proceed as before.

$$16.3 - 7.2p \le -8.18$$

$$100(16.3 - 7.2p) \le 100(-8.18) \qquad \text{Multiplying by 100}$$

$$100(16.3) - 100(7.2p) \le 100(-8.18) \qquad \text{Using the distributive law}$$

$$1630 - 720p \le -818 \qquad \text{Simplifying}$$

$$1630 - 720p - 1630 \le -818 - 1630 \qquad \text{Subtracting 1630}$$

$$-720p \le -2448 \qquad \text{Simplifying}$$

$$\frac{-720p}{-720} \ge \frac{-2448}{-720} \qquad \text{Dividing by } -720$$

The symbol must be reversed.

$$p \ge 3.4$$

The solution set is $\{p \mid p \ge 3.4\}$.

Do Exercise 19. ▷

19. Solve:
$$2.1x + 43.2 \ge 1.2 - 8.4x.$$

EXAMPLE 18 Solve: $\dfrac{2}{3}x - \dfrac{1}{6} + \dfrac{1}{2}x > \dfrac{7}{6} + 2x$.

The number 6 is the least common multiple of all the denominators. Thus we first multiply by 6 on both sides to clear fractions.

$$\frac{2}{3}x - \frac{1}{6} + \frac{1}{2}x > \frac{7}{6} + 2x$$

$$6\left(\frac{2}{3}x - \frac{1}{6} + \frac{1}{2}x\right) > 6\left(\frac{7}{6} + 2x\right) \qquad \text{Multiplying by 6 on both sides}$$

$$6 \cdot \frac{2}{3}x - 6 \cdot \frac{1}{6} + 6 \cdot \frac{1}{2}x > 6 \cdot \frac{7}{6} + 6 \cdot 2x \qquad \text{Using the distributive law}$$

$$4x - 1 + 3x > 7 + 12x \qquad \text{Simplifying}$$

$$7x - 1 > 7 + 12x \qquad \text{Collecting like terms}$$

$$7x - 1 - 7x > 7 + 12x - 7x \qquad \text{Subtracting } 7x. \text{ The coefficient of the } x\text{-term will be positive.}$$

$$-1 > 7 + 5x \qquad \text{Simplifying}$$

$$-1 - 7 > 7 + 5x - 7 \qquad \text{Subtracting 7}$$

$$-8 > 5x \qquad \text{Simplifying}$$

$$\frac{-8}{5} > \frac{5x}{5} \qquad \text{Dividing by 5}$$

$$-\frac{8}{5} > x$$

The solution set is $\left\{x \mid -\frac{8}{5} > x\right\}$, or $\left\{x \mid x < -\frac{8}{5}\right\}$.

Do Exercise 20. ▷

20. Solve:
$$\frac{3}{4} + x < \frac{7}{8}x - \frac{1}{4} + \frac{1}{2}x.$$

Answers

19. $\{x \mid x \ge -4\}$ **20.** $\left\{x \mid x > \frac{8}{3}\right\}$

For Extra Help

MyMathLab®

MathXL®   PRACTICE   WATCH   READ   REVIEW

✓ Reading Check

Classify each pair of inequalities as "equivalent" or "not equivalent."

RC1. $x + 10 \geq 12$; $x \leq 2$

RC2. $3x - 5 \leq -x + 1$; $2x \leq 6$

RC3. $-\dfrac{3}{4}y < 6$; $y > -8$

RC4. $2 - t > -3t + 4$; $2t > 2$

---

a Determine whether each number is a solution of the given inequality.

1. $x > -4$
   a) 4
   b) 0
   c) -4
   d) 6
   e) 5.6

2. $x \leq 5$
   a) 0
   b) 5
   c) -1
   d) -5
   e) $7\dfrac{1}{4}$

3. $x \geq 6.8$
   a) -6
   b) 0
   c) 6
   d) 8
   e) $-3\dfrac{1}{2}$

4. $x < 8$
   a) 8
   b) -10
   c) 0
   d) 11
   e) -4.7

---

b Graph on the number line.

5. $x > 4$

6. $x < 0$

7. $t < -3$

8. $y > 5$

9. $m \geq -1$

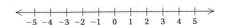

10. $x \leq -2$

11. $-3 < x \leq 4$

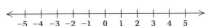

12. $-5 \leq x < 2$

13. $0 < x < 3$

14. $-5 \leq x \leq 0$

c Solve using the addition principle. Then graph.

15. $x + 7 > 2$

16. $x + 5 > 2$

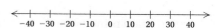

17. $x + 8 \leq -10$

18. $x + 8 \leq -11$

Solve using the addition principle.

19. $y - 7 > -12$

20. $y - 9 > -15$

21. $2x + 3 > x + 5$

22. $2x + 4 > x + 7$

23. $3x + 9 \leq 2x + 6$

24. $3x + 18 \leq 2x + 16$

25. $5x - 6 < 4x - 2$

26. $9x - 8 < 8x - 9$

27. $-9 + t > 5$

28. $-8 + p > 10$

29. $y + \dfrac{1}{4} \leq \dfrac{1}{2}$

30. $x - \dfrac{1}{3} \leq \dfrac{5}{6}$

31. $x - \dfrac{1}{3} > \dfrac{1}{4}$

32. $x + \dfrac{1}{8} > \dfrac{1}{2}$

d Solve using the multiplication principle. Then graph.

33. $5x < 35$

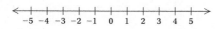

34. $8x \geq 32$

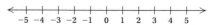

35. $-12x > -36$

36. $-16x > -64$

Solve using the multiplication principle.

37. $5y \geq -2$

38. $3x < -4$

39. $-2x \leq 12$

40. $-3x \leq 15$

41. $-4y \geq -16$ **42.** $-7x < -21$ **43.** $-3x < -17$ **44.** $-5y > -23$

45. $-2y > \dfrac{1}{7}$ **46.** $-4x \leq \dfrac{1}{9}$ **47.** $-\dfrac{6}{5} \leq -4x$ **48.** $-\dfrac{7}{9} > 63x$

e   Solve using the addition principle and the multiplication principle.

49. $4 + 3x < 28$ **50.** $3 + 4y < 35$ **51.** $3x - 5 \leq 13$

52. $5y - 9 \leq 21$ **53.** $13x - 7 < -46$ **54.** $8y - 6 < -54$

55. $30 > 3 - 9x$ **56.** $48 > 13 - 7y$ **57.** $4x + 2 - 3x \leq 9$

58. $15x + 5 - 14x \leq 9$ **59.** $-3 < 8x + 7 - 7x$ **60.** $-8 < 9x + 8 - 8x - 3$

61. $6 - 4y > 4 - 3y$ **62.** $9 - 8y > 5 - 7y + 2$ **63.** $5 - 9y \leq 2 - 8y$

64. $6 - 18x \leq 4 - 12x - 5x$ **65.** $19 - 7y - 3y < 39$ **66.** $18 - 6y - 4y < 63 + 5y$

67. $0.9x + 19.3 > 5.3 - 2.6x$ **68.** $0.96y - 0.79 \leq 0.21y + 0.46$ **69.** $\dfrac{x}{3} - 2 \leq 1$

70. $\dfrac{2}{3} + \dfrac{x}{5} < \dfrac{4}{15}$ **71.** $\dfrac{y}{5} + 1 \leq \dfrac{2}{5}$ **72.** $\dfrac{3x}{4} - \dfrac{7}{8} \geq -15$

73. $3(2y - 3) < 27$ **74.** $4(2y - 3) > 28$ **75.** $2(3 + 4m) - 9 \geq 45$

76. $3(5 + 3m) - 8 \leq 88$ **77.** $8(2t + 1) > 4(7t + 7)$ **78.** $7(5y - 2) > 6(6y - 1)$

79. $3(r - 6) + 2 < 4(r + 2) - 21$ **80.** $5(x + 3) + 9 \leq 3(x - 2) + 6$

81. $0.8(3x + 6) \geq 1.1 - (x + 2)$ **82.** $0.4(2x + 8) \geq 20 - (x + 5)$

83. $\dfrac{5}{3} + \dfrac{2}{3}x < \dfrac{25}{12} + \dfrac{5}{4}x + \dfrac{3}{4}$ **84.** $1 - \dfrac{2}{3}y \geq \dfrac{9}{5} - \dfrac{y}{5} + \dfrac{3}{5}$

Skill Maintenance

Add or subtract.   [10.3a], [10.4a]

85. $-\dfrac{3}{4} + \dfrac{1}{8}$ **86.** $8.12 - 9.23$

87. $-2.3 - 7.1$ **88.** $-\dfrac{3}{4} - \dfrac{1}{8}$

Simplify.

89. $5 - 3^2 + (8 - 2)^2 \cdot 4$ [10.8d]   **90.** $10 \div 2 \cdot 5 - 3^2 + (-5)^2$ [10.8d]

91. $5(2x - 4) - 3(4x + 1)$ [10.8b]   **92.** $9(3 + 5x) - 4(7 + 2x)$ [10.8b]

Synthesis

93. Determine whether each number is a solution of the inequality $|x| < 3$.

   a) 0   **b)** -2
   c) -3 **d)** 4
   e) 3   **f)** 1.7
   g) -2.8

94. Graph $|x| < 3$ on the number line.

Solve.

95. $x + 3 < 3 + x$ **96.** $x + 4 > 3 + x$

OBJECTIVES

a Translate number sentences to inequalities.

b Solve applied problems using inequalities.

SKILL TO REVIEW

Objective 11.7d: Solve inequalities using the multiplication principle.

Solve.

1. $-8x \leq -512$
2. $-300 > 15y$

Translate.

1. Sara worked no fewer than 15 hr last week.

2. The price of that Volkswagen Beetle convertible is at most $31,210.

3. The time of the test was between 45 min and 55 min.

4. Camila's weight is less than 110 lb.

5. That number is more than -2.

6. The costs of production of that marketing video cannot exceed $12,500.

7. At most 1250 people attended the concert.

8. Yesterday, at least 23 people got tickets for speeding.

Answers

Skill to Review:
1. $\{x \mid x \geq 64\}$ 2. $\{y \mid y < -20\}$

Margin Exercises:
1. $h \geq 15$ 2. $p \leq 31,210$ 3. $45 < t < 55$
4. $w < 110$ 5. $n > -2$ 6. $c \leq 12,500$
7. $p \leq 1250$ 8. $s \geq 23$

The five steps for problem solving can be used for problems involving inequalities.

a TRANSLATING TO INEQUALITIES

Before solving problems that involve inequalities, we list some important phrases to look for. Sample translations are listed as well.

IMPORTANT WORDS	SAMPLE SENTENCE	TRANSLATION
is at least	Bill is at least 21 years old.	$b \geq 21$
is at most	At most 5 students dropped the course.	$n \leq 5$
cannot exceed	To qualify, earnings cannot exceed $12,000.	$r \leq 12,000$
must exceed	The speed must exceed 15 mph.	$s > 15$
is less than	Tucker's weight is less than 50 lb.	$w < 50$
is more than	Nashville is more than 200 mi away.	$d > 200$
is between	The film is between 90 min and 100 min long.	$90 < t < 100$
no more than	Cooper weighs no more than 90 lb.	$w \leq 90$
no less than	Sofia scored no less than 8.3.	$s \geq 8.3$

The following phrases deserve special attention.

TRANSLATING "AT LEAST" AND "AT MOST"

A quantity x is at least some amount q: $x \geq q$.
  (If x is at least q, it cannot be less than q.)

A quantity x is at most some amount q: $x \leq q$.
  (If x is at most q, it cannot be more than q.)

◀ Do Margin Exercises 1–8.

b SOLVING PROBLEMS

EXAMPLE 1 *Catering Costs.* To cater a company's annual lobster-bake cookout, Jayla's Catering charges a $325 setup fee plus $18.50 per person. The cost cannot exceed $3200. How many people can attend the cookout?

1. **Familiarize.** Suppose that 130 people were to attend the cookout. The cost would then be $325 + $18.50(130), or $2730. This shows that more than 130 people could attend the picnic without exceeding $3200. Instead of making another guess, we let $n =$ the number of people in attendance.

2. Translate. Our guess shows us how to translate. The cost of the cookout will be the $325 setup fee plus $18.50 times the number of people attending. We translate to an inequality:

Rewording:	The setup fee	plus	the cost of the meals	cannot exceed	$3200.
Translating:	325	+	18.50n	≤	3200.

3. Solve. We solve the inequality for n:

$$325 + 18.50n \le 3200$$

$$325 + 18.50n - 325 \le 3200 - 325 \qquad \text{Subtracting 325}$$

$$18.50n \le 2875 \qquad \text{Simplifying}$$

$$\frac{18.50n}{18.50} \le \frac{2875}{18.50} \qquad \text{Dividing by 18.50}$$

$$n \le 155.4. \qquad \text{Rounding to the nearest tenth}$$

4. Check. Although the solution set of the inequality is all numbers less than or equal to about 155.4, since n = the number of people in attendance, we round *down* to 155 people. If 155 people attend, the cost will be $325 + $18.50(155), or $3192.50. If 156 attend, the cost will exceed $3200.

5. State. At most, 155 people can attend the lobster-bake cookout.

Do Exercise 9. ▶

.... **Caution!** ....

Solutions of problems should always be checked using the original wording of the problem. In some cases, answers might need to be whole numbers or integers or rounded off in a particular direction.

....

EXAMPLE 2 *Nutrition.* The U.S. Department of Agriculture recommends that for a typical 2000-calorie daily diet, no more than 20 g of saturated fat be consumed. In the first three days of a four-day vacation, Ethan consumed 26 g, 17 g, and 22 g of saturated fat. Determine (in terms of an inequality) how many grams of saturated fat Ethan can consume on the fourth day if he is to average no more than 20 g of saturated fat per day.

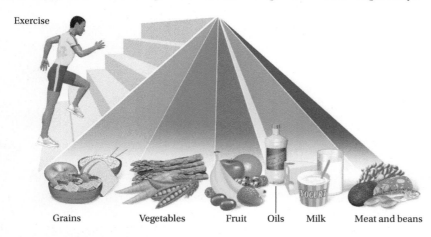

Exercise

Grains   Vegetables   Fruit   Oils   Milk   Meat and beans

SOURCES: U.S. Department of Health and Human Services; U.S. Department of Agriculture

GS Translate to an inequality and solve.

9. *Butter Temperatures.* Butter stays solid at Fahrenheit temperatures below 88°. The formula

$$F = \tfrac{9}{5}C + 32$$

can be used to convert Celsius temperatures C to Fahrenheit temperatures F. Determine (in terms of an inequality) those Celsius temperatures for which butter stays solid.

Translate and *Solve*:

$$F < 88$$

$$\frac{9}{5}C + 32 < 88$$

$$\frac{9}{5}C + 32 - 32 < 88 - \boxed{}$$

$$\frac{9}{5}C < 56$$

$$\boxed{} \cdot \frac{9}{5}C < \frac{5}{9} \cdot 56$$

$$C < \frac{\boxed{}}{9}$$

$$C < 31\tfrac{1}{9}.$$

Butter stays solid at Celsius temperatures less than $31\tfrac{1}{9}°$—that is, $\{C \,|\, C < 31\tfrac{1}{9}°\}$.

Answer

9. $\frac{9}{5}C + 32 < 88$; $\{C \,|\, C < 31\tfrac{1}{9}°\}$

Guided Solution:

9. $32, \frac{5}{9}, 280$

1. **Familiarize.** Suppose Ethan consumed 19 g of saturated fat on the fourth day. His daily average for the vacation would then be

$$\frac{26\text{ g} + 17\text{ g} + 22\text{ g} + 19\text{ g}}{4} = \frac{84\text{ g}}{4} = 21\text{ g}.$$

This shows that Ethan cannot consume 19 g of saturated fat on the fourth day, if he is to average no more than 20 g of fat per day. We let $x =$ the number of grams of fat that Ethan can consume on the fourth day.

2. **Translate.** We reword the problem and translate to an inequality as follows:

Rewording: $\underbrace{\text{The average consumption of saturated fat}}$ $\underbrace{\text{should be no more than}}$ $\underbrace{\text{20 g.}}$

Translating: $\dfrac{26 + 17 + 22 + x}{4}$ $\quad \leq \quad$ $20.$

3. **Solve.** Because of the fraction expression, it is convenient to use the multiplication principle first to clear the fraction:

$$\frac{26 + 17 + 22 + x}{4} \leq 20$$

$$4\left(\frac{26 + 17 + 22 + x}{4}\right) \leq 4 \cdot 20 \qquad \text{Multiplying by 4}$$

$$26 + 17 + 22 + x \leq 80$$

$$65 + x \leq 80 \qquad \text{Simplifying}$$

$$x \leq 15. \qquad \text{Subtracting 65}$$

4. **Check.** As a partial check, we show that Ethan can consume 15 g of saturated fat on the fourth day and not exceed a 20-g average for the four days:

$$\frac{26 + 17 + 22 + 15}{4} = \frac{80}{4} = 20.$$

5. **State.** Ethan's average intake of saturated fat for the vacation will not exceed 20 g per day if he consumes no more than 15 g of saturated fat on the fourth day.

◀ **Do Exercise 10.**

Translate to an inequality and solve.

10. *Test Scores.* A pre-med student is taking a chemistry course in which four tests are given. To get an A, she must average at least 90 on the four tests. The student got scores of 91, 86, and 89 on the first three tests. Determine (in terms of an inequality) what scores on the last test will allow her to get an A.

Answer

10. $\dfrac{91 + 86 + 89 + s}{4} \geq 90; \{s \mid s \geq 94\}$

✓ Reading Check

Match each sentence with one of the following.

$$q < r \qquad q \le r \qquad r < q \qquad r \le q$$

RC1. r is at most q.

RC2. q is no more than r.

RC3. r is less than q.

RC4. r is at least q.

RC5. q exceeds r.

RC6. q is no less than r.

a   Translate to an inequality.

1. A number is at least 7.

2. A number is greater than or equal to 5.

3. The baby weighs more than 2 kilograms (kg).

4. Between 75 and 100 people attended the concert.

5. The speed of the train was between 90 mph and 110 mph.

6. The attendance was no more than 180.

7. Brianna works no more than 20 hr per week.

8. The amount of acid must exceed 40 liters (L).

9. The cost of gasoline is no less than $3.20 per gallon.

10. The temperature is at most $-2°$.

11. A number is greater than 8.

12. A number is less than 5.

13. A number is less than or equal to -4.

14. A number is greater than or equal to 18.

15. The number of people is at least 1300.

16. The cost is at most $4857.95.

17. The amount of water is not to exceed 500 liters.

18. The cost of ground beef is no less than $3.19 per pound.

19. Two more than three times a number is less than 13.

20. Five less than one-half a number is greater than 17.

b Solve.

21. *Test Scores.* Xavier is taking a geology course in which four tests are given. To get a B, he must average at least 80 on the four tests. He got scores of 82, 76, and 78 on the first three tests. Determine (in terms of an inequality) what scores on the last test will allow him to get at least a B.

22. *Test Scores.* Chloe is taking a French class in which five quizzes are given. Her first four quiz grades are 73, 75, 89, and 91. Determine (in terms of an inequality) what scores on the last quiz will allow her to get an average quiz grade of at least 85.

23. *Gold Temperatures.* Gold stays solid at Fahrenheit temperatures below 1945.4°. Determine (in terms of an inequality) those Celsius temperatures for which gold stays solid. Use the formula given in Margin Exercise 9.

24. *Body Temperatures.* The human body is considered to be fevered when its temperature is higher than 98.6°F. Using the formula given in Margin Exercise 9, determine (in terms of an inequality) those Celsius temperatures for which the body is fevered.

25. *World Records in the 1500-m Run.* The formula

$$R = -0.075t + 3.85$$

can be used to predict the world record in the 1500-m run t years after 1930. Determine (in terms of an inequality) those years for which the world record will be less than 3.5 min.

26. *World Records in the 200-m Dash.* The formula

$$R = -0.028t + 20.8$$

can be used to predict the world record in the 200-m dash t years after 1920. Determine (in terms of an inequality) those years for which the world record will be less than 19.0 sec.

27. *Blueprints.* To make copies of blueprints, Vantage Reprographics charges a $5 setup fee plus $4 per copy. Myra can spend no more than $65 for copying her blueprints. What numbers of copies will allow her to stay within budget?

28. *Banquet Costs.* The Shepard College women's volleyball team can spend at most $750 for its awards banquet at a local restaurant. If the restaurant charges an $80 setup fee plus $16 per person, at most how many can attend?

29. *Envelope Size.* For a direct-mail campaign, Hollcraft Advertising determines that any envelope with a fixed width of $3\frac{1}{2}$ in. and an area of at least $17\frac{1}{2}$ in^2 can be used. Determine (in terms of an inequality) those lengths that will satisfy the company constraints.

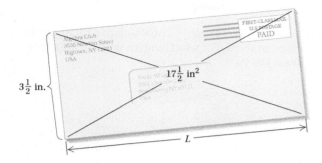

$3\frac{1}{2}$ in.

$17\frac{1}{2}$ in^2

L

30. *Package Sizes.* Logan Delivery Service accepts packages of up to 165 in. in length and girth combined. (Girth is the distance around the package.) A package has a fixed girth of 53 in. Determine (in terms of an inequality) those lengths for which a package is acceptable.

L   Girth = 53 in.

31. *Phone Costs.* Simon claims that it costs him at least $3.00 every time he calls an overseas customer. If his typical call costs 75¢ plus 45¢ for each minute, how long do his calls typically last? (*Hint*: 75¢ = $0.75.)

32. *Parking Costs.* Laura is certain that every time she parks in the municipal garage it costs her at least $6.75. If the garage charges $1.50 plus 75¢ for each half hour, for how long is Laura's car generally parked?

33. *College Tuition.* Angelica's financial aid stipulates that her tuition cannot exceed $1000. If her local community college charges a $35 registration fee plus $375 per course, what is the greatest number of courses for which Angelica can register?

34. *Furnace Repairs.* RJ's Plumbing and Heating charges $45 for a service call plus $30 per hour for emergency service. Gary remembers being billed over $150 for an emergency call. How long was RJ's there?

35. *Nutrition.* Following the guidelines of the Food and Drug Administration, Dale tries to eat at least 5 servings of fruits or vegetables each day. For the first six days of one week, he had 4, 6, 7, 4, 6, and 4 servings. How many servings of fruits or vegetables should Dale eat on Saturday in order to average at least 5 servings per day for the week?

36. *College Course Load.* To remain on financial aid, Millie needs to complete an average of at least 7 credits per quarter each year. In the first three quarters of 2013, Millie completed 5, 7, and 8 credits. How many credits of course work must Millie complete in the fourth quarter if she is to remain on financial aid?

37. *Perimeter of a Rectangle.* The width of a rectangle is fixed at 8 ft. What lengths will make the perimeter at least 200 ft? at most 200 ft?

38. *Perimeter of a Triangle.* One side of a triangle is 2 cm shorter than the base. The other side is 3 cm longer than the base. What lengths of the base will allow the perimeter to be greater than 19 cm?

39. *Area of a Rectangle.* The width of a rectangle is fixed at 4 cm. For what lengths will the area be less than 86 cm^2?

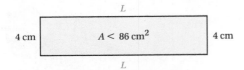

40. *Area of a Rectangle.* The width of a rectangle is fixed at 16 yd. For what lengths will the area be at least 264 yd^2?

41. *Insurance-Covered Repairs.* Most insurance companies will replace a vehicle if an estimated repair exceeds 80% of the "blue-book" value of the vehicle. Rachel's insurance company paid $8500 for repairs to her Toyota after an accident. What can be concluded about the blue-book value of the car?

42. *Insurance-Covered Repairs.* Following an accident, Jeff's Ford pickup was replaced by his insurance company because the damage was so extensive. Before the damage, the blue-book value of the truck was $21,000. How much would it have cost to repair the truck? (See Exercise 41.)

43. *Reduced-Fat Foods.* In order for a food to be labeled "reduced fat," it must have at least 25% less fat than the regular item. One brand of reduced-fat peanut butter contains 12 g of fat per serving. What can you conclude about the fat content in a serving of the brand's regular peanut butter?

44. *Reduced-Fat Foods.* One brand of reduced-fat chocolate chip cookies contains 5 g of fat per serving. What can you conclude about the fat content of the brand's regular chocolate chip cookies? (See Exercise 43.)

45. *Area of a Triangular Flag.* As part of an outdoor education course at Baxter YMCA, Wendy needs to make a bright-colored triangular flag with an area of at least 3 ft^2. What heights can the triangle be if the base is $1\frac{1}{2}$ ft?

46. *Area of a Triangular Sign.* Zoning laws in Harrington prohibit displaying signs with areas exceeding 12 ft^2. If Flo's Marina is ordering a triangular sign with an 8-ft base, how tall can the sign be?

47. *Pond Depth.* On July 1, Garrett's Pond was 25 ft deep. Since that date, the water level has dropped $\frac{2}{3}$ ft per week. For what dates will the water level not exceed 21 ft?

48. *Weight Gain.* A 3-lb puppy is gaining weight at a rate of $\frac{3}{4}$ lb per week. When will the puppy's weight exceed $22\frac{1}{2}$ lb?

49. *Electrician Visits.* Dot's Electric made 17 customer calls last week and 22 calls this week. How many calls must be made next week in order to maintain a weekly average of at least 20 calls for the three-week period?

50. *Volunteer Work.* George and Joan do volunteer work at a hospital. Joan worked 3 more hr than George, and together they worked more than 27 hr. What possible numbers of hours did each work?

Skill Maintenance

Solve.

51. $-13 + x = 27$ [11.1b]

52. $-6y = 132$ [11.2a]

53. $4a - 3 = 45$ [11.3a]

54. $8x + 3x = 66$ [11.3b]

55. $-\frac{1}{2} + x = x - \frac{1}{2}$ [11.3c]

56. $9x - 1 + 11x - 18 = 3x - 15 + 4 + 17x$ [11.3c]

Solve. [11.5a]

57. What percent of 200 is 15?

58. What is 10% of 310?

59. 25 is 2% of what number?

60. 80 is what percent of 96?

Synthesis

Solve.

61. *Ski Wax.* Green ski wax works best between 5° and 15° Fahrenheit. Determine those Celsius temperatures for which green ski wax works best. Use the formula given in Margin Exercise 9.

62. *Parking Fees.* Mack's Parking Garage charges $4.00 for the first hour and $2.50 for each additional hour. For how long has a car been parked when the charge exceeds $16.50?

63. *Low-Fat Foods.* In order for a food to be labeled "low fat," it must have fewer than 3 g of fat per serving. One brand of reduced-fat tortilla chips contains 60% less fat than regular nacho cheese tortilla chips, but still cannot be labeled low fat. What can you conclude about the fat content of a serving of nacho cheese tortilla chips?

64. *Parking Fees.* When asked how much the parking charge is for a certain car, Mack replies "between 14 and 24 dollars." For how long has the car been parked? (See Exercise 62.)

Vocabulary Reinforcement

Complete each statement with the correct word or words from the column on the right. Some of the choices may not be used.

addition principle
multiplication principle
solution
equivalent
equation
inequality

1. Any replacement for the variable that makes an equation true is called a(n) _____ of the equation.   [11.1a]

2. The _____ for equations states that for any real numbers a, b, and c, $a = b$ is equivalent to $a + c = b + c$.   [11.1b]

3. The _____ for equations states that for any real numbers a, b, and c, $a = b$ is equivalent to $a \cdot c = b \cdot c$.   [11.2a]

4. An _____ is a number sentence with $<$, $\leq$, $>$, or $\geq$ as its verb.   [11.7a]

5. Equations with the same solutions are called _____ equations.   [11.1b]

Concept Reinforcement

Determine whether each statement is true or false.

_____ **1.** Some equations have no solution.   [11.3c]

_____ **2.** For any number n, $n \geq n$.   [11.7a]

_____ **3.** $2x - 7 < 11$ and $x < 2$ are equivalent inequalities.   [11.7c]

_____ **4.** If $x > y$, then $-x < -y$.   [11.7d]

Study Guide

Objective 11.3a   Solve equations using both the addition principle and the multiplication principle.

Objective 11.3b   Solve equations in which like terms may need to be collected.

Objective 11.3c   Solve equations by first removing parentheses and collecting like terms.

Example   Solve: $6y - 2(2y - 3) = 12$.

$6y - 2(2y - 3) = 12$

$6y - 4y + 6 = 12$ Removing parentheses

$2y + 6 = 12$ Collecting like terms

$2y + 6 - 6 = 12 - 6$ Subtracting 6

$2y = 6$

$\dfrac{2y}{2} = \dfrac{6}{2}$ Dividing by 2

$y = 3$

The solution is 3.

Practice Exercise

1. Solve: $4(x - 3) = 6(x + 2)$.

Objective 11.3c Solve equations with an infinite number of solutions and equations with no solutions.

Example Solve: $8 + 2x - 4 = 6 + 2(x - 1)$.

$$8 + 2x - 4 = 6 + 2(x - 1)$$
$$8 + 2x - 4 = 6 + 2x - 2$$
$$2x + 4 = 2x + 4$$
$$2x + 4 - 2x = 2x + 4 - 2x$$
$$4 = 4$$

Every real number is a solution of the equation $4 = 4$, so all real numbers are solutions of the original equation. The equation has infinitely many solutions.

Example Solve: $2 + 5(x - 1) = -6 + 5x + 7$.

$$2 + 5(x - 1) = -6 + 5x + 7$$
$$2 + 5x - 5 = -6 + 5x + 7$$
$$5x - 3 = 5x + 1$$
$$5x - 3 - 5x = 5x + 1 - 5x$$
$$-3 = 1$$

This is a false equation, so the original equation has no solution.

Practice Exercises

2. Solve: $4 + 3y - 7 = 3 + 3(y - 2)$.

3. Solve: $4(x - 3) + 7 = -5 + 4x + 10$.

Objective 11.4b Solve a formula for a specified letter.

Example Solve for n: $M = \dfrac{m + n}{5}$.

$$M = \frac{m + n}{5}$$
$$5 \cdot M = 5\left(\frac{m + n}{5}\right)$$
$$5M = m + n$$
$$5M - m = m + n - m$$
$$5M - m = n$$

Practice Exercise

4. Solve for b: $A = \dfrac{1}{2}bh$.

Objective 11.7b Graph an inequality on the number line.

Example Graph each inequality: **(a)** $x < 2$; **(b)** $x \geq -3$.

a) The solutions of $x < 2$ are all numbers less than 2. We shade all points to the left of 2, and we use a parenthesis at 2 to indicate that 2 *is not* part of the graph.

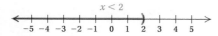

b) The solutions of $x \geq -3$ are all numbers greater than -3 and the number -3 as well. We shade all points to the right of -3, and we use a bracket at -3 to indicate that -3 *is* part of the graph.

Practice Exercises

5. Graph: $x > 1$.

6. Graph: $x \leq -1$.

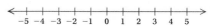

Objective 11.7e Solve inequalities using the addition principle and the multiplication principle together.

Example Solve: $8y - 7 \leq 5y + 2$.

$$8y - 7 \leq 5y + 2$$
$$8y - 7 - 8y \leq 5y + 2 - 8y$$
$$-7 \leq -3y + 2$$
$$-7 - 2 \leq -3y + 2 - 2$$
$$-9 \leq -3y$$
$$\frac{-9}{-3} \geq \frac{-3y}{-3} \quad \text{Reversing the symbol}$$
$$3 \geq y$$

The solution set is $\{y \mid 3 \geq y\}$, or $\{y \mid y \leq 3\}$.

Practice Exercise

7. Solve: $6y + 5 > 3y - 7$.

Review Exercises

Solve. [11.1b]

1. $x + 5 = -17$

2. $n - 7 = -6$

3. $x - 11 = 14$

4. $y - 0.9 = 9.09$

Solve. [11.2a]

5. $-\frac{2}{3}x = -\frac{1}{6}$

6. $-8x = -56$

7. $-\frac{x}{4} = 48$

8. $15x = -35$

9. $\frac{4}{5}y = -\frac{3}{16}$

Solve. [11.3a]

10. $5 - x = 13$

11. $\frac{1}{4}x - \frac{5}{8} = \frac{3}{8}$

Solve. [11.3b, c]

12. $5t + 9 = 3t - 1$

13. $7x - 6 = 25x$

14. $14y = 23y - 17 - 10$

15. $0.22y - 0.6 = 0.12y + 3 - 0.8y$

16. $\frac{1}{4}x - \frac{1}{8}x = 3 - \frac{1}{16}x$

17. $14y + 17 + 7y = 9 + 21y + 8$

18. $4(x + 3) = 36$

19. $3(5x - 7) = -66$

20. $8(x - 2) - 5(x + 4) = 20 + x$

21. $-5x + 3(x + 8) = 16$

22. $6(x - 2) - 16 = 3(2x - 5) + 11$

Determine whether the given number is a solution of the inequality $x \leq 4$. [11.7a]

23. -3 **24.** 7 **25.** 4

Solve. Write set notation for the answers.   [11.7c, d, e]

26. $y + \dfrac{2}{3} \geq \dfrac{1}{6}$

27. $9x \geq 63$

28. $2 + 6y > 14$

29. $7 - 3y \geq 27 + 2y$

30. $3x + 5 < 2x - 6$

31. $-4y < 28$

32. $4 - 8x < 13 + 3x$

33. $-4x \leq \dfrac{1}{3}$

Graph on the number line.   [11.7b, e]

34. $4x - 6 < x + 3$

35. $-2 < x \leq 5$

36. $y > 0$

Solve.   [11.4b]

37. $C = \pi d$, for d

38. $V = \dfrac{1}{3}Bh$, for B

39. $A = \dfrac{a + b}{2}$, for a

40. $y = mx + b$, for x

Solve.   [11.6a]

41. *Dimensions of Wyoming.*   The state of Wyoming is roughly in the shape of a rectangle whose perimeter is 1280 mi. The length is 90 mi more than the width. Find the dimensions.

42. *Interstate Mile Markers.*   The sum of two consecutive mile markers on I-5 in California is 691. Find the numbers on the markers.

43. An entertainment center sold for $2449 in June. This was $332 more than the cost in February. What was the cost in February?

44. Ty is paid a commission of $4 for each magazine subscription he sells. One week, he received $108 in commissions. How many subscriptions did he sell?

45. The measure of the second angle of a triangle is 50° more than that of the first angle. The measure of the third angle is 10° less than twice the first angle. Find the measures of the angles.

Solve.   [11.5a]

46. What number is 20% of 75?

47. Fifteen is what percent of 80?

48. 18 is 3% of what number?

49. Black Bears. The population of black bears in the 26 U.S. states that border on or are east of the Mississippi River increased from 87,872 in 1988 to 164,440 in 2008. What is the percent increase?

Source: Hank Hristienko, Manitoba Conservation Wildlife and Ecosystem Protection Branch

50. First-Class Mail. The volume of first-class mail decreased from 102.4 billion pieces in 2002 to only 73.5 billion pieces in 2011. What is the percent decrease?

Source: United States Postal Service

Solve.   [11.6a]

51. After a 30% reduction, a bread maker is on sale for $154. What was the marked price (the price before the reduction)?

52. A restaurant manager's salary is $78,300, which is an 8% increase over the previous year's salary. What was the previous salary?

53. A tax-exempt organization received a bill of $145.90 for janitorial supplies. The bill incorrectly included sales tax of 5%. How much does the organization actually owe?

Solve.   [11.8b]

54. Test Scores. Noah's test grades are 71, 75, 82, and 86. What is the lowest grade that he can get on the next test and still have an average test score of at least 80?

55. The length of a rectangle is 43 cm. What widths will make the perimeter greater than 120 cm?

56. The solution of the equation
$$4(3x - 5) + 6 = 8 + x$$
is which of the following?   [11.3c]

A. Less than -1 **B.** Between -1 and 1
C. Between 1 and 5   **D.** Greater than 5

57. Solve for y: $3x + 4y = P$.   [11.4b]

A. $y = \dfrac{P - 3x}{4}$ **B.** $y = \dfrac{P + 3x}{4}$

C. $y = P - \dfrac{3x}{4}$ **D.** $y = \dfrac{P}{4} - 3x$

Synthesis

Solve.

58. $2|x| + 4 = 50$ [10.2e], [11.3a]

59. $|3x| = 60$ [10.2e], [11.2a]

60. $y = 2a - ab + 3$, for a [11.4b]

Understanding Through Discussion and Writing

1. Would it be better to receive a 5% raise and then an 8% raise or the other way around? Why?   [11.5a]

2. Erin returns a tent that she bought during a storewide 25%-off sale that has ended. She is offered store credit for 125% of what she paid (not to be used on sale items). Is this fair to Erin? Why or why not?   [11.5a]

3. Are the inequalities $x > -5$ and $-x < 5$ equivalent? Why or why not?   [11.7d]

4. Explain in your own words why it is necessary to reverse the inequality symbol when multiplying on both sides of an inequality by a negative number.   [11.7d]

5. If f represents Fran's age and t represents Todd's age, write a sentence that would translate to $t + 3 < f$.   [11.8a]

6. Explain how the meanings of "Five more than a number" and "Five is more than a number" differ.   [11.8a]

CHAPTER

11   Test

For Extra Help   For step-by-step test solutions, access the Chapter Test Prep Videos in
MyMathLab® or on You Tube (search "BittingerAlgebraFoundations" and click
on "Channels").

Solve.

1. $x + 7 = 15$

2. $t - 9 = 17$

3. $3x = -18$

4. $-\dfrac{4}{7}x = -28$

5. $3t + 7 = 2t - 5$

6. $\dfrac{1}{2}x - \dfrac{3}{5} = \dfrac{2}{5}$

7. $8 - y = 16$

8. $-\dfrac{2}{5} + x = -\dfrac{3}{4}$

9. $3(x + 2) = 27$

10. $-3x - 6(x - 4) = 9$

11. $0.4p + 0.2 = 4.2p - 7.8 - 0.6p$

12. $4(3x - 1) + 11 = 2(6x + 5) - 8$

13. $-2 + 7x + 6 = 5x + 4 + 2x$

Solve. Write set notation for the answers.

14. $x + 6 \le 2$

15. $14x + 9 > 13x - 4$

16. $12x \le 60$

17. $-2y \ge 26$

18. $-4y \le -32$

19. $-5x \ge \dfrac{1}{4}$

20. $4 - 6x > 40$

21. $5 - 9x \ge 19 + 5x$

Graph on the number line.

22. $y \le 9$

23. $6x - 3 < x + 2$

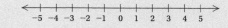

24. $-2 \le x \le 2$

Solve.

25. What number is 24% of 75?

26. 15.84 is what percent of 96?

27. 800 is 2% of what number?

28. *Bottled Water.* Annual bottled water consumption
per person in the United States increased from 18.2 gal
in 2001 to 29.2 gal in 2011. What is the percent
increase?

Source: Beverage Marketing Corporation

29. Perimeter of a Photograph. The perimeter of a rectangular photograph is 36 cm. The length is 4 cm greater than the width. Find the width and the length.

30. Cost of Raising a Child. It is estimated that $41,500 will be spent for child care and K–12 education for a child to age 17. This number represents approximately 18% of the total cost of raising a child to age 17. What is the total cost of raising a child to age 17?

Sources: U.S. Department of Agriculture/Upromise

31. Raffle Tickets. The numbers on three raffle tickets are consecutive integers whose sum is 7530. Find the integers.

32. Savings Account. Money is invested in a savings account at 5% simple interest. After 1 year, there is $924 in the account. How much was originally invested?

33. Board Cutting. An 8-m board is cut into two pieces. One piece is 2 m longer than the other. How long are the pieces?

34. Lengths of a Rectangle. The width of a rectangle is 96 yd. Find all possible lengths such that the perimeter of the rectangle will be at least 540 yd.

35. Budgeting. Jason has budgeted an average of $95 per month for entertainment. For the first five months of the year, he has spent $98, $89, $110, $85, and $83. How much can Jason spend in the sixth month without exceeding his average budget?

36. Copy Machine Rental. A catalog publisher needs to lease a copy machine for use during a special project that they anticipate will take 3 months. It costs $225 per month plus 3.2¢ per copy to rent the machine. The company must stay within a budget of $4500 for copies. Determine (in terms of an inequality) the number of copies they can make and still remain within budget.

37. Solve $A = 2\pi rh$ for r.

38. Solve $y = 8x + b$ for x.

39. Senior Population. The number of Americans ages 65 and older is projected to grow from 40.4 million to 70.3 million between 2011 and 2030. Find the percent increase.

Source: U.S. Census Bureau

A. 42.5% **B.** 47%
C. 57.5% **D.** 74%

Synthesis

40. Solve $c = \dfrac{1}{a - d}$ for d.

41. Solve: $3|w| - 8 = 37$.

42. A movie theater had a certain number of tickets to give away. Five people got the tickets. The first got one-third of the tickets, the second got one-fourth of the tickets, and the third got one-fifth of the tickets. The fourth person got eight tickets, and there were five tickets left for the fifth person. Find the total number of tickets given away.

Graphs of Linear Equations

STUDYING FOR SUCCESS   *Preparing for a Test*

☐ Make up your own test questions as you study.
☐ Do an overall review of the chapter, focusing on the objectives and the examples.
☐ Do the exercises in the mid-chapter review and in the summary and review at the end of the chapter.
☐ Take the chapter test at the end of the chapter.

12.1   Introduction to Graphing

OBJECTIVES

a Plot points associated with ordered pairs of numbers; determine the quadrant in which a point lies.

b Find the coordinates of a point on a graph.

c Determine whether an ordered pair is a solution of an equation with two variables.

SKILL TO REVIEW

Objective 11.1a: Determine whether a given number is a solution of a given equation.

Determine whether −3 is a solution of each equation.

1. $8(w - 3) = 0$
2. $15 = -2y + 9$

You probably have seen bar graphs like the following in newspapers and magazines. Note that a straight line can be drawn along the tops of the bars. Such a line is a *graph of a linear equation*. In this chapter, we study how to graph linear equations and consider properties such as slope and intercepts. Many applications of these topics will also be considered.

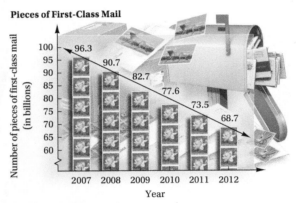

Pieces of First-Class Mail

Number of pieces of first-class mail (in billions)

96.3, 90.7, 82.7, 77.6, 73.5, 68.7

2007 2008 2009 2010 2011 2012

Year

SOURCE: U. S. Postal Service

a   PLOTTING ORDERED PAIRS

In Chapter 11, we graphed numbers and inequalities in one variable on a line. To enable us to graph an equation that contains two variables, we now learn to graph number pairs on a plane.

On the number line, each point is the graph of a number. On a plane, each point is the graph of a number pair. To form the plane, we use two perpendicular number lines called **axes**. They cross at a point called the **origin**. The arrows show the positive directions.

Consider the **ordered pair** $(3, 4)$, shown in the figure at left. The numbers in an ordered pair are called **coordinates**. In $(3, 4)$, the **first coordinate** (the **abscissa**) is 3 and the **second coordinate** (the **ordinate**) is 4. To plot $(3, 4)$, we start at the origin and move *horizontally* to the 3. Then we move up *vertically* 4 units and make a "dot."

The point $(4, 3)$ is also plotted at left. Note that $(3, 4)$ and $(4, 3)$ represent different points. The order of the numbers in the pair is important. We use the term *ordered* pairs because it makes a difference which number comes first. The coordinates of the origin are $(0, 0)$.

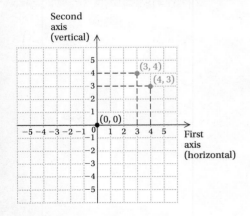

Second axis (vertical)

$(3, 4)$
$(4, 3)$
$(0, 0)$

First axis (horizontal)

Answers

Skill to Review:
1. −3 is not a solution.   **2.** −3 is a solution.

EXAMPLE 1 Plot the point $(-5, 2)$.

The first number, -5, is negative. Starting at the origin, we move -5 units in the horizontal direction (5 units to the left). The second number, 2, is positive. We move 2 units in the vertical direction (up).

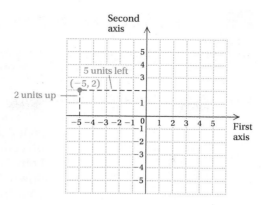

Plot these points on the gird below.

1. $(4, 5)$ 2. $(5, 4)$

3. $(-2, 5)$ 4. $(-3, -4)$

5. $(5, -3)$ 6. $(-2, -1)$

7. $(0, -3)$ 8. $(2, 0)$

Caution!

The *first* coordinate of an ordered pair is always graphed in a *horizontal* direction, and the *second* coordinate is always graphed in a *vertical* direction.

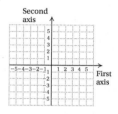

Do Exercises 1–8. ▶

The following figure shows some points and their coordinates. In region I (the *first quadrant*), both coordinates of any point are positive. In region II (the *second quadrant*), the first coordinate is negative and the second positive. In region III (the *third quadrant*), both coordinates are negative. In region IV (the *fourth quadrant*), the first coordinate is positive and the second is negative.

EXAMPLE 2 In which quadrant, if any, are the points $(-4, 5)$, $(5, -5)$, $(2, 4)$, $(-2, -5)$, and $(-5, 0)$ located?

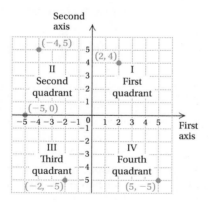

The point $(-4, 5)$ is in the second quadrant. The point $(5, -5)$ is in the fourth quadrant. The point $(2, 4)$ is in the first quadrant. The point $(-2, -5)$ is in the third quadrant. The point $(-5, 0)$ is on an axis and is *not in any quadrant*.

9. What can you say about the coordinates of a point in the third quadrant?

10. What can you say about the coordinates of a point in the fourth quadrant?

In which quadrant, if any, is each point located?

11. $(5, 3)$ 12. $(-6, -4)$

13. $(10, -14)$ 14. $(-13, 9)$

15. $(0, -3)$ 16. $\left(-\dfrac{1}{2}, \dfrac{1}{4}\right)$

Do Exercises 9–16. ▶

Answers

1.–8.

Second axis

9. First, negative; second, negative   10. First, positive; second, negative   11. I   12. III   13. IV   14. II   15. On an axis, not in any quadrant   16. II

b FINDING COORDINATES

To find the coordinates of a point, we see how far to the right or to the left of the origin it is located and how far up or down from the origin.

EXAMPLE 3 Find the coordinates of points A, B, C, D, E, F, and G.

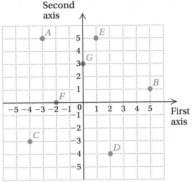

Point A is 3 units to the left (horizontal direction) and 5 units up (vertical direction). Its coordinates are $(-3, 5)$. Point D is 2 units to the right and 4 units down. Its coordinates are $(2, -4)$. The coordinates of the other points are as follows:

B: $(5, 1)$; C: $(-4, -3)$;

E: $(1, 5)$; F: $(-2, 0)$;

G: $(0, 3)$.

17. Find the coordinates of points A, B, C, D, E, F, and G on the graph below.

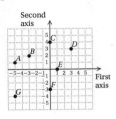

◀ Do Exercise 17.

c SOLUTIONS OF EQUATIONS

Now we begin to learn how graphs can be used to represent solutions of equations. When an equation contains two variables, the solutions of the equation are *ordered pairs* in which each number in the pair corresponds to a letter in the equation. Unless stated otherwise, to determine whether a pair is a solution, we use the first number in each pair to replace the variable that occurs first *alphabetically*.

EXAMPLE 4 Determine whether each of the following pairs is a solution of $4q - 3p = 22$: $(2, 7)$ and $(-1, 6)$.

For $(2, 7)$, we substitute 2 for p and 7 for q (using alphabetical order of variables):

$$\begin{array}{c|c} \multicolumn{2}{c}{4q - 3p = 22} \\ \hline 4 \cdot 7 - 3 \cdot 2 \;?\; 22 \\ 28 - 6 \\ 22 & \text{TRUE} \end{array}$$

Thus, $(2, 7)$ *is* a solution of the equation.

For $(-1, 6)$, we substitute -1 for p and 6 for q:

$$\begin{array}{c|c} \multicolumn{2}{c}{4q - 3p = 22} \\ \hline 4 \cdot 6 - 3 \cdot (-1) \;?\; 22 \\ 24 + 3 \\ 27 & \text{FALSE} \end{array}$$

Thus, $(-1, 6)$ *is not* a solution of the equation.

◀ Do Exercises 18 and 19.

18. Determine whether $(2, -4)$ is a solution of $4q - 3p = 22$. **GS**

$$\begin{array}{c|c} \multicolumn{2}{c}{4q - 3p = 22} \\ \hline 4 \cdot (\boxed{}) - 3 \cdot \boxed{} \;?\; 22 \\ -16 - \boxed{} \\ \boxed{} & \text{FALSE} \end{array}$$

Thus, $(2, -4)$ _____ a solution.
        is/is not

19. Determine whether $(2, -4)$ is a solution of $7a + 5b = -6$.

EXAMPLE 5 Show that the pairs $(3, 7)$, $(0, 1)$, and $(-3, -5)$ are solutions of $y = 2x + 1$. Then graph the three points and use the graph to determine another pair that is a solution.

To show that a pair is a solution, we substitute, replacing x with the first coordinate and y with the second coordinate of each pair:

$$\begin{array}{l} y = 2x + 1 \\ \hline 7 \;?\; 2 \cdot 3 + 1 \\ \quad\;\; 6 + 1 \\ \quad\;\; 7 \qquad \text{TRUE} \end{array} \qquad \begin{array}{l} y = 2x + 1 \\ \hline 1 \;?\; 2 \cdot 0 + 1 \\ \quad\;\; 0 + 1 \\ \quad\;\; 1 \qquad \text{TRUE} \end{array}$$

$$\begin{array}{l} y = 2x + 1 \\ \hline -5 \;?\; 2(-3) + 1 \\ \quad\;\;\; -6 + 1 \\ \quad\;\;\; -5 \qquad \text{TRUE} \end{array}$$

In each of the three cases, the substitution results in a true equation. Thus the pairs are all solutions.

We plot the points as shown at right. The order of the points follows the alphabetical order of the variables. That is, x is before y, so x-values are first coordinates and y-values are second coordinates. Similarly, we also label the horizontal axis as the x-axis and the vertical axis as the y-axis.

Note that the three points appear to "line up." That is, they appear to be on a straight line. Will other points that line up with these points also represent solutions of $y = 2x + 1$? To find out, we use a straightedge and sketch a line passing through $(3, 7)$, $(0, 1)$, and $(-3, -5)$.

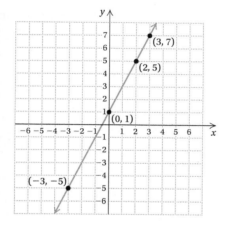

The line appears to pass through $(2, 5)$ as well. Let's see if this pair is a solution of $y = 2x + 1$:

$$\begin{array}{l} y = 2x + 1 \\ \hline 5 \;?\; 2 \cdot 2 + 1 \\ \quad\;\; 4 + 1 \\ \quad\;\; 5 \qquad \text{TRUE} \end{array}$$

Thus, $(2, 5)$ is a solution.

Do Exercise 20.

Example 5 leads us to suspect that any point on the line that passes through $(3, 7)$, $(0, 1)$, and $(-3, -5)$ represents a solution of $y = 2x + 1$. In fact, every solution of $y = 2x + 1$ is represented by a point on that line and every point on that line represents a solution. The line is the *graph* of the equation.

GRAPH OF AN EQUATION

The **graph** of an equation is a drawing that represents all of its solutions.

20. Use the graph in Example 5 to find at least two more points that are solutions of $y = 2x + 1$.

Answer

20. $(-2, -3)$, $(1, 3)$; answers may vary

✓ Reading Check

Determine whether each statement is true or false.

RC1. In the ordered pair $(-6, 2)$, the first coordinate, -6, is also called the abscissa.

RC2. The point $(1, 0)$ is in quadrant I and in quadrant IV.

RC3. The ordered pairs $(4, -7)$ and $(-7, 4)$ name the same point.

RC4. To plot the point $(-3, 5)$, start at the origin and move horizontally to -3. Then move up vertically 5 units and make a "dot."

a

1. Plot these points.

$(2, 5)$ $(-1, 3)$ $(3, -2)$ $(-2, -4)$

$(0, 4)$ $(0, -5)$ $(5, 0)$ $(-5, 0)$

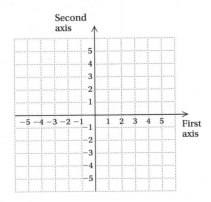

2. Plot these points.

$(4, 4)$ $(-2, 4)$ $(5, -3)$ $(-5, -5)$

$(0, 2)$ $(0, -4)$ $(3, 0)$ $(-4, 0)$

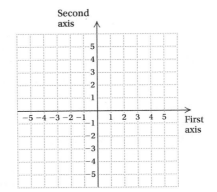

In which quadrant, if any, is each point located?

3. $(-5, 3)$

4. $(1, -12)$

5. $(100, -1)$

6. $(-2.5, 35.6)$

7. $(-6, -29)$

8. $(3.6, 105.9)$

9. $(3.8, 0)$

10. $(0, -492)$

11. $\left(-\dfrac{1}{3}, \dfrac{15}{7}\right)$

12. $\left(-\dfrac{2}{3}, -\dfrac{9}{8}\right)$

13. $\left(12\dfrac{7}{8}, -1\dfrac{1}{2}\right)$

14. $\left(23\dfrac{5}{8}, 81.74\right)$

Complete the table regarding the signs of coordinates in certain quadrants.

	Quadrant	First Coordinates	Second Coordinates
15.		Positive	Positive
16.	III		Negative
17.	II	Negative	
18.		Positive	Negative

In which quadrant(s) can the point described be located?

19. The first coordinate is negative and the second coordinate is positive.

20. The first and second coordinates are positive.

21. The first coordinate is positive.

22. The second coordinate is negative.

23. The first and second coordinates are equal.

24. The first coordinate is the additive inverse of the second coordinate.

 Find the coordinates of points *A, B, C, D,* and *E.*

25.

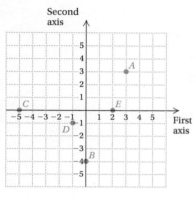

26.

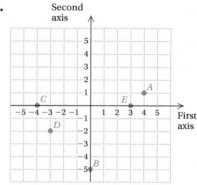

 Determine whether the given ordered pair is a solution of the equation.

27. $(2, 9)$; $y = 3x - 1$

28. $(1, 7)$; $y = 2x + 5$

29. $(4, 2)$; $2x + 3y = 12$

30. $(0, 5)$; $5x - 3y = 15$

31. $(3, -1)$; $3a - 4b = 13$

32. $(-5, 1)$; $2p - 3q = -13$

In each of Exercises 33–38, an equation and two ordered pairs are given. Show that each pair is a solution of the equation. Then use the graph of the equation to determine another solution. Answers may vary.

33. $y = x - 5$; $(4, -1)$ and $(1, -4)$

34. $y = x + 3$; $(-1, 2)$ and $(3, 6)$

35. $y = \frac{1}{2}x + 3$; $(4, 5)$ and $(-2, 2)$

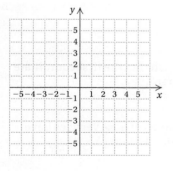

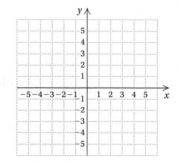

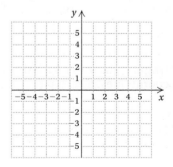

36. $3x + y = 7$; $(2, 1)$ and $(4, -5)$ **37.** $4x - 2y = 10$; $(0, -5)$ and $(4, 3)$ **38.** $6x - 3y = 3$; $(1, 1)$ and $(-1, -3)$

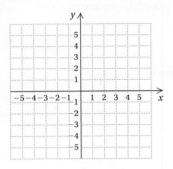   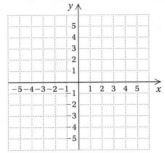   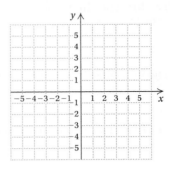

Skill Maintenance

Solve.   [11.3c]

39. $6(z - 5) = 2(z + 1)$

40. $8 - 4(q + 2) = -7$

41. $-5 + x = x - 5$

42. $-\dfrac{3}{4} + x = x + \dfrac{1}{4}$

43. $4t - 2 = 3(5 - 2t)$

44. $2b - 5 + b = 6(b - 1)$

Graph on the number line.   [11.7b]

45. $-2 < y \le 1$

<------+----+----+----+----+----+----+----+----+----+----+------>
      -5   -4   -3   -2   -1   0   1   2   3   4   5

46. $p \ge -3$

<------+----+----+----+----+----+----+----+----+----+----+------>
      -5   -4   -3   -2   -1   0   1   2   3   4   5

Solve.   [11.5a]

47. What is 15% of $23.80?

48. $7.29 is 15% of what number?

49. 75 is what percent of 500?

Subtract.   [10.4a]

50. $\dfrac{1}{4} - \dfrac{2}{5}$

51. $-3.1 - (-3.1)$

52. $-403 - 52$

Synthesis

53. The points $(-1, 1)$, $(4, 1)$, and $(4, -5)$ are three vertices of a rectangle. Find the coordinates of the fourth vertex.

54. Three parallelograms share the vertices $(-2, -3)$, $(-1, 2)$, and $(4, -3)$. Find the fourth vertex of each parallelogram.

55. Graph eight points such that the sum of the coordinates in each pair is 6.

56. Graph eight points such that the first coordinate minus the second coordinate is 1.

57. Find the perimeter of a rectangle whose vertices have coordinates $(5, 3)$, $(5, -2)$, $(-3, -2)$, and $(-3, 3)$.

58. Find the area of a triangle whose vertices have coordinates $(0, 9)$, $(0, -4)$, and $(5, -4)$.

Graphing Linear Equations

a   GRAPHS OF LINEAR EQUATIONS

Equations like $y = 2x + 1$ and $4q - 3p = 22$ are said to be **linear** because the graph of each equation is a straight line. In general, any equation equivalent to one of the form $y = mx + b$ or $Ax + By = C$, where m, b, A, B, and C are constants (not variables) and A and B are not both 0, is linear.

To graph a linear equation:

1. Select a value for one variable and calculate the corresponding value of the other variable. Form an ordered pair using alphabetical order as indicated by the variables.

2. Repeat step (1) to obtain at least two other ordered pairs. Two points are essential to determine a straight line. A third point serves as a check.

3. Plot the ordered pairs and draw a straight line passing through the points.

In general, calculating three (or more) ordered pairs is not difficult for equations of the form $y = mx + b$. We simply substitute values for x and calculate the corresponding values for y.

EXAMPLE 1   Graph: $y = 2x$.

First, we find some ordered pairs that are solutions. We choose *any* number for x and then determine y by substitution. Since $y = 2x$, we find y by doubling x. Suppose that we choose 3 for x. Then

$$y = 2x = 2 \cdot 3 = 6.$$

We get a solution: the ordered pair $(3, 6)$.
   Suppose that we choose 0 for x. Then

$$y = 2x = 2 \cdot 0 = 0.$$

We get another solution: the ordered pair $(0, 0)$.
   For a third point, we make a negative choice for x. If x is -3, we have

$$y = 2x = 2 \cdot (-3) = -6.$$

This gives us the ordered pair $(-3, -6)$.
   We now have enough points to plot the line, but if we wish, we can compute more. If a number takes us off the graph paper, we either do not use it or we use larger paper or rescale the axes. Continuing in this manner, we create a table like the one shown on the following page.

CALCULATOR CORNER

Finding Solutions of Equations   A table of values representing ordered pairs that are solutions of an equation can be displayed on a graphing calculator. To do this for the equation in Example 1, $y = 2x$, we first access the equation-editor screen. Then we clear any equations that are present. Next, we enter the equation, display the table set-up screen, and set both **INDPNT** and **DEPEND** to **AUTO**.
   We will display a table of values that starts with $x = -2$ (**TBLSTART**) and adds 1 (Δ**TBL**) to the preceding x-value.

X	Y₁
-2	-4
-1	-2
0	0
1	2
2	4
3	6
4	8

X = -2

EXERCISE:

1. Create a table of ordered pairs that are solutions of the equations in Examples 2 and 3.

Answers

Skill to Review:
1. II   2. IV

Complete each table and graph.

Now we plot these points. Then we draw the line, or graph, with a straightedge and label it $y = 2x$.

1. $y = -2x$ GS

x	y	(x, y)
-3	6	$(-3, 6)$
-1	☐	$(-1, ☐)$
0	0	$(0, 0)$
1	☐	$(☐, -2)$
3	☐	$(3, ☐)$

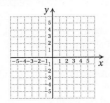

x	y $y = 2x$	(x, y)
3	6	$(3, 6)$
1	2	$(1, 2)$
0	0	$(0, 0)$
-2	-4	$(-2, -4)$
-3	-6	$(-3, -6)$

(1) Choose x.
(2) Compute y.
(3) Form the pair (x, y).
(4) Plot the points.

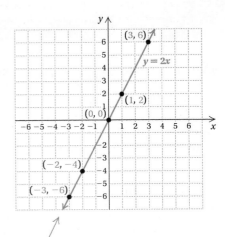

·········· **Caution!** ··········

Keep in mind that you can choose *any* number for x and then compute y. Our choice of certain numbers in the examples does not dictate those that you must choose.
···

◀ **Do Exercises 1 and 2.**

EXAMPLE 2 Graph: $y = -3x + 1$.

We select a value for x, compute y, and form an ordered pair. Then we repeat the process for other choices of x.

If $x = 2$,   then $y = -3 \cdot 2 + 1 = -5$,   and $(2, -5)$ is a solution.
If $x = 0$,   then $y = -3 \cdot 0 + 1 = 1$,   and $(0, 1)$ is a solution.
If $x = -1$, then $y = -3 \cdot (-1) + 1 = 4$,   and $(-1, 4)$, is a solution.

Results are listed in the table below. The points corresponding to each pair are then plotted.

2. $y = \frac{1}{2}x$

x	y	(x, y)
4		
2		
0		
-2		
-4		
-1		

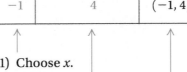

x	y $y = -3x + 1$	(x, y)
2	-5	$(2, -5)$
0	1	$(0, 1)$
-1	4	$(-1, 4)$

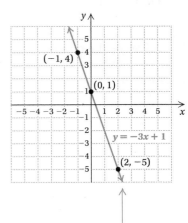

(1) Choose x.
(2) Compute y.
(3) Form the pair (x, y).
(4) Plot the points.

Answers

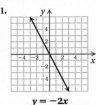
$y = -2x$

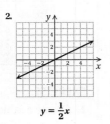
$y = \frac{1}{2}x$

Guided Solution:
1. $2, 2, -2, 1, -6, -6$

Note that all three points line up. If they did not, we would know that we had made a mistake. When only two points are plotted, a mistake is harder to detect. We use a ruler or another straightedge to draw a line through the points. Every point on the line represents a solution of $y = -3x + 1$.

Do Exercises 3 and 4. ▶

In Example 1, we saw that $(0, 0)$ is a solution of $y = 2x$. It is also the point at which the graph crosses the y-axis. Similarly, in Example 2, we saw that $(0, 1)$ is a solution of $y = -3x + 1$. It is also the point at which the graph crosses the y-axis. A generalization can be made: If x is replaced with 0 in the equation $y = mx + b$, then the corresponding y-value is $m \cdot 0 + b$, or b. Thus any equation of the form $y = mx + b$ has a graph that passes through the point $(0, b)$. Since $(0, b)$ is the point at which the graph crosses the y-axis, it is called the **y-intercept**. Sometimes, for convenience, we simply refer to b as the y-intercept.

---

y-INTERCEPT

The graph of the equation $y = mx + b$ passes through the **y-intercept** $(0, b)$.

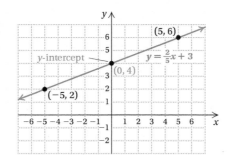

EXAMPLE 3 Graph $y = \frac{2}{5}x + 4$ and identify the y-intercept.

We select a value for x, compute y, and form an ordered pair. Then we repeat the process for other choices of x. In this case, using multiples of 5 avoids fractions. We try to avoid graphing ordered pairs with fractions because they are difficult to graph accurately.

If $x = 0$,   then $y = \dfrac{2}{5} \cdot 0 + 4 = 4$,   and $(0, 4)$ is a solution.

If $x = 5$,   then $y = \dfrac{2}{5} \cdot 5 + 4 = 6$,   and $(5, 6)$ is a solution.

If $x = -5$,   then $y = \dfrac{2}{5} \cdot (-5) + 4 = 2$,   and $(-5, 2)$ is a solution.

The following table lists these solutions. Next, we plot the points and see that they form a line. Finally, we draw and label the line.

x	y $y = \frac{2}{5}x + 4$	(x, y)
0	4	$(0, 4)$
5	6	$(5, 6)$
-5	2	$(-5, 2)$

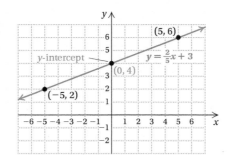

Complete each table and graph.

3. $y = 2x + 3$

x	y	(x, y)

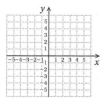

4. $y = -\dfrac{1}{2}x - 3$

x	y	(x, y)

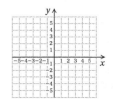

Answers

3.
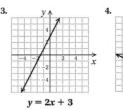
$y = 2x + 3$

4.
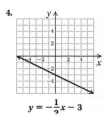
$y = -\dfrac{1}{2}x - 3$

Graph each equation and identify the y-intercept.

5. $y = \dfrac{3}{5}x + 2$

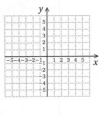

6. $y = -\dfrac{3}{5}x - 1$

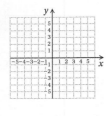

We see that $(0, 4)$ is a solution of $y = \frac{2}{5}x + 4$. It is the y-intercept. Because the equation is in the form $y = mx + b$, we can read the y-intercept directly from the equation as follows:

$$y = \frac{2}{5}x + \underset{\uparrow}{4} \qquad (0, 4) \text{ is the } y\text{-intercept.}$$

◀ **Do Exercises 5 and 6.**

Calculating ordered pairs is generally easiest when y is isolated on one side of the equation, as in $y = mx + b$. To graph an equation in which y is not isolated, we can use the addition and multiplication principles to solve for y. (See Sections 2.3 and 2.4.)

EXAMPLE 4 Graph $3y + 5x = 0$ and identify the y-intercept.

To find an equivalent equation in the form $y = mx + b$, we solve for y:

$$\begin{aligned}
3y + 5x &= 0 \\
3y + 5x - 5x &= 0 - 5x \qquad &\text{Subtracting } 5x \\
3y &= -5x \qquad &\text{Collecting like terms} \\
\frac{3y}{3} &= \frac{-5x}{3} \qquad &\text{Dividing by 3} \\
y &= -\frac{5}{3}x.
\end{aligned}$$

Because all the equations above are equivalent, we can use $y = -\frac{5}{3}x$ to draw the graph of $3y + 5x = 0$. To graph $y = -\frac{5}{3}x$, we select x-values and compute y-values. In this case, if we select multiples of 3, we can avoid fractions.

$$\text{If } x = 0, \quad \text{then } y = -\frac{5}{3} \cdot 0 = 0.$$

$$\text{If } x = 3, \quad \text{then } y = -\frac{5}{3} \cdot 3 = -5.$$

$$\text{If } x = -3, \quad \text{then } y = -\frac{5}{3} \cdot (-3) = 5.$$

We list these solutions in a table. Next, we plot the points and see that they form a line. Finally, we draw and label the line. The y-intercept is $(0, 0)$.

Graph each equation and identify the y-intercept.

7. $5y + 4x = 0$

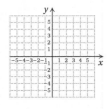

8. $4y = 3x$

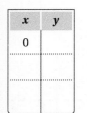 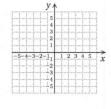

x	y	
0	0	← y-intercept
3	−5	
−3	5	

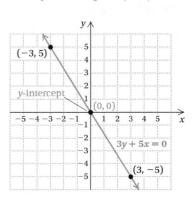

◀ **Do Exercises 7 and 8.**

Answers

Answers to Margin Exercises 5–8 are on p. 183.

EXAMPLE 5 Graph $4y + 3x = -8$ and identify the y-intercept.

To find an equivalent equation in the form $y = mx + b$, we solve for y:

$$4y + 3x = -8$$
$$4y + 3x - 3x = -8 - 3x \qquad \text{Subtracting } 3x$$
$$4y = -3x - 8 \qquad \text{Simplifying}$$
$$\frac{1}{4} \cdot 4y = \frac{1}{4} \cdot (-3x - 8) \qquad \begin{array}{l}\text{Multiplying by } \frac{1}{4} \text{ or}\\ \text{dividing by } 4\end{array}$$
$$y = \frac{1}{4} \cdot (-3x) - \frac{1}{4} \cdot 8 \qquad \text{Using the distributive law}$$
$$y = -\frac{3}{4}x - 2. \qquad \text{Simplifying}$$

Thus, $4y + 3x = -8$ is equivalent to $y = -\frac{3}{4}x - 2$. The y-intercept is $(0, -2)$. We find two other pairs using multiples of 4 for x to avoid fractions. We then complete and label the graph as shown.

x	y
0	-2	← y-intercept
4	-5
-4	1

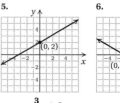

Do Exercises 9 and 10. ▶

b APPLICATIONS OF LINEAR EQUATIONS

Mathematical concepts become more understandable through visualization. Throughout this text, you will occasionally see the heading Algebraic–Graphical Connection, as in Example 6, which follows. In this feature, the algebraic approach is enhanced and expanded with a graphical connection. Relating a solution of an equation to a graph can often give added meaning to the algebraic solution.

EXAMPLE 6 *World Population.* The world population, in billions, is estimated and projected by the equation

$$y = 0.072x + 4.593,$$

where x is the number of years since 1980. That is, $x = 0$ corresponds to 1980, $x = 12$ corresponds to 1992, and so on.

Source: U.S. Census Bureau

Graph each equation and identify the y-intercept.

 9. $5y - 3x = -10$

x	y
0	☐	← y-intercept, $(0, \boxed{})$
5	1
-5	☐

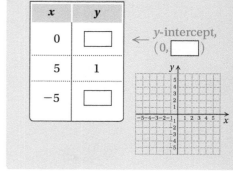

10. $5y + 3x = 20$

x	y

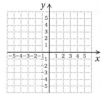

a) Estimate the world population in 1980 and in 2005. Then project the population in 2030.

b) Graph the equation and then use the graph to estimate the world population in 2015.

c) In what year could we project the world population to be 7.761 billion?

a) The years 1980, 2005, and 2030 correspond to $x = 0$, $x = 25$, and $x = 50$, respectively. We substitute 0, 25, and 50 for x and then calculate y:

$$y = 0.072(0) + 4.593 = 0 + 4.593 = 4.593;$$
$$y = 0.072(25) + 4.593 = 1.8 + 4.593 = 6.393;$$
$$y = 0.072(50) + 4.593 = 3.6 + 4.593 = 8.193.$$

The world population in 1980, in 2005, and in 2030 is estimated and projected to be 4.593 billion, 6.393 billion, and 8.193 billion, respectively.

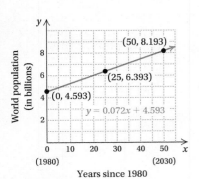

FIGURE 1

ALGEBRAIC ▶◀ **GRAPHICAL CONNECTION**

b) We have three ordered pairs from part (a). We plot these points and see that they line up. Thus our calculations are probably correct. Since we are considering only the year 2015 and the number of years since 1980 ($x \geq 0$) and since the population, in billions, for those years will be positive ($y > 0$), we need only the first quadrant for the graph. We use the three points we have plotted to draw a straight line. (See Figure 1.)

To use the graph to estimate world population in 2015, we first note in Figure 2 that this year corresponds to $x = 35$. We need to determine which y-value is paired with $x = 35$. We locate the point on the graph by moving up vertically from $x = 35$, and then find the value on the y-axis that corresponds to that point. It appears that the world population in 2015 will be about 7.1 billion.

To find a more accurate value, we can simply substitute into the equation:

$$y = 0.072(35) + 4.593 = 7.113.$$

The world population in 2015 is estimated to be 7.113 billion.

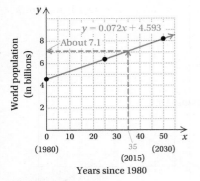

FIGURE 2

c) We substitute 7.761 for y and solve for x:

$$y = 0.072x + 4.593$$
$$7.761 = 0.072x + 4.593$$
$$3.168 = 0.072x$$
$$44 = x.$$

In 44 years after 1980, or in 2024, the world population is projected to be approximately 7.761 billion.

◀ **Do Exercise 11 on the following page.**

Many equations in two variables have graphs that are not straight lines. Three such nonlinear graphs are shown below. We will cover some such graphs in the optional Calculator Corners throughout the text and in Chapter 20.

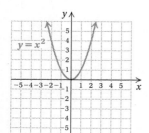 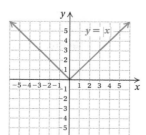 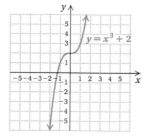

CALCULATOR CORNER

Graphing Equations   Equations must be solved for y before they can be graphed on the TI-84 Plus. Consider the equation $3x + 2y = 6$. Solving for y, we get $y = \dfrac{6 - 3x}{2}$. We enter this equation as $y_1 = (6 - 3x)/2$ on the equation-editor screen. Then we select the standard viewing window and display the graph.

$$y = (6 - 3x)/2$$

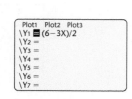 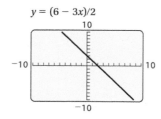

EXERCISES: Graph each equation in the standard viewing window $[-10, 10, -10, 10]$, with Xscl $= 1$ and Yscl $= 1$.

1. $y = -5x + 3$ **2.** $y = 4x - 5$

3. $4x - 5y = -10$ **4.** $5y + 5 = -3x$

---

11. *Milk Consumption.*   Milk consumption per capita (per person) in the United States can be estimated by

$$M = -0.183t + 27.776,$$

where M is the consumption, in gallons, t years since 1985.

Source: U.S. Department of Agriculture

a) Find the per capita consumption of milk in 1985, in 1995, and in 2015.

b) Graph the equation and use the graph to estimate milk consumption in 2010.

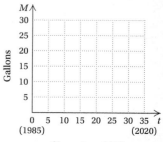

c) In which year would the per capita consumption of milk be 21.737 gal?

Answers

11. **(a)** 27.776 gal; 25.946 gal; 22.286 gal;
(b) about 23.2 gal;
(c) 33 years after 1985, or in 2018

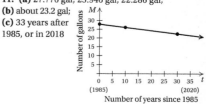

| **12.2** | **Exercise Set** |

For Extra Help

MyMathLab®   MathXL° PRACTICE WATCH READ REVIEW

✓ Reading Check

For each of the following equations, choose from the column on the right an equivalent equation.

RC1. $2y + 5x = -10$

RC2. $5y - 2x = 0$

RC3. $2x + 5y = 0$

RC4. $5x - 2y = -5$

a) $y = \dfrac{5}{2}x + \dfrac{5}{2}$

b) $y = \dfrac{2}{5}x$

c) $y = -\dfrac{5}{2}x - 5$

d) $y = -\dfrac{2}{5}x$

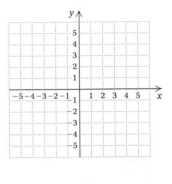

a   Graph each equation and identify the *y*-intercept.

1. $y = x + 1$

x	y
−2	
−1	
0	
1	
2	
3	

2. $y = x - 1$

x	y
−2	
−1	
0	
1	
2	
3	

3. $y = x$

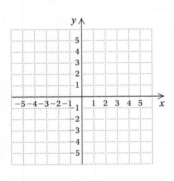

x	y
−2	
−1	
0	
1	
2	
3	

4. $y = -x$

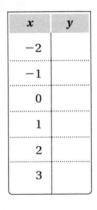
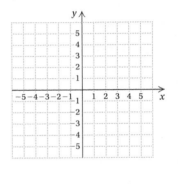

x	y
−2	
−1	
0	
1	
2	
3	

5. $y = \dfrac{1}{2}x$

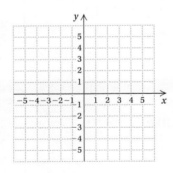

x	y
−2	
0	
4	

6. $y = \dfrac{1}{3}x$

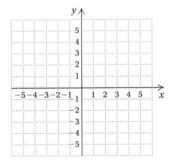

x	y
−6	
0	
3	

7. $y = x - 3$

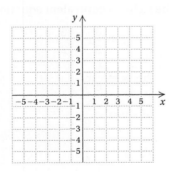

x	y

8. $y = x + 3$

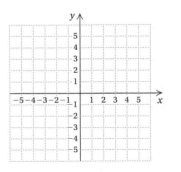

x	y

9. $y = 3x - 2$

x	y

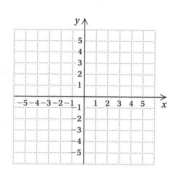

10. $y = 2x + 2$

x	y

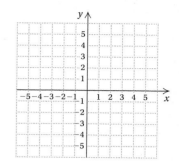

11. $y = \dfrac{1}{2}x + 1$

x	y

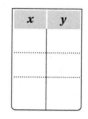

12. $y = \dfrac{1}{3}x - 4$

x	y

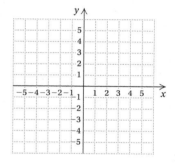

13. $x + y = -5$

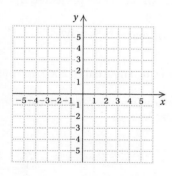

14. $x + y = 4$

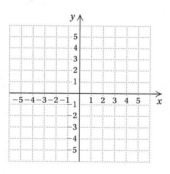

15. $y = \dfrac{5}{3}x - 2$

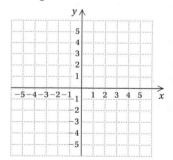

16. $y = \dfrac{5}{2}x + 3$

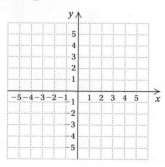

17. $x + 2y = 8$

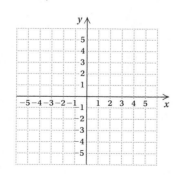

18. $x + 2y = -6$

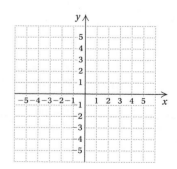

19. $y = \dfrac{3}{2}x + 1$

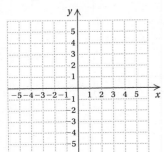

20. $y = -\dfrac{1}{2}x - 3$

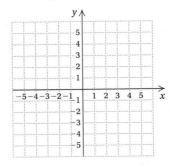

21. $8x - 2y = -10$

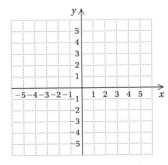

22. $6x - 3y = 9$

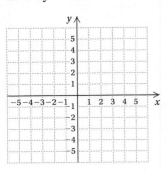

23. $8y + 2x = -4$

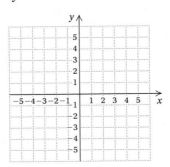

24. $6y + 2x = 8$

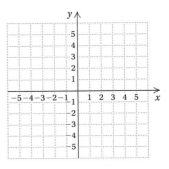

b   Solve.

25. *Online Advertising.*   Spending for online advertising is increasing. The amount A, in billions of dollars, spent worldwide for online advertising can be approximated and projected by

$$A = 12.83t + 68.38,$$

where t is the number of years since 2010.

a) Find the amount spent for online advertising in 2010, in 2014, and in 2015.

b) Graph the equation and use the graph to estimate the amount spent for online advertising in 2012.

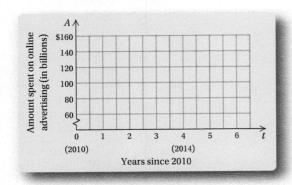

c) In what year will online advertising spending be about $158.19 billion?

26. *Price of a New Car.*   The average price P of a new car, in dollars, can be approximated and projected by

$$P = 426t + 25{,}710$$

where t is the number of years since 2002.

Source: Edmunds.com

a) Find the average price of a new car in 2002, in 2006, and in 2011.

b) Graph the equation and use the graph to estimate the price of a new car in 2010.

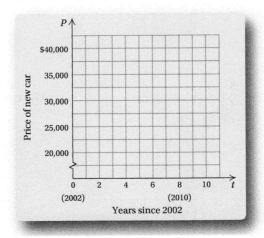

c) In what year will the average price of a new car be approximately $35,000?

27. Sheep and Lambs. The number of sheep and lambs S, in millions, on farms in the United States has declined in recent years and can be approximated and projected by

$$S = -0.125t + 6.898,$$

where t is the number of years since 2000.

Sources: National Agricultural Statistics; U.S. Department of Agriculture

a) Find the number of sheep and lambs in 2000, in 2007, and in 2012.

b) Graph the equation and use the graph to estimate the number of sheep and lambs in 2010.

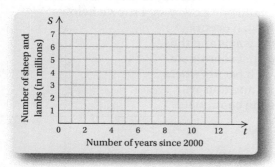

c) At this rate of decline, in what year will the number of sheep and lambs be 4.898 million?

28. Record Temperature Drop. On 22 January 1943, the temperature T, in degrees Fahrenheit, in Spearfish, South Dakota, could be approximated by

$$T = -2.15m + 54,$$

where m is the number of minutes since 9:00 that morning.

Source: Information Please Almanac

a) Find the temperature at 9:01 A.M., at 9:08 A.M., and at 9:20 A.M.

b) Graph the equation and use the graph to estimate the temperature at 9:15 A.M.

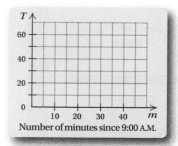

c) The temperature stopped dropping when it reached $-4°F$. At what time did this occur?

Skill Maintenance

Find the absolute value.   [10.2e]

29. $|-12|$

30. $|4.89|$

31. $|0|$

32. $\left|-\dfrac{4}{5}\right|$

Solve.   [11.3a]

33. $2x - 14 = 29$

34. $\dfrac{1}{3}t + 6 = -12$

35. $-10 = 1.2y + 2$

36. $4 - 5w = -16$

Solve.   [11.6a]

37. Books in Libraries. The Library of Congress houses 33.5 million books. This number is 0.3 million more than twice the number of books in the New York Public Library. How many books are in the New York Public Library?

Source: American Library Association

38. Busiest Orchestras. In 2011, the orchestras who performed the greatest number of concerts were the San Francisco Symphony and the Chicago Symphony. The Chicago Symphony had 133 concerts. This number is 24 fewer than the number of concerts by the San Francisco Symphony. How many concerts did the San Francisco Symphony have?

Source: www.bachtrack.com

Calculate.

39. $-\dfrac{3}{5} \div 5$ [10.6c]

40. $2.8 - (-0.2)$ [10.4a]

41. $-\dfrac{9}{16} + \left(-\dfrac{3}{8}\right)$ [10.3a]

42. $4.2 \times (-100)$ [10.5a]

43. $-\dfrac{8}{7} \div \left(-\dfrac{1}{4}\right)$ [10.6c]

44. $23.3 - 32.3$ [10.4a]

12.3

More with Graphing and Intercepts

OBJECTIVES

a Find the intercepts of a linear equation, and graph using intercepts.

b Graph equations equivalent to those of the type $x = a$ and $y = b$.

SKILL TO REVIEW

Objective 11.3a: Solve equations using both the addition principle and the multiplication principle.

Solve.

1. $5x - 7 = -10$

2. $-20 = \dfrac{7}{4}x + 8$

---

1. Look at the graph shown below.

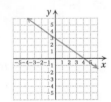

a) Find the coordinates of the y-intercept.

b) Find the coordinates of the x-intercept.

---

a GRAPHING USING INTERCEPTS

In Section 12.2, we graphed linear equations of the form $Ax + By = C$ by first solving for y to find an equivalent equation in the form $y = mx + b$. We did so because it is then easier to calculate the y-value that corresponds to a given x-value. Another convenient way to graph $Ax + By = C$ is to use **intercepts**. Look at the graph of $-2x + y = 4$ shown below.

The y-intercept is $(0, 4)$. It occurs where the line crosses the y-axis and thus will always have 0 as the first coordinate. The x-intercept is $(-2, 0)$. It occurs where the line crosses the x-axis and thus will always have 0 as the second coordinate.

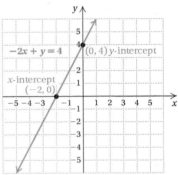

◀ Do Margin Exercise 1.

We find intercepts as follows.

INTERCEPTS

The **y-intercept** is $(0, b)$. To find b, let $x = 0$ and solve the equation for y.

The **x-intercept** is $(a, 0)$. To find a, let $y = 0$ and solve the equation for x.

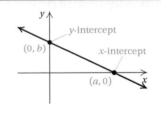

Now let's draw a graph using intercepts.

EXAMPLE 1 Consider $4x + 3y = 12$. Find the intercepts. Then graph the equation using the intercepts.

To find the y-intercept, we let $x = 0$. Then we solve for y:

$$4 \cdot 0 + 3y = 12$$
$$3y = 12$$
$$y = 4.$$

Thus, $(0, 4)$ is the y-intercept. Note that finding this intercept involves covering up the x-term and solving the rest of the equation for y.

To find the x-intercept, we let $y = 0$. Then we solve for x:

$$4x + 3 \cdot 0 = 12$$
$$4x = 12$$
$$x = 3.$$

Answers

Skill to Review:

1. $-\dfrac{3}{5}$ **2.** -16

Margin Exercises:

1. (a) $(0, 3)$; **(b)** $(4, 0)$

Thus, $(3, 0)$ is the x-intercept. Note that finding this intercept involves covering up the y-term and solving the rest of the equation for x.

We plot these points and draw the line, or graph.

x	y	
3	0	← x-intercept
0	4	← y-intercept
-2	$6\frac{2}{3}$	← Check point

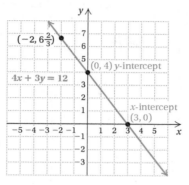

A third point should be used as a check. We substitute any convenient value for x and solve for y. In this case, we choose $x = -2$. Then

$$4(-2) + 3y = 12 \qquad \text{Substituting } -2 \text{ for } x$$
$$-8 + 3y = 12$$
$$3y = 20 \qquad \text{Adding 8 on both sides}$$
$$y = \frac{20}{3}, \text{ or } 6\frac{2}{3}. \qquad \text{Solving for } y$$

It appears that the point $\left(-2, 6\frac{2}{3}\right)$ is on the graph, though graphing fraction values can be inexact. The graph is probably correct.

Do Exercises 2 and 3. ▶

Graphs of equations of the type $y = mx$ pass through the origin. Thus the x-intercept and the y-intercept are the same, $(0, 0)$. In such cases, we must calculate another point in order to complete the graph. A third point would also need to be calculated if a check is desired.

EXAMPLE 2 Graph: $y = 3x$.

We know that $(0, 0)$ is both the x-intercept and the y-intercept. We calculate values at two other points and complete the graph, knowing that it passes through the origin $(0, 0)$.

x	y	
-1	-3	x-intercept
0	0	y-intercept
1	3	

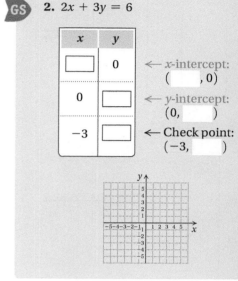

Do Exercises 4 and 5 on the following page. ▶

For each equation, find the intercepts. Then graph the equation using the intercepts.

GS **2.** $2x + 3y = 6$

x	y	
☐	0	← x-intercept: (☐ , 0)
0	☐	← y-intercept: (0, ☐)
-3	☐	← Check point: $(-3,$ ☐ $)$

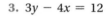

3. $3y - 4x = 12$

x	y	
		← x-intercept
		← y-intercept
		← Check point

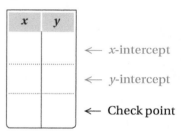

Answers

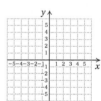

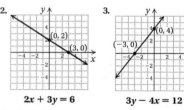

$2x + 3y = 6$ $3y - 4x = 12$

Guided Solution:
2. 3, 2, 4, 3, 2, 4

Graph.

4. $y = 2x$

x	y
-1	
0	
1	

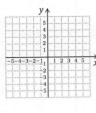

5. $y = -\dfrac{2}{3}x$

x	y

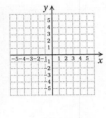

CALCULATOR CORNER

Viewing the Intercepts Knowing the intercepts of a linear equation helps us to determine a good viewing window for the graph of the equation. For example, when we graph the equation $y = -x + 15$ in the standard window, we see only a small portion of the graph in the upper right-hand corner of the screen, as shown on the left below.

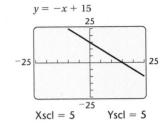

Using algebra, as we did in Example 1, we find that the intercepts of the graph of this equation are $(0, 15)$ and $(15, 0)$. This tells us that, if we are to see more of the graph than is shown on the left above, both Xmax and Ymax should be greater than 15. We can try different window settings until we find one that suits us. One good choice is $[-25, 25, -25, 25]$, with Xscl $= 5$ and Yscl $= 5$, shown on the right above.

EXERCISES: Find the intercepts of each equation algebraically. Then graph the equation on a graphing calculator, choosing window settings that allow the intercepts to be seen clearly. (Settings may vary.)

1. $y = -7.5x - 15$ 　　**2.** $y - 2.15x = 43$

3. $6x - 5y = 150$ 　　**4.** $y = 0.2x - 4$

5. $y = 1.5x - 15$ 　　**6.** $5x - 4y = 2$

b EQUATIONS WHOSE GRAPHS ARE HORIZONTAL LINES OR VERTICAL LINES

EXAMPLE 3 Graph: $y = 3$.

The equation $y = 3$ tells us that y must be 3, but it doesn't give us any information about x. We can also think of this equation as $0 \cdot x + y = 3$. No matter what number we choose for x, we find that y is 3. We make up a table with all 3's in the y-column.

x	y
	3
	3
	3

Choose any number for x. →

y must be 3.

x	y
-2	3
0	3
4	3

← y-intercept

Answers

4.

$y = 2x$

5.

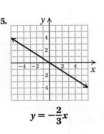

$y = -\dfrac{2}{3}x$

When we plot the ordered pairs $(-2, 3)$, $(0, 3)$, and $(4, 3)$ and connect the points, we obtain a horizontal line. Any ordered pair $(x, 3)$ is a solution. So the line is parallel to the x-axis with y-intercept $(0, 3)$.

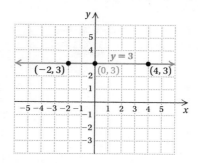

EXAMPLE 4   Graph: $x = -4$.

Consider $x = -4$. We can also think of this equation as $x + 0 \cdot y = -4$. We make up a table with all -4's in the x-column.

x	y
-4	
-4	
-4	
-4	

x must be -4.

x	y
-4	-5
-4	1
-4	3
-4	0

← Choose any number for y.

x-intercept →

When we plot the ordered pairs $(-4, -5)$, $(-4, 1)$, $(-4, 3)$, and $(-4, 0)$ and connect the points, we obtain a vertical line. Any ordered pair $(-4, y)$ is a solution. So the line is parallel to the y-axis with x-intercept $(-4, 0)$.

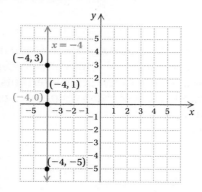

HORIZONTAL LINES AND VERTICAL LINES

The graph of $y = b$ is a **horizontal line**. The y-intercept is $(0, b)$.

The graph of $x = a$ is a **vertical line**. The x-intercept is $(a, 0)$.

Do Exercises 6–9. ▶

Graph.

 6. $x = 5$

x	y
5	-4
	0
	3

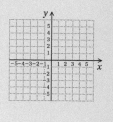

7. $y = -2$

x	y
-1	-2
0	
2	

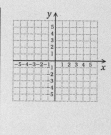

8. $x = -3$

x	y

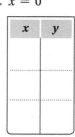

9. $x = 0$

x	y

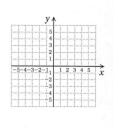

Answers

Answers to Margin Exercises 6–9 and Guided Solutions 6 and 7 are on p. 194.

The following is a general procedure for graphing linear equations.

GRAPHING LINEAR EQUATIONS

1. If the equation is of the type $x = a$ or $y = b$, the graph will be a line parallel to an axis; $x = a$ is vertical and $y = b$ is horizontal.

   Examples.

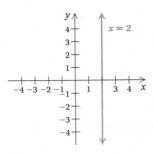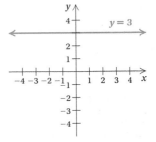

2. If the equation is of the type $y = mx$, both intercepts are the origin, $(0, 0)$. Plot $(0, 0)$ and two other points.

   Example.

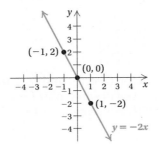

3. If the equation is of the type $y = mx + b$, plot the y-intercept $(0, b)$ and two other points.

   Example.

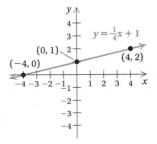

4. If the equation is of the type $Ax + By = C$, but not of the type $x = a$ or $y = b$, then either solve for y and proceed as with the equation $y = mx + b$, or graph using intercepts. If the intercepts are too close together, choose another point or points farther from the origin.

   Examples.

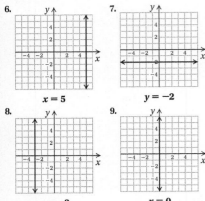

Visualizing
for Success

A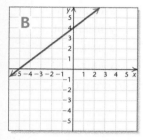

B

C

D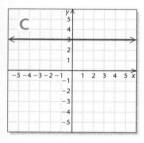

E

Match each equation with its graph.

1. $5y + 20 = 4x$

2. $y = 3$

3. $3x + 5y = 15$

4. $5y + 4x = 20$

5. $5y = 10 - 2x$

6. $4x + 5y + 20 = 0$

7. $5x - 4y = 20$

8. $4y + 5x + 20 = 0$

9. $5y - 4x = 20$

10. $x = -3$

Answers on page A-21

F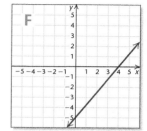

G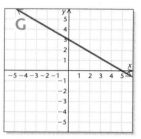

H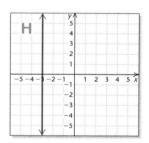

I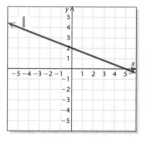

J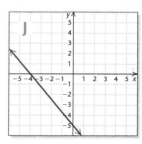

✓ Reading Check

Choose from the column on the right the word or the expression that best completes each statement. Not every choice will be used.

RC1. The graph of $y = -3$ is a(n) _____ line with $(0, -3)$ as its _____.

RC2. The x-intercept occurs when a line crosses the _____.

RC3. To find the x-intercept, let _____.

RC4. In the graph of $y = 2x$, the point _____ is both the x-intercept and the y-intercept.

RC5. To find the y-intercept, let _____.

RC6. The graph of $x = 4$ is a(n) _____ line with $(4, 0)$ as its _____.

$(0, 2)$
$(0, 0)$
horizontal
vertical
$x = 0$
$y = 0$
x-intercept
y-intercept
x-axis
y-axis

a For each of Exercises 1–4, find **(a)** the coordinates of the y-intercept and **(b)** the coordinates of the x-intercept.

1.
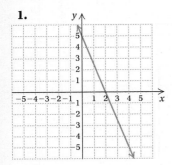

2.
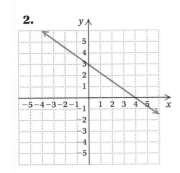

3.
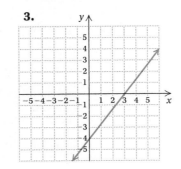

4.
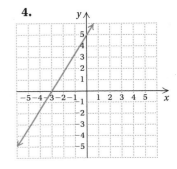

For each of Exercises 5–12, find **(a)** the coordinates of the y-intercept and **(b)** the coordinates of the x-intercept. Do not graph.

5. $3x + 5y = 15$

6. $5x + 2y = 20$

7. $7x - 2y = 28$

8. $3x - 4y = 24$

9. $-4x + 3y = 10$

10. $-2x + 3y = 7$

11. $6x - 3 = 9y$

12. $4y - 2 = 6x$

For each equation, find the intercepts. Then use the intercepts to graph the equation.

13. $x + 3y = 6$

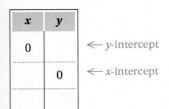

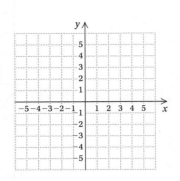

14. $x + 2y = 2$

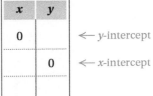

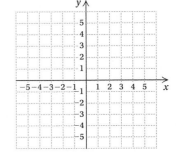

15. $-x + 2y = 4$

x	y
0	← y-intercept
	0 ← x-intercept

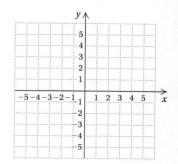

16. $-x + y = 5$

x	y
0	← y-intercept
	0 ← x-intercept

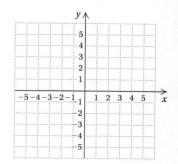

17. $3x + y = 6$

x	y
0	← y-intercept
	0 ← x-intercept

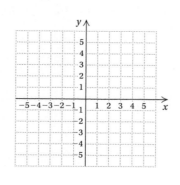

18. $2x + y = 6$

x	y
0	← y-intercept
	0 ← x-intercept

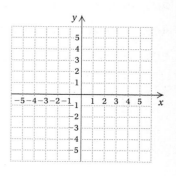

19. $2y - 2 = 6x$

x	y
	← y-intercept
	← x-intercept

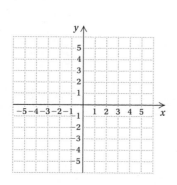

20. $3y - 6 = 9x$

x	y
	← y-intercept
	← x-intercept

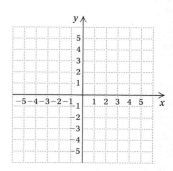

21. $3x - 9 = 3y$

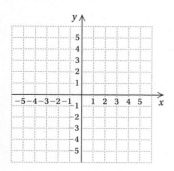

22. $5x - 10 = 5y$

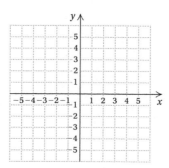

23. $2x - 3y = 6$

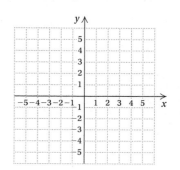

24. $2x - 5y = 10$

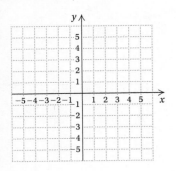

25. $4x + 5y = 20$

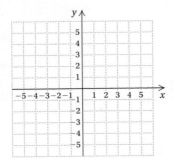

26. $2x + 6y = 12$

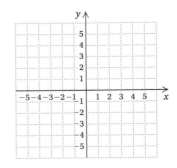

27. $2x + 3y = 8$

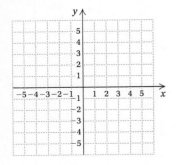

28. $x - 1 = y$

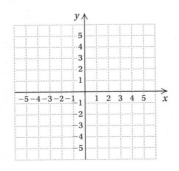

29. $3x + 4y = 5$

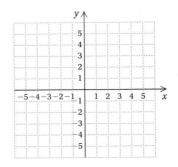

30. $2x - 1 = y$

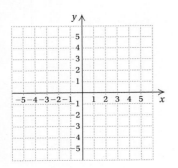

31. $3x - 2 = y$

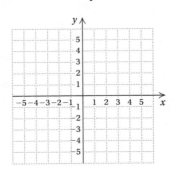

32. $4x - 3y = 12$

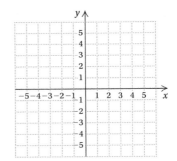

33. $6x - 2y = 12$

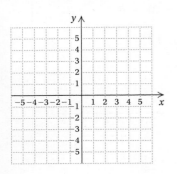

34. $7x + 2y = 6$

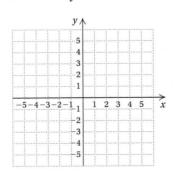

35. $y = -3 - 3x$

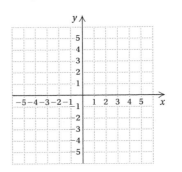

36. $-3x = 6y - 2$

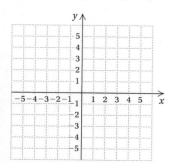

37. $y - 3x = 0$

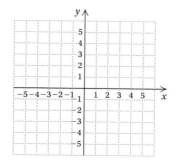

38. $x + 2y = 0$

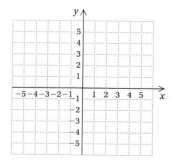

b Graph.

39. $x = -2$

x	y
-2	
-2	
-2	

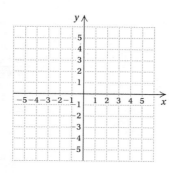

40. $x = 1$

x	y
1	
1	
1	

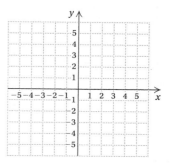

41. $y = 2$

x	y
	2
	2
	2

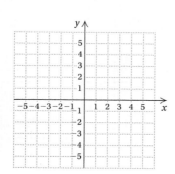

42. $y = -4$

x	y
	-4
	-4
	-4

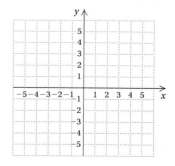

43. $x = 2$

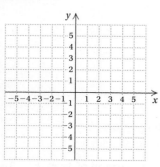

44. $x = 3$

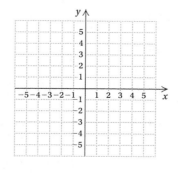

45. $y = 0$

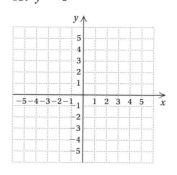

46. $y = -1$

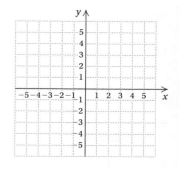

47. $x = \dfrac{3}{2}$

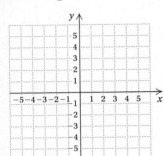

48. $x = -\dfrac{5}{2}$

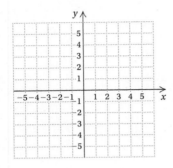

49. $3y = -5$

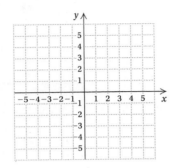

50. $12y = 45$

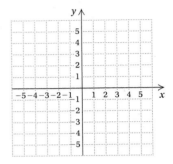

51. $4x + 3 = 0$

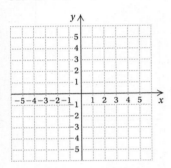

52. $-3x + 12 = 0$

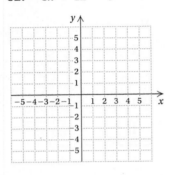

53. $48 - 3y = 0$

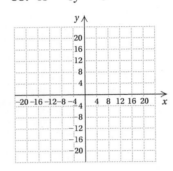

54. $63 + 7y = 0$

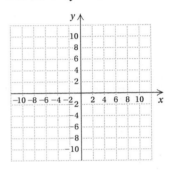

Write an equation for the graph shown.

55.

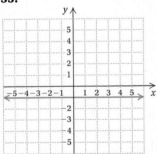

56.

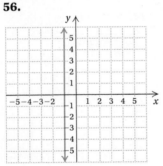

57.

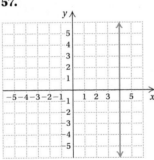

58.

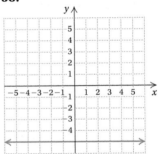

Skill Maintenance

Solve. [11.7e]

59. $x + (x - 1) < (x + 2) - (x + 1)$

60. $6 - 18x \leq 4 - 12x - 5x$

61. $\dfrac{2x}{7} - 4 \leq -2$

62. $\dfrac{1}{4} + \dfrac{x}{3} > \dfrac{7}{12}$

Synthesis

63. Write an equation of a line parallel to the x-axis and passing through $(-3, -4)$.

64. Find the value of m such that the graph of $y = mx + 6$ has an x-intercept of $(2, 0)$.

65. Find the value of k such that the graph of $3x + k = 5y$ has an x-intercept of $(-4, 0)$.

66. Find the value of k such that the graph of $4x = k - 3y$ has a y-intercept of $(0, -8)$.

Mid-Chapter Review

Concept Reinforcement

Determine whether each statement is true or false.

_____ **1.** In quadrant II, the first coordinate of all points is less than the second coordinate. [12.1a]

_____ **2.** The y-intercept of the graph of $2 - y = 3x$ is $(0, -3)$. [12.3a]

_____ **3.** The y-intercept of $Ax + By = C, B \neq 0$, is $\left(0, \dfrac{C}{B}\right)$. [12.3a]

_____ **4.** Both coordinates of points in quadrant IV are negative. [12.1a]

Guided Solutions

 5. Given the graph of the line below, fill in the letters and numbers that create correct statements. [12.3a]

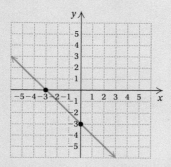

    a) The [ ]-intercept is ([ ] , −3).

    b) The [ ]-intercept is ([ ] , 0).

6. Given the graph of the line below, fill in the letters and numbers that create correct statements. [12.3a]

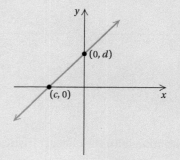

    a) The x-intercept is ([ ] , [ ]).

    b) The y-intercept is ([ ] , [ ]).

Mixed Review

7. Determine the coordinates of points A, B, C, D, and E. [12.1b]

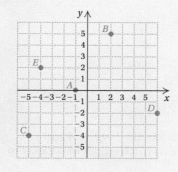

8. Determine the coordinates of points F, G, H, I, and J. [12.1b]

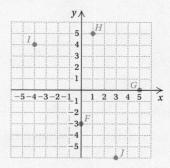

Determine whether the given ordered pair is a solution of the equation. [12.1c]

9. $(8, -5)$; $-2q - 7p = 19$

10. $\left(-1, \dfrac{2}{3}\right)$; $6y = -3x + 1$

Find the coordinates of the *x*-intercept and the *y*-intercept. [12.3a]

11. $-3x + 2y = 18$

12. $x - \dfrac{1}{2} = 10y$

13. $x - 40 = 20y$

14. $\dfrac{1}{3}y - \dfrac{5}{6}x = 35$

Graph. [12.2a], [12.3a, b]

15. $-2x + y = -3$

16. $y = -\dfrac{3}{2}$

17. $y = -x + 4$

18. $x = 0$

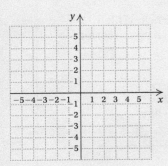
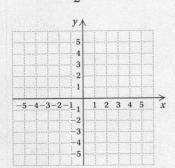
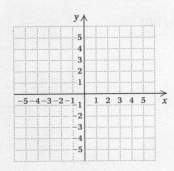
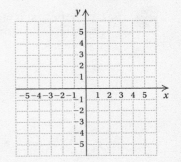

Match each equation with the characteristics listed on the right. [12.3a, b]

19. $y = -1$

20. $x = 1$

21. $y = -x - 1$

22. $y = x - 1$

23. $y = x + 1$

A. The *y*-intercept is $(0, 1)$ and the *x*-intercept is $(-1, 0)$.
B. The *x*-intercept is $(-1, 0)$ and the *y*-intercept is $(0, -1)$.
C. The line is vertical and the *x*-intercept is $(1, 0)$.
D. The line is horizontal and the *y*-intercept is $(0, -1)$.
E. The *y*-intercept is $(0, -1)$ and the *x*-intercept is $(1, 0)$.

Understanding Through Discussion and Writing

24. Do all graphs of linear equations have *y*-intercepts? Why or why not? [12.3b]

25. The equations $3x + 4y = 8$ and $y = -\frac{3}{4}x + 2$ are equivalent. Which equation is easier to graph and why? [12.2a], [12.3a]

26. If the graph of the equation $Ax + By = C$ is a horizontal line, what can you conclude about *A*? Why? [12.3b]

27. Explain in your own words why the graph of $x = 7$ is a vertical line. [12.3b]

Slope and Applications

12.4

a SLOPE

OBJECTIVES

a Given the coordinates of two points on a line, find the slope of the line, if it exists.

b Find the slope of a line from an equation.

c Find the slope, or rate of change, in an applied problem involving slope.

We have considered two forms of a linear equation, $Ax + By = C$ and $y = mx + b$. We found that from the form of the equation $y = mx + b$, we know that the y-intercept of the line is $(0, b)$.

$$y = mx + b.$$
$$? \qquad \longrightarrow \text{The } y\text{-intercept is } (0, b).$$

What about the constant m? Does it give us information about the line? Look at the graphs in the margin and see if you can make any connection between the constant m and the "slant" of the line.

The graphs of some linear equations slant upward from left to right. Others slant downward. Some are vertical and some are horizontal. Some slant more steeply than others. We now look for a way to describe such possibilities with numbers.

Consider a line with two points marked P and Q. As we move from P to Q, the y-coordinate changes from 1 to 3 and the x-coordinate changes from 2 to 6. The change in y is $3 - 1$, or 2. The change in x is $6 - 2$, or 4.

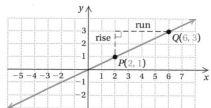

SKILL TO REVIEW

Objective 10.4a: Subtract real numbers.

Subtract.
1. $-4 - 20$
2. $-21 - (-5)$

We call the change in y the **rise** and the change in x the **run**. The ratio rise/run is the same for any two points on a line. We call this ratio the **slope** of the line. Slope describes the slant of a line. The slope of the line in the graph above is given by

$$\frac{\text{rise}}{\text{run}} = \frac{\text{the change in } y}{\text{the change in } x}, \text{ or } \frac{2}{4}, \text{ or } \frac{1}{2}.$$

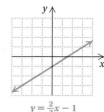

$y = \frac{2}{3}x - 1$

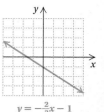

$y = -\frac{2}{3}x - 1$

SLOPE

The **slope** of a line containing points (x_1, y_1) and (x_2, y_2) is given by

$$m = \frac{\text{rise}}{\text{run}} = \frac{\text{the change in } y}{\text{the change in } x} = \frac{y_2 - y_1}{x_2 - x_1}.$$

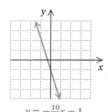

$y = -\frac{10}{3}x - 1$

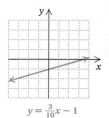

$y = \frac{3}{10}x - 1$

Answers

Skill to Review:
1. -24 2. -16

In the preceding definition, (x_1, y_1) and (x_2, y_2)—read "x sub-one, y sub-one and x sub-two, y sub-two"—represent two different points on a line. It does not matter which point is considered (x_1, y_1) and which is considered (x_2, y_2) so long as coordinates are subtracted in the same order in both the numerator and the denominator:

$$\frac{y_2 - y_1}{x_2 - x_1} = \frac{y_1 - y_2}{x_1 - x_2}.$$

EXAMPLE 1 Graph the line containing the points $(-4, 3)$ and $(2, -6)$ and find the slope.

The graph is shown below. We consider (x_1, y_1) to be $(-4, 3)$ and (x_2, y_2) to be $(2, -6)$. From $(-4, 3)$ and $(2, -6)$, we see that the change in y, or the rise, is $-6 - 3$, or -9. The change in x, or the run, is $2 - (-4)$, or 6.

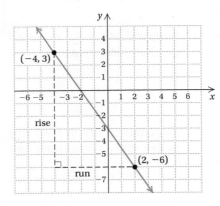

$$\text{Slope} = \frac{\text{rise}}{\text{run}} = \frac{\text{change in } y}{\text{change in } x}$$
$$= \frac{y_2 - y_1}{x_2 - x_1}$$
$$= \frac{-6 - 3}{2 - (-4)}$$
$$= \frac{-9}{6} = -\frac{9}{6}, \text{ or } -\frac{3}{2}$$

When we use the formula

$$m = \frac{y_2 - y_1}{x_2 - x_1},$$

we must remember to subtract the x-coordinates in the same order that we subtract the y-coordinates. **Let's redo Example 1, where we consider (x_1, y_1)** to be $(2, -6)$ and (x_2, y_2) to be $(-4, 3)$:

$$\text{Slope} = \frac{\text{change in } y}{\text{change in } x} = \frac{3 - (-6)}{-4 - 2} = \frac{9}{-6} = -\frac{9}{6} = -\frac{3}{2}.$$

◀ **Do Exercises 1 and 2.**

The slope of a line tells how it slants. A line with positive slope slants up from left to right. The larger the slope, the steeper the slant. A line with negative slope slants downward from left to right.

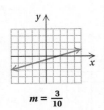 $m = \frac{3}{10}$

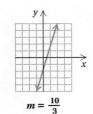 $m = \frac{10}{3}$

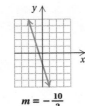 $m = -\frac{10}{3}$

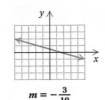 $m = -\frac{3}{10}$

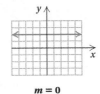 $m = 0$

 m is not defined.

Later in this section, in Examples 7 and 8, we will discuss the slope of a horizontal line and of a vertical line.

Graph the line containing the points and find the slope in two different ways.

1. $(-2, 3)$ and $(3, 5)$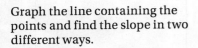

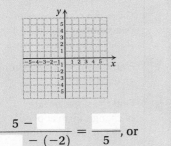

$$\frac{5 - \boxed{}}{\boxed{} - (-2)} = \frac{\boxed{}}{5}, \text{ or}$$

$$\frac{3 - \boxed{}}{\boxed{} - 3} = \frac{-2}{\boxed{}} = \frac{2}{\boxed{}}$$

2. $(0, -3)$ and $(-3, 2)$

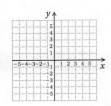

Answers

Answers to Margin Exercises 1 and 2 are on p. 205.

b FINDING THE SLOPE FROM AN EQUATION

It is possible to find the slope of a line from its equation. Let's consider the equation $y = 2x + 3$, which is in the form $y = mx + b$. The graph of this equation is shown at right. We can find two points by choosing convenient values for x—say, 0 and 1—and substituting to find the corresponding y-values. We find the two points on the line to be $(0, 3)$ and $(1, 5)$. The slope of the line is found using the definition of slope:

$$m = \frac{\text{change in } y}{\text{change in } x} = \frac{5 - 3}{1 - 0} = \frac{2}{1} = 2.$$

The slope is 2. Note that this is also the coefficient of the x-term in the equation $y = 2x + 3$.

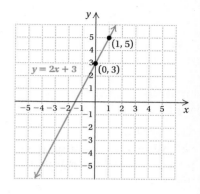

DETERMINING SLOPE FROM THE EQUATION $y = mx + b$

The slope of the line $y = mx + b$ is m. To find the slope of a non-vertical line, solve the linear equation in x and y for y and get the resulting equation in the form $y = mx + b$. The coefficient of the x-term, m, is the slope of the line.

EXAMPLES Find the slope of each line.

2. $y = -3x + \dfrac{2}{9}$

$\quad\quad\quad \longrightarrow m = -3 = \text{Slope}$

3. $y = \dfrac{4}{5}x$

$\quad\quad\quad \longrightarrow m = \dfrac{4}{5} = \text{Slope}$

4. $y = x + 6$

$\quad\quad\quad \longrightarrow m = 1 = \text{Slope}$

5. $y = -0.6x - 3.5$

$\quad\quad\quad \longrightarrow m = -0.6 = \text{Slope}$

Do Exercises 3–6. ▶

To find slope from an equation, we may need to first find an equivalent form of the equation.

EXAMPLE 6 Find the slope of the line $2x + 3y = 7$.

We solve for y to get the equation in the form $y = mx + b$:

$$2x + 3y = 7$$
$$3y = -2x + 7$$
$$y = \frac{1}{3}(-2x + 7)$$
$$y = -\frac{2}{3}x + \frac{7}{3}. \quad \text{This is } y = mx + b.$$

The slope is $-\frac{2}{3}$.

Do Exercises 7 and 8. ▶

Find the slope of each line.

3. $y = 4x + 11$

4. $y = -17x + 8$

5. $y = -x + \dfrac{1}{2}$

6. $y = \dfrac{2}{3}x - 1$

Find the slope of each line.

7. $4x + 4y = 7$

GS **8.** $5x - 4y = 8$

$$5x = \boxed{} + 8$$
$$5x - \boxed{} = 4y$$
$$\frac{5x - 8}{\boxed{}} = \frac{4y}{4}$$
$$\boxed{} \cdot x - 2 = y, \text{ or}$$
$$y = \boxed{} \cdot x - 2$$
$$\downarrow$$
$$\text{Slope is } \boxed{}.$$

Answers

1. $\dfrac{2}{5}$ 2. $-\dfrac{5}{3}$

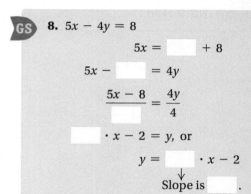

3. 4 4. -17 5. -1 6. $\dfrac{2}{3}$

7. -1 8. $\dfrac{5}{4}$

Guided Solutions

1. 3, 3, 2; 5, -2, -5, 5 8. $4y$, 8, 4, $\dfrac{5}{4}, \dfrac{5}{4}, \dfrac{5}{4}$

What about the slope of a horizontal line or a vertical line?

EXAMPLE 7 Find the slope of the line $y = 5$.

We can think of $y = 5$ as $y = 0x + 5$. Then from this equation, we see that $m = 0$. Consider the points $(-3, 5)$ and $(4, 5)$, which are on the line. The change in $y = 5 - 5$, or 0. The change in $x = -3 - 4$, or -7. We have

$$m = \frac{5 - 5}{-3 - 4}$$

$$= \frac{0}{-7}$$

$$= 0.$$

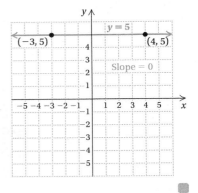

Any two points on a horizontal line have the same y-coordinate. The change in y is 0. Thus the slope of a horizontal line is 0.

EXAMPLE 8 Find the slope of the line $x = -4$.

Consider the points $(-4, 3)$ and $(-4, -2)$, which are on the line. The change in $y = 3 - (-2)$, or 5. The change in $x = -4 - (-4)$, or 0. We have

$$m = \frac{3 - (-2)}{-4 - (-4)}$$

$$= \frac{5}{0}. \qquad \text{Not defined}$$

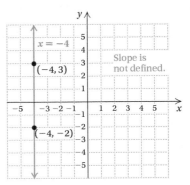

Since division by 0 is not defined, the slope of this line is not defined. The answer in this example is "The slope of this line is not defined."

SLOPE 0; SLOPE NOT DEFINED

The slope of a horizontal line is 0.

The slope of a vertical line is not defined.

◀ **Do Exercises 9 and 10.**

Find the slope, if it exists, of each line.

9. $x = 7$

10. $y = -5$

Answers

9. Not defined **10.** 0

c APPLICATIONS OF SLOPE; RATES OF CHANGE

Slope has many real-world applications. For example, numbers like 2%, 3%, and 6% are often used to represent the *grade* of a road, a measure of how steep a road on a hill or mountain is. For example, a 3% grade ($3\% = \frac{3}{100}$) means that for every horizontal distance of 100 ft, the road rises 3 ft, and a −3% grade means that for every horizontal distance of 100 ft, the road drops 3 ft. (Road signs do not include negative signs.)

Road grade = $\frac{a}{b}$
(expressed as a percent)

   The concept of grade also occurs in skiing or snowboarding, where a 7% grade is considered very tame, but a 70% grade is considered extremely steep.

EXAMPLE 9 *Dubai Ski Run.* Dubai Ski Resort has the fifth longest indoor ski run in the world. It drops 197 ft over a horizontal distance of 1297 ft. Find the grade of the ski run.

197 ft

1297 ft

The grade of the ski run is its slope, expressed as a percent:

$$m = \frac{197}{1297} \quad \begin{array}{l} \leftarrow \text{Vertical distance} \\ \leftarrow \text{Horizontal distance} \end{array}$$

$$\approx 0.15$$

$$\approx 15\%$$

Do Exercise 11. ▶

11. *Grade of a Treadmill.* During a stress test, a physician may change the grade, or slope, of a treadmill to measure its effect on heart rate (number of beats per minute). Find the grade, or slope, of the treadmill shown below.

0.4 ft

5 ft

Answer

11. 8%

Slope can also be considered as a **rate of change**.

EXAMPLE 10 *Car Assembly Line.* Cameron, a supervisor in a car assembly plant, prepared the following graph to display data from a recent day's work. Use the graph to determine the slope, or the rate of change of the number of cars that came off an assembly line with respect to time.

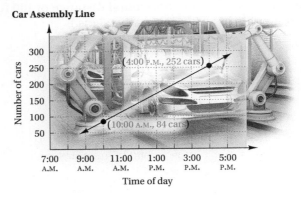

Car Assembly Line

The vertical axis of the graph shows the number of cars, and the horizontal axis shows the time, in units of one hour. We can describe the rate of change of the number of cars with respect to time as

$$\frac{\text{Cars}}{\text{Hours}}, \quad \text{or} \quad \text{number of cars per hour.}$$

This value is the slope of the line. We determine two ordered pairs on the graph—in this case,

$$(10{:}00 \text{ A.M., } 84 \text{ cars}) \quad \text{and} \quad (4{:}00 \text{ P.M., } 252 \text{ cars}).$$

This tells us that in the 6 hr between 10:00 A.M. and 4:00 P.M., $252 - 84$, or 168, cars came off the assembly line. Thus,

$$\text{Rate of change} = \frac{252 \text{ cars } - 84 \text{ cars}}{4{:}00 \text{ P.M. } - 10{:}00 \text{ A.M.}}$$
$$= \frac{168 \text{ cars}}{6 \text{ hours}}$$
$$= 28 \text{ cars per hour.}$$

◀ **Do Exercise 12.**

12. *Masonry.* Daryl, a mason, graphed data from a recent day's work. Use the following graph to determine the slope, or the rate of change of the number of bricks he can lay with respect to time.

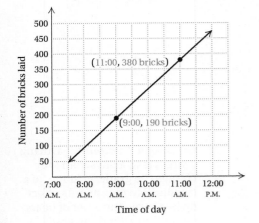

EXAMPLE 11 *Advertising Revenue.* Print-newspaper advertising revenue has been decreasing since 2005. Use the following graph to determine the slope, or rate of change in the advertising revenue with respect to time.

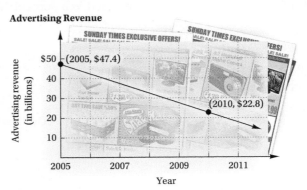

Advertising Revenue

SOURCE: Research Department, Newspaper Association of America

Answer

12. 95 bricks per hour

768 CHAPTER 12 Graphs of Linear Equations

The vertical axis of the graph shows the advertising revenue, in billions of dollars, and the horizontal axis shows the years. We can describe the rate of change in the advertising revenue with respect to time as

$$\frac{\text{Change in advertising revenue}}{\text{Years}}, \quad \text{or} \quad \text{change in advertising revenue per year.}$$

This value is the slope of the line. We determine two ordered pairs on the graph—in this case,

$$(2005, \$47.4) \quad \text{and} \quad (2010, \$22.8).$$

This tells us that in the 5 years from 2005 to 2010, newspaper advertising revenue dropped from $47.4 billion to $22.8 billion. Thus,

$$\text{Rate of change} = \frac{\$22.8 - \$47.4}{2010 - 2005}$$

$$= \frac{-\$24.6}{5} \approx -\$4.9 \text{ billion per year.}$$

Do Exercise 13. ▶

13. *Sunday Newspapers.* Use the following graph to determine the rate of change in the circulation of Sunday newspapers since 2005.

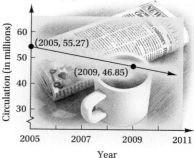

Sunday Newspaper

SOURCE: *Editor & Publisher International Yearbook,* 2010

Answer

13. About −2.11 million Sunday newspapers per year

For Extra Help

MyMathLab®   MathXL®   PRACTICE   WATCH   READ   REVIEW

12.4 | Exercise Set

✓ Reading Check

Match each expression with an appropriate description or value from the column on the right.

RC1. Slope of a horizontal line

RC2. y-intercept of $y = mx + b$

RC3. Change in x

RC4. Slope of a vertical line

RC5. Slope

RC6. Change in y

a) Rise
b) Run
c) Rise/run
d) 0
e) Not defined
f) $(0, b)$

---

a   Find the slope, if it exists, of each line.

1.

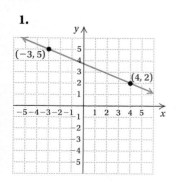

$(-3, 5)$ $(4, 2)$

2.

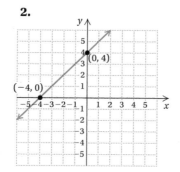

$(0, 4)$ $(-4, 0)$

3.

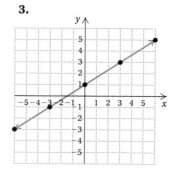

4.

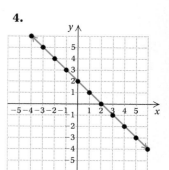

5.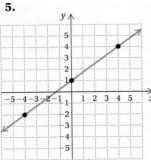

6.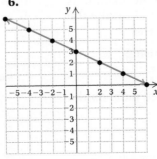

7.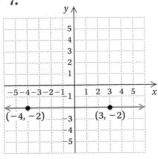

8.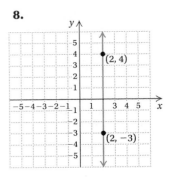

Graph the line containing the given pair of points and find the slope.

9. $(-2, 4), (3, 0)$

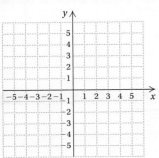

10. $(2, -4), (-3, 2)$

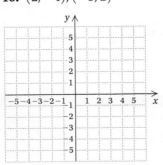

11. $(-4, 0), (-5, -3)$

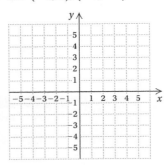

12. $(-3, 0), (-5, -2)$

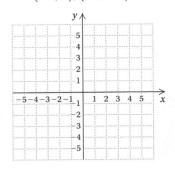

13. $(-4, 1), (2, -3)$

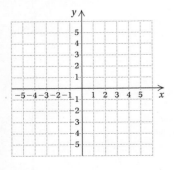

14. $(-3, 5), (4, -3)$

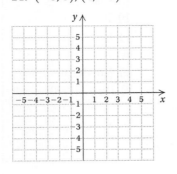

15. $(5, 3), (-3, -4)$

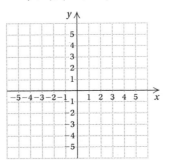

16. $(-4, -3), (2, 5)$

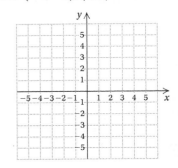

Find the slope, if it exists, of the line containing the given pair of points.

17. $\left(2, -\frac{1}{2}\right), \left(5, \frac{3}{2}\right)$

18. $\left(\frac{2}{3}, -1\right), \left(\frac{5}{3}, 2\right)$

19. $(4, -2), (4, 3)$

20. $(4, -3), (-2, -3)$

21. $(-11, 7), (15, -3)$

22. $(-13, 22), (8, -17)$

23. $\left(-\frac{1}{2}, \frac{3}{11}\right), \left(\frac{5}{4}, \frac{3}{11}\right)$

24. $(0.2, 4), (0.2, -0.04)$

b Find the slope, if it exists, of each line.

25. $y = -10x$

26. $y = \frac{10}{3}x$

27. $y = 3.78x - 4$

28. $y = -\frac{3}{5}x + 28$

29. $3x - y = 4$

30. $-2x + y = 8$

31. $x + 5y = 10$

32. $x - 4y = 8$

33. $3x + 2y = 6$

34. $2x - 4y = 8$

35. $x = \frac{2}{15}$

36. $y = -\frac{1}{3}$

37. $y = 2 - x$

38. $y = \frac{3}{4} + x$

39. $9x = 3y + 5$

40. $4y = 9x - 7$

41. $5x - 4y + 12 = 0$

42. $16 + 2x - 8y = 0$

43. $y = 4$

44. $x = -3$

45. $x = \frac{3}{4}y - 2$

46. $3x - \frac{1}{5}y = -4$

47. $\frac{2}{3}y = -\frac{7}{4}x$

48. $-x = \frac{2}{11}y$

c In each of Exercises 49–52, find the slope (or rate of change).

49. Find the slope (or pitch) of the roof.

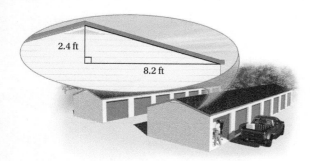

2.4 ft

8.2 ft

50. Find the slope (or grade) of the road.

148.8 m

2400 m

51. *Slope of a River.* When a river flows, the strength or force of the river depends on how far the river falls vertically compared to how far it flows horizontally. Find the slope of the river shown below.

56 ft

258 ft

52. *Constructing Stairs.* Carpenters use slope when designing and building stairs. Public buildings normally include steps with 7-in. risers and 11-in. treads. Find the grade of such a stairway.

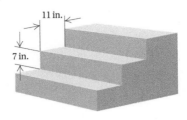

11 in.

7 in.

53. *Grade of a Transit System.* The maximum grade allowed between two stations in a rapid-transit rail system is 3.5%. Between station A and station B, which are 280 ft apart, the tracks rise $8\frac{1}{2}$ ft. What is the grade of the tracks between these two stations? Round the answer to the nearest tenth of a percent. Does this grade meet the rapid-transit rail standards?

Source: Brian Burell, *Merriam Webster's Guide to Everyday Math*, Merriam-Webster, Inc., Springfield MA

54. *Slope of Long's Peak.* From a base elevation of 9600 ft, Long's Peak in Colorado rises to a summit elevation of 14,256 ft over a horizontal distance of 15,840 ft. Find the grade of Long's Peak.

In each of Exercises 55–58, use the graph to calculate a rate of change in which the units of the horizontal axis are used in the denominator.

55. *Kindergarten in China.* The number of children enrolled in kindergarten in China is projected to increase from 26.58 million in 2009 to 34 million in 2015. Use the following graph to find the rate of change, rounded to the nearest hundredth of a million, in the number of children enrolled in kindergarten in China with respect to time.

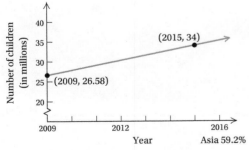

SOURCE: *China Daily*, Lin Qi and Guo Shuhan report, June 23, 2011, p.18

56. *Injuries on Farms.* In 1998, there were 37,775 injuries on farms to people under age 20. Since then, this number has steadily decreased. Using the following graph, find the rate of change, rounded to the nearest ten, in the number of farm injuries to people under age 20 with respect to time.

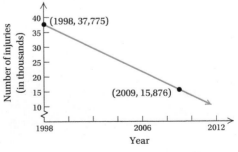

SOURCE: National Institute for Occupational Health and Safety

57. *Population Decrease of New Orleans.* The change in the population of New Orleans, Louisiana, is illustrated in the following graph. Find the rate of change, to the nearest hundred, in the population with respect to time.

Population Decrease of New Orleans

(2000, 484,949)
New Orleans
Louisiana
(2010, 343,829)

Population (in thousands): 100, 200, 300, 400, 500
Year: 2000, 2006, 2010

SOURCE: Decennial Census, U. S. Census Bureau

58. *Population Increase of Charlotte.* The change in the population of Charlotte, North Carolina, is illustrated in the following graph. Find the rate of change, to the nearest hundred, in the population with respect to time.

Population Increase of Charlotte

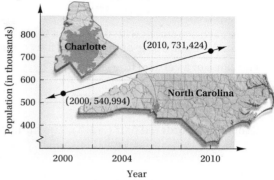

Charlotte
(2010, 731,424)
North Carolina
(2000, 540,994)

Population (in thousands): 400, 500, 600, 700, 800
Year: 2000, 2004, 2010

SOURCE: Decennial Census, U. S. Census Bureau

59. *Production of Blueberries.* U.S. production of blueberries is continually increasing. In 2006, 358,000,000 lb of blueberries were produced. By 2011, this amount had increased to 511,000,000 lb. Find the rate of change in the production of blueberries with respect to time.

Source: U.S. Department of Agriculture

60. *Bottled Water.* Bottled water consumption per person per year in the United States has increased from 16.7 gal in 2000 to 29.2 gal in 2011. Find the rate of change, rounded to the nearest tenth, in the number of gallons of bottled water consumed annually per person per year.

Sources: Beverage Marketing Corporation; International Bottled Water Association

Skill Maintenance

Solve.   [11.3a]

61. $2x - 11 = 4$

62. $5 - \frac{1}{2}x = 11$

Collect like terms.   [10.7e]

63. $\frac{1}{3}p - p$

64. $t - 6 + 4t + 5$

Synthesis

In each of Exercises 65–68, find an equation for the graph shown.

65.

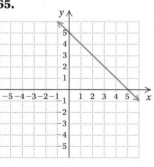

66.

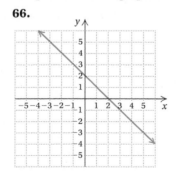

67.

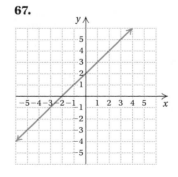

68.

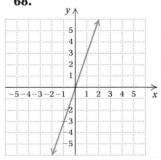

Vocabulary Reinforcement

Complete each statement with the correct term or expression from the column on the right. Some of the choices may not be used.

1. The slope of a vertical line is _____. [12.4b]

2. The graph of $y = b$ is a(n) _____ line. The y-intercept is _____. [12.3b]

3. Consider the ordered pair $(-5, 3)$. The numbers -5 and 3 are called _____. [12.1a]

4. The _____ occurs when a line crosses the y-axis and thus will always have 0 as the first coordinate. [12.3a]

5. The graph of $x = a$ is a(n) _____ line. The x-intercept is _____. [12.3b]

6. The slope of a horizontal line is _____. [12.4b]

$(0, b)$

$(b, 0)$

$(a, 0)$

$(0, a)$

0

not defined

x-intercept

y-intercept

vertical

horizontal

coordinates

axes

Concept Reinforcement

Determine whether each statement is true or false.

_____ 1. The x- and y-intercepts of $y = mx$ are both $(0, 0)$. [12.3a]

_____ 2. A slope of $-\frac{3}{4}$ is steeper than a slope of $-\frac{5}{2}$. [12.4a]

_____ 3. The slope of the line that passes through (a, b) and (c, d) is $\frac{d - b}{c - a}$. [12.4a]

_____ 4. The second coordinate of all points in quadrant III is negative. [12.1a]

_____ 5. The x-intercept of $Ax + By = C$, $C \neq 0$, is $\left(\frac{A}{C}, 0\right)$. [12.3a]

_____ 6. The slope of the line that passes through $(0, t)$ and $(-t, 0)$ is $\frac{1}{t}$. [12.4a]

Study Guide

Objective 12.1b Find the coordinates of a point on a graph.

Example Find the coordinates of points Q, R, and S.

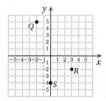

Point Q is 2 units left of the origin and 5 units up. Its coordinates are $(-2, 5)$.

Point R is 3 units right of the origin and 2 units down. Its coordinates are $(3, -2)$.

Point S is 0 units left or right of the origin and 4 units down. Its coordinates are $(0, -4)$.

Practice Exercise

1. Find the coordinates of points F, G, and H.

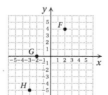

Objective 12.2a   Graph linear equations of the type $y = mx + b$ and $Ax + By = C$, identifying the y-intercept.

Example   Graph $2y + 2 = -3x$ and identify the y-intercept.

    To find an equivalent equation in the form $y = mx + b$, we solve for y: $y = -\frac{3}{2}x - 1$. The y-intercept is $(0, -1)$.

    We then find two other points using multiples of 2 for x to avoid fractions.

x	y
0	-1	$\leftarrow$ y-intercept
-2	2
2	-4

Practice Exercise

2. Graph $x + 2y = 8$ and identify the y-intercept.

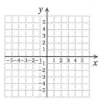

Objective 12.3a   Find the intercepts of a linear equation, and graph using intercepts.

Example   For $2x - y = -6$, find the intercepts. Then use the intercepts to graph the equation.

    To find the y-intercept, we let $x = 0$ and solve for y:

$$2 \cdot 0 - y = -6 \quad \text{and} \quad y = 6.$$

The y-intercept is $(0, 6)$.

    To find the x-intercept, we let $y = 0$ and solve for x:

$$2x - 0 = -6 \quad \text{and} \quad x = -3.$$

The x-intercept is $(-3, 0)$.

    We find a third point as a check.

x	y
0	6	$\leftarrow$ y-intercept
-3	0	$\leftarrow$ x-intercept
-1	4

Practice Exercise

3. For $y - 2x = -4$, find the intercepts. Then use the intercepts to graph the equation.

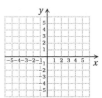

Objective 12.3b   Graph equations equivalent to those of the type $x = a$ and $y = b$.

Example   Graph: $y = 1$ and $x = -\dfrac{3}{2}$.

    For $y = 1$, no matter what number we choose for x, $y = 1$. The graph is a horizontal line. For $x = -\frac{3}{2}$, no matter what number we choose for y, $x = -\frac{3}{2}$. The graph is a vertical line.

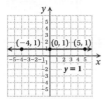 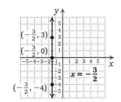

Practice Exercises

Graph.

4. $y = -\dfrac{5}{2}$ **5.** $x = 2$

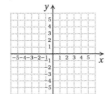

Objective 12.4a Given the coordinates of two points on a line, find the slope of the line, if it exists.

Example Find the slope, if it exists, of the line containing the given points.

$(-9, 3)$ and $(5, -6)$: $m = \dfrac{-6 - 3}{5 - (-9)} = \dfrac{-9}{14} = -\dfrac{9}{14}$;

$\left(7, \dfrac{1}{2}\right)$ and $\left(-13, \dfrac{1}{2}\right)$: $m = \dfrac{\frac{1}{2} - \frac{1}{2}}{-13 - 7} = \dfrac{0}{-20} = 0$;

$(0.6, 1.5)$ and $(0.6, -1.5)$: $m = \dfrac{-1.5 - 1.5}{0.6 - 0.6} = \dfrac{-3}{0}$,

m is not defined.

Practice Exercises

Find the slope, if it exists, of the line containing the given points.

6. $(-8, 20), (-8, 14)$

7. $(2, -1), (16, 20)$

8. $(0.5, 2.8), (1.5, 2.8)$

Objective 12.4b Find the slope of a line from an equation.

Example Find the slope, if it exists, of each line.

a) $5x - 20y = -10$

We first solve for y: $y = \frac{1}{4}x + \frac{1}{2}$. The slope is $\frac{1}{4}$.

b) $y = -\frac{4}{5}$

Think: $y = 0 \cdot x - \frac{4}{5}$. This line is horizontal. The slope is 0.

c) $x = 6$

This line is vertical. The slope is not defined.

Practice Exercises

Find the slope, if it exists, of the line.

9. $x = 0.25$

10. $7y + 14x = -28$

11. $y = -5$

Objective 12.4c Find the slope, or rate of change, in an applied problem involving slope.

Example In 2000, the population of Cincinnati, Ohio, was approximately 331,410. By 2010, the population had decreased to 296,943. Find the rate of change, to the nearest hundred, in the population with respect to time.

Source: U.S. Census Bureau

Rate of change $= \dfrac{296{,}943 - 331{,}410}{2010 - 2000}$

$= \dfrac{-34{,}467}{10} \approx -3400$ people per year

Practice Example

12. In 2000, the population of Idaho, was 1,293,953. By 2010, the population had increased to 1,567,582. Find the rate of change, to the nearest hundred, in the population with respect to time.

Source: U.S. Census Bureau

Review Exercises

Plot each point. [12.1a]

1. $(2, 5)$

2. $(0, -3)$

3. $(-4, -2)$

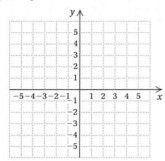

Find the coordinates of each point. [12.1b]

4. A

5. B

6. C

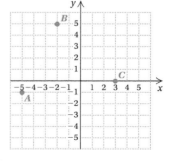

In which quadrant is each point located? [12.1a]

7. $(3, -8)$ **8.** $(-20, -14)$ **9.** $(4.9, 1.3)$

Determine whether each ordered pair is a solution of $2y - x = 10$. [12.1c]

10. $(2, -6)$ **11.** $(0, 5)$

12. Show that the ordered pairs $(0, -3)$ and $(2, 1)$ are solutions of the equation $2x - y = 3$. Then use the graph of the equation to determine another solution. Answers may vary. [12.1c]

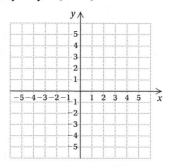

Graph each equation, identifying the y-intercept. [12.2a]

13. $y = 2x - 5$

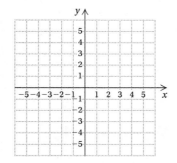

14. $y = -\dfrac{3}{4}x$

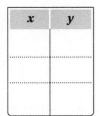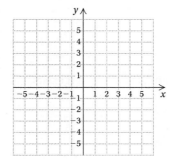

15. $y = -x + 4$

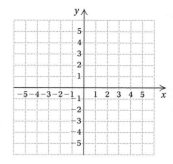

16. $y = 3 - 4x$

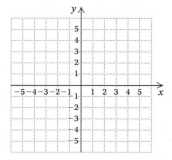

Solve. [12.2b]

17. *Kitchen Design.* Kitchen designers recommend that a refrigerator be selected on the basis of the number of people n in the household. The appropriate size S, in cubic feet, is given by

$$S = \frac{3}{2}n + 13.$$

a) Determine the recommended size of a refrigerator if the number of people is 1, 2, 5, and 10.

b) Graph the equation and use the graph to estimate the recommended size of a refrigerator for 4 people sharing an apartment.

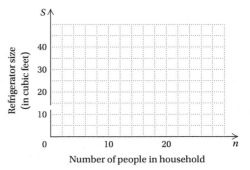

c) A refrigerator is 22 ft^3. For how many residents is it the recommended size?

Find the intercepts of each equation. Then graph the equation. [12.3a]

18. $x - 2y = 6$

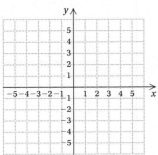

19. $5x - 2y = 10$

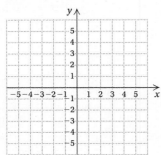

Graph each equation. [12.3b]

20. $y = 3$

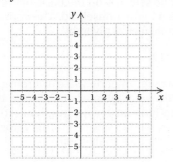

21. $5x - 4 = 0$

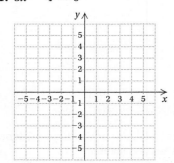

Find the slope. [12.4a]

22.

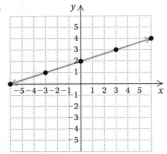

23.

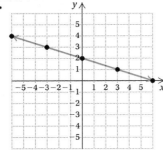

Graph the line containing the given pair of points and find the slope. [12.4a]

24. $(-5, -2), (5, 4)$

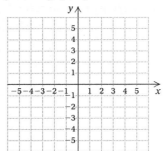

25. $(-5, 5), (4, -4)$

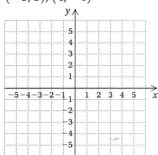

Find the slope, if it exists. [12.4b]

26. $y = -\dfrac{5}{8}x - 3$

27. $2x - 4y = 8$

28. $x = -2$

29. $y = 9$

30. $x - 10y = 20$

31. $6x - 5 = 4y$

32. *Snow Removal.* By 3:00 P.M., Erin had plowed 7 driveways and by 5:30 P.M., she had completed 13 driveways. [12.4c]
  a) Find Erin's plowing rate, in number of driveways per hour.
  b) Find Erin's plowing rate, in number of minutes per driveway.

33. *Road Grade.* At one point, Beartooth Highway in Yellowstone National Park rises 315 ft over a horizontal distance of 4500 ft. Find the slope, or grade, of the road. [12.4c]

34. *Organic Food.* Each year in the United States, the amount of sales in the organic food industry increases. Use the following graph to determine the slope, or rate of change, in the amount of sales, in billions of dollars, of organic food products with respect to time. [12.4a]

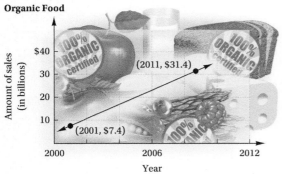

Organic Food

(2011, $31.4)

(2001, $7.4)

SOURCE: Organic Trade Association

35. Which of the following is the x-intercept of the graph of $5x - y = -15$? [12.3a]
  A. $(0, -3)$ **B.** $(-15, 0)$
  C. $(0, 15)$ **D.** $(-3, 0)$

36. What is the slope, if it exists, of the line containing the points $(-8, 8)$ and $(8, -8)$? [12.4a]
  A. 1 **B.** 0
  C. -1 **D.** Not defined

Synthesis

37. Find the area and the perimeter of a rectangle for which $(-2, 2)$, $(7, 2)$, and $(7, -3)$ are three of the vertices. [12.1a]

38. *Gondola Aerial Lift.* In Telluride, Colorado, there is a free gondola ride that provides a spectacular view of the town and the surrounding mountains. The gondolas that begin in the town at an elevation of 8725 ft travel 5750 ft to Station St. Sophia, whose altitude is 10,550 ft. They then continue 3913 ft to Mountain Village, whose elevation is 9500 ft. [12.3c]

Station St. Sophia · Elevation: 10,550 ft · 5750 ft · 3913 ft · Elevation: 9500 ft · Town · Elevation: 8725 ft · Mountain Village

A visitor departs from the town at 11:55 A.M. and with no stop at Station St. Sophia reaches Mountain Village at 12:07 P.M.
  a) Find the gondola's average rate of ascent and descent, in number of feet per minute.
  b) Find the gondola's average rate of ascent and descent, in number of minutes per foot.

Understanding Through Discussion and Writing

1. Explain why the slant of a line with slope $\frac{5}{3}$ is steeper than the slant of a line with slope $\frac{4}{3}$. [12.4a]

2. Do all graphs of linear equations have x-intercepts? Explain. [12.3b]

3. Explain why the first coordinate of the y-intercept is always 0. [12.2a]

4. Explain why the graph of $y = -2$ is a horizontal line. [12.3b]

CHAPTER
12 | Test

For Extra Help For step-by-step test solutions, access the Chapter Test Prep Videos in MyMathLab® or on YouTube (search "BittingerAlgebraFoundations" and click on "Channels").

In which quadrant is each point located?

1. $\left(-\frac{1}{2}, 7\right)$

2. $(-5, -6)$

Find the coordinates of each point.

3. *A* **4.** *B*

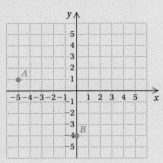

5. Show that the ordered pairs $(-4, -3)$ and $(-1, 3)$ are solutions of the equation $y - 2x = 5$. Then use the graph of the straight line containing the two points to determine another solution. Answers may vary.

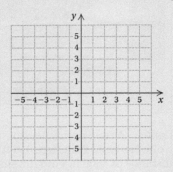

Graph each equation. Identify the *y*-intercept.

6. $y = 2x - 1$

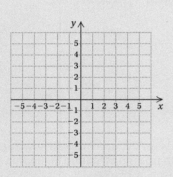

x	y

7. $y = -\frac{3}{2}x$

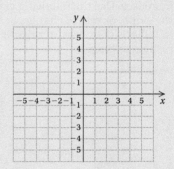

x	y

Find the intercepts of each equation. Then graph the equation.

8. $2x - 4y = -8$

x	y
		← *x*-intercept
		← *y*-intercept

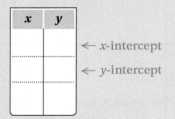

9. $2x - y = 3$

x	y
		← *x*-intercept
		← *y*-intercept

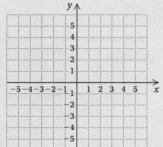
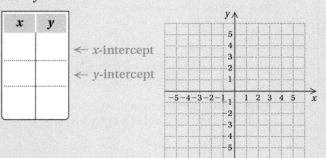

Graph each equation.

10. $2x + 8 = 0$

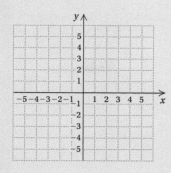

11. $y = 5$

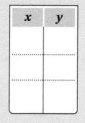

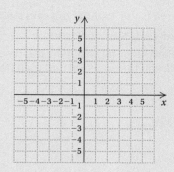

12. Find the slope.

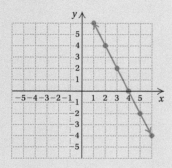

13. Graph the line containing $(-3, 1)$ and $(5, 4)$ and find the slope.

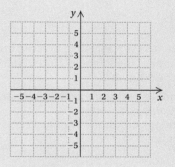

Find the slope, if it exists, of each line.

14. $2x - 5y = 10$

15. $x = -2$

16. $3y = \dfrac{1}{9}$

17. $y = -11x + 6$

18. *Navigation.* Capital Rapids drops 54 ft vertically over a horizontal distance of 1080 ft. What is the slope of the rapids?

19. *Health Insurance Cost.* The total annual cost, employer plus employee, of health insurance can be approximated by

$$C = 606t + 8593,$$

where t is the number of years since 2007. That is, $t = 0$ corresponds to 2007, $t = 3$ corresponds to 2010, and so on.

Source: TW/NBGH Value Purchasing Survey

a) Find the total cost of health insurance in 2007, in 2009, and in 2012.

b) Graph the equation and then use the graph to estimate the cost of health insurance in 2016.

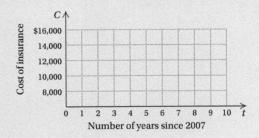

c) Predict the year in which the cost of health insurance will be $15,259.

20. *Elevators.* At 2:38, Serge entered an elevator on the 34th floor of the Regency Hotel. At 2:40, he stepped off at the 5th floor.

a) Find the elevator's average rate of travel, in number of floors per minute.

b) Find the elevator's average rate of travel, in seconds per floor.

21. *Train Travel.* The following graph shows data concerning a recent train ride from Denver to Kansas City. At what rate did the train travel?

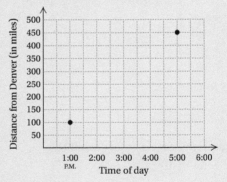

22. Which of the following is the correct description of the line $6x - 1 = 3y + 2$?

A. The slope is -6, and the x-intercept is $\left(-\frac{1}{2}, 0\right)$.

B. The slope is 2, and the y-intercept is $(0, -1)$.

C. The slope is -2, and the x-intercept is $\left(\frac{1}{2}, 0\right)$.

D. The slope is 6, and the y-intercept is $(0, -3)$.

Synthesis

23. Write an equation of a line whose graph is parallel to the x-axis and 3 units above it.

24. A diagonal of a square connects the points $(-3, -1)$ and $(2, 4)$. Find the area and the perimeter of the square.

Polynomials: Operations

From Chapter 13 of *Algebra Foundations: Basic Mathematics, Introductory Algebra, and Intermediate Algebra,* First Edition.
Marvin J. Bittinger, Judith A. Beecher, and Barbara L. Johnson. Copyright © 2015 by Pearson Education, Inc. All rights reserved.

13.1   Integers as Exponents

OBJECTIVES

a Tell the meaning of exponential notation.

b Evaluate exponential expressions with exponents of 0 and 1.

c Evaluate algebraic expressions containing exponents.

d Use the product rule to multiply exponential expressions with like bases.

e Use the quotient rule to divide exponential expressions with like bases.

f Express an exponential expression involving negative exponents with positive exponents.

SKILL TO REVIEW

Objective 10.1a: Evaluate algebraic expressions by substitution.

1. Evaluate $6y$ when $y = 4$.

2. Evaluate $\dfrac{m}{n}$ when $m = 48$ and $n = 8$.

a EXPONENTIAL NOTATION

An exponent of 2 or greater tells how many times the base is used as a factor. For example, $a \cdot a \cdot a \cdot a = a^4$. In this case, the **exponent** is 4 and the **base** is a. An expression for a power is called **exponential notation**.

This is the base. $\longrightarrow a^n \longleftarrow$ This is the exponent.

EXAMPLE 1   What is the meaning of 3^5? of n^4? of $(2n)^3$? of $50x^2$? of $(-n)^3$? of $-n^3$?

3^5 means $3 \cdot 3 \cdot 3 \cdot 3 \cdot 3$; $\qquad n^4$ means $n \cdot n \cdot n \cdot n$;

$(2n)^3$ means $2n \cdot 2n \cdot 2n$; $\qquad 50x^2$ means $50 \cdot x \cdot x$;

$(-n)^3$ means $(-n) \cdot (-n) \cdot (-n)$; $\qquad -n^3$ means $-1 \cdot n \cdot n \cdot n$

Do Margin Exercises 1–6 on the following page.

We read a^n as the **nth power of a**, or simply **a to the nth**, or **a to the n**. We often read x^2 as "**x-squared**" because the area of a square of side x is $x \cdot x$, or x^2. We often read x^3 as "**x-cubed**" because the volume of a cube with length, width, and height x is $x \cdot x \cdot x$, or x^3.

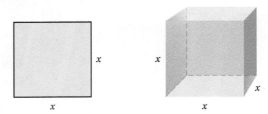

b ONE AND ZERO AS EXPONENTS

Look for a pattern in the following:

On each side, we **divide** by 8 at each step.	$8 \cdot 8 \cdot 8 \cdot 8 = 8^4$	On this side, the exponents **decrease** by 1 at each step.
	$8 \cdot 8 \cdot 8 = 8^3$	
	$8 \cdot 8 = 8^2$	
	$8 = 8^?$	
	$1 = 8^?$.	

To continue the pattern, we would say that $8 = 8^1$ and $1 = 8^0$.

Answers

Skill to Review:
1. 24   2. 6

We make the following definition.

What is the meaning of each of the following?

1. 5^4 **2.** x^5

3. $(3t)^2$ **4.** $3t^2$

5. $(-x)^4$ **6.** $-y^3$

> **EXPONENTS OF 0 AND 1**
>
> $a^1 = a$, for any number a;
>
> $a^0 = 1$, for any nonzero number a

We consider 0^0 to be not defined. We will explain why later in this section.

EXAMPLE 2 Evaluate 5^1, $(-8)^1$, 3^0, and $(-749.21)^0$.

$$5^1 = 5; \qquad (-8)^1 = -8;$$
$$3^0 = 1; \qquad (-749.21)^0 = 1$$

Do Exercises 7–12. ▷

Evaluate.

7. 6^1 **8.** 7^0

9. $(8.4)^1$ **10.** 8654^0

11. $(-1.4)^1$ **12.** 0^1

c   EVALUATING ALGEBRAIC EXPRESSIONS

Algebraic expressions can involve exponential notation. For example, the following are algebraic expressions:

$$x^4, \qquad (3x)^3 - 2, \qquad a^2 + 2ab + b^2.$$

We evaluate algebraic expressions by replacing variables with numbers and following the rules for order of operations.

EXAMPLE 3 Evaluate $1000 - x^4$ when $x = 5$.

$$
\begin{aligned}
1000 - x^4 &= 1000 - 5^4 && \text{Substituting} \\
&= 1000 - 625 && \text{Evaluating } 5^4 \\
&= 375
\end{aligned}
$$

EXAMPLE 4 *Area of a Circular Region.* The Richat Structure is a circular eroded geologic dome with a radius of 20 km. Find the area of the structure.

$$
\begin{aligned}
A &= \pi r^2 && \text{Using the formula for the area of a circle} \\
&= \pi (20 \text{ km})^2 && \text{Substituting} \\
&= \pi \cdot 20 \text{ km} \cdot 20 \text{ km} \\
&\approx 3.14 \times 400 \text{ km}^2 && \text{Using 3.14 as an approximation for } \pi \\
&= 1256 \text{ km}^2
\end{aligned}
$$

In Example 4, "km^2" means "square kilometers" and "$\approx$" means "is approximately equal to."

EXAMPLE 5 Evaluate $(5x)^3$ when $x = -2$.

When we evaluate with a negative number, we often use extra parentheses to show the substitution.

$$
\begin{aligned}
(5x)^3 &= [5 \cdot (-2)]^3 && \text{Substituting} \\
&= [-10]^3 && \text{Multiplying within brackets first} \\
&= [-10] \cdot [-10] \cdot [-10] \\
&= -1000 && \text{Evaluating the power}
\end{aligned}
$$

Answers

1. $5 \cdot 5 \cdot 5 \cdot 5$ **2.** $x \cdot x \cdot x \cdot x \cdot x$
3. $3t \cdot 3t$ **4.** $3 \cdot t \cdot t$
5. $(-x) \cdot (-x) \cdot (-x) \cdot (-x)$ **6.** $-1 \cdot y \cdot y \cdot y$
7. 6   **8.** 1   **9.** 8.4   **10.** 1
11. -1.4 **12.** 0

13. Evaluate t^3 when $t = 5$.

14. Evaluate $-5x^5$ when $x = -2$.

15. Find the area of a circle when $r = 32$ cm. Use 3.14 for π.

16. Evaluate $200 - a^4$ when $a = 3$.

17. Evaluate $t^1 - 4$ and $t^0 - 4$ when $t = 7$.

18. **a)** Evaluate $(4t)^2$ when $t = -3$.

b) Evaluate $4t^2$ when $t = -3$.

c) Determine whether $(4t)^2$ and $4t^2$ are equivalent.

a) $(4t)^2 = \left[4 \cdot \left(\right) \right]^2$

$= \left[\right]^2$

$= $

b) $4t^2 = 4 \cdot \left(\right)^2$

$= 4 \cdot \left(\right)$

$= $

c) Since $144 \neq 36$, the expressions _____ equivalent.

are/are not

Multiply and simplify.

19. $3^5 \cdot 3^5$

20. $x^4 \cdot x^6$

21. $p^4 p^{12} p^8$

22. $x \cdot x^4$

23. $(a^2 b^3)(a^7 b^5)$

EXAMPLE 6 Evaluate $5x^3$ when $x = -2$.

$$5x^3 = 5 \cdot (-2)^3 \qquad \text{Substituting}$$
$$= 5 \cdot (-2) \cdot (-2) \cdot (-2) \qquad \text{Evaluating the power first}$$
$$= 5(-8) \qquad\qquad (-2)(-2)(-2) = -8$$
$$= -40$$

Recall that two expressions are equivalent if they have the same value for all meaningful replacements. Note that Examples 5 and 6 show that $(5x)^3$ and $5x^3$ are *not* equivalent—that is, $(5x)^3 \neq 5x^3$.

◀ Do Exercises 13–18.

d MULTIPLYING POWERS WITH LIKE BASES

We can multiply powers with like bases by adding exponents. For example,

$$a^3 \cdot a^2 = (a \cdot a \cdot a)(a \cdot a) = a \cdot a \cdot a \cdot a \cdot a = a^5.$$

3 factors   2 factors   5 factors

Note that the exponent in a^5 is the sum of those in $a^3 \cdot a^2$. That is, $3 + 2 = 5$. Likewise,

$$b^4 \cdot b^3 = (b \cdot b \cdot b \cdot b)(b \cdot b \cdot b) = b^7, \quad \text{where} \quad 4 + 3 = 7.$$

Adding the exponents gives the correct result.

THE PRODUCT RULE

For any number a and any positive integers m and n,

$$a^m \cdot a^n = a^{m+n}.$$

(When multiplying with exponential notation, if the bases are the same, keep the base and add the exponents.)

EXAMPLES Multiply and simplify.

7. $5^6 \cdot 5^2 = 5^{6+2}$ Adding exponents: $a^m \cdot a^n = a^{m+n}$
$ = 5^8$

8. $m^5 m^{10} m^3 = m^{5+10+3} = m^{18}$

9. $x \cdot x^8 = x^1 \cdot x^8$ Writing x as x^1
$ = x^{1+8}$
$ = x^9$

10. $(a^3 b^2)(a^3 b^5) = (a^3 a^3)(b^2 b^5)$
$ = a^6 b^7$

11. $(4y)^6 (4y)^3 = (4y)^{6+3} = (4y)^9$

◀ Do Exercises 19–23.

Answers

13. 125   14. 160   15. 3215.36 cm²   16. 119
17. 3; −3   18. (a) 144; (b) 36; (c) no
19. 3^{10} 20. x^{10} 21. p^{24} 22. x^5 23. $a^9 b^8$

Guided Solution:
18. (a) −3, −12, 144; (b) −3, 9, 36; (c) are not

e DIVIDING POWERS WITH LIKE BASES

The following suggests a rule for dividing powers with like bases, such as a^5/a^2:

$$\frac{a^5}{a^2} = \frac{a \cdot a \cdot a \cdot a \cdot a}{a \cdot a} = \frac{a \cdot a \cdot a \cdot a \cdot a}{1 \cdot a \cdot a} = \frac{a \cdot a \cdot a}{1} \cdot \frac{a \cdot a}{a \cdot a}$$

$$= \frac{a \cdot a \cdot a}{1} \cdot 1 = a \cdot a \cdot a = a^3.$$

Note that the exponent in a^3 is the difference of those in $a^5 \div a^2$. That is, $5 - 2 = 3$. In a similar way, we have

$$\frac{t^9}{t^4} = \frac{t \cdot t \cdot t \cdot t \cdot t \cdot t \cdot t \cdot t \cdot t}{t \cdot t \cdot t \cdot t} = t^5, \quad \text{where} \quad 9 - 4 = 5.$$

Subtracting exponents gives the correct answer.

THE QUOTIENT RULE

For any nonzero number a and any positive integers m and n,

$$\frac{a^m}{a^n} = a^{m-n}.$$

(When dividing with exponential notation, if the bases are the same, keep the base and subtract the exponent of the denominator from the exponent of the numerator.)

EXAMPLES Divide and simplify.

12. $\dfrac{6^5}{6^3} = 6^{5-3}$ Subtracting exponents

$= 6^2$

13. $\dfrac{x^8}{x} = \dfrac{x^8}{x^1} = x^{8-1}$

$= x^7$

14. $\dfrac{(3t)^{12}}{(3t)^2} = (3t)^{12-2}$

$= (3t)^{10}$

15. $\dfrac{p^5 q^7}{p^2 q^5} = \dfrac{p^5}{p^2} \cdot \dfrac{q^7}{q^5} = p^{5-2} q^{7-5}$

$= p^3 q^2$

The quotient rule can also be used to explain the definition of 0 as an exponent. Consider the expression a^4/a^4, where a is nonzero:

$$\frac{a^4}{a^4} = \frac{a \cdot a \cdot a \cdot a}{a \cdot a \cdot a \cdot a} = 1.$$

This is true because the numerator and the denominator are the same. Now suppose we apply the rule for dividing powers with the same base:

$$\frac{a^4}{a^4} = a^{4-4} = a^0.$$

Since $a^4/a^4 = 1$ and $a^4/a^4 = a^0$, it follows that $a^0 = 1$, when $a \neq 0$.

We can explain why we do not define 0^0 using the quotient rule. We know that 0^0 is 0^{1-1}. But 0^{1-1} is also equal to $0^1/0^1$, or $0/0$. We have already seen that division by 0 is not defined, so 0^0 is also not defined.

Do Exercises 24–27. ▶

Divide and simplify.

24. $\dfrac{4^5}{4^2}$

25. $\dfrac{y^6}{y^2}$

26. $\dfrac{p^{10}}{p}$

27. $\dfrac{a^7 b^6}{a^3 b^4}$

Answers

24. 4^3 **25.** y^4 **26.** p^9 **27.** $a^4 b^2$

f NEGATIVE INTEGERS AS EXPONENTS

We can use the rule for dividing powers with like bases to lead us to a definition of exponential notation when the exponent is a negative integer. Consider $5^3/5^7$ and first simplify it using procedures we have learned for working with fractions:

$$\frac{5^3}{5^7} = \frac{5 \cdot 5 \cdot 5}{5 \cdot 5 \cdot 5 \cdot 5 \cdot 5 \cdot 5 \cdot 5} = \frac{5 \cdot 5 \cdot 5 \cdot 1}{5 \cdot 5 \cdot 5 \cdot 5 \cdot 5 \cdot 5 \cdot 5}$$

$$= \frac{5 \cdot 5 \cdot 5}{5 \cdot 5 \cdot 5} \cdot \frac{1}{5 \cdot 5 \cdot 5 \cdot 5} = \frac{1}{5^4}.$$

Now we apply the rule for dividing exponential expressions with the same bases. Then

$$\frac{5^3}{5^7} = 5^{3-7} = 5^{-4}.$$

From these two expressions for $5^3/5^7$, it follows that

$$5^{-4} = \frac{1}{5^4}.$$

This leads to our definition of negative exponents.

> ### NEGATIVE EXPONENT
>
> For any real number a that is nonzero and any integer n,
>
> $$a^{-n} = \frac{1}{a^n}.$$

In fact, the numbers a^n and a^{-n} are reciprocals because

$$a^n \cdot a^{-n} = a^n \cdot \frac{1}{a^n} = \frac{a^n}{a^n} = 1.$$

The following is another way to arrive at the definition of negative exponents.

On each side, we **divide** by 5 at each step.		On this side, the exponents **decrease** by 1 at each step.
	$5 \cdot 5 \cdot 5 \cdot 5 = 5^4$	
	$5 \cdot 5 \cdot 5 = 5^3$	
	$5 \cdot 5 = 5^2$	
	$5 = 5^1$	
	$1 = 5^0$	
	$\frac{1}{5} = 5^?$	
	$\frac{1}{25} = 5^?$	

To continue the pattern, it should follow that

$$\frac{1}{5} = \frac{1}{5^1} = 5^{-1} \quad \text{and} \quad \frac{1}{25} = \frac{1}{5^2} = 5^{-2}.$$

EXAMPLES Express using positive exponents. Then simplify.

16. $4^{-2} = \dfrac{1}{4^2} = \dfrac{1}{16}$

17. $(-3)^{-2} = \dfrac{1}{(-3)^2} = \dfrac{1}{(-3)(-3)} = \dfrac{1}{9}$

18. $m^{-3} = \dfrac{1}{m^3}$

19. $ab^{-1} = a\left(\dfrac{1}{b^1}\right) = a\left(\dfrac{1}{b}\right) = \dfrac{a}{b}$

20. $\dfrac{1}{x^{-3}} = x^{-(-3)} = x^3$

21. $3c^{-5} = 3\left(\dfrac{1}{c^5}\right) = \dfrac{3}{c^5}$

Example 20 might also be done as follows:

$$\dfrac{1}{x^{-3}} = \dfrac{1}{\dfrac{1}{x^3}} = 1 \cdot \dfrac{x^3}{1} = x^3.$$

··········· **Caution!** ···········

As shown in Examples 16 and 17, a negative exponent does not necessarily mean that an expression is negative.

Do Exercises 28–33. ▶

The rules for multiplying and dividing powers with like bases hold when exponents are 0 or negative.

EXAMPLES Simplify. Write the result using positive exponents.

22. $7^{-3} \cdot 7^6 = 7^{-3+6}$ Adding exponents

$\quad = 7^3$

23. $x^4 \cdot x^{-3} = x^{4+(-3)} = x^1 = x$

24. $\dfrac{5^4}{5^{-2}} = 5^{4-(-2)}$ Subtracting exponents

$\quad = 5^{4+2} = 5^6$

25. $\dfrac{x}{x^7} = x^{1-7} = x^{-6} = \dfrac{1}{x^6}$

26. $\dfrac{b^{-4}}{b^{-5}} = b^{-4-(-5)}$

$\quad = b^{-4+5} = b^1 = b$

27. $y^{-4} \cdot y^{-8} = y^{-4+(-8)}$

$\quad = y^{-12} = \dfrac{1}{y^{12}}$

Do Exercises 34–38. ▶

The following is a summary of the definitions and rules for exponents that we have considered in this section.

DEFINITIONS AND RULES FOR EXPONENTS
···

1 as an exponent: $a^1 = a$

0 as an exponent: $a^0 = 1, a \neq 0$

Negative integers as exponents: $a^{-n} = \dfrac{1}{a^n}, \dfrac{1}{a^{-n}} = a^n; a \neq 0$

Product Rule: $a^m \cdot a^n = a^{m+n}$

Quotient Rule: $\dfrac{a^m}{a^n} = a^{m-n}, a \neq 0$

Express with positive exponents. Then simplify.

28. 4^{-3}

29. 5^{-2}

30. 2^{-4}

31. $(-2)^{-3}$

32. $\dfrac{1}{x^{-2}}$

 33. $4p^{-3}$

$= 4\left(\dfrac{1}{\boxed{}}\right) = \dfrac{4}{\boxed{}}$

Simplify.

34. $5^{-2} \cdot 5^4$

35. $x^{-3} \cdot x^{-4}$

36. $\dfrac{7^{-2}}{7^3}$

37. $\dfrac{b^{-2}}{b^{-3}}$

38. $\dfrac{t}{t^{-5}}$

Answers

28. $\dfrac{1}{4^3} = \dfrac{1}{64}$ **29.** $\dfrac{1}{5^2} = \dfrac{1}{25}$ **30.** $\dfrac{1}{2^4} = \dfrac{1}{16}$
31. $\dfrac{1}{(-2)^3} = -\dfrac{1}{8}$ **32.** x^2 **33.** $\dfrac{4}{p^3}$ **34.** 5^2
35. $\dfrac{1}{x^7}$ **36.** $\dfrac{1}{7^5}$ **37.** b **38.** t^6

Guided Solution:
33. p^3, p^3

✓ Reading Check

Match each expression with the appropriate value from the column on the right. Choices may be used more than once or not at all.

RC1. _____ y^1

RC2. _____ $y^0, y \neq 0$

RC3. _____ $y^1 \cdot y^1$

RC4. _____ $\dfrac{y^9}{y^8}$

RC5. _____ $\dfrac{y^8}{y^9}$

RC6. _____ $\dfrac{1}{y^{-1}}$

a) 1

b) 0

c) y

d) $\dfrac{1}{y}$

e) y^2

a What is the meaning of each of the following?

1. 3^4

2. 4^3

3. $(-1.1)^5$

4. $(87.2)^6$

5. $\left(\dfrac{2}{3}\right)^4$

6. $\left(-\dfrac{5}{8}\right)^3$

7. $(7p)^2$

8. $(11c)^3$

9. $8k^3$

10. $17x^2$

11. $-6y^4$

12. $-q^5$

b Evaluate.

13. $a^0, a \neq 0$

14. $t^0, t \neq 0$

15. b^1

16. c^1

17. $\left(\dfrac{2}{3}\right)^0$

18. $\left(-\dfrac{5}{8}\right)^0$

19. $(-7.03)^1$

20. $\left(\dfrac{4}{5}\right)^1$

21. 8.38^0

22. 8.38^1

23. $(ab)^1$

24. $(ab)^0, a \neq 0, b \neq 0$

25. $ab^0, b \neq 0$

26. ab^1

c Evaluate.

27. m^3, when $m = 3$

28. x^6, when $x = 2$

29. p^1, when $p = 19$

30. x^{19}, when $x = 0$

31. $-x^4$, when $x = -3$

32. $-2y^7$, when $y = 2$

33. x^4, when $x = 4$

34. y^{15}, when $y = 1$

35. $y^2 - 7$, when $y = -10$

36. $z^5 + 5$, when $z = -2$

37. $161 - b^2$, when $b = 5$

38. $325 - v^3$, when $v = -3$

39. $x^1 + 3$ and $x^0 + 3$, when $x = 7$

40. $y^0 - 8$ and $y^1 - 8$, when $y = -3$

41. Find the area of a circle when $r = 34$ ft. Use 3.14 for π.

42. The area A of a square with sides of length s is given by $A = s^2$. Find the area of a square with sides of length 24 m.

f Express using positive exponents. Then simplify.

43. 3^{-2}

44. 2^{-3}

45. 10^{-3}

46. 5^{-4}

47. a^{-3}

48. x^{-2}

49. $\dfrac{1}{8^{-2}}$

50. $\dfrac{1}{2^{-5}}$

51. $\dfrac{1}{y^{-4}}$

52. $\dfrac{1}{t^{-7}}$

53. $5z^{-4}$

54. $6n^{-5}$

55. xy^{-2}

56. ab^{-3}

Express using negative exponents.

57. $\dfrac{1}{4^3}$

58. $\dfrac{1}{5^2}$

59. $\dfrac{1}{x^3}$

60. $\dfrac{1}{y^2}$

61. $\dfrac{1}{a^5}$

62. $\dfrac{1}{b^7}$

d , **f** Multiply and simplify.

63. $2^4 \cdot 2^3$

64. $3^5 \cdot 3^2$

65. $9^{17} \cdot 9^{21}$

66. $7^{22} \cdot 7^{15}$

67. $x^4 \cdot x$

68. $y \cdot y^9$

69. $x^{14} \cdot x^3$

70. $x^9 \cdot x^4$

71. $(3y)^4(3y)^8$

72. $(2t)^8(2t)^{17}$

73. $(7y)^1(7y)^{16}$

74. $(8x)^0(8x)^1$

75. $3^{-5} \cdot 3^8$

76. $5^{-8} \cdot 5^9$

77. $x^{-2} \cdot x^2$

78. $x \cdot x^{-1}$

79. $x^{-7} \cdot x^{-6}$ **80.** $y^{-5} \cdot y^{-8}$ **81.** $a^{11} \cdot a^{-3} \cdot a^{-18}$ **82.** $a^{-11} \cdot a^{-3} \cdot a^{-7}$

83. $(x^4 y^7)(x^2 y^8)$ **84.** $(a^5 c^2)(a^3 c^9)$ **85.** $(s^2 t^3)(s t^4)$ **86.** $(m^4 n)(m^2 n^7)$

e , **f**   Divide and simplify.

87. $\dfrac{7^5}{7^2}$ **88.** $\dfrac{5^8}{5^6}$ **89.** $\dfrac{y^9}{y}$ **90.** $\dfrac{x^{11}}{x}$

91. $\dfrac{16^2}{16^8}$ **92.** $\dfrac{7^2}{7^9}$ **93.** $\dfrac{m^6}{m^{12}}$ **94.** $\dfrac{a^3}{a^4}$

95. $\dfrac{(8x)^6}{(8x)^{10}}$ **96.** $\dfrac{(8t)^4}{(8t)^{11}}$ **97.** $\dfrac{x}{x^{-1}}$ **98.** $\dfrac{t^8}{t^{-3}}$

99. $\dfrac{z^{-6}}{z^{-2}}$ **100.** $\dfrac{x^{-9}}{x^{-3}}$ **101.** $\dfrac{x^{-5}}{x^{-8}}$ **102.** $\dfrac{y^{-2}}{y^{-9}}$

103. $\dfrac{m^{-9}}{m^{-9}}$ **104.** $\dfrac{x^{-7}}{x^{-7}}$ **105.** $\dfrac{a^5 b^3}{a^2 b}$ **106.** $\dfrac{s^8 t^4}{s t^3}$

Matching. In Exercises 107 and 108, match each item in the first column with the appropriate item in the second column by drawing connecting lines. Items in the second column may be used more than once.

107.

5^2	$-\dfrac{1}{10}$
5^{-2}	$\dfrac{1}{10}$
$\left(\dfrac{1}{5}\right)^2$	$-\dfrac{1}{25}$
$\left(\dfrac{1}{5}\right)^{-2}$	10
-5^2	25
$(-5)^2$	-25
$-\left(-\dfrac{1}{5}\right)^2$	$\dfrac{1}{25}$
$\left(-\dfrac{1}{5}\right)^{-2}$	-10

108.

$-\left(\dfrac{1}{8}\right)^2$	16
$\left(\dfrac{1}{8}\right)^{-2}$	-16
8^{-2}	64
8^2	-64
-8^2	$\dfrac{1}{64}$
$(-8)^2$	$-\dfrac{1}{64}$
$\left(-\dfrac{1}{8}\right)^{-2}$	$-\dfrac{1}{16}$
$\left(-\dfrac{1}{8}\right)^2$	$\dfrac{1}{16}$

Skill Maintenance

Solve.

109. A 12-in. submarine sandwich is cut into two pieces. One piece is twice as long as the other. How long are the pieces?   [11.6a]

110. The first angle of a triangle is 24° more than the second. The third angle is twice the first. Find the measures of the angles of the triangle.   [11.6a]

111. A warehouse stores 1800 lb of peanuts, 1500 lb of cashews, and 700 lb of almonds. What percent of the total is peanuts? cashews? almonds?   [11.5a]

112. The width of a rectangle is fixed at 10 ft. For what lengths will the area be less than 25 ft²?   [11.8b]

Solve.

113. $2x - 4 - 5x + 8 = x - 3$ [11.3b]

114. $8x + 7 - 9x = 12 - 6x + 5$ [11.3b]

115. $-6(2 - x) + 10(5x - 7) = 10$ [11.3c]

116. $-10(x - 4) = 5(2x + 5) - 7$ [11.3c]

Synthesis

Determine whether each of the following equations is true.

117. $(x + 1)^2 = x^2 + 1$

118. $(x - 1)^2 = x^2 - 2x + 1$

119. $(5x)^0 = 5x^0$

120. $\dfrac{x^3}{x^5} = x^2$

Simplify.

121. $(y^{2x})(y^{3x})$

122. $a^{5k} \div a^{3k}$

123. $\dfrac{a^{6t}(a^{7t})}{a^{9t}}$

124. $\dfrac{\left(\frac{1}{2}\right)^4}{\left(\frac{1}{2}\right)^5}$

125. $\dfrac{(0.8)^5}{(0.8)^3(0.8)^2}$

126. $\dfrac{(x - 3)^5}{x - 3}$

Use >, <, or = for ☐ to write a true sentence.

127. 3^5 ☐ 3^4

128. 4^2 ☐ 4^3

129. 4^3 ☐ 5^3

130. 4^3 ☐ 3^4

Evaluate.

131. $\dfrac{1}{-z^4}$, when $z = -10$

132. $\dfrac{1}{-z^5}$, when $z = -0.1$

133. Determine whether $(a + b)^2$ and $a^2 + b^2$ are equivalent. (*Hint*: Choose values for a and b and evaluate.)

13.2

Exponents and Scientific Notation

OBJECTIVES

a · Use the power rule to raise powers to powers.

b · Raise a product to a power and a quotient to a power.

c · Convert between scientific notation and decimal notation.

d · Multiply and divide using scientific notation.

e · Solve applied problems using scientific notation.

SKILL TO REVIEW

Objective 10.5a: Multiply real numbers.

Multiply.

1. $-5 \cdot 8$ **2.** $(-3)(-5)$

Simplify. Express the answers using positive exponents.

1. $(3^4)^5$ 2. $(x^{-3})^4$

3. $(y^{-5})^{-3}$ 4. $(x^4)^{-8}$

We now consider three rules used to simplify exponential expressions. We then apply our knowledge of exponents to *scientific notation*.

a · RAISING POWERS TO POWERS

Consider an expression like $(3^2)^4$. We are raising 3^2 to the fourth power:

$$(3^2)^4 = (3^2)(3^2)(3^2)(3^2)$$
$$= (3 \cdot 3)(3 \cdot 3)(3 \cdot 3)(3 \cdot 3)$$
$$= 3 \cdot 3 \cdot 3 \cdot 3 \cdot 3 \cdot 3 \cdot 3 \cdot 3$$
$$= 3^8.$$

Note that in this case we could have multiplied the exponents:

$$(3^2)^4 = 3^{2 \cdot 4} = 3^8.$$

THE POWER RULE

For any real number a and any integers m and n,

$$(a^m)^n = a^{mn}.$$

(To raise a power to a power, multiply the exponents.)

EXAMPLES Simplify. Express the answers using positive exponents.

1. $(3^5)^4 = 3^{5 \cdot 4}$ Multiplying
$= 3^{20}$ exponents

2. $(2^2)^5 = 2^{2 \cdot 5} = 2^{10}$

3. $(y^{-5})^7 = y^{-5 \cdot 7} = y^{-35} = \dfrac{1}{y^{35}}$

4. $(x^4)^{-2} = x^{4(-2)} = x^{-8} = \dfrac{1}{x^8}$

5. $(a^{-4})^{-6} = a^{(-4)(-6)} = a^{24}$

◀ Do Margin Exercises 1–4.

b · RAISING A PRODUCT OR A QUOTIENT TO A POWER

When an expression inside parentheses is raised to a power, the inside expression is the base. Let's compare $2a^3$ and $(2a)^3$:

$$2a^3 = 2 \cdot a \cdot a \cdot a;$$ The base is a.

$$(2a)^3 = (2a)(2a)(2a)$$ The base is $2a$.
$$= (2 \cdot 2 \cdot 2)(a \cdot a \cdot a)$$ Using the associative and commutative laws of multiplication to regroup the factors
$$= 2^3 a^3$$
$$= 8a^3.$$

We see that $2a^3$ and $(2a)^3$ are *not* equivalent. We also see that we can evaluate the power $(2a)^3$ by raising each factor to the power 3. This leads us to a rule for raising a product to a power.

Answers

Skill to Review:
1. -40 2. 15

Margin Exercises:
1. 3^{20} 2. $\dfrac{1}{x^{12}}$ 3. y^{15} 4. $\dfrac{1}{x^{32}}$

RAISING A PRODUCT TO A POWER

For any real numbers a and b and any integer n,

$$(ab)^n = a^n b^n.$$

(To raise a product to the nth power, raise each factor to the nth power.)

EXAMPLES   Simplify.

6. $(4x^2)^3 = (4^1 x^2)^3$ $4 = 4^1$

$ = (4^1)^3 \cdot (x^2)^3$ Raising *each* factor to the third power

$ = 4^3 \cdot x^6 = 64x^6$ Using the power rule and simplifying

7. $(5x^3 y^5 z^2)^4 = 5^4 (x^3)^4 (y^5)^4 (z^2)^4$ Raising *each* factor to the fourth power

$ = 625 x^{12} y^{20} z^8$

8. $(-5x^4 y^3)^3 = (-5)^3 (x^4)^3 (y^3)^3$

$ = -125 x^{12} y^9$

9. $[(-x)^{25}]^2 = (-x)^{50}$ Using the power rule

$\phantom{[(-x)^{25}]^2} = (-1 \cdot x)^{50}$ Using the property of -1: $-x = -1 \cdot x$

$\phantom{[(-x)^{25}]^2} = (-1)^{50} x^{50}$ Raising each factor to the fiftieth power

$\phantom{[(-x)^{25}]^2} = 1 \cdot x^{50}$ The product of an even number of negative factors is positive.

$\phantom{[(-x)^{25}]^2} = x^{50}$

10. $(3x^3 y^{-5} z^2)^4 = 3^4 (x^3)^4 (y^{-5})^4 (z^2)^4 = 81 x^{12} y^{-20} z^8 = \dfrac{81 x^{12} z^8}{y^{20}}$

11. $(-x^4)^{-3} = (-1 \cdot x^4)^{-3} = (-1)^{-3} \cdot (x^4)^{-3} = (-1)^{-3} \cdot x^{-12}$

$\phantom{(-x^4)^{-3}} = \dfrac{1}{(-1)^3} \cdot \dfrac{1}{x^{12}} = \dfrac{1}{-1} \cdot \dfrac{1}{x^{12}} = -\dfrac{1}{x^{12}}$

12. $(-2x^{-5} y^4)^{-4} = (-2)^{-4} (x^{-5})^{-4} (y^4)^{-4} = \dfrac{1}{(-2)^4} \cdot x^{20} \cdot y^{-16}$

$\phantom{(-2x^{-5} y^4)^{-4}} = \dfrac{1}{16} \cdot x^{20} \cdot \dfrac{1}{y^{16}} = \dfrac{x^{20}}{16 y^{16}}$

Do Exercises 5–11. ▶

Simplify.

5. $(2x^5 y^{-3})^4$

6. $(5x^5 y^{-6} z^{-3})^2$

7. $[(-x)^{37}]^2$

8. $(3y^{-2} x^{-5} z^8)^3$

9. $(-y^8)^{-3}$

GS **10.** $(-2x^4)^{-2}$

$= (-2)^{-2} (\boxed{})^{-2}$

$= \dfrac{1}{(-2)^{\boxed{}}} \cdot x^{\boxed{}}$

$= \dfrac{1}{\boxed{}} \cdot \dfrac{1}{x^{\boxed{}}}$

$= \dfrac{1}{\boxed{}}$

11. $(-3x^2 y^{-5})^{-3}$

There is a similar rule for raising a quotient to a power.

RAISING A QUOTIENT TO A POWER

For any real numbers a and b, $b \neq 0$, and any integer n,

$$\left(\frac{a}{b}\right)^n = \frac{a^n}{b^n}.$$

(To raise a quotient to the nth power, raise both the numerator and the denominator to the nth power.)

Answers

5. $\dfrac{16x^{20}}{y^{12}}$ **6.** $\dfrac{25x^{10}}{y^{12} z^6}$ **7.** x^{74} **8.** $\dfrac{27z^{24}}{y^6 x^{15}}$

9. $-\dfrac{1}{y^{24}}$ **10.** $\dfrac{1}{4x^8}$ **11.** $-\dfrac{y^{15}}{27x^6}$

Guided Solution:
10. $x^4, 2, -8, 4, 8, 4x^8$

Simplify.

12. $\left(\dfrac{x^6}{5}\right)^2$

13. $\left(\dfrac{2t^5}{w^4}\right)^3$

14. $\left(\dfrac{a^4}{3b^{-2}}\right)^3$

15. $\left(\dfrac{x^4}{3}\right)^{-2}$ GS

Do this two ways.

$$\left(\frac{x^4}{3}\right)^{-2} = \frac{(x^4)^{\boxed{}}}{3^{-2}} = \frac{x^{\boxed{}}}{3^{-2}}$$

$$= \frac{\dfrac{1}{x^{\boxed{}}}}{\dfrac{1}{3^2}} = \frac{1}{x^8} \div \frac{1}{3^2}$$

$$= \frac{1}{x^8} \cdot \frac{3^2}{\boxed{}} = \frac{9}{\boxed{}}$$

This can be done a second way.

$$\left(\frac{x^4}{3}\right)^{-2} = \left(\frac{3}{x^4}\right)^{\boxed{}}$$

$$= \frac{3^2}{(x^4)^{\boxed{}}} = \frac{9}{\boxed{}}$$

EXAMPLES Simplify.

13. $\left(\dfrac{x^2}{4}\right)^3 = \dfrac{(x^2)^3}{4^3} = \dfrac{x^6}{64}$ Raising *both* the numerator and the denominator to the third power

14. $\left(\dfrac{3a^4}{b^3}\right)^2 = \dfrac{(3a^4)^2}{(b^3)^2} = \dfrac{3^2(a^4)^2}{b^{3\cdot2}} = \dfrac{9a^8}{b^6}$

15. $\left(\dfrac{y^2}{2z^{-5}}\right)^4 = \dfrac{(y^2)^4}{(2z^{-5})^4} = \dfrac{(y^2)^4}{2^4(z^{-5})^4} = \dfrac{y^8}{16z^{-20}} = \dfrac{y^8z^{20}}{16}$

16. $\left(\dfrac{y^3}{5}\right)^{-2} = \dfrac{(y^3)^{-2}}{5^{-2}} = \dfrac{y^{-6}}{5^{-2}} = \dfrac{\dfrac{1}{y^6}}{\dfrac{1}{5^2}} = \dfrac{1}{y^6} \div \dfrac{1}{5^2} = \dfrac{1}{y^6} \cdot \dfrac{5^2}{1} = \dfrac{25}{y^6}$

The following can often be used to simplify a quotient that is raised to a negative power.

> For $a \neq 0$ and $b \neq 0$,
> $$\left(\frac{a}{b}\right)^{-n} = \left(\frac{b}{a}\right)^n.$$

Example 16 might also be completed as follows:

$$\left(\frac{y^3}{5}\right)^{-2} = \left(\frac{5}{y^3}\right)^2 = \frac{5^2}{(y^3)^2} = \frac{25}{y^6}.$$

◀ **Do Exercises 12–15.**

c SCIENTIFIC NOTATION

We can write numbers using different types of notation, such as fraction notation, decimal notation, and percent notation. Another type, **scientific notation**, makes use of exponential notation. Scientific notation is especially useful when calculations involve very large or very small numbers. The following are examples of scientific notation.

The number of flamingos in Africa's Great Rift Valley:
$4 \times 10^6 = 4{,}000{,}000$

The length of an *E.coli* bacterium:
$2 \times 10^{-6}\,\text{m} = 0.000002\,\text{m}$

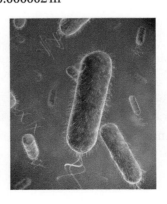

Answers

12. $\dfrac{x^{12}}{25}$ **13.** $\dfrac{8t^{15}}{w^{12}}$ **14.** $\dfrac{a^{12}b^6}{27}$ **15.** $\dfrac{9}{x^8}$

Guided Solution:
15. $-2, -8, 8, 1, x^8; 2, 2, x^8$

SCIENTIFIC NOTATION

Scientific notation for a number is an expression of the type

$$M \times 10^n,$$

where n is an integer, M is greater than or equal to 1 and less than 10 ($1 \leq M < 10$), and M is expressed in decimal notation. 10^n is also considered to be scientific notation when $M = 1$.

You should try to make conversions to scientific notation mentally as often as possible. Here is a handy mental device.

A positive exponent in scientific notation indicates a large number (greater than or equal to 10) and a negative exponent indicates a small number (between 0 and 1).

EXAMPLES Convert to scientific notation.

17. $78{,}000 = 7.8 \times 10^4$

    7.8,000.   Large number, so the exponent is positive
        4 places

18. $0.0000057 = 5.7 \times 10^{-6}$

    0.000005.7   Small number, so the exponent is negative
        6 places

Do Exercises 16 and 17. ▶

EXAMPLES Convert mentally to decimal notation.

19. $7.893 \times 10^5 = 789{,}300$

    7.89300.   Positive exponent, so the answer is a large number
        5 places

20. $4.7 \times 10^{-8} = 0.000000047$

    .00000004.7   Negative exponent, so the answer is a small number
        8 places

Do Exercises 18 and 19. ▶

d MULTIPLYING AND DIVIDING USING SCIENTIFIC NOTATION

Multiplying

Consider the product

$$400 \cdot 2000 = 800{,}000.$$

In scientific notation, this is

$$(4 \times 10^2) \cdot (2 \times 10^3) = (4 \cdot 2)(10^2 \cdot 10^3) = 8 \times 10^5.$$

Caution!

Each of the following is *not* scientific notation.

$$\underset{\uparrow}{12.46} \times 10^7$$

This number is greater than 10.

$$\underset{\uparrow}{0.347} \times 10^{-5}$$

This number is less than 1.

Convert to scientific notation.

16. 0.000517

17. 523,000,000

Convert to decimal notation.

18. 6.893×10^{11}

19. 5.67×10^{-5}

Answers
16. 5.17×10^{-4} **17.** 5.23×10^8
18. 689,300,000,000   **19.** 0.0000567

Multiply and write scientific notation for the result.

20. $(1.12 \times 10^{-8})(5 \times 10^{-7})$

21. $(9.1 \times 10^{-17})(8.2 \times 10^3)$

CALCULATOR CORNER

To find the product in Example 21 and express the result in scientific notation on a graphing calculator, we first set the calculator in Scientific mode using **MODE**. Then we go to the home screen and enter the computation by pressing ① · ⑧ **2ND** **EE** ⑥ **×** ② · ③ **2ND** **EE** (-) ④ **ENTER**. (EE is the second operation associated with the **,** key.) The decimal portion of a number written in scientific notation appears before a small E and the exponent follows the E.

```
1.8E6*2.3E−4
              4.14E2
```

EXERCISES Multiply or divide and express the answer in scientific notation.

1. $(3.15 \times 10^7)(4.3 \times 10^{-12})$

2. $(8 \times 10^9)(4 \times 10^{-5})$

3. $\dfrac{4.5 \times 10^6}{1.5 \times 10^{12}}$

4. $\dfrac{4 \times 10^{-9}}{5 \times 10^{16}}$

Divide and write scientific notation for the result.

22. $\dfrac{4.2 \times 10^5}{2.1 \times 10^2}$

23. $\dfrac{1.1 \times 10^{-4}}{2.0 \times 10^{-7}}$

Answers

20. 5.6×10^{-15} **21.** 7.462×10^{-13}
22. 2.0×10^3 **23.** 5.5×10^2

EXAMPLE 21 Multiply: $(1.8 \times 10^6) \cdot (2.3 \times 10^{-4})$.

We apply the commutative and associative laws to get

$$(1.8 \times 10^6) \cdot (2.3 \times 10^{-4}) = (1.8 \cdot 2.3) \times (10^6 \cdot 10^{-4})$$
$$= 4.14 \times 10^{6+(-4)}$$
$$= 4.14 \times 10^2.$$

We get 4.14 by multiplying 1.8 and 2.3. We get 10^2 by adding the exponents 6 and -4.

EXAMPLE 22 Multiply: $(3.1 \times 10^5) \cdot (4.5 \times 10^{-3})$.

$(3.1 \times 10^5) \cdot (4.5 \times 10^{-3}) = (3.1 \times 4.5)(10^5 \cdot 10^{-3})$

$= 13.95 \times 10^2$ Not scientific notation; 13.95 is greater than 10.

$= (1.395 \times 10^1) \times 10^2$ Substituting 1.395×10^1 for 13.95

$= 1.395 \times (10^1 \times 10^2)$ Associative law

$= 1.395 \times 10^3$ The answer is now in scientific notation.

◀ **Do Exercises 20 and 21.**

Dividing

Consider the quotient $800{,}000 \div 400 = 2000$. In scientific notation, this is

$$(8 \times 10^5) \div (4 \times 10^2) = \frac{8 \times 10^5}{4 \times 10^2} = \frac{8}{4} \times \frac{10^5}{10^2} = 2 \times 10^3.$$

EXAMPLE 23 Divide: $(3.41 \times 10^5) \div (1.1 \times 10^{-3})$.

$(3.41 \times 10^5) \div (1.1 \times 10^{-3}) = \dfrac{3.41 \times 10^5}{1.1 \times 10^{-3}} = \dfrac{3.41}{1.1} \times \dfrac{10^5}{10^{-3}}$

$= 3.1 \times 10^{5-(-3)}$

$= 3.1 \times 10^8$

EXAMPLE 24 Divide: $(6.4 \times 10^{-7}) \div (8.0 \times 10^6)$.

$(6.4 \times 10^{-7}) \div (8.0 \times 10^6) = \dfrac{6.4 \times 10^{-7}}{8.0 \times 10^6}$

$= \dfrac{6.4}{8.0} \times \dfrac{10^{-7}}{10^6}$

$= 0.8 \times 10^{-7-6}$

$= 0.8 \times 10^{-13}$ Not scientific notation; 0.8 is less than 1.

$= (8.0 \times 10^{-1}) \times 10^{-13}$ Substituting 8.0×10^{-1} for 0.8

$= 8.0 \times (10^{-1} \times 10^{-13})$ Associative law

$= 8.0 \times 10^{-14}$ Adding exponents

◀ **Do Exercises 22 and 23.**

e APPLICATIONS WITH SCIENTIFIC NOTATION

EXAMPLE 25 *Distance from the Sun to Earth.* Light from the sun traveling at a rate of 300,000 kilometers per second (km/s) reaches Earth in 499 sec. Find the distance, expressed in scientific notation, from the sun to Earth.

The time t that it takes for light to reach Earth from the sun is 4.99×10^2 sec (s). The speed is 3.0×10^5 km/s. Recall that distance can be expressed in terms of speed and time as

$$\text{Distance} = \text{Speed} \cdot \text{Time}$$
$$d = rt.$$

We substitute 3.0×10^5 for r and 4.99×10^2 for t:

$$
\begin{aligned}
d &= rt \\
&= (3.0 \times 10^5)(4.99 \times 10^2) \qquad \text{Substituting} \\
&= 14.97 \times 10^7 \\
&= (1.497 \times 10^1) \times 10^7 \\
&= 1.497 \times (10^1 \times 10^7) \qquad \text{Converting to scientific notation} \\
&= 1.497 \times 10^8 \text{ km.}
\end{aligned}
$$

Thus the distance from the sun to Earth is 1.497×10^8 km.

Do Exercise 24. ▶

EXAMPLE 26 *Media Usage.* In January 2013, the 800 million active YouTube users viewed 120 billion videos on the site. On average, how many videos did each user view?

Source: YouTube

In order to find the average number of YouTube videos that each user viewed, we divide the total number viewed by the number of users. We first write each number using scientific notation:

$$800 \text{ million} = 800{,}000{,}000 = 8 \times 10^8,$$

$$120 \text{ billion} = 120{,}000{,}000{,}000 = 1.2 \times 10^{11}.$$

---

24. *Niagara Falls Water Flow.* On the Canadian side, the amount of water that spills over Niagara Falls in 1 min during the summer is about

$$1.3088 \times 10^8 \text{ L.}$$

How much water spills over the falls in one day? Express the answer in scientific notation.

Answer
24. 1.884672×10^{11} L

25. *DNA.* The width of a DNA (deoxyribonucleic acid) double helix is about 2×10^{-9} m. If its length, fully stretched, is 5×10^{-2} m, how many times longer is the helix than it is wide?

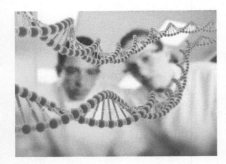

We then divide 1.2×10^{11} by 8×10^8:

$$\frac{1.2 \times 10^{11}}{8 \times 10^8} = \frac{1.2}{8} \times \frac{10^{11}}{10^8}$$
$$= 0.15 \times 10^3 = (1.5 \times 10^{-1}) \times 10^3 = 1.5 \times 10^2.$$

On average, each user viewed 1.5×10^2 videos.

◀ Do Exercise 25.

The following is a summary of the definitions and rules for exponents that we have considered in this section and the preceding one.

Answer

25. The length of the helix is 2.5×10^7 times its width.

DEFINITIONS AND RULES FOR EXPONENTS

Exponent of 1:	$a^1 = a$
Exponent of 0:	$a^0 = 1, a \neq 0$
Negative exponents:	$a^{-n} = \dfrac{1}{a^n}, \dfrac{1}{a^{-n}} = a^n, a \neq 0$
Product Rule:	$a^m \cdot a^n = a^{m+n}$
Quotient Rule:	$\dfrac{a^m}{a^n} = a^{m-n}, a \neq 0$
Power Rule:	$(a^m)^n = a^{mn}$
Raising a product to a power:	$(ab)^n = a^n b^n$
Raising a quotient to a power:	$\left(\dfrac{a}{b}\right)^n = \dfrac{a^n}{b^n}, b \neq 0;$
	$\left(\dfrac{a}{b}\right)^{-n} = \left(\dfrac{b}{a}\right)^n, b \neq 0, a \neq 0$
Scientific notation:	$M \times 10^n$, where $1 \leq M < 10$

13.2 | **Exercise Set**

For Extra Help

MyMathLab®   MathXL®
PRACTICE   WATCH   READ   REVIEW

✓ Reading Check

Choose from the column on the right the appropriate word to complete each statement. Not every word will be used.

RC1. To raise a power to a power, _____ the exponents.

RC2. To raise a product to the nth power, raise each factor to the _____ power.

RC3. To convert a number less than 1 to scientific notation, move the decimal point to the _____.

RC4. A _____ exponent in scientific notation indicates a number greater than or equal to 10.

add
left
multiply
negative
nth
positive
right

a , **b**   Simplify.

1. $(2^3)^2$

2. $(5^2)^4$

3. $(5^2)^{-3}$

4. $(7^{-3})^5$

5. $(x^{-3})^{-4}$

6. $(a^{-5})^{-6}$

7. $(a^{-2})^9$

8. $(x^{-5})^6$

9. $(t^{-3})^{-6}$

10. $(a^{-4})^{-7}$

11. $(t^4)^{-3}$

12. $(t^5)^{-2}$

13. $(x^{-2})^{-4}$

14. $(t^{-6})^{-5}$

15. $(ab)^3$

16. $(xy)^2$

17. $(ab)^{-3}$

18. $(xy)^{-6}$

19. $(mn^2)^{-3}$

20. $(x^3y)^{-2}$

21. $(4x^3)^2$

22. $4(x^3)^2$

23. $(3x^{-4})^2$

24. $(2a^{-5})^3$

25. $(x^4y^5)^{-3}$

26. $(t^5x^3)^{-4}$

27. $(x^{-6}y^{-2})^{-4}$

28. $(x^{-2}y^{-7})^{-5}$

29. $(a^{-2}b^7)^{-5}$

30. $(q^5r^{-1})^{-3}$

31. $(5r^{-4}t^3)^2$

32. $(4x^5y^{-6})^3$

33. $(a^{-5}b^7c^{-2})^3$

34. $(x^{-4}y^{-2}z^9)^2$

35. $(3x^3y^{-8}z^{-3})^2$

36. $(2a^2y^{-4}z^{-5})^3$

37. $(-4x^3y^{-2})^2$

38. $(-8x^3y^{-2})^3$

39. $(-a^{-3}b^{-2})^{-4}$

40. $(-p^{-4}q^{-3})^{-2}$

41. $\left(\dfrac{y^3}{2}\right)^2$

42. $\left(\dfrac{a^5}{3}\right)^3$

43. $\left(\dfrac{a^2}{b^3}\right)^4$

44. $\left(\dfrac{x^3}{y^4}\right)^5$

45. $\left(\dfrac{y^2}{2}\right)^{-3}$

46. $\left(\dfrac{a^4}{3}\right)^{-2}$

47. $\left(\dfrac{7}{x^{-3}}\right)^2$

48. $\left(\dfrac{3}{a^{-2}}\right)^3$

49. $\left(\dfrac{x^2y}{z}\right)^3$

50. $\left(\dfrac{m}{n^4p}\right)^3$

51. $\left(\dfrac{a^2b}{cd^3}\right)^{-2}$

52. $\left(\dfrac{2a^2}{3b^4}\right)^{-3}$

c   Convert to scientific notation.

53. 28,000,000,000

54. 4,900,000,000,000

55. 907,000,000,000,000,000

56. 168,000,000,000,000

57. 0.00000304

58. 0.000000000865

59. 0.000000018

60. 0.00000000002

61. 100,000,000,000

62. 0.0000001

63. *Population of the United States.*   It is estimated that the population of the United States will be 419,854,000 in 2050. Convert 419,854,000 to scientific notation.

Source: U.S. Census Bureau

64. *Microprocessors.*   The minimum feature size of a microprocessor is the transistor gate length. In 2011, the transistor gate length for a new microprocessor was about 0.000000028 m. Convert 0.000000028 to scientific notation.

65. *Wavelength of Light.*   The wavelength of red light is 0.00000068 m. Convert 0.00000068 to scientific notation.

66. *Olympics.*   Great Britain spent about $15,000,000,000 to stage the 2012 Summer Olympics. Convert 15,000,000,000 to scientific notation.

Source: cnn.com

Convert to decimal notation.

67. 8.74×10^7

68. 1.85×10^8

69. 5.704×10^{-8}

70. 8.043×10^{-4}

71. 10^7

72. 10^6

73. 10^{-5}

74. 10^{-8}

d   Multiply or divide and write scientific notation for the result.

75. $(3 \times 10^4)(2 \times 10^5)$

76. $(3.9 \times 10^8)(8.4 \times 10^{-3})$

77. $(5.2 \times 10^5)(6.5 \times 10^{-2})$

78. $(7.1 \times 10^{-7})(8.6 \times 10^{-5})$

79. $(9.9 \times 10^{-6})(8.23 \times 10^{-8})$

80. $(1.123 \times 10^4) \times 10^{-9}$

81. $\dfrac{8.5 \times 10^8}{3.4 \times 10^{-5}}$

82. $\dfrac{5.6 \times 10^{-2}}{2.5 \times 10^5}$

83. $(3.0 \times 10^6) \div (6.0 \times 10^9)$

84. $(1.5 \times 10^{-3}) \div (1.6 \times 10^{-6})$

85. $\dfrac{7.5 \times 10^{-9}}{2.5 \times 10^{12}}$

86. $\dfrac{4.0 \times 10^{-3}}{8.0 \times 10^{20}}$

 Solve.

87. *River Discharge.* The average discharge at the mouths of the Amazon River is 4,200,000 cubic feet per second. How much water is discharged from the Amazon River in 1 year? Express the answer in scientific notation.

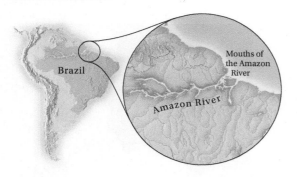

Brazil · Mouths of the Amazon River · Amazon River

88. *Coral Reefs.* There are 10 million bacteria per square centimeter of coral in a coral reef. The coral reefs near the Hawaiian Islands cover 14,000 km². How many bacteria are there in Hawaii's coral reefs?

Sources: livescience.com; U.S. Geological Survey

89. *Stars.* It is estimated that there are 10 billion trillion stars in the known universe. Express the number of stars in scientific notation. (1 billion $= 10^9$; 1 trillion $= 10^{12}$)

90. *Water Contamination.* Americans who change their own motor oil generate about 150 million gallons of used oil annually. If this oil is not disposed of properly, it can contaminate drinking water and soil. One gallon of used oil can contaminate one million gallons of drinking water. How many gallons of drinking water can 150 million gallons of oil contaminate? Express the answer in scientific notation. (1 million $= 10^6$).

Source: *New Car Buying Guide*

91. *Earth vs. Jupiter.* The mass of Earth is about 6×10^{21} metric tons. The mass of Jupiter is about 1.908×10^{24} metric tons. About how many times the mass of Earth is the mass of Jupiter? Express the answer in scientific notation.

92. *Office Supplies.* A ream of copier paper weighs 2.25 kg. How much does a sheet of copier paper weigh?

93. *Information Technology.* In 2012, about 2.5×10^{18} bytes of information were generated each day by the worldwide online population of 2×10^9 people. Find the average amount of information generated per person per day in 2012.

Sources: IBM; internetworldstats.com

94. *Computer Technology.* Intel Corporation has developed silicon-based connections that use lasers to move data at a rate of 50 gigabytes per second. The printed collection of the U.S. Library of Congress contains 10 terabytes of information. How long would it take to copy the Library of Congress using these connections? *Note:* 1 gigabyte $= 10^9$ bytes and 1 terabyte $= 10^{12}$ bytes.

Sources: spie.org; newworldencyclopedia.org

95. *Gold Leaf.* Gold can be milled into a very thin film called gold leaf. This film is so thin that it took only 43 oz of gold to cover the dome of Georgia's state capitol building. The gold leaf used was 5×10^{-6} m thick. In contrast, a U.S. penny is 1.55×10^{-3} m thick. How many sheets of gold leaf are in a stack that is the height of a penny?

Source: georgiaencyclopedia.org

96. *Relative Size.* An influenza virus is about 1.2×10^{-7} m in diameter. A staphylococcus bacterium is about 1.5×10^{-6} m in diameter. How many influenza viruses would it take, laid side by side, to equal the diameter of the bacterium?

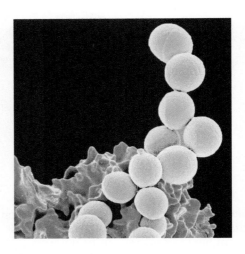

Space Travel. Use the following information for Exercises 97 and 98.

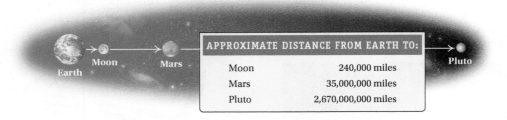

APPROXIMATE DISTANCE FROM EARTH TO:

Moon	240,000 miles
Mars	35,000,000 miles
Pluto	2,670,000,000 miles

97. *Time to Reach Mars.* Suppose that it takes about 3 days for a space vehicle to travel from Earth to the moon. About how long would it take the same space vehicle traveling at the same speed to reach Mars? Express the answer in scientific notation.

98. *Time to Reach Pluto.* Suppose that it takes about 3 days for a space vehicle to travel from Earth to the moon. About how long would it take the same space vehicle traveling at the same speed to reach the dwarf planet Pluto? Express the answer in scientific notation.

Skill Maintenance

Graph.

99. $y = x - 5$ [12.2a]

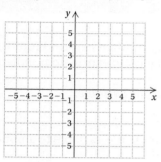

100. $2x + y = 4$ [12.2a]

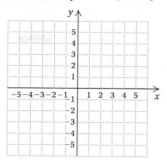

101. $3x - y = 3$ [12.2a]

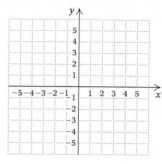

102. $y = -x$ [12.2a]

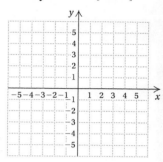

103. $2x = -10$ [12.3b]

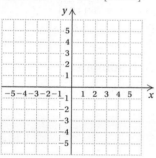

104. $y = -4$ [12.3b]

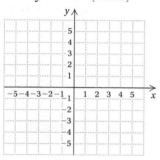

105. $8y - 16 = 0$ [12.3b]

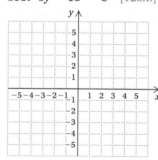

106. $x = 4$ [12.3b]

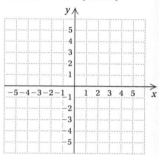

Synthesis

107. ▦ Carry out the indicated operations. Express the result in scientific notation.

$$\frac{(5.2 \times 10^6)(6.1 \times 10^{-11})}{1.28 \times 10^{-3}}$$

108. Find the reciprocal and express it in scientific notation.

$$6.25 \times 10^{-3}$$

Simplify.

109. $\dfrac{(5^{12})^2}{5^{25}}$

110. $\dfrac{a^{22}}{(a^2)^{11}}$

111. $\dfrac{(3^5)^4}{3^5 \cdot 3^4}$

112. $\left(\dfrac{5x^{-2}}{3y^{-2}z}\right)^0$

113. $\dfrac{49^{18}}{7^{35}}$

114. $\left(\dfrac{1}{a}\right)^{-n}$

115. $\dfrac{(0.4)^5}{[(0.4)^3]^2}$

116. $\left(\dfrac{4a^3b^{-2}}{5c^{-3}}\right)^1$

Determine whether each equation is true or false for all pairs of integers m and n and all positive numbers x and y.

117. $x^m \cdot y^n = (xy)^{mn}$

118. $x^m \cdot y^m = (xy)^{2m}$

119. $(x - y)^m = x^m - y^m$

120. $-x^m = (-x)^m$

121. $(-x)^{2m} = x^{2m}$

122. $x^{-m} = \dfrac{-1}{x^m}$

OBJECTIVES

a Evaluate a polynomial for a given value of the variable.

b Identify the terms of a polynomial and classify a polynomial by its number of terms.

c Identify the coefficient and the degree of each term of a polynomial and the degree of the polynomial.

d Collect the like terms of a polynomial.

e Arrange a polynomial in descending order, or collect the like terms and then arrange in descending order.

f Identify the missing terms of a polynomial.

SKILL TO REVIEW

Objective 10.7e: Collect like terms.

Collect like terms.

1. $3x - 4y + 5x + y$

2. $2a - 7b + 6 - 3a + 4b - 1$

1. Write three polynomials.

Answers

Skill to Review:

1. $8x - 3y$ **2.** $-a - 3b + 5$

Margin Exercise:

1. $4x^2 - 3x + \dfrac{5}{4}$; $15y^3$; $-7x^3 + 1.1$;

answers may vary

We have already learned to evaluate and to manipulate certain kinds of algebraic expressions. We will now consider algebraic expressions called *polynomials*.

The following are examples of *monomials in one variable*:

$$3x^2, \qquad 2x, \qquad -5, \qquad 37p^4, \qquad 0.$$

Each expression is a constant or a constant times some variable to a non-negative integer power.

MONOMIAL

A **monomial** is an expression of the type ax^n, where a is a real-number constant and n is a nonnegative integer.

Algebraic expressions like the following are **polynomials**:

$$\tfrac{3}{4}y^5, \quad -2, \quad 5y + 3, \quad 3x^2 + 2x - 5, \quad -7a^3 + \tfrac{1}{2}a, \quad 6x, \quad 37p^4, \quad x, \quad 0.$$

POLYNOMIAL

A **polynomial** is a monomial or a combination of sums and/or differences of monomials.

The following algebraic expressions are *not* polynomials:

$$(1) \ \frac{x+3}{x-4}, \qquad (2) \ 5x^3 - 2x^2 + \frac{1}{x}, \qquad (3) \ \frac{1}{x^3 - 2}.$$

Expressions (1) and (3) are not polynomials because they represent quotients, not sums or differences. Expression (2) is not a polynomial because

$$\frac{1}{x} = x^{-1},$$

and this is not a monomial because the exponent is negative.

◀ Do Margin Exercise 1.

a EVALUATING POLYNOMIALS AND APPLICATIONS

When we replace the variable in a polynomial with a number, the polynomial then represents a number called a **value** of the polynomial. Finding that number, or value, is called **evaluating the polynomial**. We evaluate a polynomial using the rules for order of operations.

EXAMPLE 1 Evaluate the polynomial when $x = 2$.

a) $3x + 5 = 3 \cdot 2 + 5$
$= 6 + 5$
$= 11$

b) $2x^2 - 7x + 3 = 2 \cdot 2^2 - 7 \cdot 2 + 3$
$= 2 \cdot 4 - 7 \cdot 2 + 3$
$= 8 - 14 + 3$
$= -3$

EXAMPLE 2 Evaluate the polynomial when $x = -4$.

a) $2 - x^3 = 2 - (-4)^3 = 2 - (-64)$
$$= 2 + 64 = 66$$

b) $-x^2 - 3x + 1 = -(-4)^2 - 3(-4) + 1$
$$= -16 + 12 + 1 = -3$$

Do Exercises 2–5. ▶

ALGEBRAIC ▶◀ **GRAPHICAL CONNECTION**

An equation like $y = 2x - 2$, which has a polynomial on one side and only y on the other, is called a **polynomial equation**. For such an equation, determining y is the same as evaluating the polynomial. Once the graph of such an equation has been drawn, we can evaluate the polynomial for a given x-value by finding the y-value that is paired with it on the graph.

EXAMPLE 3 Use *only* the given graph of $y = 2x - 2$ to evaluate the polynomial $2x - 2$ when $x = 3$.

First, we locate 3 on the x-axis. From there we move vertically to the graph of the equation and then horizontally to the y-axis. There we locate the y-value that is paired with 3. It appears that the y-value 4 is paired with 3. Thus the value of $2x - 2$ is 4 when $x = 3$. We can check this by evaluating $2x - 2$ when $x = 3$.

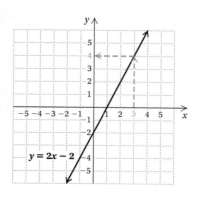

Do Exercise 6. ▶

Polynomial equations can be used to model many real-world situations.

EXAMPLE 4 *Games in a Sports League.* In a sports league of x teams in which each team plays every other team twice, the total number of games N to be played is given by the polynomial equation

$$N = x^2 - x.$$

A women's slow-pitch softball league has 10 teams and each team plays every other team twice. What is the total number of games to be played?

We evaluate the polynomial when $x = 10$:

$$N = x^2 - x = 10^2 - 10 = 100 - 10 = 90.$$

The league plays 90 games.

Do Exercise 7. ▶

Evaluate each polynomial when $x = 3$.

2. $-4x - 7$

3. $-5x^3 + 7x + 10$

Evaluate each polynomial when $x = -5$.

4. $5x + 7$

 5. $2x^2 + 5x - 4$
$$= 2(\quad)^2 + 5(\quad) - 4$$
$$= 2(\quad) + (\quad) - 4$$
$$= 50 - \quad - 4$$
$$= \quad$$

6. Use *only* the graph shown in Example 3 to evaluate the polynomial $2x - 2$ when $x = 4$ and when $x = -1$.

7. Refer to Example 4. Determine the total number of games to be played in a league of 12 teams in which each team plays every other team twice.

Answers

2. -19 **3.** -104 **4.** -18 **5.** 21
6. 6; -4 **7.** 132 games

Guided Solution:
5. $-5, -5, 25, -25, 25, 21$

CALCULATOR CORNER

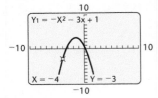

To use a graphing calculator to evaluate the polynomial in Example 2(b), $-x^2 - 3x + 1$, when $x = -4$, we can first graph $y_1 = -x^2 - 3x + 1$ in a window that includes the x-value -4. Then we can use the Value feature from the **CALC** menu, supplying the desired x-value and pressing **ENTER** to see X = −4, Y = −3 at the bottom of the screen. Thus, when $x = -4$, the value of $-x^2 - 3x + 1$ is -3.

```
            10
 Y1 = −X² − 3x + 1

−10 ┤├┼┼┼┼┼┼┼┼┼┼┼┼┤ 10
        X
  X = −4      Y = −3
           −10
```

EXERCISES: Use the Value feature to evaluate each polynomial for the given values of x.

1. $-x^2 - 3x + 1$, when $x = -2$, when $x = -0.5$, and when $x = 4$

2. $3x^2 - 5x + 2$, when $x = -3$, when $x = 1$, and when $x = 2.6$

---

8. *Medical Dosage.* Refer to Example 5.

  a) Determine the concentration after 3 hr by evaluating the polynomial when $t = 3$.

  b) Use *only* the graph showing medical dosage to check the value found in part (a).

9. *Medical Dosage.* Refer to Example 5. Use *only* the graph showing medical dosage to estimate the value of the polynomial when $t = 26$.

EXAMPLE 5 *Medical Dosage.* The concentration C, in parts per million, of a certain antibiotic in the bloodstream after t hours is given by the polynomial equation

$$C = -0.05t^2 + 2t + 2.$$

Find the concentration after 2 hr.

To find the concentration after 2 hr, we evaluate the polynomial when $t = 2$:

$$
\begin{aligned}
C &= -0.05t^2 + 2t + 2 \\
&= -0.05(2)^2 + 2(2) + 2 &&\text{Substituting 2 for } t \\
&= -0.05(4) + 2(2) + 2 &&\text{Carrying out the calculation using} \\
& &&\text{the rules for order of operations} \\
&= -0.2 + 4 + 2 \\
&= 3.8 + 2 \\
&= 5.8.
\end{aligned}
$$

The concentration after 2 hr is 5.8 parts per million.

ALGEBRAIC ▶◀ GRAPHICAL CONNECTION

The polynomial equation in Example 5 can be graphed if we evaluate the polynomial for several values of t. We list the values in a table and show the graph below. Note that the concentration peaks at the 20-hr mark and after slightly more than 40 hr, the concentration is 0. Since neither time nor concentration can be negative, our graph uses only the first quadrant.

	C
t	$C = -0.05t^2 + 2t + 2$
0	2
2	5.8 ← Example 5
10	17
20	22
30	17

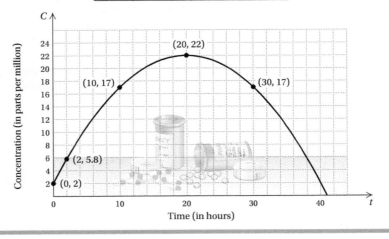

◀ **Do Exercises 8 and 9.**

Answers

8. (a) 7.55 parts per million; **(b)** When $t = 3$, $C \approx 7.5$ so the value found in part (a) appears to be correct. **9.** 20 parts per million

b IDENTIFYING TERMS AND CLASSIFYING POLYNOMIALS

For any polynomial that has some subtractions, we can find an equivalent polynomial using only additions.

EXAMPLES Find an equivalent polynomial using only additions.

6. $-5x^2 - x = -5x^2 + (-x)$

7. $4x^5 - 2x^6 + 4x - 7 = 4x^5 + (-2x^6) + 4x + (-7)$

Do Exercises 10 and 11. ▶

Find an equivalent polynomial using only additions.

10. $-9x^3 - 4x^5$

11. $-2y^3 + 3y^7 - 7y - 9$

When a polynomial is written using only additions, the monomials being added are called **terms**. In Example 6, the terms are $-5x^2$ and $-x$. In Example 7, the terms are $4x^5$, $-2x^6$, $4x$, and -7.

EXAMPLE 8 Identify the terms of the polynomial

$$4x^7 + 3x + 12 + 8x^3 + 5x.$$

Terms: $4x^7$, $3x$, 12, $8x^3$, and $5x$.

If there are subtractions, you can *think* of them as additions without rewriting.

EXAMPLE 9 Identify the terms of the polynomial

$$3t^4 - 5t^6 - 4t + 2.$$

Terms: $3t^4$, $-5t^6$, $-4t$, and 2.

Do Exercises 12 and 13. ▶

Identify the terms of each polynomial.

12. $3x^2 + 6x + \dfrac{1}{2}$

13. $-4y^5 + 7y^2 - 3y - 2$

Polynomials with just one term are called **monomials**. Polynomials with just two terms are called **binomials**. Those with just three terms are called **trinomials**. Those with more than three terms are generally not specified with a name.

EXAMPLE 10

MONOMIALS	BINOMIALS	TRINOMIALS	NONE OF THESE
$-4x^2$	$2x + 4$	$3x^5 + 4x^4 + 7x$	$4x^3 - 5x^2 + x - 8$
9	$-9x^7 - 6x$	$4x^2 - 6x - \frac{1}{2}$	$z^5 + 2z^4 - z^3 + 7z + 3$

Do Exercises 14–17. ▶

Classify each polynomial as a monomial, a binomial, a trinomial, or none of these.

14. $3x^2 + x$

15. $5x^4$

16. $4x^3 - 3x^2 + 4x + 2$

17. $3x^2 + 2x - 4$

c COEFFICIENTS AND DEGREES

The coefficient of the term $5x^3$ is 5. In the following polynomial, the red numbers are the **coefficients**, 3, -2, 5, and 4:

$$3x^5 - 2x^3 + 5x + 4.$$

Answers

10. $-9x^3 + (-4x^5)$

11. $-2y^3 + 3y^7 + (-7y) + (-9)$

12. $3x^2, 6x, \dfrac{1}{2}$ **13.** $-4y^5, 7y^2, -3y, -2$

14. Binomial **15.** Monomial

16. None of these **17.** Trinomial

EXAMPLE 11 Identify the coefficient of each term in the polynomial

$$3x^4 - 4x^3 + \frac{1}{2}x^2 + x - 8.$$

The coefficient of $3x^4$ is 3.
The coefficient of $-4x^3$ is -4.
The coefficient of $\frac{1}{2}x^2$ is $\frac{1}{2}$.
The coefficient of x (or $1x$) is 1.
The coefficient of -8 is -8.

18. Identify the coefficient of each term in the polynomial
$2x^4 - 7x^3 - 8.5x^2 - x - 4.$

◀ Do Exercise 18.

The **degree** of a term is the exponent of the variable. The degree of the term $-5x^3$ is 3.

EXAMPLE 12 Identify the degree of each term of $8x^4 - 3x + 7$.

The degree of $8x^4$ is 4.
The degree of $-3x$ (or $-3x^1$) is 1. $x = x^1$
The degree of 7 (or $7x^0$) is 0. $7 = 7 \cdot 1 = 7 \cdot x^0$, since $x^0 = 1$

Because we can write 1 as x^0, the degree of any constant term (except 0) is 0. The term 0 is a special case. We agree that it has *no* degree because we can express 0 as $0 = 0x^5 = 0x^7$, and so on, using any exponent we wish.

> The degree of any nonzero constant term is 0.

The **degree of a polynomial** is the largest of the degrees of the terms, unless it is the polynomial 0.

EXAMPLE 13 Identify the degree of the polynomial $5x^3 - 6x^4 + 7$.

$$5x^3 - 6x^4 + 7.$$ The largest degree is 4.

The degree of the polynomial is 4.

Identify the degree of each term and the degree of the polynomial.

19. $-6x^4 + 8x^2 - 2x + 9$

20. $4 - x^3 + \frac{1}{2}x^6 - x^5$

◀ Do Exercises 19 and 20.

Let's summarize the terminology that we have learned, using the polynomial $3x^4 - 8x^3 + x^2 + 7x - 6$.

TERM	COEFFICIENT	DEGREE OF THE TERM	DEGREE OF THE POLYNOMIAL
$3x^4$	3	4	
$-8x^3$	-8	3	
x^2	1	2	4
$7x$	7	1	
-6	-6	0	

d COLLECTING LIKE TERMS

When terms have the same variable and the same exponent, we say that they are **like terms**.

EXAMPLES Identify the like terms in each polynomial.

14. $4x^3 + 5x - 4x^2 + 2x^3 + x^2$

Like terms: $4x^3$ and $2x^3$ Same variable and exponent

Like terms: $-4x^2$ and x^2 Same variable and exponent

15. $6 - 3a^2 - 8 - a - 5a$

Like terms: 6 and -8 Constant terms are like terms; note that $6 = 6x^0$ and $-8 = -8x^0$.

Like terms: $-a$ and $-5a$

Do Exercises 21–23. ▶

We can often simplify polynomials by **collecting like terms**, or **combining like terms**. To do this, we use the distributive laws.

EXAMPLES Collect like terms.

16. $2x^3 - 6x^3 = (2 - 6)x^3 = -4x^3$

17. $5x^2 + 7 + 4x^4 + 2x^2 - 11 - 2x^4 = (5 + 2)x^2 + (4 - 2)x^4 + (7 - 11)$
$$= 7x^2 + 2x^4 - 4$$

Note that using the distributive laws in this manner allows us to collect like terms by adding or subtracting the coefficients. Often the middle step is omitted and we add or subtract mentally, writing just the answer. In collecting like terms, we may get 0.

EXAMPLE 18 Collect like terms: $3x^5 + 2x^2 - 3x^5 + 8$.

$3x^5 + 2x^2 - 3x^5 + 8 = (3 - 3)x^5 + 2x^2 + 8$
$$= 0x^5 + 2x^2 + 8$$
$$= 2x^2 + 8$$

Do Exercises 24–29. ▶

Expressing a term like x^2 by showing 1 as a factor, $1 \cdot x^2$, may make it easier to understand how to factor or collect like terms.

EXAMPLES Collect like terms.

19. $5x^8 - 6x^5 - x^8 = 5x^8 - 6x^5 - 1x^8$ Replacing x^8 with $1x^8$
$$= (5 - 1)x^8 - 6x^5 \quad \text{Using a distributive law}$$
$$= 4x^8 - 6x^5$$

20. $\frac{2}{3}x^4 - x^3 - \frac{1}{6}x^4 + \frac{2}{5}x^3 - \frac{3}{10}x^3$
$$= \left(\frac{2}{3} - \frac{1}{6}\right)x^4 + \left(-1 + \frac{2}{5} - \frac{3}{10}\right)x^3 \quad -x^3 = -1 \cdot x^3$$
$$= \left(\frac{4}{6} - \frac{1}{6}\right)x^4 + \left(-\frac{10}{10} + \frac{4}{10} - \frac{3}{10}\right)x^3$$
$$= \frac{3}{6}x^4 - \frac{9}{10}x^3 = \frac{1}{2}x^4 - \frac{9}{10}x^3$$

Do Exercises 30–32. ▶

Identify the like terms in each polynomial.

21. $4x^3 - x^3 + 2$

22. $4t^4 - 9t^3 - 7t^4 + 10t^3$

23. $5x^2 + 3x - 10 + 7x^2 - 8x + 11$

Collect like terms.

24. $3x^2 + 5x^2$

25. $4x^3 - 2x^3 + 2 + 5$

26. $\frac{1}{2}x^5 - \frac{3}{4}x^5 + 4x^2 - 2x^2$

27. $24 - 4x^3 - 24$

28. $5x^3 - 8x^5 + 8x^5$

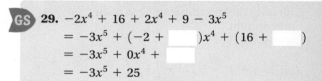

GS 29. $-2x^4 + 16 + 2x^4 + 9 - 3x^5$
$$= -3x^5 + (-2 + \boxed{})x^4 + (16 + \boxed{})$$
$$= -3x^5 + 0x^4 + \boxed{}$$
$$= -3x^5 + 25$$

Collect like terms.

30. $5x^3 - x^3 + 4$

31. $\frac{3}{4}x^3 + 4x^2 - x^3 + 7$

32. $\frac{4}{5}x^4 - x^4 + x^5 - \frac{1}{5} - \frac{1}{4}x^4 + 10$

Answers

21. $4x^3$ and $-x^3$ **22.** $4t^4$ and $-7t^4$; $-9t^3$ and $10t^3$ **23.** $5x^2$ and $7x^2$; $3x$ and $-8x$; -10 and 11

24. $8x^2$ **25.** $2x^3 + 7$ **26.** $-\frac{1}{4}x^5 + 2x^2$

27. $-4x^3$ **28.** $5x^3$ **29.** $-3x^5 + 25$

30. $4x^3 + 4$ **31.** $-\frac{1}{4}x^3 + 4x^2 + 7$

32. $x^5 - \frac{9}{20}x^4 + \frac{49}{5}$

Guided Solution:
29. 2, 9, 25

e DESCENDING ORDER

A polynomial is written in **descending order** when the term with the largest degree is written first, the term with the next largest degree is written next, and so on, in order from left to right.

EXAMPLES Arrange each polynomial in descending order.

21. $6x^5 + 4x^7 + x^2 + 2x^3 = 4x^7 + 6x^5 + 2x^3 + x^2$

22. $\frac{2}{3} + 4x^5 - 8x^2 + 5x - 3x^3 = 4x^5 - 3x^3 - 8x^2 + 5x + \frac{2}{3}$

◀ Do Exercises 33 and 34.

EXAMPLE 23 Collect like terms and then arrange in descending order:

$$2x^2 - 4x^3 + 3 - x^2 - 2x^3.$$

$$
\begin{aligned}
2x^2 - 4x^3 + 3 - x^2 - 2x^3 &= x^2 - 6x^3 + 3 && \text{Collecting like terms} \\
&= -6x^3 + x^2 + 3 && \text{Arranging in descending order}
\end{aligned}
$$

◀ Do Exercises 35 and 36.

The opposite of descending order is called **ascending order**. Generally, if an exercise is written in a certain order, we give the answer in that same order.

f MISSING TERMS

If a coefficient is 0, we generally do not write the term. We say that we have a **missing term**.

EXAMPLE 24 Identify the missing terms in the polynomial

$$8x^5 - 2x^3 + 5x^2 + 7x + 8.$$

There is no term with x^4. We say that the x^4-term is missing.

◀ Do Exercises 37–39.

We can either write missing terms with zero coefficients or leave space.

EXAMPLE 25 Write the polynomial $x^4 - 6x^3 + 2x - 1$ in two ways: with its missing term and by leaving space for it.

a) $x^4 - 6x^3 + 2x - 1 = x^4 - 6x^3 + 0x^2 + 2x - 1$ Writing with the missing x^2-term

b) $x^4 - 6x^3 + 2x - 1 = x^4 - 6x^3 \quad\quad + 2x - 1$ Leaving space for the missing x^2-term

EXAMPLE 26 Write the polynomial $y^5 - 1$ in two ways: with its missing terms and by leaving space for them.

a) $y^5 - 1 = y^5 + 0y^4 + 0y^3 + 0y^2 + 0y - 1$

b) $y^5 - 1 = y^5 \quad\quad\quad\quad\quad\quad\quad - 1$

◀ Do Exercises 40 and 41.

Arrange each polynomial in descending order.

33. $4x^2 - 3 + 7x^5 + 2x^3 - 5x^4$

34. $-14 + 7t^2 - 10t^5 + 14t^7$

Collect like terms and then arrange in descending order.

35. $3x^2 - 2x + 3 - 5x^2 - 1 - x$

36. $-x + \frac{1}{2} + 14x^4 - 7x - 1 - 4x^4$

Identify the missing term(s) in each polynomial.

37. $2x^3 + 4x^2 - 2$

38. $-3x^4$

39. $x^3 + 1$

Write each polynomial in two ways: with its missing term(s) and by leaving space for them.

40. $2x^3 + 4x^2 - 2$

41. $a^4 + 10$

Answers

33. $7x^5 - 5x^4 + 2x^3 + 4x^2 - 3$
34. $14t^7 - 10t^5 + 7t^2 - 14$ **35.** $-2x^2 - 3x + 2$
36. $10x^4 - 8x - \frac{1}{2}$ **37.** x **38.** x^3, x^2, x, x^0
39. x^2, x **40.** $2x^3 + 4x^2 + 0x - 2$;
$\quad\quad\quad 2x^3 + 4x^2 \quad\quad - 2$
41. $a^4 + 0a^3 + 0a^2 + 0a + 10$;
$\quad a^4 \quad\quad\quad\quad\quad\quad + 10$

☑ **Reading Check**

Choose from the column on the right the expression that best fits each description.

RC1. _____ The value of $x^2 - x$ when $x = -1$

RC2. _____ A polynomial written in ascending order

RC3. _____ A coefficient of $5x^4 - 3x + 7$

RC4. _____ A term of $5x^4 - 3x + 7$

RC5. _____ The degree of one of the terms of $5x^4 - 3x + 7$

RC6. _____ An example of a binomial

a) 0
b) 2
c) 5
d) $-3x$
e) $8x - 9$
f) $y + 6y^2 - 2y^8$

a   Evaluate each polynomial when $x = 4$ and when $x = -1$.

1. $-5x + 2$

2. $-8x + 1$

3. $2x^2 - 5x + 7$

4. $3x^2 + x - 7$

5. $x^3 - 5x^2 + x$

6. $7 - x + 3x^2$

Evaluate each polynomial when $x = -2$ and when $x = 0$.

7. $\frac{1}{3}x + 5$

8. $8 - \frac{1}{4}x$

9. $x^2 - 2x + 1$

10. $5x + 6 - x^2$

11. $-3x^3 + 7x^2 - 3x - 2$

12. $-2x^3 + 5x^2 - 4x + 3$

13. *Skydiving.* During the first 13 sec of a jump, the distance S, in feet, that a skydiver falls in t seconds can be approximated by the polynomial equation

$$S = 11.12t^2.$$

In 2009, 108 U.S. skydivers fell headfirst in formation from a height of 18,000 ft. How far had they fallen 10 sec after having jumped from the plane?

Source: www.telegraph.co.uk

14. *Skydiving.* For jumps that exceed 13 sec, the polynomial equation

$$S = 173t - 369$$

can be used to approximate the distance S, in feet, that a skydiver has fallen in t seconds. Approximately how far has a skydiver fallen 20 sec after having jumped from a plane?

15. Stacking Spheres. In 2004, the journal *Annals of Mathematics* accepted a proof of the so-called Kepler Conjecture: that the most efficient way to pack spheres is in the shape of a square pyramid. The number N of balls in the stack is given by the polynomial equation

$$N = \frac{1}{3}x^3 + \frac{1}{2}x^2 + \frac{1}{6}x,$$

where x is the number of layers. A square pyramid with 3 layers is illustrated below. Find the number of oranges in a pyramid with 5 layers.

Source: *The New York Times 4/6/04*

Bottom layer   Second layer   Top layer

16. SCAD Diving. The distance s, in feet, traveled by a body falling freely from rest in t seconds is approximated by the polynomial equation

$$s = 16t^2.$$

The SCAD thrill ride is a 2.5-sec free fall into a net. How far does the diver fall?

Source: "What is SCAD?", www.scadfreefall.co.uk

17. The graph of the polynomial equation $y = 5 - x^2$ is shown below. Use *only* the graph to estimate the value of the polynomial when $x = -3$, $x = -1$, $x = 0$, $x = 1.5$, and $x = 2$.

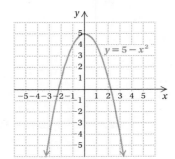

18. The graph of the polynomial equation $y = 6x^3 - 6x$ is shown below. Use *only* the graph to estimate the value of the polynomial when $x = -1$, $x = -0.5$, $x = 0.5$, $x = 1$, and $x = 1.1$.

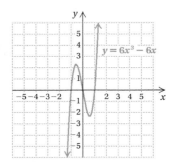

19. Solar Capacity. The average capacity C, in kilowatts (kW), of U.S. residential installations generating energy from the sun can be estimated by the polynomial equation

$$C = 0.27t + 2.97,$$

where t is the number of years since 2000.

Source: Based on data from IREC 2011 Updates & Trends

a) Use the equation to estimate the average capacity of a solar-energy residential installation in 2010.

b) Check the result of part (a) using the graph below.

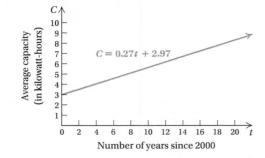

Number of years since 2000

20. Senior Population. The number N, in millions, of people in the United States ages 65 and older can be estimated by the polynomial equation

$$N = 1.24t + 28.4,$$

where t is the number of years since 2000.

Source: U.S. Census Bureau

a) Use the equation to estimate the number of people in the United States ages 65 and older in 2030.

b) Check the result of part (a) using the graph below.

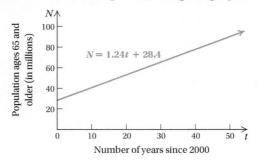

Number of years since 2000

Memorizing Words. Participants in a psychology experiment were able to memorize an average of M words in t minutes, where $M = -0.001t^3 + 0.1t^2$. Use the graph below for Exercises 21–26.

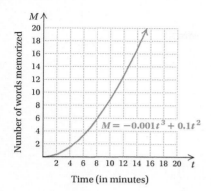

21. Estimate the number of words memorized after 10 min.

22. Estimate the number of words memorized after 14 min.

23. Find the approximate value of M for $t = 8$.

24. Find the approximate value of M for $t = 12$.

25. Estimate the value of M when t is 13.

26. Estimate the value of M when t is 7.

b   Identify the terms of each polynomial.

27. $2 - 3x + x^2$

28. $2x^2 + 3x - 4$

29. $-2x^4 + \frac{1}{3}x^3 - x + 3$

30. $-\frac{2}{5}x^5 - x^3 + 6$

Classify each polynomial as a monomial, a binomial, a trinomial, or none of these.

31. $x^2 - 10x + 25$

32. $-6x^4$

33. $x^3 - 7x^2 + 2x - 4$

34. $x^2 - 9$

35. $4x^2 - 25$

36. $2x^4 - 7x^3 + x^2 + x - 6$

37. $40x$

38. $4x^2 + 12x + 9$

c   Identify the coefficient of each term of the polynomial.

39. $-3x + 6$

40. $2x - 4$

41. $5x^2 + \frac{3}{4}x + 3$

42. $\frac{2}{3}x^2 - 5x + 2$

43. $-5x^4 + 6x^3 - 2.7x^2 + x - 2$

44. $7x^3 - x^2 - 4.2x + 5$

Identify the degree of each term of the polynomial and the degree of the polynomial.

45. $2x - 4$

46. $6 - 3x$

47. $3x^2 - 5x + 2$

48. $5x^3 - 2x^2 + 3$

49. $-7x^3 + 6x^2 + \frac{3}{5}x + 7$

50. $5x^4 + \frac{1}{4}x^2 - x + 2$

51. $x^2 - 3x + x^6 - 9x^4$

52. $8x - 3x^2 + 9 - 8x^3$

53. Complete the following table for the polynomial $-7x^4 + 6x^3 - x^2 + 8x - 2$.

TERM	COEFFICIENT	DEGREE OF THE TERM	DEGREE OF THE POLYNOMIAL
$-7x^4$			
$6x^3$	6		
		2	
$8x$		1	
	-2		

54. Complete the following table for the polynomial $3x^2 + x^5 - 46x^3 + 6x - 2.4 - \frac{1}{2}x^4$.

TERM	COEFFICIENT	DEGREE OF THE TERM	DEGREE OF THE POLYNOMIAL
		5	
$-\frac{1}{2}x^4$		4	
	-46		
$3x^2$		2	
	6		
-2.4			

d Identify the like terms in each polynomial.

55. $5x^3 + 6x^2 - 3x^2$

56. $3x^2 + 4x^3 - 2x^2$

57. $2x^4 + 5x - 7x - 3x^4$

58. $-3t + t^3 - 2t - 5t^3$

59. $3x^5 - 7x + 8 + 14x^5 - 2x - 9$

60. $8x^3 + 7x^2 - 11 - 4x^3 - 8x^2 - 29$

Collect like terms.

61. $2x - 5x$

62. $2x^2 + 8x^2$

63. $x - 9x$

64. $x - 5x$

65. $5x^3 + 6x^3 + 4$

66. $6x^4 - 2x^4 + 5$

67. $5x^3 + 6x - 4x^3 - 7x$

68. $3a^4 - 2a + 2a + a^4$

69. $6b^5 + 3b^2 - 2b^5 - 3b^2$

70. $2x^2 - 6x + 3x + 4x^2$

71. $\frac{1}{4}x^5 - 5 + \frac{1}{2}x^5 - 2x - 37$

72. $\frac{1}{3}x^3 + 2x - \frac{1}{6}x^3 + 4 - 16$

73. $6x^2 + 2x^4 - 2x^2 - x^4 - 4x^2$

74. $8x^2 + 2x^3 - 3x^3 - 4x^2 - 4x^2$

75. $\frac{1}{4}x^3 - x^2 - \frac{1}{6}x^2 + \frac{3}{8}x^3 + \frac{5}{16}x^3$

76. $\frac{1}{5}x^4 + \frac{1}{5} - 2x^2 + \frac{1}{10} - \frac{3}{15}x^4 + 2x^2 - \frac{3}{10}$

e   Arrange each polynomial in descending order.

77. $x^5 + x + 6x^3 + 1 + 2x^2$

78. $3 + 2x^2 - 5x^6 - 2x^3 + 3x$

79. $5y^3 + 15y^9 + y - y^2 + 7y^8$

80. $9p - 5 + 6p^3 - 5p^4 + p^5$

Collect like terms and then arrange in descending order.

81. $3x^4 - 5x^6 - 2x^4 + 6x^6$

82. $-1 + 5x^3 - 3 - 7x^3 + x^4 + 5$

83. $-2x + 4x^3 - 7x + 9x^3 + 8$

84. $-6x^2 + x - 5x + 7x^2 + 1$

85. $3x + 3x + 3x - x^2 - 4x^2$

86. $-2x - 2x - 2x + x^3 - 5x^3$

87. $-x + \frac{3}{4} + 15x^4 - x - \frac{1}{2} - 3x^4$

88. $2x - \frac{5}{6} + 4x^3 + x + \frac{1}{3} - 2x$

f   Identify the missing terms in each polynomial.

89. $x^3 - 27$

90. $x^5 + x$

91. $x^4 - x$

92. $5x^4 - 7x + 2$

93. $2x^3 - 5x^2 + x - 3$

94. $-6x^3$

Write each polynomial in two ways: with its missing terms and by leaving space for them.

95. $x^3 - 27$

96. $x^5 + x$

97. $x^4 - x$

98. $5x^4 - 7x + 2$

99. $2x^3 - 5x^2 + x - 3$

100. $-6x^3$

Skill Maintenance

Add. [10.3a]

101. $1 + (-20)$

102. $-\dfrac{2}{3} + \left(-\dfrac{1}{3}\right)$

103. $-4.2 + 1.95$

104. $-\dfrac{5}{8} + \dfrac{1}{4}$

Subtract. [10.4a]

105. $1 - 20$

106. $\dfrac{1}{8} - \dfrac{5}{6}$

107. $\dfrac{3}{8} - \left(-\dfrac{1}{4}\right)$

108. $5.6 - 8.2$

Multiply. [10.5a]

109. $(-6)(-3)$

110. $\left(-\dfrac{1}{2}\right)\left(\dfrac{2}{3}\right)$

111. $0.5\,(-1.2)$

112. $(-2)(-3)(-4)$

Divide, if possible. [10.6a, c]

113. $-600 \div (-30)$

114. $\dfrac{4}{5} \div \left(-\dfrac{1}{2}\right)$

115. $0 \div (-4)$

116. $\dfrac{-6.3}{-5 + 5}$

Synthesis

Collect like terms.

117. $6x^3 \cdot 7x^2 - (4x^3)^2 + (-3x^3)^2 - (-4x^2)(5x^3) - 10x^5 + 17x^6$

118. $(3x^2)^3 + 4x^2 \cdot 4x^4 - x^4(2x)^2 + ((2x)^2)^3 - 100x^2(x^2)^2$

119. Construct a polynomial in x (meaning that x is the variable) of degree 5 with four terms and coefficients that are integers.

120. What is the degree of $(5m^5)^2$?

121. A polynomial in x has degree 3. The coefficient of x^2 is 3 less than the coefficient of x^3. The coefficient of x is three times the coefficient of x^2. The remaining coefficient is 2 more than the coefficient of x^3. The sum of the coefficients is -4. Find the polynomial.

Use the CALC feature and choose VALUE on your graphing calculator to find the values in each of the following. (Refer to the Calculator Corner on p. 808.)

122. Exercise 18

123. Exercise 17

124. Exercise 22

125. Exercise 21

Addition and Subtraction of Polynomials

a ADDITION OF POLYNOMIALS

To add two polynomials, we can write a plus sign between them and then collect like terms. Depending on the situation, you may see polynomials written in descending order, ascending order, or neither. Generally, if an exercise is written in a particular order, we write the answer in that same order.

EXAMPLE 1 Add $(-3x^3 + 2x - 4)$ and $(4x^3 + 3x^2 + 2)$.

$$(-3x^3 + 2x - 4) + (4x^3 + 3x^2 + 2)$$
$$= (-3 + 4)x^3 + 3x^2 + 2x + (-4 + 2) \quad \text{Collecting like terms}$$
$$= x^3 + 3x^2 + 2x - 2$$

EXAMPLE 2 Add:

$$\left(\tfrac{2}{3}x^4 + 3x^2 - 2x + \tfrac{1}{2}\right) + \left(-\tfrac{1}{3}x^4 + 5x^3 - 3x^2 + 3x - \tfrac{1}{2}\right).$$

We have

$$\left(\tfrac{2}{3}x^4 + 3x^2 - 2x + \tfrac{1}{2}\right) + \left(-\tfrac{1}{3}x^4 + 5x^3 - 3x^2 + 3x - \tfrac{1}{2}\right)$$
$$= \left(\tfrac{2}{3} - \tfrac{1}{3}\right)x^4 + 5x^3 + (3 - 3)x^2 + (-2 + 3)x + \left(\tfrac{1}{2} - \tfrac{1}{2}\right) \quad \begin{array}{l}\text{Collecting}\\\text{like terms}\end{array}$$
$$= \tfrac{1}{3}x^4 + 5x^3 + x.$$

We can add polynomials as we do because they represent numbers. After some practice, you will be able to add mentally.

Do Margin Exercises 1–4. ▶

EXAMPLE 3 Add $(3x^2 - 2x + 2)$ and $(5x^3 - 2x^2 + 3x - 4)$.

$$(3x^2 - 2x + 2) + (5x^3 - 2x^2 + 3x - 4)$$
$$= 5x^3 + (3 - 2)x^2 + (-2 + 3)x + (2 - 4) \quad \begin{array}{l}\text{You might do this}\\\text{step mentally.}\end{array}$$
$$= 5x^3 + x^2 + x - 2 \quad \text{Then you would write only this.}$$

Do Exercises 5 and 6 on the following page. ▶

We can also add polynomials by writing like terms in columns.

EXAMPLE 4 Add $9x^5 - 2x^3 + 6x^2 + 3$ and $5x^4 - 7x^2 + 6$ and $3x^6 - 5x^5 + x^2 + 5$.

We arrange the polynomials with the like terms in columns.

$$
\begin{array}{l}
9x^5 \qquad\quad\; -\, 2x^3 + 6x^2 + \;\; 3 \\
\qquad\quad 5x^4 \qquad\quad\; -\, 7x^2 + \;\; 6 \qquad \text{We leave spaces for missing terms.} \\
\underline{3x^6 - 5x^5 \qquad\qquad\qquad\;\; +\; x^2 + \;\; 5} \\
3x^6 + 4x^5 + 5x^4 - 2x^3 \qquad\quad\; +\; 14 \qquad \text{Adding}
\end{array}
$$

We write the answer as $3x^6 + 4x^5 + 5x^4 - 2x^3 + 14$ without the space.

OBJECTIVES

a Add polynomials.

b Simplify the opposite of a polynomial.

c Subtract polynomials.

d Use polynomials to represent perimeter and area.

SKILL TO REVIEW

Objective 10.4a: Subtract real numbers and simplify combinations of additions and subtractions.

Simplify.

1. $-4 - (-8)$
2. $-5 - 6 + 4$

Add.

1. $(3x^2 + 2x - 2) + (-2x^2 + 5x + 5)$

2. $(-4x^5 + x^3 + 4) + (7x^4 + 2x^2)$

3. $(31x^4 + x^2 + 2x - 1) + (-7x^4 + 5x^3 - 2x + 2)$

4. $(17x^3 - x^2 + 3x + 4) + \left(-15x^3 + x^2 - 3x - \dfrac{2}{3}\right)$

Answers

Skill to Review:
1. 4   2. -7

Margin Exercises:
1. $x^2 + 7x + 3$
2. $-4x^5 + 7x^4 + x^3 + 2x^2 + 4$
3. $24x^4 + 5x^3 + x^2 + 1$
4. $2x^3 + \dfrac{10}{3}$

Add mentally. Try to write just the answer.

5. $(4x^2 - 5x + 3) +$
   $(-2x^2 + 2x - 4)$

6. $(3x^3 - 4x^2 - 5x + 3) +$
   $\left(5x^3 + 2x^2 - 3x - \dfrac{1}{2}\right)$

Add.

7. $\quad -2x^3 + 5x^2 - 2x + 4$
   $\quad\ \ x^4 \qquad\ + 6x^2 + 7x - 10$
   $-9x^4 + 6x^3 +\ \ x^2 \qquad\quad - 2$

8. $-3x^3 + 5x + 2$ and
   $x^3 + x^2 + 5$ and
   $x^3 - 2x - 4$

Simplify.

9. $-(4x^3 - 6x + 3)$

10. $-(5x^4 + 3x^2 + 7x - 5)$

11. $-(14x^{10} - \frac{1}{2}x^5 + 5x^3 - x^2 + 3x)$

Subtract.

12. $(7x^3 + 2x + 4) - (5x^3 - 4)$

13. $(-3x^2 + 5x - 4) -$
    $(-4x^2 + 11x - 2)$

◀ **Do Exercises 7 and 8.**

b OPPOSITES OF POLYNOMIALS

We can use the property of -1 to write an equivalent expression for an opposite. For example, the opposite of $x - 2y + 5$ can be written as

$$-(x - 2y + 5).$$

We find an equivalent expression by changing the sign of every term:

$$-(x - 2y + 5) = -x + 2y - 5.$$

We use this concept when we subtract polynomials.

OPPOSITES OF POLYNOMIALS

To find an equivalent polynomial for the **opposite**, or **additive inverse**, of a polynomial, change the sign of every term. This is the same as multiplying by -1.

EXAMPLE 5 Simplify: $-(x^2 - 3x + 4)$.

$$-(x^2 - 3x + 4) = -x^2 + 3x - 4$$

EXAMPLE 6 Simplify: $-(-t^3 - 6t^2 - t + 4)$.

$$-(-t^3 - 6t^2 - t + 4) = t^3 + 6t^2 + t - 4$$

EXAMPLE 7 Simplify: $-(-7x^4 - \frac{5}{9}x^3 + 8x^2 - x + 67)$.

$$-(-7x^4 - \tfrac{5}{9}x^3 + 8x^2 - x + 67) = 7x^4 + \tfrac{5}{9}x^3 - 8x^2 + x - 67$$

◀ **Do Exercises 9–11.**

c SUBTRACTION OF POLYNOMIALS

Recall that we can subtract a real number by adding its opposite, or additive inverse: $a - b = a + (-b)$. This allows us to subtract polynomials.

EXAMPLE 8 Subtract:

$$(9x^5 + x^3 - 2x^2 + 4) - (2x^5 + x^4 - 4x^3 - 3x^2).$$

We have

$(9x^5 + x^3 - 2x^2 + 4) - (2x^5 + x^4 - 4x^3 - 3x^2)$
$\quad = 9x^5 + x^3 - 2x^2 + 4 + [-(2x^5 + x^4 - 4x^3 - 3x^2)]$ Adding the opposite

$\quad = 9x^5 + x^3 - 2x^2 + 4 - 2x^5 - x^4 + 4x^3 + 3x^2$ Finding the opposite by changing the sign of *each* term

$\quad = 7x^5 - x^4 + 5x^3 + x^2 + 4.$ Adding (collecting like terms)

◀ **Do Exercises 12 and 13.**

We combine steps by changing the sign of each term of the polynomial being subtracted and collecting like terms. Try to do this mentally as much as possible.

EXAMPLE 9 Subtract: $(9x^5 + x^3 - 2x) - (-2x^5 + 5x^3 + 6)$.

$$(9x^5 + x^3 - 2x) - (-2x^5 + 5x^3 + 6)$$
$$= 9x^5 + x^3 - 2x + 2x^5 - 5x^3 - 6 \quad \text{Finding the opposite by changing the sign of each term}$$

$$= 11x^5 - 4x^3 - 2x - 6 \quad \text{Collecting like terms}$$

Do Exercises 14 and 15. ▶

We can use columns to subtract. We replace coefficients with their opposites, as shown in Example 9.

EXAMPLE 10 Write in columns and subtract:

$$(5x^2 - 3x + 6) - (9x^2 - 5x - 3).$$

a) $\begin{array}{l} 5x^2 - 3x + 6 \\ -(9x^2 - 5x - 3) \end{array}$ Writing like terms in columns

b) $\begin{array}{l} 5x^2 - 3x + 6 \\ -9x^2 + 5x + 3 \end{array}$ Changing signs

c) $\begin{array}{l} 5x^2 - 3x + 6 \\ \underline{-9x^2 + 5x + 3} \\ -4x^2 + 2x + 9 \end{array}$ Adding

If you can do so without error, you can arrange the polynomials in columns and write just the answer, remembering to change the signs and add.

EXAMPLE 11 Write in columns and subtract:

$$(x^3 + x^2 + 2x - 12) - (-2x^3 + x^2 - 3x).$$

$$\begin{array}{l} x^3 + x^2 + 2x - 12 \\ \underline{-(-2x^3 + x^2 - 3x)} \\ 3x^3 + 5x - 12 \end{array}$$

Leaving space for the missing term

Changing the signs and adding

Do Exercises 16 and 17. ▶

d POLYNOMIALS AND GEOMETRY

EXAMPLE 12 Find a polynomial for the sum of the areas of these four rectangles.

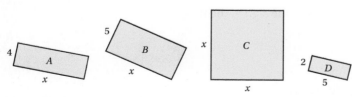

Recall that the area of a rectangle is the product of the length and the width. The sum of the areas is a sum of products. We find these products and then collect like terms.

Subtract.

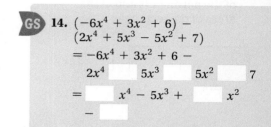

GS **14.** $(-6x^4 + 3x^2 + 6) - (2x^4 + 5x^3 - 5x^2 + 7)$

$$= -6x^4 + 3x^2 + 6 - 2x^4 \boxed{} 5x^3 \boxed{} 5x^2 \boxed{} 7$$

$$= \boxed{} x^4 - 5x^3 + \boxed{} x^2 - \boxed{}$$

15. $\left(\dfrac{3}{2}x^3 - \dfrac{1}{2}x^2 + 0.3\right) - \left(\dfrac{1}{2}x^3 + \dfrac{1}{2}x^2 + \dfrac{4}{3}x + 1.2\right)$

Write in columns and subtract.

16. $(4x^3 + 2x^2 - 2x - 3) - (2x^3 - 3x^2 + 2)$

17. $(2x^3 + x^2 - 6x + 2) - (x^5 + 4x^3 - 2x^2 - 4x)$

Answers

14. $-8x^4 - 5x^3 + 8x^2 - 1$

15. $x^3 - x^2 - \dfrac{4}{3}x - 0.9$

16. $2x^3 + 5x^2 - 2x - 5$

17. $-x^5 - 2x^3 + 3x^2 - 2x + 2$

Guided Solution:

14. $-, +, -, -8, 8, 1$

18. Find a polynomial for the sums of the perimeters and of the areas of the rectangles.

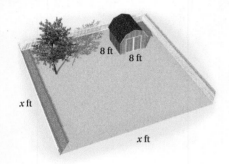

19. *Lawn Area.* An 8-ft by 8-ft shed is placed on a lawn x ft on a side. Find a polynomial for the remaining area.

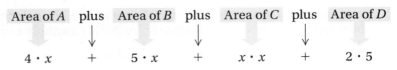

$$4 \cdot x \quad + \quad 5 \cdot x \quad + \quad x \cdot x \quad + \quad 2 \cdot 5$$

We collect like terms:

$$4x + 5x + x^2 + 10 = x^2 + 9x + 10.$$

◀ **Do Exercise 18.**

EXAMPLE 13 *Lawn Area.* A new city park is to contain a square grassy area that is x ft on a side. Within that grassy area will be a circular playground, with a radius of 15 ft, that will be mulched. To determine the amount of sod needed, find a polynomial for the grassy area.

We make a drawing, reword the problem, and write the polynomial.

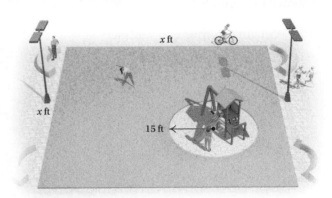

$$\underbrace{\text{Area of square}}_{x \cdot x \text{ ft}^2} \quad - \quad \underbrace{\text{Area of playground}}_{\pi \cdot 15^2 \text{ ft}^2} \quad = \quad \underbrace{\text{Area of grass}}_{\text{Area of grass}}$$

Then $(x^2 - 225\pi)$ ft^2 = Area of grass.

◀ **Do Exercise 19.**

Answers

18. Sum of perimeters: $13x$; sum of areas: $\dfrac{7}{2}x^2$

19. $(x^2 - 64)$ ft^2

13.4 Exercise Set

For Extra Help
MyMathLab®   MathXL®   PRACTICE   WATCH   READ   REVIEW

✓ Reading Check

Determine whether each statement is true or false.

RC1. To find the opposite of a polynomial, we need change only the sign of the first term.

RC2. We can subtract a polynomial by adding its opposite.

RC3. The sum of two binomials is always a binomial.

RC4. The area of a rectangle is the sum of its length and its width.

Add.

1. $(3x + 2) + (-4x + 3)$

2. $(6x + 1) + (-7x + 2)$

3. $(-6x + 2) + \left(x^2 + \frac{1}{2}x - 3\right)$

4. $\left(x^2 - \frac{5}{3}x + 4\right) + (8x - 9)$

5. $(x^2 - 9) + (x^2 + 9)$

6. $(x^3 + x^2) + (2x^3 - 5x^2)$

7. $(3x^2 - 5x + 10) + (2x^2 + 8x - 40)$

8. $(6x^4 + 3x^3 - 1) + (4x^2 - 3x + 3)$

9. $(1.2x^3 + 4.5x^2 - 3.8x) + (-3.4x^3 - 4.7x^2 + 23)$

10. $(0.5x^4 - 0.6x^2 + 0.7) + (2.3x^4 + 1.8x - 3.9)$

11. $(1 + 4x + 6x^2 + 7x^3) + (5 - 4x + 6x^2 - 7x^3)$

12. $(3x^4 - 6x - 5x^2 + 5) + (6x^2 - 4x^3 - 1 + 7x)$

13. $\left(\frac{1}{4}x^4 + \frac{2}{3}x^3 + \frac{5}{8}x^2 + 7\right) + \left(-\frac{3}{4}x^4 + \frac{3}{8}x^2 - 7\right)$

14. $\left(\frac{1}{3}x^9 + \frac{1}{5}x^5 - \frac{1}{2}x^2 + 7\right) +$
$\left(-\frac{1}{5}x^9 + \frac{1}{4}x^4 - \frac{3}{5}x^5 + \frac{3}{4}x^2 + \frac{1}{2}\right)$

15. $(0.02x^5 - 0.2x^3 + x + 0.08) +$
$(-0.01x^5 + x^4 - 0.8x - 0.02)$

16. $(0.03x^6 + 0.05x^3 + 0.22x + 0.05) +$
$\left(\frac{7}{100}x^6 - \frac{3}{100}x^3 + 0.5\right)$

17. $(9x^8 - 7x^4 + 2x^2 + 5) + (8x^7 + 4x^4 - 2x) +$
$(-3x^4 + 6x^2 + 2x - 1)$

18. $(4x^5 - 6x^3 - 9x + 1) + (6x^3 + 9x^2 + 9x) +$
$(-4x^3 + 8x^2 + 3x - 2)$

19. $\begin{array}{l} 0.15x^4 + 0.10x^3 - 0.9x^2 \\ - 0.01x^3 + 0.01x^2 + x \\ 1.25x^4 + 0.11x^2 + 0.01 \\ 0.27x^3 + 0.99 \\ \underline{-0.35x^4 + 15x^2 - 0.03} \end{array}$

20. $\begin{array}{l} 0.05x^4 + 0.12x^3 - 0.5x^2 \\ - 0.02x^3 + 0.02x^2 + 2x \\ 1.5x^4 + 0.01x^2 + 0.15 \\ 0.25x^3 + 0.85 \\ \underline{-0.25x^4 + 10x^2 - 0.04} \end{array}$

b Simplify.

21. $-(-5x)$

22. $-(x^2 - 3x)$

23. $-\left(-x^2 + \frac{3}{2}x - 2\right)$

24. $-\left(-4x^3 - x^2 - \frac{1}{4}x\right)$

25. $-(12x^4 - 3x^3 + 3)$

26. $-(4x^3 - 6x^2 - 8x + 1)$

27. $-(3x - 7)$

28. $-(-2x + 4)$

29. $-(4x^2 - 3x + 2)$

30. $-(-6a^3 + 2a^2 - 9a + 1)$

31. $-\left(-4x^4 + 6x^2 + \frac{3}{4}x - 8\right)$

32. $-(-5x^4 + 4x^3 - x^2 + 0.9)$

c Subtract.

33. $(3x + 2) - (-4x + 3)$

34. $(6x + 1) - (-7x + 2)$

35. $(-6x + 2) - (x^2 + x - 3)$

36. $(x^2 - 5x + 4) - (8x - 9)$

37. $(x^2 - 9) - (x^2 + 9)$

38. $(x^3 + x^2) - (2x^3 - 5x^2)$

39. $(6x^4 + 3x^3 - 1) - (4x^2 - 3x + 3)$

40. $(-4x^2 + 2x) - (3x^3 - 5x^2 + 3)$

41. $(1.2x^3 + 4.5x^2 - 3.8x) - (-3.4x^3 - 4.7x^2 + 23)$

42. $(0.5x^4 - 0.6x^2 + 0.7) - (2.3x^4 + 1.8x - 3.9)$

43. $\left(\frac{5}{8}x^3 - \frac{1}{4}x - \frac{1}{3}\right) - \left(-\frac{1}{8}x^3 + \frac{1}{4}x - \frac{1}{3}\right)$

44. $\left(\frac{1}{5}x^3 + 2x^2 - 0.1\right) - \left(-\frac{2}{5}x^3 + 2x^2 + 0.01\right)$

45. $(0.08x^3 - 0.02x^2 + 0.01x) - (0.02x^3 + 0.03x^2 - 1)$

46. $(0.8x^4 + 0.2x - 1) - \left(\frac{7}{10}x^4 + \frac{1}{5}x - 0.1\right)$

Subtract.

47.
$$x^2 + 5x + 6$$
$$\underline{-(x^2 + 2x)}$$

48.
$$x^3 + 1$$
$$\underline{-(x^3 + x^2)}$$

49.
$$5x^4 + 6x^3 - 9x^2$$
$$\underline{-(-6x^4 - 6x^3 + 8x + 9)}$$

50.
$$5x^4 + 6x^2 - 3x + 6$$
$$\underline{-(6x^3 + 7x^2 - 8x - 9)}$$

51.
$$x^5 - 1$$
$$\underline{-(x^5 - x^4 + x^3 - x^2 + x - 1)}$$

52.
$$x^5 + x^4 - x^3 + x^2 - x + 2$$
$$\underline{-(x^5 - x^4 + x^3 - x^2 - x + 2)}$$

 Solve.

Find a polynomial for the perimeter of each figure.

53.

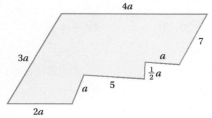

54.

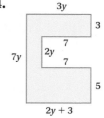

55. Find a polynomial for the sum of the areas of these rectangles.

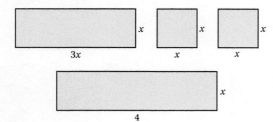

56. Find a polynomial for the sum of the areas of these circles.

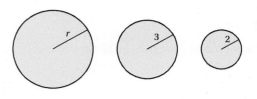

Find two algebraic expressions for the area of each figure. First, regard the figure as one large rectangle, and then regard the figure as a sum of four smaller rectangles.

57.

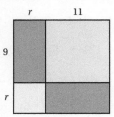

58.

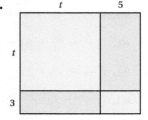

59.

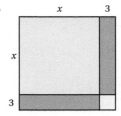

60.

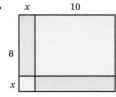

Find a polynomial for the shaded area of each figure.

61.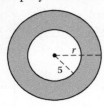

62.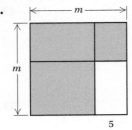

63.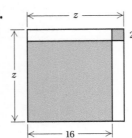

64.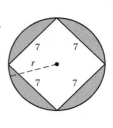

Skill Maintenance

Solve. [11.3b]

65. $8x + 3x = 66$

66. $5x - 7x = 38$

67. $\frac{3}{8}x + \frac{1}{4} - \frac{3}{4}x = \frac{11}{16} + x$

68. $5x - 4 = 26 - x$

69. $1.5x - 2.7x = 22 - 5.6x$

70. $3x - 3 = -4x + 4$

Solve. [11.3c]

71. $6(y - 3) - 8 = 4(y + 2) + 5$

72. $8(5x + 2) = 7(6x - 3)$

Solve. [11.7e]

73. $3x - 7 \leq 5x + 13$

74. $2(x - 4) > 5(x - 3) + 7$

Synthesis

Find a polynomial for the surface area of each right rectangular solid.

75.

76.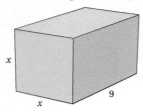

77.

78.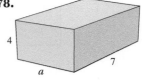

79. Find $(y - 2)^2$ using the four parts of this square.

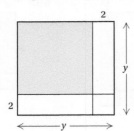

Simplify.

80. $(3x^2 - 4x + 6) - (-2x^2 + 4) + (-5x - 3)$

81. $(7y^2 - 5y + 6) - (3y^2 + 8y - 12) + (8y^2 - 10y + 3)$

82. $(-4 + x^2 + 2x^3) - (-6 - x + 3x^3) - (-x^2 - 5x^3)$

83. $(-y^4 - 7y^3 + y^2) + (-2y^4 + 5y - 2) - (-6y^3 + y^2)$

Mid-Chapter Review

Concept Reinforcement

Determine whether each statement is true or false.

_____ **1.** a^n and a^{-n} are reciprocals. [13.1f]

_____ **2.** $x^2 \cdot x^3 = x^6$ [13.1d]

_____ **3.** $-5y^4$ and $-5y^2$ are like terms. [13.3d]

_____ **4.** $4920^0 = 1$ [13.1b]

Guided Solutions

 Fill in each blank with the expression or operation sign that creates a correct statement or solution.

5. Collect like terms: $4w^3 + 6w - 8w^3 - 3w$. [13.3d]

$$4w^3 + 6w - 8w^3 - 3w = (4 - 8) \boxed{} + (6 - 3) \boxed{}$$
$$= \boxed{} w^3 + \boxed{} w$$

6. Subtract: $(3y^4 - y^2 + 11) - (y^4 - 4y^2 + 5)$. [13.4c]

$$(3y^4 - y^2 + 11) - (y^4 - 4y^2 + 5) = 3y^4 - y^2 + 11 \boxed{} y^4 \boxed{} 4y^2 \boxed{} 5$$
$$= \boxed{} y^4 + \boxed{} y^2 + \boxed{}$$

Mixed Review

Evaluate. [13.1b, c]

7. z^1

8. 4.56^0

9. a^5, when $a = -2$

10. $-x^3$, when $x = -1$

Multiply and simplify. [13.1d, f]

11. $5^3 \cdot 5^4$

12. $(3a)^2(3a)^7$

13. $x^{-8} \cdot x^5$

14. $t^4 \cdot t^{-4}$

Divide and simplify. [13.1e, f]

15. $\dfrac{7^8}{7^4}$

16. $\dfrac{x}{x^3}$

17. $\dfrac{w^5}{w^{-3}}$

18. $\dfrac{y^{-6}}{y^{-2}}$

Simplify. [13.2a, b]

19. $(3^5)^3$

20. $(x^{-3}y^2)^{-6}$

21. $\left(\dfrac{a^4}{5}\right)^6$

22. $\left(\dfrac{2y^3}{xz^2}\right)^{-2}$

Convert to scientific notation. [13.2c]

23. $25{,}430{,}000$

24. 0.00012

Convert to decimal notation. [13.2c]

25. 3.6×10^{-5}

26. 1.44×10^8

Multiply or divide and write scientific notation for the result.   [13.2d]

27. $(3 \times 10^6)(2 \times 10^{-3})$ **28.** $\dfrac{1.2 \times 10^{-4}}{2.4 \times 10^2}$

Evaluate the polynomial when $x = -3$ and when $x = 2$.   [13.3a]

29. $-3x + 7$ **30.** $x^3 - 2x + 5$

Collect like terms and then arrange in descending order.   [13.3e]

31. $3x - 2x^5 + x - 5x^2 + 2$ **32.** $4x^3 - 9x^2 - 2x^3 + x^2 + 8x^6$

Identify the degree of each term of the polynomial and the degree of the polynomial.   [13.3c]

33. $5x^3 - x + 4$ **34.** $2x - x^4 + 3x^6$

Classify the polynomial as a monomial, a binomial, a trinomial, or none of these.   [13.3b]

35. $x - 9$ **36.** $x^5 - 2x^3 + 6x^2$

Add or subtract.   [13.4a, c]

37. $(3x^2 - 1) + (5x^2 + 6)$ **38.** $(x^3 + 2x - 5) + (4x^3 - 2x^2 - 6)$

39. $(5x - 8) - (9x + 2)$ **40.** $(0.1x^2 - 2.4x + 3.6) - (0.5x^2 + x - 5.4)$

41. Find a polynomial for the sum of the areas of these rectangles.   [13.4d]

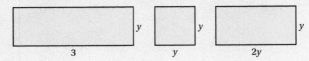

Understanding Through Discussion and Writing

42. Suppose that the length of a side of a square is three times the length of a side of a second square. How do the areas of the squares compare? Why?   [13.1d]

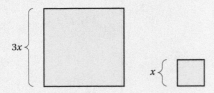

43. Suppose that the length of a side of a cube is twice the length of a side of a second cube. How do the volumes of the cubes compare? Why?   [13.1d]

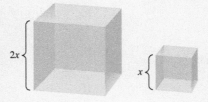

44. Explain in your own words when exponents should be added and when they should be multiplied. [13.1d], [13.2a]

45. Without performing actual computations, explain why 3^{-29} is smaller than 2^{-29}.   [13.1f]

46. Is it better to evaluate a polynomial before or after like terms have been collected? Why?   [13.3a, d]

47. Is the sum of two binomials ever a trinomial? Why or why not?   [13.3b], [13.4a]

Multiplication of Polynomials

13.5

We now multiply polynomials using techniques based, for the most part, on the distributive laws, but also on the associative and commutative laws.

a   MULTIPLYING MONOMIALS

Consider $(3x)(4x)$. We multiply as follows:

$$(3x)(4x) = 3 \cdot x \cdot 4 \cdot x \qquad \text{By the associative law of multiplication}$$
$$= 3 \cdot 4 \cdot x \cdot x \qquad \text{By the commutative law of multiplication}$$
$$= (3 \cdot 4)(x \cdot x) \qquad \text{By the associative law}$$
$$= 12x^2. \qquad \text{Using the product rule for exponents}$$

MULTIPLYING MONOMIALS

To find an equivalent expression for the product of two monomials, multiply the coefficients and then multiply the variables using the product rule for exponents.

EXAMPLES   Multiply.

1. $5x \cdot 6x = (5 \cdot 6)(x \cdot x)$ By the associative and commutative laws
$$= 30x^2 \qquad \text{Multiplying the coefficients and multiplying the variables}$$

2. $(3x)(-x) = (3x)(-1x) = (3)(-1)(x \cdot x) = -3x^2$

3. $(-7y^5)(4y^3) = (-7 \cdot 4)(y^5 \cdot y^3)$
$$= -28y^{5+3} = -28y^8 \qquad \text{Adding exponents}$$

After some practice, you will be able to multiply mentally.

Do Margin Exercises 1–8. ▶

b   MULTIPLYING A MONOMIAL AND ANY POLYNOMIAL

To multiply a monomial, such as $2x$, and a binomial, such as $5x + 3$, we use a distributive law and multiply each term of $5x + 3$ by $2x$:

$$2x(5x + 3) = (2x)(5x) + (2x)(3) \qquad \text{Using a distributive law}$$
$$= 10x^2 + 6x. \qquad \text{Multiplying the monomials}$$

OBJECTIVES

a Multiply monomials.

b Multiply a monomial and any polynomial.

c Multiply two binomials.

d Multiply any two polynomials.

SKILL TO REVIEW

Objective 10.7c: Use the distributive laws to multiply expressions like 8 and $x - y$.

Multiply.
1. $3(x - 5)$
2. $2(3y + 4z - 1)$

Multiply.

1. $(3x)(-5)$ **2.** $(-x) \cdot x$

3. $(-x)(-x)$ **4.** $(-x^2)(x^3)$

5. $3x^5 \cdot 4x^2$

6. $(4y^5)(-2y^6)$

7. $(-7y^4)(-y)$ **8.** $7x^5 \cdot 0$

Answers

Skill to Review:
1. $3x - 15$ **2.** $6y + 8z - 2$

Margin Exercises:
1. $-15x$ **2.** $-x^2$ **3.** x^2 **4.** $-x^5$
5. $12x^7$ **6.** $-8y^{11}$ **7.** $7y^5$ **8.** 0

EXAMPLE 4 Multiply: $5x(2x^2 - 3x + 4)$.

$$5x(2x^2 - 3x + 4) = (5x)(2x^2) - (5x)(3x) + (5x)(4)$$
$$= 10x^3 - 15x^2 + 20x$$

> **MULTIPLYING A MONOMIAL AND A POLYNOMIAL**
>
> To multiply a monomial and a polynomial, multiply each term of the polynomial by the monomial.

EXAMPLE 5 Multiply: $-2x^2(x^3 - 7x^2 + 10x - 4)$.

$$-2x^2(x^3 - 7x^2 + 10x - 4)$$
$$= (-2x^2)(x^3) - (-2x^2)(7x^2) + (-2x^2)(10x) - (-2x^2)(4)$$
$$= -2x^5 + 14x^4 - 20x^3 + 8x^2$$

◀ Do Exercises 9–11.

c MULTIPLYING TWO BINOMIALS

To find an equivalent expression for the product of two binomials, we use the distributive laws more than once. In Example 6, we use a distributive law three times.

EXAMPLE 6 Multiply: $(x + 5)(x + 4)$.

$$(x + 5)(x + 4) = x(x + 4) + 5(x + 4) \quad \text{Using a distributive law}$$

$$= x \cdot x + x \cdot 4 + 5 \cdot x + 5 \cdot 4 \quad \text{Using a distributive law twice}$$

$$= x^2 + 4x + 5x + 20 \quad \text{Multiplying the monomials}$$

$$= x^2 + 9x + 20 \quad \text{Collecting like terms}$$

To visualize the product in Example 6, consider a rectangle of length $x + 5$ and width $x + 4$.

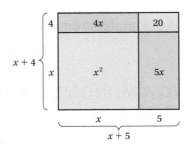

The total area can be expressed as $(x + 5)(x + 4)$ or, by adding the four smaller areas, $x^2 + 4x + 5x + 20$, or $x^2 + 9x + 20$.

◀ Do Exercises 12–14.

Multiply.

9. $4x(2x + 4)$

10. $3t^2(-5t + 2)$

11. $-5x^3(x^3 + 5x^2 - 6x + 8)$

12. a) Multiply: $(y + 2)(y + 7)$.

$(y + 2)(y + 7)$
$= y \cdot (y + 7) + 2 \cdot (\boxed{})$
$= y \cdot y + y \cdot \boxed{} + 2 \cdot y + 2 \cdot \boxed{}$
$= \boxed{} + 7y + 2y + \boxed{}$
$= y^2 + \boxed{} + 14$

b) Write an algebraic expression that represents the total area of the four smaller rectangles in the figure shown here.

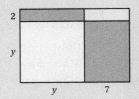

The area is $(y + 7)(y + \boxed{})$, or, from part (a), $y^2 + \boxed{} + 14$.

Multiply.

13. $(x + 8)(x + 5)$

14. $(x + 5)(x - 4)$

Answers

9. $8x^2 + 16x$ **10.** $-15t^3 + 6t^2$
11. $-5x^6 - 25x^5 + 30x^4 - 40x^3$
12. **(a)** $y^2 + 9y + 14$; **(b)** $(y + 2)(y + 7)$, or
$y^2 + 2y + 7y + 14$, or $y^2 + 9y + 14$
13. $x^2 + 13x + 40$ **14.** $x^2 + x - 20$

Guided Solution:
12. **(a)** $y + 7, 7, 7, y^2, 14, 9y$; **(b)** $2, 9y$

EXAMPLE 7 Multiply: $(4x + 3)(x - 2)$.

$(4x + 3)(x - 2) = 4x(x - 2) + 3(x - 2)$ Using a distributive law

$\qquad\qquad\quad = 4x \cdot x - 4x \cdot 2 + 3 \cdot x - 3 \cdot 2$ Using a distributive law twice

$\qquad\qquad\quad = 4x^2 - 8x + 3x - 6$ Multiplying the monomials

$\qquad\qquad\quad = 4x^2 - 5x - 6$ Collecting like terms

Do Exercises 15 and 16. ▶

Multiply.

15. $(5x + 3)(x - 4)$

16. $(2x - 3)(3x - 5)$

d MULTIPLYING ANY TWO POLYNOMIALS

Let's consider the product of a binomial and a trinomial. We use a distributive law four times. You may see ways to skip some steps and do the work mentally.

EXAMPLE 8 Multiply: $(x^2 + 2x - 3)(x^2 + 4)$.

$(x^2 + 2x - 3)(x^2 + 4) = x^2(x^2 + 4) + 2x(x^2 + 4) - 3(x^2 + 4)$

$\qquad\qquad\qquad\qquad = x^2 \cdot x^2 + x^2 \cdot 4 + 2x \cdot x^2 + 2x \cdot 4 - 3 \cdot x^2 - 3 \cdot 4$

$\qquad\qquad\qquad\qquad = x^4 + 4x^2 + 2x^3 + 8x - 3x^2 - 12$

$\qquad\qquad\qquad\qquad = x^4 + 2x^3 + x^2 + 8x - 12$

Do Exercises 17 and 18. ▶

Multiply.

17. $(x^2 + 3x - 4)(x^2 + 5)$

PRODUCT OF TWO POLYNOMIALS

To multiply two polynomials P and Q, select one of the polynomials—say, P. Then multiply each term of P by every term of Q and collect like terms.

GS 18. $(3y^2 - 7)(2y^3 - 2y + 5)$

$= 3y^2(2y^3 - 2y + 5) - \boxed{}(2y^3 - 2y + 5)$

$= 6y^5 - \boxed{} + 15y^2 - 14y^3 + \boxed{} - 35$

$= 6y^5 - \boxed{} + 15y^2 + 14y - 35$

To use columns for long multiplication, multiply each term in the top row by every term in the bottom row. We write like terms in columns, and then add the results. Such multiplication is like multiplying with whole numbers.

```
   3 2 1              300 + 20 + 1
 ×   1 2        ×            10 + 2
   6 4 2              600 + 40 + 2      Multiplying the top row by 2
   3 2 1        3000 + 200 + 10        Multiplying the top row by 10
   3 8 5 2      3000 + 800 + 50 + 2    Adding
```

EXAMPLE 9 Multiply: $(4x^3 - 2x^2 + 3x)(x^2 + 2x)$.

$$
\begin{array}{r}
4x^3 - 2x^2 + 3x \\
x^2 + 2x \\
\hline
8x^4 - 4x^3 + 6x^2 \\
4x^5 - 2x^4 + 3x^3 \\
\hline
4x^5 + 6x^4 - x^3 + 6x^2
\end{array}
$$

Multiplying the top row by $2x$

Multiplying the top row by x^2

Collecting like terms

Line up like terms in columns.

Answers

15. $5x^2 - 17x - 12$ **16.** $6x^2 - 19x + 15$
17. $x^4 + 3x^3 + x^2 + 15x - 20$
18. $6y^5 - 20y^3 + 15y^2 + 14y - 35$
Guided Solution:
18. $7, 6y^3, 14y, 20y^3$

EXAMPLE 10 Multiply: $(2x^2 + 3x - 4)(2x^2 - x + 3)$.

$$
\begin{array}{r}
2x^2 + 3x - 4 \\
2x^2 - x + 3 \\
\hline
6x^2 + 9x - 12 \\
-2x^3 - 3x^2 + 4x \\
4x^4 + 6x^3 - 8x^2 \\
\hline
4x^4 + 4x^3 - 5x^2 + 13x - 12
\end{array}
$$

Multiplying by 3
Multiplying by $-x$
Multiplying by $2x^2$
Collecting like terms

19. Multiply.

$$
\begin{array}{r}
3x^2 - 2x - 5 \\
2x^2 + x - 2
\end{array}
$$

◀ **Do Exercise 19.**

Multiply.

20. $3x^2 - 2x + 4$
$ x + 5$

21. $-5x^2 + 4x + 2$
$ -4x^2 - 8$

EXAMPLE 11 Multiply: $(5x^3 - 3x + 4)(-2x^2 - 3)$.

If terms are missing, it helps to leave spaces for them and align like terms in columns as we multiply.

$$
\begin{array}{r}
5x^3 - 3x + 4 \\
-2x^2 - 3 \\
\hline
-15x^3 + 9x - 12 \\
-10x^5 + 6x^3 - 8x^2 \\
\hline
-10x^5 - 9x^3 - 8x^2 + 9x - 12
\end{array}
$$

Multiplying by -3
Multiplying by $-2x^2$
Collecting like terms

◀ **Do Exercises 20 and 21.**

Answers

19. $6x^4 - x^3 - 18x^2 - x + 10$
20. $3x^3 + 13x^2 - 6x + 20$
21. $20x^4 - 16x^3 + 32x^2 - 32x - 16$

CALCULATOR CORNER

Checking Multiplication of Polynomials A partial check of multiplication of polynomials can be performed graphically. Consider the product $(x + 3)(x - 2) = x^2 + x - 6$. We will use two graph styles to determine whether this product is correct. First, we press **MODE** and select the **SEQUENTIAL** mode.

Next, on the Y= screen, we enter $y_1 = (x + 3)(x - 2)$ and $y_2 = x^2 + x - 6$. We will select the line-graph style for y_1 and the path style for y_2. To select these graph styles, we use ◁ to position the cursor over the icon to the left of the equation and press **ENTER** repeatedly until the desired style of icon appears, as shown below. Then we graph the equations.

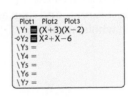

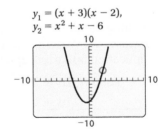

$y_1 = (x + 3)(x - 2),$
$y_2 = x^2 + x - 6$

The graphing calculator will graph y_1 first as a solid line. Then it will graph y_2 as the circular cursor traces the leading edge of the graph, allowing us to determine visually whether the graphs coincide. In this case, the graphs appear to coincide, so the factorization is probably correct.

A table of values can also be used as a check.

EXERCISES: Determine graphically whether each product is correct.

1. $(x + 5)(x + 4) = x^2 + 9x + 20$

2. $(4x + 3)(x - 2) = 4x^2 - 5x - 6$

3. $(5x + 3)(x - 4) = 5x^2 + 17x - 12$

4. $(2x - 3)(3x - 5) = 6x^2 - 19x - 15$

✓ Reading Check

Match each expression with an equivalent expression from the column on the right. Choices may be used more than once or not at all.

RC1. $8x \cdot 2x$

RC2. $(-16x)(-x)$

RC3. $2x(8x - 1)$

RC4. $(2x - 1)(8x + 1)$

a) $16x^2$
b) $-16x^2$
c) $16x^2 - 1$
d) $16x^2 - 2x$
e) $16x^2 - 6x - 1$

a   Multiply.

1. $(8x^2)(5)$

2. $(4x^2)(-2)$

3. $(-x^2)(-x)$

4. $(-x^3)(x^2)$

5. $(8x^5)(4x^3)$

6. $(10a^2)(2a^2)$

7. $(0.1x^6)(0.3x^5)$

8. $(0.3x^4)(-0.8x^6)$

9. $\left(-\frac{1}{5}x^3\right)\left(-\frac{1}{3}x\right)$

10. $\left(-\frac{1}{4}x^4\right)\left(\frac{1}{5}x^8\right)$

11. $(-4x^2)(0)$

12. $(-4m^5)(-1)$

13. $(3x^2)(-4x^3)(2x^6)$

14. $(-2y^5)(10y^4)(-3y^3)$

b   Multiply.

15. $2x(-x + 5)$

16. $3x(4x - 6)$

17. $-5x(x - 1)$

18. $-3x(-x - 1)$

19. $x^2(x^3 + 1)$

20. $-2x^3(x^2 - 1)$

21. $3x(2x^2 - 6x + 1)$

22. $-4x(2x^3 - 6x^2 - 5x + 1)$

23. $(-6x^2)(x^2 + x)$

24. $(-4x^2)(x^2 - x)$

25. $(3y^2)(6y^4 + 8y^3)$

26. $(4y^4)(y^3 - 6y^2)$

c   Multiply.

27. $(x + 6)(x + 3)$

28. $(x + 5)(x + 2)$

29. $(x + 5)(x - 2)$

30. $(x + 6)(x - 2)$

31. $(x - 1)(x + 4)$

32. $(x - 8)(x + 7)$

33. $(x - 4)(x - 3)$

34. $(x - 7)(x - 3)$

35. $(x + 3)(x - 3)$

36. $(x + 6)(x - 6)$

37. $(x - 4)(x + 4)$

38. $(x - 9)(x + 9)$

39. $(3x + 5)(x + 2)$

40. $(2x + 6)(x + 3)$

41. $(5 - x)(5 - 2x)$

42. $(3 - 4x)(2 - x)$

43. $(2x + 5)(2x + 5)$ **44.** $(3x + 4)(3x + 4)$ **45.** $(x - 3)(x - 3)$ **46.** $(x - 6)(x - 6)$

47. $\left(x - \frac{5}{2}\right)\left(x + \frac{2}{5}\right)$ **48.** $\left(x + \frac{4}{3}\right)\left(x + \frac{3}{2}\right)$ **49.** $(x - 2.3)(x + 4.7)$ **50.** $(2x + 0.13)(2x - 0.13)$

Write an algebraic expression that represents the total area of the four smaller rectangles in each figure.

51. 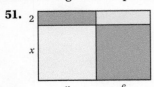   **52.** 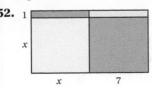   **53.** 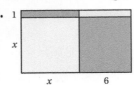   **54.**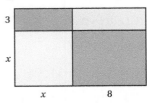

Draw and label rectangles similar to the one following Example 6 to illustrate each product.

55. $x(x + 5)$ **56.** $x(x + 2)$ **57.** $(x + 1)(x + 2)$

58. $(x + 3)(x + 1)$ **59.** $(x + 5)(x + 3)$ **60.** $(x + 4)(x + 6)$

d   Multiply.

61. $(x^2 + x + 1)(x - 1)$ **62.** $(x^2 + x - 2)(x + 2)$ **63.** $(2x + 1)(2x^2 + 6x + 1)$

64. $(3x - 1)(4x^2 - 2x - 1)$ **65.** $(y^2 - 3)(3y^2 - 6y + 2)$ **66.** $(3y^2 - 3)(y^2 + 6y + 1)$

67. $(x^3 + x^2)(x^3 + x^2 - x)$ **68.** $(x^3 - x^2)(x^3 - x^2 + x)$ **69.** $(-5x^3 - 7x^2 + 1)(2x^2 - x)$

70. $(-4x^3 + 5x^2 - 2)(5x^2 + 1)$ **71.** $(1 + x + x^2)(-1 - x + x^2)$ **72.** $(1 - x + x^2)(1 - x + x^2)$

73. $(2t^2 - t - 4)(3t^2 + 2t - 1)$ **74.** $(3a^2 - 5a + 2)(2a^2 - 3a + 4)$ **75.** $(x - x^3 + x^5)(x^2 - 1 + x^4)$

76. $(x - x^3 + x^5)(3x^2 + 3x^6 + 3x^4)$ **77.** $(x + 1)(x^3 + 7x^2 + 5x + 4)$ **78.** $(x + 2)(x^3 + 5x^2 + 9x + 3)$

79. $\left(x - \frac{1}{2}\right)\left(2x^3 - 4x^2 + 3x - \frac{2}{5}\right)$ **80.** $\left(x + \frac{1}{3}\right)\left(6x^3 - 12x^2 - 5x + \frac{1}{2}\right)$

Skill Maintenance

Simplify.

81. $x - (2x - 3)$ [10.8b]

82. $5 - 2[3 - 4(8 - 2)]$
   [10.8c]

83. $(10 - 2)(10 + 2)$
   [10.8d]

84. $10 - 2 + (-6)^2 \div 3 \cdot 2$
   [10.8d]

Factor.   [10.7d]

85. $15x - 18y + 12$

86. $16x - 24y + 36$

87. $-9x - 45y + 15$

88. $100x - 100y + 1000a$

Synthesis

Find a polynomial for the shaded area of each figure.

89.

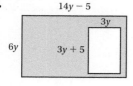

90.

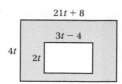

For each figure, determine what the missing number must be in order for the figure to have the given area.

91. Area $= x^2 + 8x + 15$

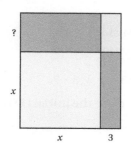

92. Area $= x^2 + 7x + 10$

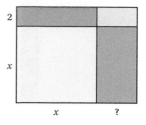

93. Find a polynomial for the volume of the solid shown below.

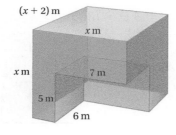

94. An open wooden box is a cube with side x cm. The box, including its bottom, is made of wood that is 1 cm thick. Find a polynomial for the interior volume of the cube.

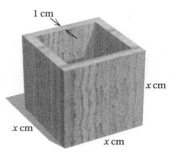

Compute and simplify.

95. $(x - 2)(x - 7) - (x - 7)(x - 2)$

96. $(x + 5)^2 - (x - 3)^2$

97. Extend the pattern and simplify:
$$(x - a)(x - b)(x - c)(x - d) \cdots (x - z).$$

98. ☑ Use a graphing calculator to check your answers to Exercises 15, 29, and 61. Use graphs, tables, or both, as directed by your instructor.

SKILL TO REVIEW

Objective 10.7e: Collect like terms.

Collect like terms.

1. $\dfrac{2}{3}x - \dfrac{2}{3}x$

2. $-12n + 12n$

We encounter certain products so often that it is helpful to have efficient methods of computing them. Such techniques are called *special products*.

a   PRODUCTS OF TWO BINOMIALS USING FOIL

To multiply two binomials, we can select one binomial and multiply each term of that binomial by every term of the other. Then we collect like terms. Consider the product $(x + 3)(x + 7)$:

$$(x + 3)(x + 7) = x(x + 7) + 3(x + 7)$$
$$= x \cdot x + x \cdot 7 + 3 \cdot x + 3 \cdot 7$$
$$= x^2 + 7x + 3x + 21$$
$$= x^2 + 10x + 21.$$

This example illustrates a special technique for finding the product of two binomials:

$$
\begin{array}{cccc}
\text{First} & \text{Outside} & \text{Inside} & \text{Last} \\
\text{terms} & \text{terms} & \text{terms} & \text{terms}
\end{array}
$$

$$(x + 3)(x + 7) = x \cdot x + 7 \cdot x + 3 \cdot x + 3 \cdot 7.$$

To remember this method of multiplying, we use the initials **FOIL**.

THE FOIL METHOD

To multiply two binomials, $A + B$ and $C + D$, multiply the First terms AC, the Outside terms AD, the Inside terms BC, and then the Last terms BD. Then collect like terms, if possible.

$$(A + B)(C + D) = AC + AD + BC + BD$$

1. Multiply First terms: AC.

2. Multiply Outside terms: AD.

3. Multiply Inside terms: BC.

4. Multiply Last terms: BD.

$$\downarrow$$

FOIL

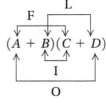

EXAMPLE 1   Multiply: $(x + 8)(x^2 - 5)$.

We have

$$
\begin{array}{cccc}
\text{F} & \text{O} & \text{I} & \text{L}
\end{array}
$$
$$(x + 8)(x^2 - 5) = x \cdot x^2 + x \cdot (-5) + 8 \cdot x^2 + 8(-5)$$
$$= x^3 - 5x + 8x^2 - 40$$
$$= x^3 + 8x^2 - 5x - 40.$$

Since each of the original binomials is in descending order, we write the product in descending order, as is customary, but this is not a "must."

Answers

Skill to Review:
1. 0   **2.** 0

Often we can collect like terms after we have multiplied.

EXAMPLES Multiply.

2. $(x + 6)(x - 6) = x^2 - 6x + 6x - 36$ Using FOIL
$$= x^2 - 36$$ Collecting like terms

3. $(x + 7)(x + 4) = x^2 + 4x + 7x + 28$
$$= x^2 + 11x + 28$$

4. $(y - 3)(y - 2) = y^2 - 2y - 3y + 6$
$$= y^2 - 5y + 6$$

5. $(x^3 - 1)(x^3 + 5) = x^6 + 5x^3 - x^3 - 5$
$$= x^6 + 4x^3 - 5$$

6. $(4t^3 + 5)(3t^2 - 2) = 12t^5 - 8t^3 + 15t^2 - 10$

Do Exercises 1–8. ▷

EXAMPLES Multiply.

7. $\left(x - \frac{2}{3}\right)\left(x + \frac{2}{3}\right) = x^2 + \frac{2}{3}x - \frac{2}{3}x - \frac{4}{9}$
$$= x^2 - \frac{4}{9}$$

8. $(x^2 - 0.3)(x^2 - 0.3) = x^4 - 0.3x^2 - 0.3x^2 + 0.09$
$$= x^4 - 0.6x^2 + 0.09$$

9. $(3 - 4x)(7 - 5x^3) = 21 - 15x^3 - 28x + 20x^4$
$$= 21 - 28x - 15x^3 + 20x^4$$

(*Note*: If the original polynomials are in ascending order, it is natural to write the product in ascending order, but this is not a "must.")

10. $(5x^4 + 2x^3)(3x^2 - 7x) = 15x^6 - 35x^5 + 6x^5 - 14x^4$
$$= 15x^6 - 29x^5 - 14x^4$$

Do Exercises 9–12. ▷

We can show the FOIL method geometrically as follows. One way to write the area of the large rectangle below is $(A + B)(C + D)$. To find another expression for the area of the large rectangle, we add the areas of the smaller rectangles.

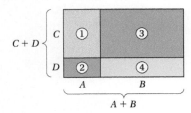

The area of rectangle ① is AC.
The area of rectangle ② is AD.
The area of rectangle ③ is BC.
The area of rectangle ④ is BD.

The area of the large rectangle is the sum of the areas of the smaller rectangles. Thus,

$$(A + B)(C + D) = AC + AD + BC + BD.$$

Multiply mentally, if possible. If you need extra steps, be sure to use them.

1. $(x + 3)(x + 4)$

2. $(x + 3)(x - 5)$

3. $(2x - 1)(x - 4)$

4. $(2x^2 - 3)(x^2 - 2)$

5. $(6x^2 + 5)(2x^3 + 1)$

6. $(y^3 + 7)(y^3 - 7)$

7. $(t + 2)(t + 3)$

8. $(2x^4 + x^2)(-x^3 + x)$

Multiply.

9. $\left(x + \frac{4}{5}\right)\left(x - \frac{4}{5}\right)$

10. $(x^3 - 0.5)(x^2 + 0.5)$

11. $(2 + 3x^2)(4 - 5x^2)$

12. $(6x^3 - 3x^2)(5x^2 - 2x)$

Answers

1. $x^2 + 7x + 12$ **2.** $x^2 - 2x - 15$
3. $2x^2 - 9x + 4$ **4.** $2x^4 - 7x^2 + 6$
5. $12x^5 + 10x^3 + 6x^2 + 5$ **6.** $y^6 - 49$
7. $t^2 + 5t + 6$ **8.** $-2x^7 + x^5 + x^3$
9. $x^2 - \dfrac{16}{25}$ **10.** $x^5 + 0.5x^3 - 0.5x^2 - 0.25$
11. $8 + 2x^2 - 15x^4$ **12.** $30x^5 - 27x^4 + 6x^3$

b MULTIPLYING SUMS AND DIFFERENCES OF TWO TERMS

Consider the product of the sum and the difference of the same two terms, such as

$$(x + 2)(x - 2).$$

Since this is the product of two binomials, we can use FOIL. This type of product occurs so often, however, that it would be valuable if we could use an even faster method. To find a faster way to compute such a product, look for a pattern in the following:

a) $(x + 2)(x - 2) = x^2 - 2x + 2x - 4$ Using FOIL
$$= x^2 - 4;$$

b) $(3x - 5)(3x + 5) = 9x^2 + 15x - 15x - 25$
$$= 9x^2 - 25.$$

◀ Do Exercises 13 and 14.

Perhaps you discovered in each case that when you multiply the two binomials, two terms are opposites, or additive inverses, which add to 0 and "drop out."

PRODUCT OF THE SUM AND THE DIFFERENCE OF TWO TERMS

The product of the sum and the difference of the same two terms is the square of the first term minus the square of the second term:

$$(A + B)(A - B) = A^2 - B^2.$$

It is helpful to memorize this rule in both words and symbols. (If you do forget it, you can, of course, use FOIL.)

EXAMPLES Multiply. (Carry out the rule and say the words as you go.)

$$(A + B)(A - B) = A^2 - B^2$$

11. $(x + 4)(x - 4) = x^2 - 4^2$ "The square of the first term, x^2, minus the square of the second, 4^2"
$$= x^2 - 16$$ Simplifying

12. $(5 + 2w)(5 - 2w) = 5^2 - (2w)^2$
$$= 25 - 4w^2$$

13. $(3x^2 - 7)(3x^2 + 7) = (3x^2)^2 - 7^2$
$$= 9x^4 - 49$$

14. $(-4x - 10)(-4x + 10) = (-4x)^2 - 10^2$
$$= 16x^2 - 100$$

15. $\left(x + \dfrac{3}{8}\right)\left(x - \dfrac{3}{8}\right) = x^2 - \left(\dfrac{3}{8}\right)^2 = x^2 - \dfrac{9}{64}$

◀ Do Exercises 15–19.

Multiply.

13. $(x + 5)(x - 5)$

14. $(2x - 3)(2x + 3)$

Multiply.

15. $(x + 8)(x - 8)$

16. $(x - 7)(x + 7)$

17. $(6 - 4y)(6 + 4y)$ GS
$$(\quad)^2 - (\quad)^2 = 36 - \boxed{}$$

18. $(2x^3 - 1)(2x^3 + 1)$

19. $\left(x - \dfrac{2}{5}\right)\left(x + \dfrac{2}{5}\right)$

Answers

13. $x^2 - 25$ **14.** $4x^2 - 9$ **15.** $x^2 - 64$
16. $x^2 - 49$ **17.** $36 - 16y^2$ **18.** $4x^6 - 1$
19. $x^2 - \dfrac{4}{25}$

Guided Solution:
17. $6, 4y, 16y^2$

c SQUARING BINOMIALS

Consider the square of a binomial, such as $(x + 3)^2$. This can be expressed as $(x + 3)(x + 3)$. Since this is the product of two binomials, we can use FOIL. But again, this type of product occurs so often that we would like to use an even faster method. Look for a pattern in the following.

a) $(x + 3)^2 = (x + 3)(x + 3)$
$= x^2 + 3x + 3x + 9$
$= x^2 + 6x + 9;$

b) $(x - 3)^2 = (x - 3)(x - 3)$
$= x^2 - 3x - 3x + 9$
$= x^2 - 6x + 9$

Do Exercises 20 and 21. ▶

Multiply.

20. $(x + 8)(x + 8)$

21. $(x - 5)(x - 5)$

When squaring a binomial, we multiply a binomial by itself. Perhaps you noticed that two terms are the same and when added give twice the product of the terms in the binomial. The other two terms are squares.

SQUARE OF A BINOMIAL

The square of a sum or a difference of two terms is the square of the first term, plus twice the product of the two terms, plus the square of the last term:

$$(A + B)^2 = A^2 + 2AB + B^2; \qquad (A - B)^2 = A^2 - 2AB + B^2.$$

It is helpful to memorize this rule in both words and symbols.

EXAMPLES Multiply. (Carry out the rule and say the words as you go.)

$(A + B)^2 = A^2 + 2 \cdot A \cdot B + B^2$

16. $(x + 3)^2 = x^2 + 2 \cdot x \cdot 3 + 3^2$ "x^2 plus 2 times x times 3 plus 3^2"
$= x^2 + 6x + 9$

$(A - B)^2 = A^2 - 2 \cdot A \cdot B + B^2$

17. $(t - 5)^2 = t^2 - 2 \cdot t \cdot 5 + 5^2$
$= t^2 - 10t + 25$

18. $(2x + 7)^2 = (2x)^2 + 2 \cdot 2x \cdot 7 + 7^2 = 4x^2 + 28x + 49$

19. $(5x - 3x^2)^2 = (5x)^2 - 2 \cdot 5x \cdot 3x^2 + (3x^2)^2 = 25x^2 - 30x^3 + 9x^4$

20. $(2.3 - 5.4m)^2 = 2.3^2 - 2(2.3)(5.4m) + (5.4m)^2$
$= 5.29 - 24.84m + 29.16m^2$

Do Exercises 22–27. ▶

Multiply.

22. $(x + 2)^2$

23. $(a - 4)^2$

24. $(2x + 5)^2$

25. $(4x^2 - 3x)^2$

26. $(7.8 + 1.2y)(7.8 + 1.2y)$

GS 27. $(3x^2 - 5)(3x^2 - 5)$
$(3x^2)^2 - 2(3x^2)() + 5^2$
$= x^4 - x^2 + 25$

Caution!

Although the square of a product is the product of the squares, the square of a sum is *not* the sum of the squares. That is, $(AB)^2 = A^2B^2$, but

The term $2AB$ is missing.
↓
$(A + B)^2 \neq A^2 + B^2.$

To illustrate this inequality, note, using the rules for order of operations, that $(7 + 5)^2 = 12^2 = 144$, whereas $7^2 + 5^2 = 49 + 25 = 74$, and $74 \neq 144$.

Answers

20. $x^2 + 16x + 64$ **21.** $x^2 - 10x + 25$
22. $x^2 + 4x + 4$ **23.** $a^2 - 8a + 16$
24. $4x^2 + 20x + 25$ **25.** $16x^4 - 24x^3 + 9x^2$
26. $60.84 + 18.72y + 1.44y^2$
27. $9x^4 - 30x^2 + 25$

We can look at the rule for finding $(A + B)^2$ geometrically as follows. The area of the large square is

$$(A + B)(A + B) = (A + B)^2.$$

This is equal to the sum of the areas of the smaller rectangles:

$$A^2 + AB + AB + B^2 = A^2 + 2AB + B^2.$$

Thus, $(A + B)^2 = A^2 + 2AB + B^2.$

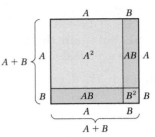

◀ Do Exercise 28.

28. In the figure at right, describe in terms of area the sum $A^2 + B^2$. How can the figure be used to verify that $(A + B)^2 \neq A^2 + B^2$?

d MULTIPLICATION OF VARIOUS TYPES

Let's now try several types of multiplications mixed together so that we can learn to sort them out. When you multiply, first see what kind of multiplication you have. Then use the best method.

> **MULTIPLYING TWO POLYNOMIALS**
> ....
>
> **1.** Is it the product of a monomial and a polynomial? If so, multiply each term of the polynomial by the monomial.
> *Example:* $5x(x + 7) = 5x \cdot x + 5x \cdot 7 = 5x^2 + 35x$
>
> **2.** Is it the product of the sum and the difference of the *same* two terms? If so, use the following:
> $$(A + B)(A - B) = A^2 - B^2.$$
> *Example:* $(x + 7)(x - 7) = x^2 - 7^2 = x^2 - 49$
>
> **3.** Is the product the square of a binomial? If so, use the following:
> $$(A + B)(A + B) = (A + B)^2 = A^2 + 2AB + B^2,$$
> or $(A - B)(A - B) = (A - B)^2 = A^2 - 2AB + B^2.$
> *Example:* $(x + 7)(x + 7) = (x + 7)^2$
> $$= x^2 + 2 \cdot x \cdot 7 + 7^2$$
> $$= x^2 + 14x + 49$$
>
> **4.** Is it the product of two binomials other than those above? If so, use FOIL.
> *Example:* $(x + 7)(x - 4) = x^2 - 4x + 7x - 28$
> $$= x^2 + 3x - 28$$
>
> **5.** Is it the product of two polynomials other than those above? If so, multiply each term of one by every term of the other. Use columns if you wish.
> *Example:*
> $$(x^2 - 3x + 2)(x + 7) = x^2(x + 7) - 3x(x + 7) + 2(x + 7)$$
> $$= x^2 \cdot x + x^2 \cdot 7 - 3x \cdot x - 3x \cdot 7$$
> $$+ 2 \cdot x + 2 \cdot 7$$
> $$= x^3 + 7x^2 - 3x^2 - 21x + 2x + 14$$
> $$= x^3 + 4x^2 - 19x + 14$$

Answers

28. $(A + B)^2$ represents the area of the large square. This includes all four sections. $A^2 + B^2$ represents the area of only two of the sections.

Guided Solution:
27. 5, 9, 30

Remember that FOIL will *always* work for two binomials. You can use it instead of either of rules (2) and (3), but those rules will make your work go faster.

EXAMPLE 21 Multiply: $(x + 3)(x - 3)$.

$$(x + 3)(x - 3) = x^2 - 3^2$$
$$= x^2 - 9$$

This is the product of the sum and the difference of the same two terms. We use $(A + B)(A - B) = A^2 - B^2$. ∎

EXAMPLE 22 Multiply: $(t + 7)(t - 5)$.

$$(t + 7)(t - 5) = t^2 + 2t - 35$$

This is the product of two binomials, but neither the square of a binomial nor the product of the sum and the difference of two terms. We use FOIL. ∎

EXAMPLE 23 Multiply: $(x + 6)(x + 6)$.

$$(x + 6)(x + 6) = x^2 + 2(6)x + 6^2$$
$$= x^2 + 12x + 36$$

This the square of a binomial. We use $(A + B)(A + B) = A^2 + 2AB + B^2$. ∎

EXAMPLE 24 Multiply: $2x^3(9x^2 + x - 7)$.

$$2x^3(9x^2 + x - 7) = 18x^5 + 2x^4 - 14x^3$$

This is the product of a monomial and a trinomial. We multiply each term of the trinomial by the monomial. ∎

EXAMPLE 25 Multiply: $(5x^3 - 7x)^2$.

$$(5x^3 - 7x)^2 = (5x^3)^2 - 2(5x^3)(7x) + (7x)^2$$
$$= 25x^6 - 70x^4 + 49x^2$$

$(A - B)^2 = A^2 - 2AB + B^2$ ∎

EXAMPLE 26 Multiply: $\left(3x + \frac{1}{4}\right)^2$.

$$\left(3x + \frac{1}{4}\right)^2 = (3x)^2 + 2(3x)\left(\frac{1}{4}\right) + \left(\frac{1}{4}\right)^2$$
$$= 9x^2 + \frac{3}{2}x + \frac{1}{16}$$

$(A + B)^2 = A^2 + 2AB + B^2$ ∎

EXAMPLE 27 Multiply: $\left(4x - \frac{3}{4}\right)\left(4x + \frac{3}{4}\right)$.

$$\left(4x - \frac{3}{4}\right)\left(4x + \frac{3}{4}\right) = (4x)^2 - \left(\frac{3}{4}\right)^2$$
$$= 16x^2 - \frac{9}{16}$$

$(A + B)(A - B) = A^2 - B^2$ ∎

EXAMPLE 28 Multiply: $(p + 3)(p^2 + 2p - 1)$.

$$\begin{array}{r} p^2 + 2p - 1 \\ p + 3 \\ \hline 3p^2 + 6p - 3 \\ p^3 + 2p^2 - p \\ \hline p^3 + 5p^2 + 5p - 3 \end{array}$$

Finding the product of two polynomials

Multiplying by 3

Multiplying by p

Do Exercises 29–36. ▶

Multiply.

29. $(x + 5)(x + 6)$

30. $(t - 4)(t + 4)$

31. $4x^2(-2x^3 + 5x^2 + 10)$

32. $(9x^2 + 1)^2$

33. $(2a - 5)(2a + 8)$

34. $\left(5x + \dfrac{1}{2}\right)^2$

35. $\left(2x - \dfrac{1}{2}\right)^2$

36. $(x^2 - x + 4)(x - 2)$

Answers

29. $x^2 + 11x + 30$ **30.** $t^2 - 16$
31. $-8x^5 + 20x^4 + 40x^2$ **32.** $81x^4 + 18x^2 + 1$
33. $4a^2 + 6a - 40$ **34.** $25x^2 + 5x + \dfrac{1}{4}$
35. $4x^2 - 2x + \dfrac{1}{4}$ **36.** $x^3 - 3x^2 + 6x - 8$

Visualizing for Success

1

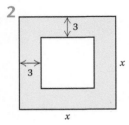

2

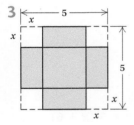

3

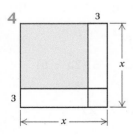

4

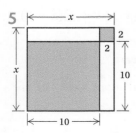

5

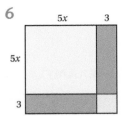

In each of Exercises 1–10, find two algebraic expressions for the shaded area of the figure from the list below.

A. $9 - 4x^2$

B. $x^2 - (x - 6)^2$

C. $(x + 3)(x - 3)$

D. $10^2 + 2^2$

E. $x^2 + 8x + 15$

F. $(x + 5)(x + 3)$

G. $x^2 - 6x + 9$

H. $(3 - 2x)^2 + 4x(3 - 2x)$

I. $(x + 3)^2$

J. $(5x + 3)^2$

K. $(5 - 2x)^2 + 4x(5 - 2x)$

L. $x^2 - 9$

M. 104

N. $x^2 - 15$

O. $12x - 36$

P. $25x^2 + 30x + 9$

Q. $(x - 5)(x - 3)$
$\quad\quad + 3(x - 5) + 5(x - 3)$

R. $(x - 3)^2$

S. $25 - 4x^2$

T. $x^2 + 6x + 9$

Answers on page A-25

6

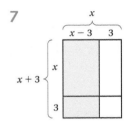

7

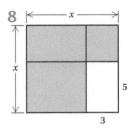

8

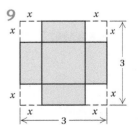

9

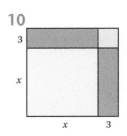

10

☑ Reading Check

Complete each statement with the appropriate word from the column on the right. A word may be used more than once or not at all.

RC1. For the FOIL multiplication method, the initials F O I L represent the words first, _____, inside, and _____.

RC2. If the polynomials being multiplied are written in descending order, we generally write the product in _____ order.

RC3. The expression $(A + B)(A - B)$ is the product of the sum and the _____ of the same two terms.

RC4. The expression $(A + B)^2$ is the _____ of a _____.

RC5. We can find the product of any two _____ using the FOIL method.

RC6. The product of the sum and the difference of the same two terms is the _____ of their squares.

ascending

binomial(s)

descending

difference

last

outside

product

square

a   Multiply. Try to write only the answer. If you need more steps, be sure to use them.

1. $(x + 1)(x^2 + 3)$

2. $(x^2 - 3)(x - 1)$

3. $(x^3 + 2)(x + 1)$

4. $(x^4 + 2)(x + 10)$

5. $(y + 2)(y - 3)$

6. $(a + 2)(a + 3)$

7. $(3x + 2)(3x + 2)$

8. $(4x + 1)(4x + 1)$

9. $(5x - 6)(x + 2)$

10. $(x - 8)(x + 8)$

11. $(3t - 1)(3t + 1)$

12. $(2m + 3)(2m + 3)$

13. $(4x - 2)(x - 1)$

14. $(2x - 1)(3x + 1)$

15. $\left(p - \frac{1}{4}\right)\left(p + \frac{1}{4}\right)$

16. $\left(q + \frac{3}{4}\right)\left(q + \frac{3}{4}\right)$

17. $(x - 0.1)(x + 0.1)$

18. $(x + 0.3)(x - 0.4)$

19. $(2x^2 + 6)(x + 1)$

20. $(2x^2 + 3)(2x - 1)$

21. $(-2x + 1)(x + 6)$

22. $(3x + 4)(2x - 4)$

23. $(a + 7)(a + 7)$

24. $(2y + 5)(2y + 5)$

25. $(1 + 2x)(1 - 3x)$

26. $(-3x - 2)(x + 1)$

27. $\left(\frac{3}{8}y - \frac{5}{6}\right)\left(\frac{3}{8}y - \frac{5}{6}\right)$

28. $\left(\frac{1}{5}x - \frac{2}{7}\right)\left(\frac{1}{5}x + \frac{2}{7}\right)$

29. $(x^2 + 3)(x^3 - 1)$

30. $(x^4 - 3)(2x + 1)$

31. $(3x^2 - 2)(x^4 - 2)$

32. $(x^{10} + 3)(x^{10} - 3)$

33. $(2.8x - 1.5)(4.7x + 9.3)$

34. $\left(x - \frac{3}{8}\right)\left(x + \frac{4}{7}\right)$

35. $(3x^5 + 2)(2x^2 + 6)$

36. $(1 - 2x)(1 + 3x^2)$

37. $(4x^2 + 3)(x - 3)$

38. $(7x - 2)(2x - 7)$

39. $(4y^4 + y^2)(y^2 + y)$

40. $(5y^6 + 3y^3)(2y^6 + 2y^3)$

b Multiply mentally, if possible. If you need extra steps, be sure to use them.

41. $(x + 4)(x - 4)$

42. $(x + 1)(x - 1)$

43. $(2x + 1)(2x - 1)$

44. $(x^2 + 1)(x^2 - 1)$

45. $(5m - 2)(5m + 2)$

46. $(3x^4 + 2)(3x^4 - 2)$

47. $(2x^2 + 3)(2x^2 - 3)$

48. $(6x^5 - 5)(6x^5 + 5)$

49. $(3x^4 - 4)(3x^4 + 4)$

50. $(t^2 - 0.2)(t^2 + 0.2)$

51. $(x^6 - x^2)(x^6 + x^2)$

52. $(2x^3 - 0.3)(2x^3 + 0.3)$

53. $(x^4 + 3x)(x^4 - 3x)$

54. $\left(\frac{3}{4} + 2x^3\right)\left(\frac{3}{4} - 2x^3\right)$

55. $(x^{12} - 3)(x^{12} + 3)$

56. $(12 - 3x^2)(12 + 3x^2)$

57. $(2y^8 + 3)(2y^8 - 3)$

58. $\left(m - \frac{2}{3}\right)\left(m + \frac{2}{3}\right)$

59. $\left(\frac{5}{8}x - 4.3\right)\left(\frac{5}{8}x + 4.3\right)$

60. $(10.7 - x^3)(10.7 + x^3)$

c Multiply mentally, if possible. If you need extra steps, be sure to use them.

61. $(x + 2)^2$

62. $(2x - 1)^2$

63. $(3x^2 + 1)^2$

64. $\left(3x + \frac{3}{4}\right)^2$

65. $\left(a - \frac{1}{2}\right)^2$

66. $\left(2a - \frac{1}{5}\right)^2$

67. $(3 + x)^2$

68. $(x^3 - 1)^2$

69. $(x^2 + 1)^2$

70. $(8x - x^2)^2$

71. $(2 - 3x^4)^2$

72. $(6x^3 - 2)^2$

73. $(5 + 6t^2)^2$

74. $(3p^2 - p)^2$

75. $\left(x - \frac{5}{8}\right)^2$

76. $(0.3y + 2.4)^2$

d Multiply mentally, if possible.

77. $(3 - 2x^3)^2$

78. $(x - 4x^3)^2$

79. $4x(x^2 + 6x - 3)$

80. $8x(-x^5 + 6x^2 + 9)$

81. $\left(2x^2 - \frac{1}{2}\right)\left(2x^2 - \frac{1}{2}\right)$

82. $(-x^2 + 1)^2$

83. $(-1 + 3p)(1 + 3p)$

84. $(-3q + 2)(3q + 2)$

85. $3t^2(5t^3 - t^2 + t)$

86. $-6x^2(x^3 + 8x - 9)$

87. $(6x^4 + 4)^2$

88. $(8a + 5)^2$

89. $(3x + 2)(4x^2 + 5)$

90. $(2x^2 - 7)(3x^2 + 9)$

91. $(8 - 6x^4)^2$

92. $\left(\frac{1}{5}x^2 + 9\right)\left(\frac{3}{5}x^2 - 7\right)$

93. $(t - 1)(t^2 + t + 1)$

94. $(y + 5)(y^2 - 5y + 25)$

Compute each of the following and compare.

95. $3^2 + 4^2$; $(3 + 4)^2$

96. $6^2 + 7^2$; $(6 + 7)^2$

97. $9^2 - 5^2$; $(9 - 5)^2$

98. $11^2 - 4^2$; $(11 - 4)^2$

Find the total area of all the shaded rectangles.

99.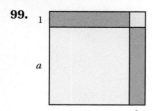

100.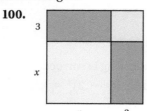

101.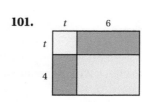

102.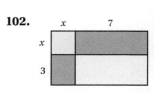

Solve. [11.3c]

103. $3x - 8x = 4(7 - 8x)$

104. $3(x - 2) = 5(2x + 7)$

105. $5(2x - 3) - 2(3x - 4) = 20$

Solve. [11.4b]

106. $3x - 2y = 12$, for y

107. $C = ab - r$, for b

108. $3a - 5d = 4$, for a

Synthesis

Multiply.

109. $5x(3x - 1)(2x + 3)$

110. $[(2x - 3)(2x + 3)](4x^2 + 9)$

111. $[(a - 5)(a + 5)]^2$

112. $(a - 3)^2(a + 3)^2$
(*Hint*: Examine Exercise 111.)

113. $(3t^4 - 2)^2(3t^4 + 2)^2$
(*Hint*: Examine Exercise 111.)

114. $[3a - (2a - 3)][3a + (2a - 3)]$

Solve.

115. $(x + 2)(x - 5) = (x + 1)(x - 3)$

116. $(2x + 5)(x - 4) = (x + 5)(2x - 4)$

117. *Factors and Sums.* To *factor* a number is to express it as a product. Since $12 = 4 \cdot 3$, we say that 12 is *factored* and that 4 and 3 are *factors* of 12. In the following table, the top number has been factored in such a way that the sum of the factors is the bottom number. For example, in the first column, 40 has been factored as $5 \cdot 8$, and $5 + 8 = 13$, the bottom number. Such thinking is important in algebra when we factor trinomials of the type $x^2 + bx + c$. Find the missing numbers in the table.

PRODUCT	40	63	36	72	−140	−96	48	168	110			
FACTOR	5									−9	−24	−3
FACTOR	8									−10	18	
SUM	13	16	−20	−38	−4	4	−14	−29	−21			18

118. Consider the rectangle below.

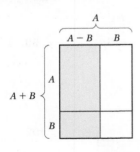

a) Find a polynomial for the area of the entire rectangle.

b) Find a polynomial for the sum of the areas of the two small unshaded rectangles.

c) Find a polynomial for the area in part (a) minus the area in part (b).

d) Find a polynomial for the area of the shaded region and compare this with the polynomial found in part (c).

Use the TABLE or GRAPH feature to check whether each of the following is correct.

119. $(x - 1)^2 = x^2 - 2x + 1$

120. $(x - 2)^2 = x^2 - 4x - 4$

121. $(x - 3)(x + 3) = x^2 - 6$

122. $(x - 3)(x + 2) = x^2 - x - 6$

Operations with Polynomials in Several Variables

13.7

The polynomials that we have been studying have only one variable. A **polynomial in several variables** is an expression like those you have already seen, but with more than one variable. Here are two examples:

$$3x + xy^2 + 5y + 4, \qquad 8xy^2z - 2x^3z - 13x^4y^2 + 15.$$

a EVALUATING POLYNOMIALS

OBJECTIVES

a Evaluate a polynomial in several variables for given values of the variables.

b Identify the coefficients and the degrees of the terms of a polynomial and the degree of a polynomial.

c Collect like terms of a polynomial.

d Add polynomials.

e Subtract polynomials.

f Multiply polynomials.

EXAMPLE 1 Evaluate the polynomial

$$4 + 3x + xy^2 + 8x^3y^3$$

when $x = -2$ and $y = 5$.

We replace x with -2 and y with 5:

$$
\begin{aligned}
4 + 3x + xy^2 + 8x^3y^3 &= 4 + 3(-2) + (-2) \cdot 5^2 + 8(-2)^3 \cdot 5^3 \\
&= 4 + 3(-2) + (-2) \cdot 25 + 8(-8)(125) \\
&= 4 - 6 - 50 - 8000 \\
&= -8052.
\end{aligned}
$$

EXAMPLE 2 *Zoology.* The weight, in kilograms, of an elephant with a girth of g centimeters at the heart, a length of l centimeters, and a footpad circumference of f centimeters can be estimated by the polynomial

$$11.5g + 7.55l + 12.5f - 4016.$$

A field zoologist finds that the girth of a 3-year-old female elephant is 231 cm, the length is 135 cm, and the footpad circumference is 86 cm. Approximately how much does the elephant weigh?

Source: "How Much Does That Elephant Weigh?" by Mark MacAllister on fieldtripearth.org

We evaluate the polynomial for $g = 231$, $l = 135$, and $f = 86$:

$$
\begin{aligned}
11.5g + 7.55l + 12.5f - 4016 &= 11.5(231) + 7.55(135) + 12.5(86) - 4016 \\
&= 734.75.
\end{aligned}
$$

The elephant weighs about 735 kg.

Do Exercises 1–3. ▷

1. Evaluate the polynomial
$$4 + 3x + xy^2 + 8x^3y^3$$
when $x = 2$ and $y = -5$.

2. Evaluate the polynomial
$$8xy^2 - 2x^3z - 13x^4y^2 + 5$$
when $x = -1$, $y = 3$, and $z = 4$.

3. *Zoology.* Refer to Example 2. A 25-year-old female elephant has a girth of 366 cm, a length of 226 cm, and a footpad circumference of 117 cm. How much does the elephant weigh?

Answers

1. −7940   **2.** −176   **3.** About 3362 kg

b COEFFICIENTS AND DEGREES

The **degree** of a term is the sum of the exponents of the variables. For example, the degree of $3x^5y^2$ is $5 + 2$, or 7. The **degree of a polynomial** is the degree of the term of highest degree.

EXAMPLE 3 Identify the coefficient and the degree of each term and the degree of the polynomial

$$9x^2y^3 - 14xy^2z^3 + xy + 4y + 5x^2 + 7.$$

TERM	COEFFICIENT	DEGREE	DEGREE OF THE POLYNOMIAL
$9x^2y^3$	9	5	
$-14xy^2z^3$	-14	6	6
xy	1	2	
$4y$	4	1	
$5x^2$	5	2	
7	7	0	

Think: $4y = 4y^1$.

Think: $7 = 7x^0$, or $7x^0y^0z^0$.

◀ Do Exercises 4 and 5.

4. Identify the coefficient of each term:
$$-3xy^2 + 3x^2y - 2y^3 + xy + 2.$$

5. Identify the degree of each term and the degree of the polynomial
$$4xy^2 + 7x^2y^3z^2 - 5x + 2y + 4.$$

c COLLECTING LIKE TERMS

Like terms have exactly the same variables with exactly the same exponents. For example,

$3x^2y^3$ and $-7x^2y^3$ are like terms;
$9x^4z^7$ and $12x^4z^7$ are like terms.

But

$13xy^5$ and $-2x^2y^5$ are *not* like terms, because the x-factors have different exponents;

and

$3xyz^2$ and $4xy$ are *not* like terms, because there is no factor of z^2 in the second expression.

Collecting like terms is based on the distributive laws.

EXAMPLES Collect like terms.

4. $5x^2y + 3xy^2 - 5x^2y - xy^2 = (5 - 5)x^2y + (3 - 1)xy^2 = 2xy^2$

5. $8a^2 - 2ab + 7b^2 + 4a^2 - 9ab - 17b^2 = 12a^2 - 11ab - 10b^2$

6. $7xy - 5xy^2 + 3xy^2 - 7 + 6x^3 + 9xy - 11x^3 + y - 1$
$= 16xy - 2xy^2 - 5x^3 + y - 8$

◀ Do Exercises 6 and 7.

Collect like terms.

6. $4x^2y + 3xy - 2x^2y$

7. $-3pt - 5ptr^3 - 12 + 8pt + 5ptr^3 + 4$ **GS**

The like terms are $-3pt$ and

[   ] , $-5ptr^3$ and [   ] ,

and -12 and [   ] .

Collecting like terms, we have

$(-3 + $ [   ] $)pt +$

$(-5 + $ [   ] $) ptr^3 +$

$(-12 + $ [   ] $)$

$= $ [   ] $ - 8.$

d ADDITION

We can find the sum of two polynomials in several variables by writing a plus sign between them and then collecting like terms.

EXAMPLE 7 Add: $(-5x^3 + 3y - 5y^2) + (8x^3 + 4x^2 + 7y^2)$.

$$(-5x^3 + 3y - 5y^2) + (8x^3 + 4x^2 + 7y^2)$$
$$= (-5 + 8)x^3 + 4x^2 + 3y + (-5 + 7)y^2$$
$$= 3x^3 + 4x^2 + 3y + 2y^2$$

EXAMPLE 8 Add:

$$(5xy^2 - 4x^2y + 5x^3 + 2) + (3xy^2 - 2x^2y + 3x^3y - 5).$$

We have

$$(5xy^2 - 4x^2y + 5x^3 + 2) + (3xy^2 - 2x^2y + 3x^3y - 5)$$
$$= (5 + 3)xy^2 + (-4 - 2)x^2y + 5x^3 + 3x^3y + (2 - 5)$$
$$= 8xy^2 - 6x^2y + 5x^3 + 3x^3y - 3.$$

Do Exercises 8–10. ▷

Add.

8. $(4x^3 + 4x^2 - 8y - 3) + (-8x^3 - 2x^2 + 4y + 5)$

9. $(13x^3y + 3x^2y - 5y) + (x^3y + 4x^2y - 3xy + 3y)$

10. $(-5p^2q^4 + 2p^2q^2 + 3q) + (6pq^2 + 3p^2q + 5)$

e SUBTRACTION

We subtract a polynomial by adding its opposite, or additive inverse. The opposite of the polynomial $4x^2y - 6x^3y^2 + x^2y^2 - 5y$ is

$$-(4x^2y - 6x^3y^2 + x^2y^2 - 5y) = -4x^2y + 6x^3y^2 - x^2y^2 + 5y.$$

EXAMPLE 9 Subtract:

$$(4x^2y + x^3y^2 + 3x^2y^3 + 6y + 10) - (4x^2y - 6x^3y^2 + x^2y^2 - 5y - 8).$$

We have

$$(4x^2y + x^3y^2 + 3x^2y^3 + 6y + 10) - (4x^2y - 6x^3y^2 + x^2y^2 - 5y - 8)$$
$$= 4x^2y + x^3y^2 + 3x^2y^3 + 6y + 10 - 4x^2y + 6x^3y^2 - x^2y^2 + 5y + 8$$

    Finding the opposite by changing the sign of each term

$$= 7x^3y^2 + 3x^2y^3 - x^2y^2 + 11y + 18.$$

    Collecting like terms. (Try to write just the answer!)

·············· **Caution!** ··············

Do *not* add exponents when collecting like terms—that is,

$$7x^3 + 8x^3 \ne 15x^6;$$ ← Adding exponents is incorrect.

$$7x^3 + 8x^3 = 15x^3.$$ ← Correct

Do Exercises 11 and 12. ▷

Subtract.

11. $(-4s^4t + s^3t^2 + 2s^2t^3) - (4s^4t - 5s^3t^2 + s^2t^2)$

12. $(-5p^4q + 5p^3q^2 - 3p^2q^3 - 7q^4 - 2) - (4p^4q - 4p^3q^2 + p^2q^3 + 2q^4 - 7)$

Answers

8. $-4x^3 + 2x^2 - 4y + 2$
9. $14x^3y + 7x^2y - 3xy - 2y$
10. $-5p^2q^4 + 2p^2q^2 + 3p^2q + 6pq^2 + 3q + 5$
11. $-8s^4t + 6s^3t^2 + 2s^2t^3 - s^2t^2$
12. $-9p^4q + 9p^3q^2 - 4p^2q^3 - 9q^4 + 5$

f MULTIPLICATION

To multiply polynomials in several variables, we can multiply each term of one by every term of the other. We can use columns for long multiplications as with polynomials in one variable. We multiply each term at the top by every term at the bottom. We write like terms in columns, and then we add.

EXAMPLE 10   Multiply: $(3x^2y - 2xy + 3y)(xy + 2y)$.

$$
\begin{array}{r}
3x^2y - 2xy + 3y \\
xy + 2y \\
\hline
6x^2y^2 - 4xy^2 + 6y^2 \\
3x^3y^2 - 2x^2y^2 + 3xy^2 \\
\hline
3x^3y^2 + 4x^2y^2 - xy^2 + 6y^2
\end{array}
$$

Multiplying by $2y$
Multiplying by xy
Adding

◀ **Do Exercises 13 and 14.**

Where appropriate, we use the special products that we have learned.

EXAMPLES   Multiply.

11. $(x^2y + 2x)(xy^2 + y^2) = x^3y^3 + x^2y^3 + 2x^2y^2 + 2xy^2$ Using FOIL

12. $(p + 5q)(2p - 3q) = 2p^2 - 3pq + 10pq - 15q^2$ Using FOIL
$$= 2p^2 + 7pq - 15q^2$$

$(A + B)^2 = A^2 + 2 \cdot A \cdot B + B^2$

13. $(3x + 2y)^2 = (3x)^2 + 2(3x)(2y) + (2y)^2 = 9x^2 + 12xy + 4y^2$

$(A - B)^2 = A^2 - 2 \cdot A \cdot B + B^2$

14. $(2y^2 - 5x^2y)^2 = (2y^2)^2 - 2(2y^2)(5x^2y) + (5x^2y)^2$
$$= 4y^4 - 20x^2y^3 + 25x^4y^2$$

$(A + B)(A - B) = A^2 - B^2$

15. $(3x^2y + 2y)(3x^2y - 2y) = (3x^2y)^2 - (2y)^2 = 9x^4y^2 - 4y^2$

16. $(-2x^3y^2 + 5t)(2x^3y^2 + 5t) = (5t - 2x^3y^2)(5t + 2x^3y^2)$
The sum and the difference of the same two terms
$$= (5t)^2 - (2x^3y^2)^2 = 25t^2 - 4x^6y^4$$

$(A - B)(A + B) = A^2 - B^2$

17. $(2x + 3 - 2y)(2x + 3 + 2y) = (2x + 3)^2 - (2y)^2$
$$= 4x^2 + 12x + 9 - 4y^2$$

Remember that FOIL will always work when you are multiplying binomials. You can use it instead of the rules for special products, but those rules will make your work go faster.

◀ **Do Exercises 15–22.**

Multiply.

13. $(x^2y^3 + 2x)(x^3y^2 + 3x)$

14. $(p^4q - 2p^3q^2 + 3q^3)(p + 2q)$

Multiply.

15. $(3xy + 2x)(x^2 + 2xy^2)$

16. $(x - 3y)(2x - 5y)$

17. $(4x + 5y)^2$

18. $(3x^2 - 2xy^2)^2$

19. $(2xy^2 + 3x)(2xy^2 - 3x)$

20. $(3xy^2 + 4y)(-3xy^2 + 4y)$

21. $(3y + 4 - 3x)(3y + 4 + 3x)$

22. $(2a + 5b + c)(2a - 5b - c)$ GS
$$= [2a + (5b + c)][2a - (\qquad)]$$
$$= (2a)^2 - (\qquad)^2$$
$$= \qquad - (25b^2 + 10bc + \qquad)$$
$$= 4a^2 - 25b^2 - 10bc - \boxed{}$$

Answers
13. $x^5y^5 + 2x^4y^2 + 3x^3y^3 + 6x^2$
14. $p^5q - 4p^3q^3 + 3pq^3 + 6q^4$
15. $3x^3y + 6x^2y^3 + 2x^3 + 4x^2y^2$
16. $2x^2 - 11xy + 15y^2$
17. $16x^2 + 40xy + 25y^2$
18. $9x^4 - 12x^3y^2 + 4x^2y^4$
19. $4x^2y^4 - 9x^2$
20. $16y^2 - 9x^2y^4$
21. $9y^2 + 24y + 16 - 9x^2$
22. $4a^2 - 25b^2 - 10bc - c^2$

Guided Solution:
22. $5b + c, 5b + c, 4a^2, c^2, c^2$

✓ Reading Check

Determine whether each sentence is true or false.

RC1. The variables in the polynomial
$8x - xy + t^2 - xy^2$ are t, x, and y.

RC2. The degree of the term $4xy$ is 4.

RC3. The terms $3x^2y$ and $3xy^2$ are like terms.

RC4. When we collect like terms, we add the exponents of the variables.

---

a   Evaluate the polynomial when $x = 3$, $y = -2$, and $z = -5$.

1. $x^2 - y^2 + xy$

2. $x^2 + y^2 - xy$

3. $x^2 - 3y^2 + 2xy$

4. $x^2 - 4xy + 5y^2$

5. $8xyz$

6. $-3xyz^2$

7. $xyz^2 - z$

8. $xy - xz + yz$

9. *Lung Capacity.* The polynomial equation
$$C = 0.041h - 0.018A - 2.69$$
can be used to estimate the lung capacity C, in liters, of a person of height h, in centimeters, and age A, in years. Find the lung capacity of a 20-year-old person who is 165 cm tall.

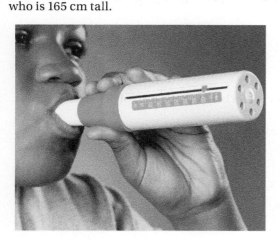

10. *Altitude of a Launched Object.* The altitude h, in meters, of a launched object is given by the polynomial equation
$$h = h_0 + vt - 4.9t^2,$$
where h_0 is the height, in meters, from which the launch occurs, v is the initial upward speed (or velocity), in meters per second (m/s), and t is the number of seconds for which the object is airborne. A rock is thrown upward from the top of the Lands End Arch, near San Lucas, Baja, Mexico, 32 m above the ground. The upward speed is 10 m/s. How high will the rock be 3 sec after it has been thrown?

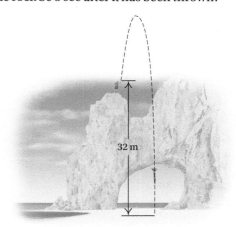

32 m

11. *Male Caloric Needs.* The number of calories needed each day by a moderately active man who weighs w kilograms, is h centimeters tall, and is a years old can be estimated by the polynomial
$$19.18w + 7h - 9.52a + 92.4.$$
Steve is moderately active, weighs 82 kg, is 185 cm tall, and is 67 years old. What is his daily caloric need?

Source: Parker, M., *She Does Math.* Mathematical Association of America

12. *Female Caloric Needs.* The number of calories needed each day by a moderately active woman who weighs w pounds, is h inches tall, and is a years old can be estimated by the polynomial
$$917 + 6w + 6h - 6a.$$
Christine is moderately active, weighs 125 lb, is 64 in. tall, and is 27 years old. What is her daily caloric need?

Source: Parker, M., *She Does Math.* Mathematical Association of America

Surface Area of a Right Circular Cylinder. The surface area S of a right circular cylinder is given by the polynomial equation $S = 2\pi rh + 2\pi r^2$, where h is the height and r is the radius of the base. Use this formula for Exercises 13 and 14.

13. A 12-oz beverage can has a height of 4.7 in. and a radius of 1.2 in. Find the surface area of the can. Use 3.14 for π.

14. A 26-oz coffee can has a height of 6.5 in. and a radius of 2.5 in. Find the surface area of the can. Use 3.14 for π.

Surface Area of a Silo. A silo is a structure that is shaped like a right circular cylinder with a half sphere on top. The surface area S of a silo of height h and radius r (including the area of the base) is given by the polynomial equation $S = 2\pi rh + \pi r^2$. Note that h is the height of the entire silo.

15. A coffee grinder is shaped like a silo, with a height of 7 in. and a radius of $1\frac{1}{2}$ in. Find the surface area " of the coffee grinder. Use 3.14 for π.

16. A $1\frac{1}{2}$-oz bottle of roll-on deodorant has a height of 4 in. and a radius of $\frac{3}{4}$ in. Find the surface area of the bottle if the bottle is shaped like a silo. Use 3.14 for π.

b Identify the coefficient and the degree of each term of the polynomial. Then find the degree of the polynomial.

17. $x^3y - 2xy + 3x^2 - 5$

18. $5x^2y^2 - y^2 + 15xy + 1$

19. $17x^2y^3 - 3x^3yz - 7$

20. $6 - xy + 8x^2y^2 - y^5$

c Collect like terms.

21. $a + b - 2a - 3b$

22. $xy^2 - 1 + y - 6 - xy^2$

23. $3x^2y - 2xy^2 + x^2$

24. $m^3 + 2m^2n - 3m^2 + 3mn^2$

25. $6au + 3av + 14au + 7av$

26. $3x^2y - 2z^2y + 3xy^2 + 5z^2y$

27. $2u^2v - 3uv^2 + 6u^2v - 2uv^2$

28. $3x^2 + 6xy + 3y^2 - 5x^2 - 10xy - 5y^2$

d   Add.

29. $(2x^2 - xy + y^2) + (-x^2 - 3xy + 2y^2)$

30. $(2zt - z^2 + 5t^2) + (z^2 - 3zt + t^2)$

31. $(r - 2s + 3) + (2r + s) + (s + 4)$

32. $(ab - 2a + 3b) + (5a - 4b) + (3a + 7ab - 8b)$

33. $(b^3a^2 - 2b^2a^3 + 3ba + 4)$
     $+ (b^2a^3 - 4b^3a^2 + 2ba - 1)$

34. $(2x^2 - 3xy + y^2) + (-4x^2 - 6xy - y^2)$
     $+ (x^2 + xy - y^2)$

e   Subtract.

35. $(a^3 + b^3) - (a^2b - ab^2 + b^3 + a^3)$

36. $(x^3 - y^3) - (-2x^3 + x^2y - xy^2 + 2y^3)$

37. $(xy - ab - 8) - (xy - 3ab - 6)$

38. $(3y^4x^2 + 2y^3x - 3y - 7)$
     $- (2y^4x^2 + 2y^3x - 4y - 2x + 5)$

39. $(-2a + 7b - c) - (-3b + 4c - 8d)$

40. Subtract $5a + 2b$ from the sum of $2a + b$ and $3a - b$.

f   Multiply.

41. $(3z - u)(2z + 3u)$

42. $(a - b)(a^2 + b^2 + 2ab)$

43. $(a^2b - 2)(a^2b - 5)$

44. $(xy + 7)(xy - 4)$

45. $(a^3 + bc)(a^3 - bc)$

46. $(m^2 + n^2 - mn)(m^2 + mn + n^2)$

47. $(y^4x + y^2 + 1)(y^2 + 1)$

48. $(a - b)(a^2 + ab + b^2)$

49. $(3xy - 1)(4xy + 2)$

50. $(m^3n + 8)(m^3n - 6)$

51. $(3 - c^2d^2)(4 + c^2d^2)$

52. $(6x - 2y)(5x - 3y)$

53. $(m^2 - n^2)(m + n)$

54. $(pq + 0.2) \times$
$(0.4pq - 0.1)$

55. $(xy + x^5y^5) \times$
$(x^4y^4 - xy)$

56. $(x - y^3)(2y^3 + x)$

57. $(x + h)^2$

58. $(y - a)^2$

59. $(3a + 2b)^2$

60. $(2ab - cd)^2$

61. $(r^3t^2 - 4)^2$

62. $(3a^2b - b^2)^2$

63. $(p^4 + m^2n^2)^2$

64. $\left(2a^3 - \frac{1}{2}b^3\right)^2$

65. $3a(a - 2b)^2$

66. $-3x(x + 8y)^2$

67. $(m + n - 3)^2$

68. $(a^2 + b + 2)^2$

69. $(a + b)(a - b)$

70. $(x - y)(x + y)$

71. $(2a - b)(2a + b)$

72. $(w + 3z)(w - 3z)$

73. $(c^2 - d)(c^2 + d)$

74. $(p^3 - 5q)(p^3 + 5q)$

75. $(ab + cd^2) \times$
$(ab - cd^2)$

76. $(xy + pq) \times$
$(xy - pq)$

77. $(x + y - 3)(x + y + 3)$

78. $(p + q + 4)(p + q - 4)$

79. $[x + y + z][x - (y + z)]$

80. $[a + b + c][a - (b + c)]$

81. $(a + b + c)(a + b - c)$

82. $(3x + 2 - 5y)(3x + 2 + 5y)$

83. $(x^2 - 4y + 2)(3x^2 + 5y - 3)$

84. $(2x^2 - 7y + 4)(x^2 + y - 3)$

Skill Maintenance

In which quadrant is each point located? [12.1a]

85. $(2, -5)$ **86.** $(-8, -9)$ **87.** $(16, 23)$ **88.** $(-3, 2)$

89. Find the absolute value: $|-39|$. [10.2e]

90. Convert $\dfrac{9}{8}$ to decimal notation. [10.2c]

91. Use either $<$ or $>$ for ☐ to write a true sentence: -17 ☐ -5. [10.2d]

92. Evaluate $-(-x)$ when $x = -3$. [10.3b]

Synthesis

Find a polynomial for each shaded area. (Leave results in terms of π where appropriate.)

93.

94.
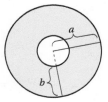

95.
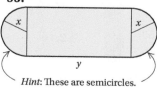

Hint: These are semicircles.

96.
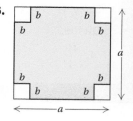

Find a formula for the surface area of each solid object. Leave results in terms of π.

97.

98.
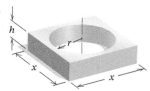

99. *Observatory Paint Costs.* The observatory at Danville University is shaped like a silo that is 40 ft high and 30 ft wide (see Exercise 15). The Heavenly Bodies Astronomy Club is to paint the exterior of the observatory using paint that covers 250 ft² per gallon. How many gallons should they purchase?

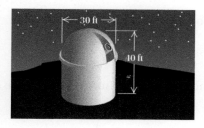

100. *Interest Compounded Annually.* An amount of money P that is invested at the yearly interest rate r grows to the amount

$$P(1 + r)^t$$

after t years. Find a polynomial that can be used to determine the amount to which P will grow after 2 years.

101. Suppose that $10,400 is invested at 3.5%, compounded annually. How much is in the account at the end of 5 years? (See Exercise 100.)

102. Multiply: $(x + a)(x - b)(x - a)(x + b)$.

OBJECTIVES

a   Divide a polynomial by a monomial.

b   Divide a polynomial by a divisor that is a binomial.

SKILL TO REVIEW

Objective 10.6a: Divide integers.

Divide.

1. $\dfrac{20}{4}$ **2.** $\dfrac{-30}{5}$

Divide.

1. $\dfrac{20x^3}{5x}$ **2.** $\dfrac{-28x^{14}}{4x^3}$

3. $\dfrac{-56p^5q^7}{2p^2q^6}$ **4.** $\dfrac{x^5}{4x}$

In this section, we consider division of polynomials. You will see that such division is similar to what is done in arithmetic.

a   DIVIDING BY A MONOMIAL

We first consider division by a monomial. When dividing a monomial by a monomial, we use the quotient rule to subtract exponents when the bases are the same. We also divide the coefficients.

EXAMPLES   Divide.

1. $\dfrac{10x^2}{2x} = \dfrac{10}{2} \cdot \dfrac{x^2}{x} = 5x^{2-1} = 5x$

2. $\dfrac{x^9}{3x^2} = \dfrac{1x^9}{3x^2} = \dfrac{1}{3} \cdot \dfrac{x^9}{x^2} = \dfrac{1}{3}x^{9-2} = \dfrac{1}{3}x^7$

3. $\dfrac{-18x^{10}}{3x^3} = \dfrac{-18}{3} \cdot \dfrac{x^{10}}{x^3} = -6x^{10-3} = -6x^7$

4. $\dfrac{42a^2b^5}{-3ab^2} = \dfrac{42}{-3} \cdot \dfrac{a^2}{a} \cdot \dfrac{b^5}{b^2} = -14a^{2-1}b^{5-2} = -14ab^3$

·········· **Caution!** ··········

The coefficients are divided, but the exponents are subtracted.

◀ **Do Margin Exercises 1–4.**

To divide a polynomial by a monomial, we note that since

$$\frac{A}{C} + \frac{B}{C} = \frac{A+B}{C},$$

it follows that

$$\frac{A+B}{C} = \frac{A}{C} + \frac{B}{C}. \quad \text{Switching the left and right sides of the equation}$$

This is actually the procedure we use when performing divisions like $86 \div 2$. Although we might write

$$\frac{86}{2} = 43,$$

we could also calculate as follows:

$$\frac{86}{2} = \frac{80+6}{2} = \frac{80}{2} + \frac{6}{2} = 40 + 3 = 43.$$

Similarly, to divide a polynomial by a monomial, we divide each term by the monomial.

EXAMPLE 5   Divide: $(9x^8 + 12x^6) \div (3x^2)$.

We have

$$(9x^8 + 12x^6) \div (3x^2) = \frac{9x^8 + 12x^6}{3x^2}$$

$$= \frac{9x^8}{3x^2} + \frac{12x^6}{3x^2}. \quad \text{To see this, add and get the original expression.}$$

We now perform the separate divisions:

$$\frac{9x^8}{3x^2} + \frac{12x^6}{3x^2} = \frac{9}{3} \cdot \frac{x^8}{x^2} + \frac{12}{3} \cdot \frac{x^6}{x^2}$$

$$= 3x^{8-2} + 4x^{6-2}$$

$$= 3x^6 + 4x^4.$$

·············· **Caution!** ··············

The coefficients are *divided*, but the exponents are *subtracted*.

·······································

To check, we multiply the quotient, $3x^6 + 4x^4$, by the divisor, $3x^2$:

$$3x^2(3x^6 + 4x^4) = (3x^2)(3x^6) + (3x^2)(4x^4) = 9x^8 + 12x^6.$$

This is the polynomial that was being divided, so our answer is $3x^6 + 4x^4$.

Do Exercises 5–7. ▶

EXAMPLE 6 Divide and check: $(10a^5b^4 - 2a^3b^2 + 6a^2b) \div (-2a^2b)$.

$$\frac{10a^5b^4 - 2a^3b^2 + 6a^2b}{-2a^2b} = \frac{10a^5b^4}{-2a^2b} - \frac{2a^3b^2}{-2a^2b} + \frac{6a^2b}{-2a^2b}$$

$$= \frac{10}{-2} \cdot a^{5-2}b^{4-1} - \frac{2}{-2} \cdot a^{3-2}b^{2-1} + \frac{6}{-2}$$

$$= -5a^3b^3 + ab - 3$$

Check: $-2a^2b(-5a^3b^3 + ab - 3) = (-2a^2b)(-5a^3b^3) + (-2a^2b)(ab) - (-2a^2b)(3)$

$$= 10a^5b^4 - 2a^3b^2 + 6a^2b$$

Our answer, $-5a^3b^3 + ab - 3$, checks. ▪

> To divide a polynomial by a monomial, divide each term of the polynomial by the monomial.

Do Exercises 8 and 9. ▶

b DIVIDING BY A BINOMIAL

Let's first consider long division as it is performed in arithmetic. When we divide, we repeat the procedure at right.

We review this by considering the division $3711 \div 8$.

$$\begin{array}{r} 4 \\ 8 \overline{)\ 3\ 7\ 1\ 1} \\ \underline{3\ 2} \\ 5\ 1 \end{array}$$

① Divide: $37 \div 8 \approx 4$.

② Multiply: $4 \times 8 = 32$.

③ Subtract: $37 - 32 = 5$.

④ Bring down the 1.

$$\begin{array}{r} 4\ 6\ 3 \\ 8 \overline{)\ 3\ 7\ 1\ 1} \\ \underline{3\ 2} \\ 5\ 1 \\ \underline{4\ 8} \\ 3\ 1 \\ \underline{2\ 4} \\ 7 \end{array}$$

Next, we repeat the process two more times. We obtain the complete division as shown on the right above. The quotient is 463. The remainder is 7, expressed as R = 7. We write the answer as

$$463\, \text{R}\, 7 \quad \text{or} \quad 463 + \frac{7}{8} = 463\frac{7}{8}.$$

Divide. Check the result.

5. $(28x^7 + 32x^5) \div (4x^3)$

$$\frac{28x^7 + 32x^5}{4x^3} = \frac{28x^7}{\boxed{}} + \frac{32x^5}{\boxed{}}$$

$$= \frac{28}{4}x^{7-\boxed{}} + \frac{32}{\boxed{}}x^{5-3}$$

$$= 7x^{\boxed{}} + \boxed{}x^2$$

6. $(2x^3 + 6x^2 + 4x) \div (2x)$

7. $(6x^2 + 3x - 2) \div 3$

Divide and check.

8. $(8x^2 - 3x + 1) \div (-2)$

9. $\dfrac{2x^4y^6 - 3x^3y^4 + 5x^2y^3}{x^2y^2}$

> **To carry out long division:**
> 1. Divide,
> 2. Multiply,
> 3. Subtract, and
> 4. Bring down the next number or term.

Answers

5. $7x^4 + 8x^2$ **6.** $x^2 + 3x + 2$

7. $2x^2 + x - \dfrac{2}{3}$ **8.** $-4x^2 + \dfrac{3}{2}x - \dfrac{1}{2}$

9. $2x^2y^4 - 3xy^2 + 5y$

Guided Solution:

5. $4x^3, 4x^3, 3, 4, 4, 8$

We check the answer, 463 R 7, by multiplying the quotient, 463, by the divisor, 8, and adding the remainder, 7:

$$8 \cdot 463 + 7 = 3704 + 7 = 3711.$$

Now let's look at long division with polynomials. We use this procedure when the divisor is not a monomial. We write polynomials in descending order and then write in missing terms, if necessary.

EXAMPLE 7   Divide $x^2 + 5x + 6$ by $x + 2$.

$$
\begin{array}{r}
x \\
x + 2 \overline{)x^2 + 5x + 6} \\
x^2 + 2x \\
\hline
3x
\end{array}
$$

— Divide the first term by the first term: $x^2/x = x$. Ignore the term 2 for this step.

— Multiply x above by the divisor, $x + 2$.

— Subtract: $(x^2 + 5x) - (x^2 + 2x) = x^2 + 5x - x^2 - 2x$
    $= 3x$.

We now "bring down" the next term of the dividend—in this case, 6.

$$
\begin{array}{r}
x + 3 \\
x + 2 \overline{)x^2 + 5x + 6} \\
x^2 + 2x \\
\hline
3x + 6 \\
3x + 6 \\
\hline
0
\end{array}
$$

— Divide the first term of $3x + 6$ by the first term of the divisor: $3x/x = 3$.

— The 6 has been "brought down."

— Multiply 3 above by the divisor, $x + 2$.

— Subtract: $(3x + 6) - (3x + 6) = 3x + 6 - 3x - 6 = 0$.

The quotient is $x + 3$. The remainder is 0. A remainder of 0 is generally not included in an answer.

To check, we multiply the quotient by the divisor and add the remainder, if any, to see if we get the dividend:

Divisor	Quotient	Remainder		Dividend
$(x + 2) \cdot$	$(x + 3) +$	0	$=$	$x^2 + 5x + 6$.

The division checks.

◀ **Do Exercise 10.**

EXAMPLE 8   Divide and check: $(x^2 + 2x - 12) \div (x - 3)$.

$$
\begin{array}{r}
x \\
x - 3 \overline{)x^2 + 2x - 12} \\
x^2 - 3x \\
\hline
5x
\end{array}
$$

— Divide the first term by the first term: $x^2/x = x$.

— Multiply x above by the divisor, $x - 3$.

— Subtract: $(x^2 + 2x) - (x^2 - 3x) = x^2 + 2x - x^2 + 3x$
    $= 5x$.

We now "bring down" the next term of the dividend—in this case, -12.

$$
\begin{array}{r}
x + 5 \\
x - 3 \overline{)x^2 + 2x - 12} \\
x^2 - 3x \\
\hline
5x - 12 \\
5x - 15 \\
\hline
3
\end{array}
$$

— Divide the first term of $5x - 12$ by the first term of the divisor: $5x/x = 5$.

— Bring down the -12.

— Multiply 5 above by the divisor, $x - 3$.

— Subtract: $(5x - 12) - (5x - 15) = 5x - 12 - 5x + 15$
    $= 3$.

10. Divide and check:
$(x^2 + x - 6) \div (x + 3)$.

$$
\begin{array}{r}
\boxed{} - \boxed{} \\
x + 3 \overline{)x^2 + x - 6} \\
x^2 + \boxed{} \\
\hline
\boxed{} - 6 \\
-2x - \boxed{} \\
\hline
\boxed{}
\end{array}
$$

The answer is $x + 5$ with R $= 3$, or

$$\underbrace{x + 5}_{\text{Quotient}} + \underbrace{\frac{3}{x - 3}}_{} \cdot$$

Quotient — $x + 5$ $\frac{3}{x-3}$ → Remainder → Divisor

(This is the way answers will be given at the back of the book.)

Check: We can check by multiplying the divisor by the quotient and adding the remainder, as follows:

$$(x - 3)(x + 5) + 3 = x^2 + 2x - 15 + 3$$
$$= x^2 + 2x - 12.$$

When dividing, an answer may "come out even" (that is, have a remainder of 0, as in Example 7), or it may not (as in Example 8). **If a remainder is not 0, we continue dividing until the degree of the remainder is less than the degree of the divisor.**

Do Exercises 11 and 12. ▶

Divide and check.

11. $x - 2 \overline{\smash{)}x^2 + 2x - 8}$

12. $x + 3 \overline{\smash{)}x^2 + 7x + 10}$

EXAMPLE 9 Divide and check: $(x^3 + 1) \div (x + 1)$.

$$
\begin{array}{r}
x^2 - x + 1 \\
x + 1 \overline{\smash{)}x^3 + 0x^2 + 0x + 1} \quad \leftarrow \text{Fill in the missing terms. (See Section 13.3.)} \\
\underline{x^3 + x^2} \quad \text{Subtract: } x^3 - (x^3 + x^2) = -x^2. \\
-x^2 + 0x \\
\underline{-x^2 - x} \quad \text{Subtract: } -x^2 - (-x^2 - x) = x. \\
x + 1 \\
\underline{x + 1} \quad \text{Subtract: } (x + 1) - (x + 1) = 0. \\
0
\end{array}
$$

The answer is $x^2 - x + 1$. The check is left to the student.

EXAMPLE 10 Divide and check: $(9x^4 - 7x^2 - 4x + 13) \div (3x - 1)$.

$$
\begin{array}{r}
3x^3 + x^2 - 2x - 2 \\
3x - 1 \overline{\smash{)}9x^4 + 0x^3 - 7x^2 - 4x + 13} \quad \leftarrow \text{Fill in the missing term.} \\
\underline{9x^4 - 3x^3} \quad \text{Subtract: } 9x^4 - (9x^4 - 3x^3) = 3x^3. \\
3x^3 - 7x^2 \\
\underline{3x^3 - x^2} \quad \text{Subtract:} \\
-6x^2 - 4x \quad (3x^3 - 7x^2) - (3x^3 - x^2) = -6x^2. \\
\underline{-6x^2 + 2x} \quad \text{Subtract:} \\
-6x + 13 \quad (-6x^2 - 4x) - (-6x^2 + 2x) = -6x. \\
\underline{-6x + 2} \quad \text{Subtract:} \\
11 \quad (-6x + 13) - (-6x + 2) = 11.
\end{array}
$$

The answer is $3x^3 + x^2 - 2x - 2$ with R $= 11$, or

$$3x^3 + x^2 - 2x - 2 + \frac{11}{3x - 1} \cdot$$

Check: $(3x - 1)(3x^3 + x^2 - 2x - 2) + 11$
$$= 9x^4 + 3x^3 - 6x^2 - 6x - 3x^3 - x^2 + 2x + 2 + 11$$
$$= 9x^4 - 7x^2 - 4x + 13$$

Do Exercises 13 and 14. ▶

Divide and check.

13. $(x^3 - 1) \div (x - 1)$

14. $(8x^4 + 10x^2 + 2x + 9) \div (4x + 2)$

Answers

11. $x + 4$ 12. $x + 4$ with R $= -2$, or
$x + 4 + \dfrac{-2}{x + 3}$ 13. $x^2 + x + 1$
14. $2x^3 - x^2 + 3x - 1$ with R $= 11$, or
$2x^3 - x^2 + 3x - 1 + \dfrac{11}{4x + 2}$

For Extra Help

MyMathLab®

 MathXL®
 PRACTICE
 WATCH
READ
 REVIEW

✓ Reading Check

Complete each statement with the appropriate word(s) from the column on the right. A word may be used more than once.

RC1. When dividing a monomial by a monomial, we _____ exponents and _____ coefficients.

add

subtract

multiply

divide

RC2. To divide a polynomial by a monomial, we _____ each term by the monomial.

RC3. To carry out long division, we repeat the following process: divide, _____, _____, and bring down the next term.

RC4. To check division, we _____ the divisor and the quotient, and then _____ the remainder.

a Divide and check.

1. $\dfrac{24x^4}{8}$

2. $\dfrac{-2u^2}{u}$

3. $\dfrac{25x^3}{5x^2}$

4. $\dfrac{16x^7}{-2x^2}$

5. $\dfrac{-54x^{11}}{-3x^8}$

6. $\dfrac{-75a^{10}}{3a^2}$

7. $\dfrac{64a^5b^4}{16a^2b^3}$

8. $\dfrac{-34p^{10}q^{11}}{-17pq^9}$

9. $\dfrac{24x^4 - 4x^3 + x^2 - 16}{8}$

10. $\dfrac{12a^4 - 3a^2 + a - 6}{6}$

11. $\dfrac{u - 2u^2 - u^5}{u}$

12. $\dfrac{50x^5 - 7x^4 + x^2}{x}$

13. $(15t^3 + 24t^2 - 6t) \div (3t)$

14. $(25t^3 + 15t^2 - 30t) \div (5t)$

15. $(20x^6 - 20x^4 - 5x^2) \div (-5x^2)$

16. $(24x^6 + 32x^5 - 8x^2) \div (-8x^2)$

17. $(24x^5 - 40x^4 + 6x^3) \div (4x^3)$

18. $(18x^6 - 27x^5 - 3x^3) \div (9x^3)$

19. $\dfrac{18x^2 - 5x + 2}{2}$

20. $\dfrac{15x^2 - 30x + 6}{3}$

21. $\dfrac{12x^3 + 26x^2 + 8x}{2x}$

22. $\dfrac{2x^4 - 3x^3 + 5x^2}{x^2}$

23. $\dfrac{9r^2s^2 + 3r^2s - 6rs^2}{3rs}$

24. $\dfrac{4x^4y - 8x^6y^2 + 12x^8y^6}{4x^4y}$

b Divide.

25. $(x^2 + 4x + 4) \div (x + 2)$

26. $(x^2 - 6x + 9) \div (x - 3)$

27. $(x^2 - 10x - 25) \div (x - 5)$

28. $(x^2 + 8x - 16) \div (x + 4)$

29. $(x^2 + 4x - 14) \div (x + 6)$

30. $(x^2 + 5x - 9) \div (x - 2)$

31. $\dfrac{x^2 - 9}{x + 3}$

32. $\dfrac{x^2 - 25}{x - 5}$

33. $\dfrac{x^5 + 1}{x + 1}$

34. $\dfrac{x^4 - 81}{x - 3}$

35. $\dfrac{8x^3 - 22x^2 - 5x + 12}{4x + 3}$

36. $\dfrac{2x^3 - 9x^2 + 11x - 3}{2x - 3}$

37. $(x^6 - 13x^3 + 42) \div (x^3 - 7)$

38. $(x^6 + 5x^3 - 24) \div (x^3 - 3)$

39. $(t^3 - t^2 + t - 1) \div (t - 1)$

40. $(y^3 + 3y^2 - 5y - 15) \div (y + 3)$

41. $(y^3 - y^2 - 5y - 3) \div (y + 2)$

42. $(t^3 - t^2 + t - 1) \div (t + 1)$

43. $(15x^3 + 8x^2 + 11x + 12) \div (5x + 1)$

44. $(20x^4 - 2x^3 + 5x + 3) \div (2x - 3)$

45. $(12y^3 + 42y^2 - 10y - 41) \div (2y + 7)$

46. $(15y^3 - 27y^2 - 35y + 60) \div (5y - 9)$

Skill Maintenance

Solve.

47. $-13 = 8d - 5$ [11.3a]

48. $x + \dfrac{1}{2}x = 5$ [11.3b]

49. $4(x - 3) = 5(2 - 3x) + 1$ [11.3c]

50. $3(r + 1) - 5(r + 2) \geq 15 - (r + 7)$ [11.7e]

51. The number of patients with the flu who were treated at Riverview Clinic increased from 25 one week to 60 the next week. What was the percent increase?   [11.5a]

52. Todd's quiz grades are 82, 88, 93, and 92. Determine (in terms of an inequality) what scores on the last quiz will allow him to get an average quiz grade of at least 90.   [11.8b]

53. The perimeter of a rectangle is 640 ft. The length is 15 ft more than the width. Find the area of the rectangle.   [11.6a]

54. *Book Pages.*   The sum of the page numbers on the facing pages of a book is 457. Find the page numbers.   [11.6a]

Synthesis

Divide.

55. $(x^4 + 9x^2 + 20) \div (x^2 + 4)$

56. $(y^4 + a^2) \div (y + a)$

57. $(5a^3 + 8a^2 - 23a - 1) \div (5a^2 - 7a - 2)$

58. $(15y^3 - 30y + 7 - 19y^2) \div (3y^2 - 2 - 5y)$

59. $(6x^5 - 13x^3 + 5x + 3 - 4x^2 + 3x^4) \div (3x^3 - 2x - 1)$

60. $(5x^7 - 3x^4 + 2x^2 - 10x + 2) \div (x^2 - x + 1)$

61. $(a^6 - b^6) \div (a - b)$

62. $(x^5 + y^5) \div (x + y)$

If the remainder is 0 when one polynomial is divided by another, the divisor is a *factor* of the dividend. Find the value(s) of c for which $x - 1$ is a factor of the polynomial.

63. $x^2 + 4x + c$

64. $2x^2 + 3cx - 8$

65. $c^2x^2 - 2cx + 1$

Vocabulary Reinforcement

Complete each statement with the correct word from the column on the right. Some of the choices may not be used.

1. In the expression 7^5, the number 5 is the _____. [13.1a]

2. The _____ rule asserts that when multiplying with exponential notation, if the bases are the same, we keep the base and add the exponent. [13.1d]

3. An expression of the type ax^n, where a is a real-number constant and n is a nonnegative integer, is a(n) _____. [13.3a, b]

4. A(n) _____ is a polynomial with three terms, such as $5x^4 - 7x^2 + 4$. [13.3b]

5. The _____ rule asserts that when dividing with exponential notation, if the bases are the same, we keep the base and subtract the exponent of the denominator from the exponent of the numerator. [13.1e]

6. If the exponents in a polynomial decrease from left to right, the polynomial is arranged in _____ order. [13.3e]

7. The _____ of a term is the sum of the exponents of the variables. [13.7b]

8. The number 2.3×10^{-5} is written in _____ notation. [13.2c]

ascending
descending
degree
fraction
scientific
base
exponent
product
quotient
monomial
binomial
trinomial

Concept Reinforcement

Determine whether each statement is true or false.

_____ 1. All trinomials are polynomials. [13.3b]

_____ 2. $(x + y)^2 = x^2 + y^2$ [13.6c]

_____ 3. The square of the difference of two expressions is the difference of the squares of the two expressions. [13.6c]

_____ 4. The product of the sum and the difference of the same two expressions is the difference of the squares of the expressions. [13.6b]

Study Guide

Objective 13.1d Use the product rule to multiply exponential expressions with like bases.

Example Multiply and simplify: $x^3 \cdot x^4$.
$$x^3 \cdot x^4 = x^{3+4} = x^7$$

Practice Exercise

1. Multiply and simplify: $z^5 \cdot z^3$.

Objective 13.1e Use the quotient rule to divide exponential expressions with like bases.

Example Divide and simplify: $\dfrac{x^6 y^5}{xy^3}$.

$$\dfrac{x^6 y^5}{xy^3} = \dfrac{x^6}{x} \cdot \dfrac{y^5}{y^3}$$
$$= x^{6-1} y^{5-3} = x^5 y^2$$

Practice Exercise

2. Divide and simplify: $\dfrac{a^4 b^7}{a^2 b}$.

Objective 13.1f Express an exponential expression involving negative exponents with positive exponents.

Objective 13.2a Use the power rule to raise powers to powers.

Objective 13.2b Raise a product to a power and a quotient to a power.

Example Simplify: $\left(\dfrac{2a^3 b^{-2}}{c^4} \right)^5$.

$$\left(\dfrac{2a^3 b^{-2}}{c^4} \right)^5 = \dfrac{(2a^3 b^{-2})^5}{(c^4)^5}$$
$$= \dfrac{2^5 (a^3)^5 (b^{-2})^5}{(c^4)^5} = \dfrac{32 a^{3 \cdot 5} b^{-2 \cdot 5}}{c^{4 \cdot 5}}$$
$$= \dfrac{32 a^{15} b^{-10}}{c^{20}} = \dfrac{32 a^{15}}{b^{10} c^{20}}$$

Practice Exercise

3. Simplify: $\left(\dfrac{x^{-4} y^2}{3z^3} \right)^3$.

Objective 13.2c Convert between scientific notation and decimal notation.

Example Convert 0.00095 to scientific notation.

0.0009.5

4 places

The number is small, so the exponent is negative.
$$0.00095 = 9.5 \times 10^{-4}$$

Example Convert 3.409×10^6 to decimal notation.

3.409000.

6 places

The exponent is positive, so the number is large.
$$3.409 \times 10^6 = 3,409,000$$

Practice Exercises

4. Convert to scientific notation: 763,000.

5. Convert to decimal notation: 3×10^{-4}.

Objective 13.2d Multiply and divide using scientific notation.

Example Multiply and express the result in scientific notation: $(5.3 \times 10^9) \cdot (2.4 \times 10^{-5})$.

$$(5.3 \times 10^9) \cdot (2.4 \times 10^{-5}) = (5.3 \cdot 2.4) \times (10^9 \cdot 10^{-5})$$
$$= 12.72 \times 10^4$$

We convert 12.72 to scientific notation and simplify:
$$12.72 \times 10^4 = (1.272 \times 10) \times 10^4$$
$$= 1.272 \times (10 \times 10^4)$$
$$= 1.272 \times 10^5.$$

Practice Exercise

6. Divide and express the result in scientific notation:

$$\dfrac{3.6 \times 10^3}{6.0 \times 10^{-2}}.$$

Objective 13.3d Collect the like terms of a polynomial.

Example Collect like terms:

$4x^3 - 2x^2 + 5 + 3x^2 - 12$.

$4x^3 - 2x^2 + 5 + 3x^2 - 12$

$= 4x^3 + (-2 + 3)x^2 + (5 - 12)$

$= 4x^3 + x^2 - 7$

Practice Exercise

7. Collect like terms: $5x^4 - 6x^2 - 3x^4 + 2x^2 - 3$.

Objective 13.4a Add polynomials.

Example Add: $(4x^3 + x^2 - 8) + (2x^3 - 5x + 1)$.

$(4x^3 + x^2 - 8) + (2x^3 - 5x + 1)$

$= (4 + 2)x^3 + x^2 - 5x + (-8 + 1)$

$= 6x^3 + x^2 - 5x - 7$

Practice Exercise

8. Add: $(3x^4 - 5x^2 - 4) + (x^3 + 3x^2 + 6)$.

Objective 13.5d Multiply any two polynomials.

Example Multiply: $(z^2 - 2z + 3)(z - 1)$.

We use columns. First, we multiply the top row by -1 and then by z, placing like terms of the product in the same column. Finally, we collect like terms.

$$\begin{array}{r} z^2 - 2z + 3 \\ z - 1 \\ \hline -z^2 + 2z - 3 \\ z^3 - 2z^2 + 3z \phantom{{}- 3} \\ \hline z^3 - 3z^2 + 5z - 3 \end{array}$$

Practice Exercise

9. Multiply: $(x^4 - 3x^2 + 2)(x^2 - 3)$.

Objective 13.6a Multiply two binomials mentally using the FOIL method.

Example Multiply: $(3x + 5)(x - 1)$.

$\qquad\qquad$ F $\qquad$ O $\qquad$ I $\qquad$ L

$(3x + 5)(x - 1) = 3x \cdot x + 3x \cdot (-1) + 5 \cdot x + 5 \cdot (-1)$

$= 3x^2 - 3x + 5x - 5$

$= 3x^2 + 2x - 5$

Practice Exercise

10. Multiply: $(y + 4)(2y + 3)$.

Objective 13.6b Multiply the sum and the difference of the same two terms mentally.

Example Multiply: $(3y + 2)(3y - 2)$.

$(3y + 2)(3y - 2) = (3y)^2 - 2^2$

$= 9y^2 - 4$

Practice Exercise

11. Multiply: $(x + 5)(x - 5)$.

Objective 13.6c Square a binomial mentally.

Example Multiply: $(2x - 3)^2$.

$(2x - 3)^2 = (2x)^2 - 2 \cdot 2x \cdot 3 + 3^2$

$= 4x^2 - 12x + 9$

Practice Exercise

12. Multiply: $(3w + 4)^2$.

Objective 13.7e Subtract polynomials.

Example Subtract:
$(m^4n + 2m^3n^2 - m^2n^3) - (3m^4n + 2m^3n^2 - 4m^2n^2).$
$(m^4n + 2m^3n^2 - m^2n^3) - (3m^4n + 2m^3n^2 - 4m^2n^2)$
$= m^4n + 2m^3n^2 - m^2n^3 - 3m^4n - 2m^3n^2 + 4m^2n^2$
$= -2m^4n - m^2n^3 + 4m^2n^2$

Practice Exercise

13. Subtract:
$(a^3b^2 - 5a^2b + 2ab) - (3a^3b^2 - ab^2 + 4ab).$

Objective 13.8a Divide a polynomial by a monomial.

Example Divide: $(6x^3 - 8x^2 + 15x) \div (3x).$
$$\frac{6x^3 - 8x^2 + 15x}{3x} = \frac{6x^3}{3x} - \frac{8x^2}{3x} + \frac{15x}{3x}$$
$$= \frac{6}{3}x^{3-1} - \frac{8}{3}x^{2-1} + \frac{15}{3}x^{1-1}$$
$$= 2x^2 - \frac{8}{3}x + 5$$

Practice Exercise

14. Divide: $(5y^2 - 20y + 8) \div 5.$

Objective 13.8b Divide a polynomial by a divisor that is a binomial.

Example Divide $x^2 - 3x + 7$ by $x + 1$.

$$
\begin{array}{r}
x - 4 \\
x + 1 \overline{) x^2 - 3x + 7} \\
\underline{x^2 + x} \\
-4x + 7 \\
\underline{-4x - 4} \\
11
\end{array}
$$

The answer is $x - 4 + \dfrac{11}{x + 1}$.

Practice Exercise

15. Divide: $(x^2 - 4x + 3) \div (x + 5).$

Review Exercises

Multiply and simplify.   [13.1d, f]

1. $7^2 \cdot 7^{-4}$ **2.** $y^7 \cdot y^3 \cdot y$

3. $(3x)^5 \cdot (3x)^9$ **4.** $t^8 \cdot t^0$

Divide and simplify.   [13.1e, f]

5. $\dfrac{4^5}{4^2}$ **6.** $\dfrac{a^5}{a^8}$ **7.** $\dfrac{(7x)^4}{(7x)^4}$

Simplify.

8. $(3t^4)^2$ [13.2a, b]   **9.** $(2x^3)^2(-3x)^2$
[13.1d], [13.2a, b]

10. $\left(\dfrac{2x}{y}\right)^{-3}$ [13.2b]

11. Express using a negative exponent: $\dfrac{1}{t^5}$.   [13.1f]

12. Express using a positive exponent: y^{-4}.   [13.1f]

13. Convert to scientific notation: 0.0000328.   [13.2c]

14. Convert to decimal notation: 8.3×10^6.   [13.2c]

Multiply or divide and write scientific notation for the result.   [13.2d]

15. $(3.8 \times 10^4)(5.5 \times 10^{-1})$ **16.** $\dfrac{1.28 \times 10^{-8}}{2.5 \times 10^{-4}}$

17. *Pizza Consumption.* Each man, woman, and child in the United States eats an average of 46 slices of pizza per year. The U.S. population is projected to be about 340 million in 2020. At this rate, how many slices of pizza would be consumed in 2020? Express the answer in scientific notation. [13.2e]

Sources: Packaged Facts; U.S. Census Bureau

18. Evaluate the polynomial $x^2 - 3x + 6$ when $x = -1$. [13.3a]

19. Identify the terms of the polynomial $-4y^5 + 7y^2 - 3y - 2$. [13.3b]

20. Identify the missing terms in $x^3 + x$. [13.3f]

21. Identify the degree of each term and the degree of the polynomial $4x^3 + 6x^2 - 5x + \frac{5}{3}$. [13.3c]

Classify the polynomial as a monomial, a binomial, a trinomial, or none of these. [13.3b]

22. $4x^3 - 1$

23. $4 - 9t^3 - 7t^4 + 10t^2$

24. $7y^2$

Collect like terms and then arrange in descending order. [13.3e]

25. $3x^2 - 2x + 3 - 5x^2 - 1 - x$

26. $-x + \frac{1}{2} + 14x^4 - 7x^2 - 1 - 4x^4$

Add. [13.4a]

27. $(3x^4 - x^3 + x - 4) + (x^5 + 7x^3 - 3x^2 - 5) + (-5x^4 + 6x^2 - x)$

28. $(3x^5 - 4x^4 + x^3 - 3) + (3x^4 - 5x^3 + 3x^2) + (-5x^5 - 5x^2) + (-5x^4 + 2x^3 + 5)$

Subtract. [13.4c]

29. $(5x^2 - 4x + 1) - (3x^2 + 1)$

30. $(3x^5 - 4x^4 + 3x^2 + 3) - (2x^5 - 4x^4 + 3x^3 + 4x^2 - 5)$

31. Find a polynomial for the perimeter and for the area. [13.4d], [13.5b]

$w + 3$

w

32. Find two algebraic expressions for the area of this figure. First, regard the figure as one large rectangle, and then regard the figure as a sum of four smaller rectangles. [13.4d]

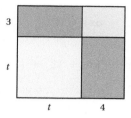

3

t

t

4

Multiply.

33. $\left(x + \frac{2}{3}\right)\left(x + \frac{1}{2}\right)$ [13.6a] **34.** $(7x + 1)^2$ [13.6c]

35. $(4x^2 - 5x + 1)(3x - 2)$ [13.5d] **36.** $(3x^2 + 4)(3x^2 - 4)$ [13.6b]

37. $5x^4(3x^3 - 8x^2 + 10x + 2)$ [13.5b]

38. $(x + 4)(x - 7)$ [13.6a]

39. $(3y^2 - 2y)^2$ [13.6c] **40.** $(2t^2 + 3)(t^2 - 7)$ [13.6a]

41. Evaluate the polynomial
$$2 - 5xy + y^2 - 4xy^3 + x^6$$
when $x = -1$ and $y = 2$. [13.7a]

42. Identify the coefficient and the degree of each term of the polynomial
$$x^5y - 7xy + 9x^2 - 8.$$
Then find the degree of the polynomial. [13.7b]

Collect like terms. [13.7c]

43. $y + w - 2y + 8w - 5$

44. $m^6 - 2m^2n + m^2n^2 + n^2m - 6m^3 + m^2n^2 + 7n^2m$

45. Add: [13.7d]
$$(5x^2 - 7xy + y^2) + (-6x^2 - 3xy - y^2) + (x^2 + xy - 2y^2).$$

46. Subtract: [13.7e]
$$(6x^3y^2 - 4x^2y - 6x) - (-5x^3y^2 + 4x^2y + 6x^2 - 6).$$

Multiply. [13.7f]

47. $(p - q)(p^2 + pq + q^2)$ **48.** $(3a^4 - \frac{1}{3}b^3)^2$

Divide.

49. $(10x^3 - x^2 + 6x) \div (2x)$ [13.8a]

50. $(6x^3 - 5x^2 - 13x + 13) \div (2x + 3)$ [13.8b]

51. The graph of the polynomial equation $y = 10x^3 - 10x$ is shown below. Use *only* the graph to estimate the value of the polynomial when $x = -1$, $x = -0.5$, $x = 0.5$, and $x = 1$. [13.3a]

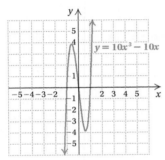

52. Subtract: $(2x^2 - 3x + 4) - (x^2 + 2x)$.   [13.4c]
   A. $x^2 - 3x - 2$ **B.** $x^2 - 5x + 4$
   C. $x^2 - x + 4$ **D.** $3x^2 - x + 4$

53. Multiply: $(x - 1)^2$.   [13.6c]
   A. $x^2 - 1$ **B.** $x^2 + 1$
   C. $x^2 - 2x - 1$ **D.** $x^2 - 2x + 1$

Synthesis

Find a polynomial for each shaded area.   [13.4d], [13.6b]

54.

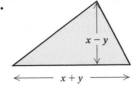

55.

56. Collect like terms: [13.1d], [13.2a], [13.3d]
$-3x^5 \cdot 3x^3 - x^6(2x)^2 + (3x^4)^2 + (2x^2)^4 - 40x^2(x^3)^2$.

57. Solve: [11.3b], [13.6a]
$(x - 7)(x + 10) = (x - 4)(x - 6)$.

58. The product of two polynomials is $x^5 - 1$. One of the polynomials is $x - 1$. Find the other.   [13.8b]

59. A rectangular garden is twice as long as it is wide and is surrounded by a sidewalk that is 4 ft wide. The area of the sidewalk is 1024 ft^2. Find the dimensions of the garden.   [11.3b], [13.4d], [13.5a], [13.6a]

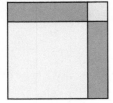

Understanding Through Discussion and Writing

1. Explain why the expression 578.6×10^{-7} is not written in scientific notation.   [13.2c]

2. Explain why an understanding of the rules for order of operations is essential when evaluating polynomials.   [13.3a]

3. How can the following figure be used to show that $(x + 3)^2 \neq x^2 + 9$?   [13.5c]

4. On an assignment, Emma *incorrectly* writes
$$\frac{12x^3 - 6x}{3x} = 4x^2 - 6x.$$
What mistake do you think she is making and how might you convince her that a mistake has been made?   [13.8a]

5. Can the sum of two trinomials in several variables be a trinomial in one variable? Why or why not?   [13.7d]

6. Is it possible for a polynomial in four variables to have a degree less than 4? Why or why not?   [13.7b]

CHAPTER

13 **Test**

For Extra Help For step-by-step test solutions, access the Chapter Test Prep Videos in MyMathLab® or on YouTube (search "BittingerAlgebraFoundations" and click on "Channels").

Multiply and simplify.

1. $6^{-2} \cdot 6^{-3}$

2. $x^6 \cdot x^2 \cdot x$

3. $(4a)^3 \cdot (4a)^8$

Divide and simplify.

4. $\dfrac{3^5}{3^2}$

5. $\dfrac{x^3}{x^8}$

6. $\dfrac{(2x)^5}{(2x)^5}$

Simplify.

7. $(x^3)^2$

8. $(-3y^2)^3$

9. $(2a^3b)^4$

10. $\left(\dfrac{ab}{c}\right)^3$

11. $(3x^2)^3(-2x^5)^3$

12. $3(x^2)^3(-2x^5)^3$

13. $2x^2(-3x^2)^4$

14. $(2x)^2(-3x^2)^4$

15. Express using a positive exponent: 5^{-3}.

16. Express using a negative exponent: $\dfrac{1}{y^8}$.

17. Convert to scientific notation: 3,900,000,000.

18. Convert to decimal notation: 5×10^{-8}.

Multiply or divide and write scientific notation for the answer.

19. $\dfrac{5.6 \times 10^6}{3.2 \times 10^{-11}}$

20. $(2.4 \times 10^5)(5.4 \times 10^{16})$

21. *Earth vs. Saturn.* The mass of Earth is about 6×10^{21} metric tons. The mass of Saturn is about 5.7×10^{23} metric tons. About how many times the mass of Earth is the mass of Saturn? Express the answer in scientific notation.

22. Evaluate the polynomial $x^5 + 5x - 1$ when $x = -2$.

23. Identify the coefficient of each term of the polynomial $\frac{1}{3}x^5 - x + 7$.

24. Identify the degree of each term and the degree of the polynomial $2x^3 - 4 + 5x + 3x^6$.

25. Classify the polynomial $7 - x$ as a monomial, a binomial, a trinomial, or none of these.

Collect like terms.

26. $4a^2 - 6 + a^2$

27. $y^2 - 3y - y + \dfrac{3}{4}y^2$

28. Collect like terms and then arrange in descending order:
$$3 - x^2 + 2x^3 + 5x^2 - 6x - 2x + x^5.$$

Add.

29. $(3x^5 + 5x^3 - 5x^2 - 3) +$
$(x^5 + x^4 - 3x^3 - 3x^2 + 2x - 4)$

30. $\left(x^4 + \dfrac{2}{3}x + 5\right) + \left(4x^4 + 5x^2 + \dfrac{1}{3}x\right)$

Subtract.

31. $(2x^4 + x^3 - 8x^2 - 6x - 3) - (6x^4 - 8x^2 + 2x)$

32. $(x^3 - 0.4x^2 - 12) - (x^5 + 0.3x^3 + 0.4x^2 + 9)$

Multiply.

33. $-3x^2(4x^2 - 3x - 5)$

34. $\left(x - \dfrac{1}{3}\right)^2$

35. $(3x + 10)(3x - 10)$

36. $(3b + 5)(b - 3)$

37. $(x^6 - 4)(x^8 + 4)$

38. $(8 - y)(6 + 5y)$

39. $(2x + 1)(3x^2 - 5x - 3)$

40. $(5t + 2)^2$

41. Collect like terms:
$x^3y - y^3 + xy^3 + 8 - 6x^3y - x^2y^2 + 11.$

42. Subtract:
$(8a^2b^2 - ab + b^3) - (-6ab^2 - 7ab - ab^3 + 5b^3).$

43. Multiply: $(3x^5 - 4y^5)(3x^5 + 4y^5).$

Divide.

44. $(12x^4 + 9x^3 - 15x^2) \div (3x^2)$

45. $(6x^3 - 8x^2 - 14x + 13) \div (3x + 2)$

46. The graph of the polynomial equation
$$y = x^3 - 5x - 1$$
is shown at right. Use *only* the graph to estimate the value of the polynomial when $x = -1$, $x = -0.5$, $x = 0.5$, $x = 1$, and $x = 1.1$.

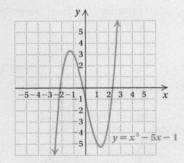

47. Find two algebraic expressions for the area of this figure. First, regard the figure as one large rectangle, and then regard the figure as a sum of four smaller rectangles.

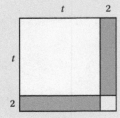

48. Which of the following is a polynomial for the surface area of this right rectangular solid?

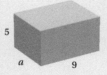

A. $28a$ **B.** $28a + 90$
C. $14a + 45$ **D.** $45a$

Synthesis

49. The height of a box is 1 less than its length, and the length is 2 more than its width. Find the volume in terms of the length.

50. Solve: $(x - 5)(x + 5) = (x + 6)^2.$

CHAPTER

16

Graphs, Functions, and Applications

From Chapter 16 of *Algebra Foundations: Basic Mathematics, Introductory Algebra, and Intermediate Algebra,* First Edition.
Marvin J. Bittinger, Judith A. Beecher, and Barbara L. Johnson. Copyright © 2015 by Pearson Education, Inc. All rights reserved.

16.1   Functions and Graphs

OBJECTIVES

a Determine whether a correspondence is a function.

b Given a function described by an equation, find function values (outputs) for specified values (inputs).

c Draw the graph of a function.

d Determine whether a graph is that of a function using the vertical-line test.

e Solve applied problems involving functions and their graphs.

SKILL TO REVIEW

Objective 10.1a: Evaluate algebraic expressions by substitution.

Evaluate.

1. $-\dfrac{1}{4}x$, when $x = 40$

2. $y^2 - 2y + 6$, when $y = -1$

a   IDENTIFYING FUNCTIONS

Consider the equation $y = 2x - 3$. If we substitute a value for x—say, 5—we get a value for y, 7:

$$y = 2x - 3 = 2(5) - 3 = 10 - 3 = 7.$$

The equation $y = 2x - 3$ is an example of a *function*, one of the most important concepts in mathematics.

In much the same way that ordered pairs form correspondences between first and second coordinates, a *function* is a correspondence from one set to another. For example:

To each student in a college, there corresponds his or her student ID.

To each item in a store, there corresponds its price.

To each real number, there corresponds the cube of that number.

In each case, the first set is called the **domain** and the second set is called the **range**. Each of these correspondences is a **function**, because given a member of the domain, there is *just one* member of the range to which it corresponds. Given a student, there is *just one* ID. Given an item, there is *just one* price. Given a real number, there is *just one* cube.

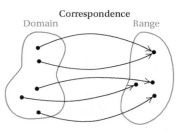
Correspondence

EXAMPLE 1   Determine whether the correspondence is a function.

Domain	Range
1 ⟶	$107.40
f: 2 ⟶	$ 34.10
3 ⟶	$ 29.60
4 ⟶	$ 19.60

Domain	Range
3 ⟶	5
g: 4 ⟶	9
5 ⟶	−7
6 ⟶	

Domain	Range
Chicago ⟨	Cubs
	White Sox
h: Baltimore ⟶	Orioles
San Diego ⟶	Padres

Domain	Range
Cubs ⟶	Chicago
p: White Sox ⟶	
Orioles ⟶	Baltimore
Padres ⟶	San Diego

Answers

Skill to Review:
1. −10   **2.** 9

The correspondence *f is* a function because each member of the domain is matched to *only one* member of the range.

The correspondence *g is* a function because each member of the domain is matched to *only one* member of the range. Note that a function allows two or more members of the domain to correspond to the same member of the range.

The correspondence *h is not* a function because one member of the domain, Chicago, is matched to *more than one* member of the range.

The correspondence *p is* a function because each member of the domain is matched to *only one* member of the range.

FUNCTION; DOMAIN; RANGE

A **function** is a correspondence between a first set, called the **domain**, and a second set, called the **range**, such that each member of the domain corresponds to exactly one member of the range.

Do Exercises 1–4. ▶

EXAMPLE 2 Determine whether each correspondence is a function.

Domain	**Correspondence**	*Range*
a) The integers	Each number's square	A set of nonnegative integers
b) A set of presidents (listed below)	Each president's appointees to the Supreme Court	A set of Supreme Court Justices (listed below)

APPOINTING PRESIDENT	SUPREME COURT JUSTICE
George H. W. Bush	Samuel A. Alito, Jr.
William Jefferson Clinton	Stephen G. Breyer
	Ruth Bader Ginsburg
George W. Bush	Elena Kagan
	John G. Roberts, Jr.
Barack H. Obama	Sonia M. Sotomayor
	Clarence Thomas

a) The correspondence *is* a function because each integer has *only one* square.

b) This correspondence *is not* a function because there is at least one member of the domain who is paired with more than one member of the range (William Jefferson Clinton with Stephen G. Breyer and Ruth Bader Ginsburg; George W. Bush with Samuel A. Alito, Jr., and John G. Roberts, Jr.; Barack H. Obama with Elena Kagan and Sonia M. Sotomayor).

Do Exercises 5–7 on the following page. ▶

Determine whether each correspondence is a function.

1.

Domain		*Range*
Cheetah	⟶	70 mph
Human	⟶	28 mph
Lion	⟶	50 mph
Chicken	⟶	9 mph

2.

Domain   *Range*

A   a
B   b
C   c
D   d
   e

3.

Domain   *Range*

−2
2   4
−3
3   9
0 ⟶ 0

4.

Domain   *Range*

4   −2
   2
9   −3
   3
0 ⟶ 0

Determine whether each correspondence is a function.

5. *Domain*
   A set of numbers

   Correspondence
   Square each number and subtract 10.

   Range
   A set of numbers

6. *Domain*
   A set of polygons

   Correspondence
   Find the perimeter of each polygon.

   Range
   A set of numbers

7. Determine whether the correspondence is a function.

   Domain
   A set of numbers

   Correspondence
   The area of a rectangle

   Range
   A set of rectangles

············· **Caution!** ·············

The notation $f(x)$ *does not mean "f times x"* and should not be read that way.

···································

When a correspondence between two sets is not a function, it is still an example of a **relation**.

RELATION

A **relation** is a correspondence between a first set, called the **domain**, and a second set, called the **range**, such that each member of the domain corresponds to **at least one** member of the range.

Thus, although the correspondences of Examples 1 and 2 are not all functions, they *are* all relations. A function is a special type of relation—one in which each member of the domain is paired with *exactly one* member of the range.

b FINDING FUNCTION VALUES

Most functions considered in mathematics are described by equations like $y = 2x + 3$ or $y = 4 - x^2$. We graph the function $y = 2x + 3$ by first performing calculations like the following:

for $x = 4$, $y = 2x + 3 = 2 \cdot 4 + 3 = 8 + 3 = 11$;

for $x = -5$, $y = 2x + 3 = 2 \cdot (-5) + 3 = -10 + 3 = -7$;

for $x = 0$, $y = 2x + 3 = 2 \cdot 0 + 3 = 0 + 3 = 3$;   and so on.

For $y = 2x + 3$, the **inputs** (members of the domain) are values of x substituted into the equation. The **outputs** (members of the range) are the resulting values of y. If we call the function f, we can use x to represent an arbitrary *input* and $f(x)$—read "f of x," or "f at x," or "the value of f at x"—to represent the corresponding *output*. In this notation, the function given by $y = 2x + 3$ is written as $f(x) = 2x + 3$ and the calculations above can be written more concisely as follows:

$$y = f(4) = 2 \cdot 4 + 3 = 8 + 3 = 11;$$
$$y = f(-5) = 2 \cdot (-5) + 3 = -10 + 3 = -7;$$
$$y = f(0) = 2 \cdot 0 + 3 = 0 + 3 = 3;   \text{and so on.}$$

Thus instead of writing "when $x = 4$, the value of y is 11," we can simply write "$f(4) = 11$," which can also be read as "f of 4 is 11" or "for the input 4, the output of f is 11."

We can think of a function as a machine. Think of $f(4) = 11$ as putting 4, a member of the domain (an input), into the machine. The machine knows the correspondence $f(x) = 2x + 3$, multiplies 4 by 2 and adds 3, and produces 11, a member of the range (the output).

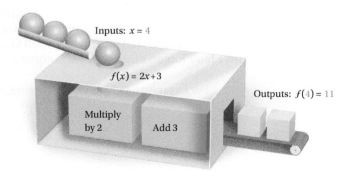

Inputs: $x = 4$

$f(x) = 2x+3$

Multiply by 2

Add 3

Outputs: $f(4) = 11$

Answers

5. Yes   **6.** Yes   **7.** No

EXAMPLE 3 A function f is given by $f(x) = 3x^2 - 2x + 8$. Find each of the indicated function values.

a) $f(0)$ **b)** $f(-5)$ **c)** $f(7a)$

One way to find function values when a formula is given is to think of the formula with blanks, or placeholders, replacing the variable as follows:

$$f(\square) = 3\square^2 - 2\square + 8.$$

To find an output for a given input, we think: "Whatever goes in the blank on the left goes in the blank(s) on the right." With this in mind, let's complete the example.

a) $f(0) = 3 \cdot 0^2 - 2 \cdot 0 + 8 = 8$

b) $f(-5) = 3(-5)^2 - 2 \cdot (-5) + 8 = 3 \cdot 25 + 10 + 8 = 75 + 10 + 8 = 93$

c) $f(7a) = 3(7a)^2 - 2(7a) + 8 = 3 \cdot 49a^2 - 14a + 8 = 147a^2 - 14a + 8$

Do Exercise 8. ▷

EXAMPLE 4 Find the indicated function value.

a) $f(5)$, for $f(x) = 3x + 2$ **b)** $g(-2)$, for $g(x) = 7$

c) $F(a + 1)$, for $F(x) = 5x - 8$ **d)** $f(a + h)$, for $f(x) = -2x + 1$

a) $f(5) = 3 \cdot 5 + 2 = 15 + 2 = 17$

b) For the function given by $g(x) = 7$, all inputs share the same output, 7. Thus, $g(-2) = 7$. The function g is an example of a **constant function**.

c) $F(a + 1) = 5(a + 1) - 8 = 5a + 5 - 8 = 5a - 3$

d) $f(a + h) = -2(a + h) + 1 = -2a - 2h + 1$

Do Exercise 9. ▷

8. Find the indicated function values for the function
$$f(x) = 2x^2 + 3x - 4.$$

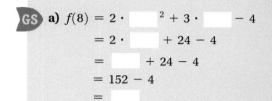

GS **a)** $f(8) = 2 \cdot \boxed{}^2 + 3 \cdot \boxed{} - 4$

$\qquad = 2 \cdot \boxed{} + 24 - 4$

$\qquad = \boxed{} + 24 - 4$

$\qquad = 152 - 4$

$\qquad = \boxed{}$

b) $f(0)$

c) $f(-5)$

d) $f(2a)$

9. Find the indicated function value.

a) $f(-6)$, for $f(x) = 5x - 3$

b) $g(55)$, for $g(x) = -3$

c) $F(a + 2)$, for $F(x) = -5x + 8$

d) $f(a - h)$, for $f(x) = 6x - 7$

Answers

8. **(a)** 148; **(b)** −4; **(c)** 31; **(d)** $8a^2 + 6a - 4$
9. **(a)** −33; **(b)** −3; **(c)** −5a − 2; **(d)** 6a − 6h − 7

Guided Solution:
8. **(a)** 8, 8, 64, 128, 148

CALCULATOR CORNER

Finding Function Values We can find function values using a graphing calculator. One method is to substitute inputs directly into the formula. Consider the function $f(x) = x^2 + 3x - 4$. We find that $f(-5) = 6$. See Figure 1.

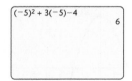

FIGURE 1

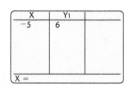

FIGURE 2

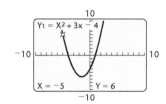

FIGURE 3

After we have entered the function as $y_1 = x^2 + 3x - 4$ on the equation-editor screen, there are other methods that we can use to find function values. We can use a table set in ASK mode and enter $x = -5$. We see that the function value, y_1, is 6. See Figure 2. We can also use the VALUE feature to evaluate the function. To do this, we first graph the function. Then we press **2ND** **CALC** **1** to access the VALUE feature. Next, we supply the desired x-value. Finally, we press **ENTER** to see X = −5, Y = 6 at the bottom of the screen. See Figure 3. Again we see that the function value is 6. Note that when the VALUE feature is used to find a function value, the x-value must be in the viewing window.

EXERCISES: Find each function value.

1. $f(-5.1)$, for $f(x) = -3x + 2$ **2.** $f(3)$, for $f(x) = 4x^2 + x - 5$

To graph a function, we find ordered pairs (x, y) or $(x, f(x))$, plot them, and connect the points. Note that y and $f(x)$ are used interchangeably—that is, $y = f(x)$—when we are working with functions and their graphs.

EXAMPLE 5 Graph: $f(x) = x + 2$.

A list of some function values is shown in the following table. We plot the points and connect them. The graph is a straight line. The "y" on the vertical axis could also be labeled "$f(x)$."

x	$f(x)$
-4	-2
-3	-1
-2	0
-1	1
0	2
1	3
2	4
3	5
4	6

10. Graph: $f(x) = x - 4$.

x	$f(x)$

◀ Do Exercise 10.

EXAMPLE 6 Graph: $g(x) = 4 - x^2$.

We calculate some function values, plot the corresponding points, and draw the curve.

$$g(0) = 4 - 0^2 = 4 - 0 = 4,$$
$$g(-1) = 4 - (-1)^2 = 4 - 1 = 3,$$
$$g(2) = 4 - 2^2 = 4 - 4 = 0,$$
$$g(-3) = 4 - (-3)^2 = 4 - 9 = -5$$

x	$g(x)$
-3	-5
-2	0
-1	3
0	4
1	3
2	0
3	-5

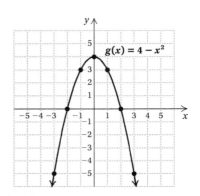

◀ Do Exercise 11.

11. Graph: $g(x) = 5 - x^2$.

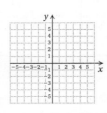

x	$g(x)$

Answers

10.

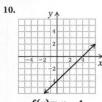

$f(x) = x - 4$

11.

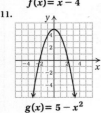

$g(x) = 5 - x^2$

EXAMPLE 7 Graph: $h(x) = |x|$.

A list of some function values is shown in the following table. We plot the points and connect them. The graph is a V-shaped "curve" that rises on either side of the vertical axis.

x	$h(x)$
-3	3
-2	2
-1	1
0	0
1	1
2	2
3	3

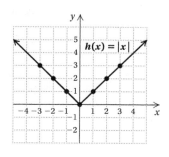

Do Exercise 12.

d THE VERTICAL-LINE TEST

Consider the graph of the function f described by $f(x) = x^2 - 5$ shown at right. It is also the graph of the equation $y = x^2 - 5$.

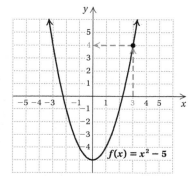

To find a function value, like $f(3)$, from a graph, we locate the input on the horizontal axis, move directly up or down to the graph of the function, and then move left or right to find the output on the vertical axis. Thus, $f(3) = 4$. Keep in mind that members of the domain are found on the horizontal axis, members of the range are found on the vertical axis, and the y on the vertical axis could also be labeled $f(x)$.

When one member of the domain is paired with two or more different members of the range, the correspondence is not a function. Thus, when a graph contains two or more different points with the same first coordinate, the graph cannot represent a function. Points sharing a common first coordinate are vertically above or below each other. (See the following graph.) This observation leads to the *vertical-line test*.

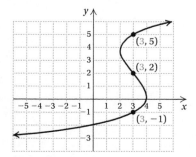

Since 3 is paired with more than one member of the range, the graph does not represent a function.

THE VERTICAL-LINE TEST

If it is possible for a vertical line to cross a graph more than once, then the graph is *not* the graph of a function.

---

12. Graph: $t(x) = 3 - |x|$.

x	$t(x)$

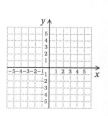

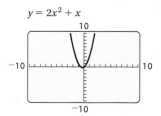
Answer

12.

$t(x) = 3 - |x|$

Determine whether each of the following is the graph of a function.

13.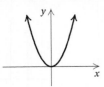

14.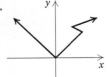

15.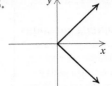

16.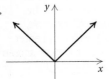

EXAMPLE 8 Determine whether each of the following is the graph of a function.

a)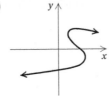

b)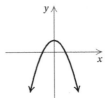

a) The graph *is not* that of a function because a vertical line can cross the graph at more than one point.

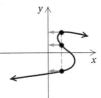

b) The graph *is* that of a function because no vertical line can cross the graph more than once.

◀ **Do Exercises 13–16.**

e APPLICATIONS OF FUNCTIONS AND THEIR GRAPHS

Functions are often described by graphs, whether or not an equation is given. To use a graph in an application, we note that each point on the graph represents a pair of values.

EXAMPLE 9 *IRS Instruction Booklet.* The following graph represents the number of pages in the IRS 1040 instruction booklet for years from 1965 through 2012. The number of pages is a function of the year. Note that no equation is given for the function.

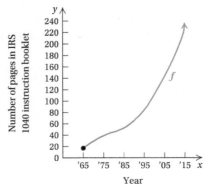

SOURCES: National Taxpayers Union; Statista.com; Internal Revenue Service

a) How many pages are in the 1975 IRS 1040 instruction booklet? That is, find $f(1975)$.

b) How many pages are in the 2010 IRS 1040 instruction booklet? That is, find $f(2010)$.

Answers

13. Yes   **14.** No   **15.** No   **16.** Yes

a) To estimate the number of pages in the 1975 booklet, we locate 1975 on the horizontal axis and move directly up until we reach the graph. Then we move across to the vertical axis. We come to a point that is about 40, so we estimate that the number of pages in the 1975 booklet is 40.

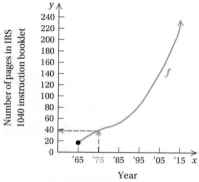

SOURCES: National Taxpayers Union;
Statista.com; Internal Revenue Service

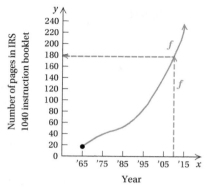

SOURCES: National Taxpayers Union;
Statista.com; Internal Revenue Service

b) To estimate the number of pages in the 2010 booklet, we locate 2010 on the horizontal axis and move directly up until we reach the graph. Then we move across to the vertical axis. We come to a point that is about 180, so we estimate that the number of pages in the 2010 booklet is 180.

Do Exercises 17 and 18. ▶

Refer to the graph in Example 9 for Margin Exercises 17 and 18.

17. How many pages are in the 2005 IRS 1040 instruction booklet?

18. How many pages are in the 2012 IRS 1040 instruction booklet?

Answers

17. About 140 pages   **18.** About 190 pages

16.1 Exercise Set

✓ Reading Check

Use the graph at right to find the given function value by locating the input on the horizontal axis, moving directly up or down to the graph of the function, and then moving left or right to find the output on the vertical axis. As an example, finding $f(4) = -5$ is illustrated.

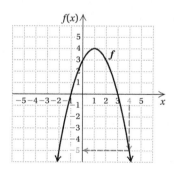

RC1. $f(2)$

RC2. $f(0)$

RC3. $f(-2)$

RC4. $f(3)$

a Determine whether each correspondence is a function.

1. Domain   Range

2 ⟶ 9
5 ⟶ 8
19

2. Domain   Range

5 ⟶ 3
−3 ⟶ 7
7
−7

3. Domain   Range

−5 ⟶ 1
5
8

4. Domain   Range

6 ⟶ −6
7 ⟶ −7
3 ⟶ −3

5. Domain   Range

9 ⟶ 3
   ⟶ −3
16 ⟶ 4
   ⟶ −4
25 ⟶ 5
   ⟶ −5

6. Domain

The Color Purple, 1982 (Pulitzer Prize 1983)
East of Eden, 1952
Fahrenheit 451, 1953
The Good Earth, 1931 (Pulitzer Prize 1932)
For Whom the Bell Tolls, 1940
The Grapes of Wrath, 1939 (Pulitzer Prize 1940)
To Kill a Mockingbird, 1960 (Pulitzer Prize 1961)
The Old Man and the Sea, 1952 (Pulitzer Prize 1953)

Range

Ray Bradbury
Pearl Buck
Ernest Hemingway
Harper Lee
John Steinbeck
Alice Walker

7. Domain   Range

Florida ⟨ Florida State University
University of Florida
University of Miami

Kansas ⟨ Baker University
Kansas State University
University of Kansas

8. Domain   Range

Colorado State University
University of Colorado ⟶ Colorado
University of Denver
Gonzaga University
University of Washington ⟶ Washington
Washington State University

Domain	Correspondence	Range
9. A set of numbers	The area of a triangle	A set of triangles
10. A family	Each person's height, in inches	A set of positive numbers
11. The set of U.S. Senators	The state that a Senator represents	The set of all states
12. The set of all states	Each state's members of the U.S. Senate	The set of U.S. Senators

UNITED STATES
Missouri
Claire McCaskill
Roy Blunt

b Find the function values.

13. $f(x) = x + 5$
  a) $f(4)$ **b)** $f(7)$
  c) $f(-3)$ **d)** $f(0)$
  e) $f(2.4)$ **f)** $f\left(\frac{2}{3}\right)$

14. $g(t) = t - 6$
  a) $g(0)$ **b)** $g(6)$
  c) $g(13)$ **d)** $g(-1)$
  e) $g(-1.08)$ **f)** $g\left(\frac{7}{8}\right)$

15. $h(p) = 3p$
  a) $h(-7)$ **b)** $h(5)$
  c) $h\left(\frac{2}{3}\right)$ **d)** $h(0)$
  e) $h(6a)$ **f)** $h(a + 1)$

16. $f(x) = -4x$
  a) $f(6)$ **b)** $f\left(-\frac{1}{2}\right)$
  c) $f(0)$ **d)** $f(-1)$
  e) $f(3a)$ **f)** $f(a - 1)$

17. $g(s) = 3s + 4$
  a) $g(1)$ **b)** $g(-7)$
  c) $g\left(\frac{2}{3}\right)$ **d)** $g(0)$
  e) $g(a - 2)$ **f)** $g(a + h)$

18. $h(x) = 19$, a constant function
  a) $h(4)$ **b)** $h(-6)$
  c) $h(12.5)$ **d)** $h(0)$
  e) $h\left(\frac{2}{3}\right)$ **f)** $h(a + 3)$

19. $f(x) = 2x^2 - 3x$
  a) $f(0)$ **b)** $f(-1)$
  c) $f(2)$ **d)** $f(10)$
  e) $f(-5)$ **f)** $f(4a)$

20. $f(x) = 3x^2 - 2x + 1$
  a) $f(0)$ **b)** $f(1)$
  c) $f(-1)$ **d)** $f(10)$
  e) $f(-3)$ **f)** $f(2a)$

21. $f(x) = |x| + 1$
  a) $f(0)$ **b)** $f(-2)$
  c) $f(2)$ **d)** $f(-10)$
  e) $f(a - 1)$ **f)** $f(a + h)$

22. $g(t) = |t - 1|$
  a) $g(4)$ **b)** $g(-2)$
  c) $g(-1)$ **d)** $g(100)$
  e) $g(5a)$ **f)** $g(a + 1)$

23. $f(x) = x^3$
  a) $f(0)$ **b)** $f(-1)$
  c) $f(2)$ **d)** $f(10)$
  e) $f(-5)$ **f)** $f(-3a)$

24. $f(x) = x^4 - 3$
  a) $f(1)$ **b)** $f(-1)$
  c) $f(0)$ **d)** $f(2)$
  e) $f(-2)$ **f)** $f(-a)$

25. *Average Age of Senators.* The function $A(s)$ given by
$$A(s) = 0.044s + 59$$
can be used to estimate the average age of senators in the U.S. Senate in the years 1945 to 2013. Let $A(s) =$ the average age of the senators and $s =$ the number of years since 1945—that is, $s = 0$ for 1945, $s = 20$ for 1965, and so on. What was the average age of U.S. Senators in 1980? in 2013?

Sources: www.slate.com/; "Democracy or Gerontocracy," Brian Palmar, January 2, 2013; Congressional Research Service

26. *Average Age of House Members.* The function $A(h)$ given by
$$A(h) = 0.059h + 53$$
can be used to estimate the average age of House members in the U.S. House of Representatives in the years 1945 to 2013. Let $A(h) =$ the average age of the House members and $h =$ the number of years since 1945. What is the average age of U.S. House members in 1980? in 2013?

Sources: www.slate.com/; "Democracy or Gerontocracy," Brian Palmar, January 2, 2013; Congressional Research Service

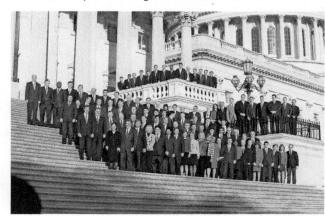

27. *Pressure at Sea Depth.*   The function $P(d) = 1 + (d/33)$ gives the pressure, in *atmospheres* (atm), at a depth of d feet in the sea. Note that $P(0) = 1$ atm, $P(33) = 2$ atm, and so on. Find the pressure at 20 ft, 30 ft, and 100 ft.

28. *Temperature as a Function of Depth.*   The function $T(d) = 10d + 20$ gives the temperature, in degrees Celsius, inside the earth as a function of the depth d, in kilometers. Find the temperature at 5 km, 20 km, and 1000 km.

29. *Melting Snow.*   The function $W(d) = 0.112d$ approximates the amount of water, in centimeters, that results from d centimeters of snow melting. Find the amount of water that results from snow melting from depths of 16 cm, 25 cm, and 100 cm.

30. *Temperature Conversions.*   The function $C(F) = \frac{5}{9}(F - 32)$ determines the Celsius temperature that corresponds to F degrees Fahrenheit. Find the Celsius temperature that corresponds to 62°F, 77°F, and 23°F.

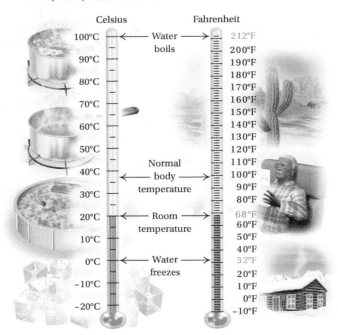

---

c   Graph each function.

31. $f(x) = -2x$

x	$f(x)$

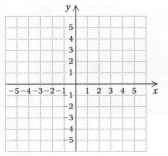

32. $g(x) = 3x$

x	$g(x)$

33. $g(x) = 3x - 1$

x	$g(x)$

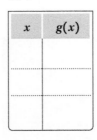

34. $f(x) = 2x + 5$

x	$f(x)$

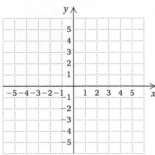

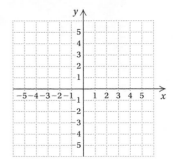

35. $g(x) = -2x + 3$

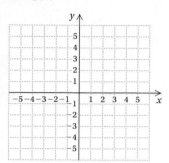

36. $f(x) = -\frac{1}{2}x + 2$

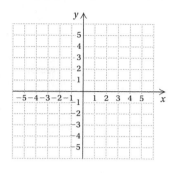

37. $f(x) = \frac{1}{2}x + 1$

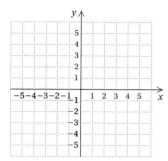

38. $f(x) = -\frac{3}{4}x - 2$

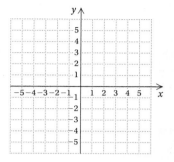

39. $f(x) = 2 - |x|$

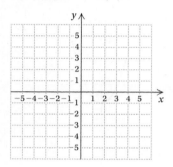

40. $f(x) = |x| - 4$

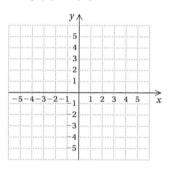

41. $g(x) = |x - 1|$

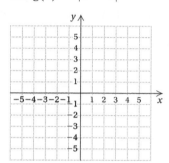

42. $g(x) = |x + 3|$

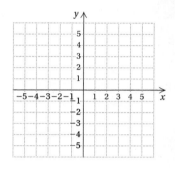

43. $g(x) = x^2 + 2$

x	g(x)
-2	
-1	
0	
1	
2	

44. $f(x) = x^2 + 1$

x	f(x)
-2	
-1	
0	
1	
2	

45. $f(x) = x^2 - 2x - 3$

x	f(x)
-2	
-1	
0	
1	
2	
3	
4	

46. $g(x) = x^2 + 6x + 5$

x	g(x)
-6	
-5	
-4	
-3	
-2	
-1	
0	

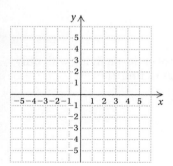

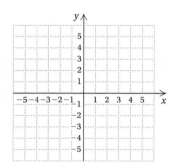

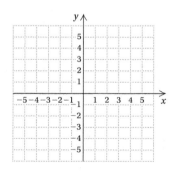

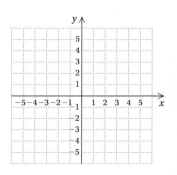

47. $f(x) = -x^2 + 1$

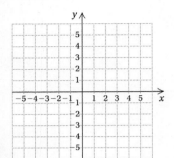

48. $f(x) = -x^2 + 2$

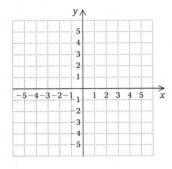

49. $f(x) = x^3 + 1$

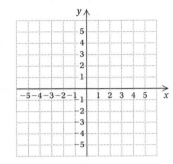

50. $f(x) = x^3 - 2$

d Determine whether each of the following is the graph of a function.

51.

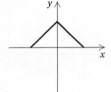

52.

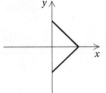

53.

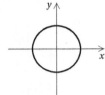

54.

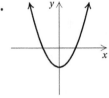

55.

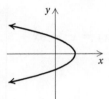

56.

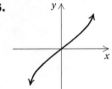

57.

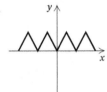

58.

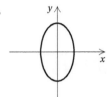

e Solve.

Living with Grandparents. The following graph approximates the number of children in the United States who lived with only their grandparents in the years from 1991 through 2009. The number of children is a function f of the year x.

59. Approximate the number of children living with only grandparents in 2009. That is, find $f(2009)$.

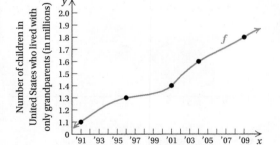

SOURCE: U.S. Census 2010

60. Approximate the number of children living with only grandparents in 1996. That is, find $f(1996)$.

Pharmacists. The following graph approximates the number of pharmacists in the United States in the years from 2002 through 2012. The number of pharmacists is a function g of the year x.

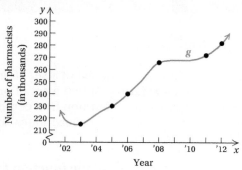

SOURCE: IDC; statista.com

61. Approximate the number of pharmacists in 2005.

62. Approximate the number of pharmacists in 2012.

Skill Maintenance

Solve.

63. $-\dfrac{5}{3} + y = -\dfrac{1}{12} - \dfrac{5}{6}$ [11.3b]

64. $6x - 31 = 11 + 6(x - 7)$ [11.3c]

65. $4 - 7y > 2y - 32$ [11.7e]

66. $x^2 + 2x - 80 = 0$ [14.8b]

Factor.

67. $121t^2 - 25$ [14.5d]

68. $w^3 - 1000$ [14.6a]

Multiply and simplify.   [13.1d, f]

69. $x^{-4} \cdot x^9$

70. $(s^2t)(s^5t^8)$

Subtract.   [13.4c]

71. $(y^2 - 2y - 7) - (y + 4)$

72. $(x^4 - 1) - (6x^4 + x^2 - 5x + 1)$

Synthesis

73. Suppose that for some function g, $g(x - 6) = 10x - 1$. Find $g(-2)$.

74. Suppose that for some function h, $h(x + 5) = x^2 - 4$. Find $h(3)$.

For Exercises 75 and 76, let $f(x) = 3x^2 - 1$ and $g(x) = 2x + 5$.

75. Find $f(g(-4))$ and $g(f(-4))$.

76. Find $f(g(-1))$ and $g(f(-1))$.

77. Suppose that a function g is such that $g(-1) = -7$ and $g(3) = 8$. Find a formula for g if $g(x)$ is of the form $g(x) = mx + b$, where m and b are constants.

OBJECTIVE

a Find the domain and the range of a function.

SKILL TO REVIEW

Objective 11.3a: Solve equations using both the addition principle and the multiplication principle.

Solve.

1. $6x - 3 = 51$

2. $15 - 2x = 0$

1. Find the domain and the range of the function f whose graph is shown below.

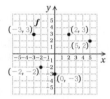

a FINDING DOMAIN AND RANGE

The solutions of an equation in two variables consist of a set of ordered pairs. A set of ordered pairs is called a **relation**. When a set of ordered pairs is such that no two different pairs share a common first coordinate, we have a **function**. The **domain** is the set of all first coordinates, and the **range** is the set of all second coordinates.

EXAMPLE 1 Find the domain and the range of the function f whose graph is shown below.

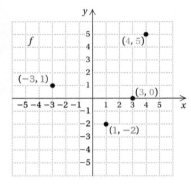

This function contains just four ordered pairs and it can be written as

$$\{(-3, 1), (1, -2), (3, 0), (4, 5)\}.$$

We can determine the domain and the range by reading the x- and y-values directly from the graph.

The domain is the set of all first coordinates, or x-values, $\{-3, 1, 3, 4\}$. The range is the set of all second coordinates, or y-values, $\{1, -2, 0, 5\}$.

◀ Do Margin Exercise 1.

EXAMPLE 2 For the function f whose graph is shown at right, determine each of the following.

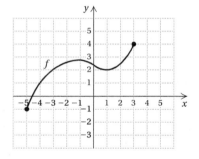

a) The number in the range that is paired with 1 from the domain. That is, find $f(1)$.

b) The domain of f

c) The numbers in the domain that are paired with 1 from the range. That is, find all x such that $f(x) = 1$.

d) The range of f

a) To determine which number in the range is paired with 1 in the domain, we locate 1 on the horizontal axis. Next, we find the point on the graph of f for which 1 is the first coordinate. From that point, we can look to the vertical axis to find the corresponding y-coordinate, 2. The input 1 has the output 2—that is, $f(1) = 2$.

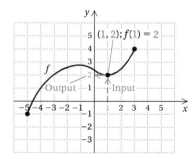

Answers

Skill to Review:

1. 9 **2.** $\frac{15}{2}$, or 7.5

Margin Exercise:

1. Domain $= \{-3, -2, 0, 2, 5\}$; range $= \{-3, -2, 2, 3\}$

b) The domain of the function is the set of all *x*-values, or inputs, of the points on the graph. These extend from −5 to 3 and can be viewed as the curve's shadow, or projection, onto the *x*-axis. Thus the domain is $\{x \mid -5 \le x \le 3\}$.

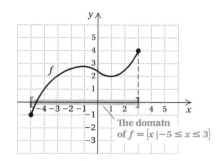

c) To determine which numbers in the domain are paired with 1 in the range, we locate 1 on the vertical axis. From there, we look left and right to the graph of *f* to find any points for which 1 is the second coordinate (output). One such point exists, $(-4, 1)$. For this function, we note that $x = -4$ is the only member of the domain paired with 1. For other functions, there might be more than one member of the domain paired with a member of the range.

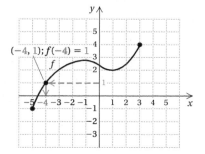

d) The range of the function is the set of all *y*-values, or outputs, of the points on the graph. These extend from −1 to 4 and can be viewed as the curve's shadow, or projection, onto the *y*-axis. Thus the range is $\{y \mid -1 \le y \le 4\}$.

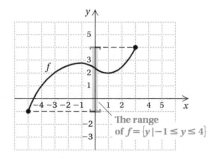

Do Exercise 2. ▶

EXAMPLE 3 Find the domain and the range of the function *h* whose graph is shown below.

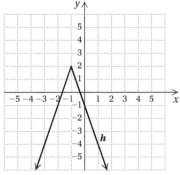

 Since no endpoints are indicated, the graph extends indefinitely horizontally. Thus the domain, or the set of inputs, is the set of all real numbers. The range, or the set of outputs, is the set of all *y*-values of the points on the graph. Thus the range is $\{y \mid y \le 2\}$.

Do Exercise 3. ▶

2. For the function *f* whose graph is shown below, determine each of the following.

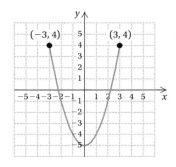

a) The number in the range that is paired with the input 1. That is, find $f(1)$.

b) The domain of *f*

c) The numbers in the domain that are paired with 4

d) The range of *f*

3. Find the domain and the range of the function *f* whose graph is shown below.

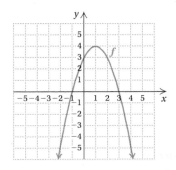

Answers

2. (a) −4; **(b)** $\{x \mid -3 \le x \le 3\}$;
(c) −3, 3; **(d)** $\{y \mid -5 \le y \le 4\}$
3. Domain: all real numbers; range: $\{y \mid y \le 4\}$

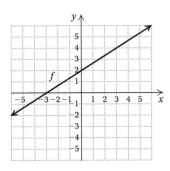

4. Find the domain and the range of the function f whose graph is shown below.

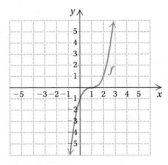

Find the domain.

5. $f(x) = x^3 - |x|$

6. $f(x) = \dfrac{4}{3x + 2}$ **GS**

Set the denominator equal to 0 and solve for x:

$3x + 2 = \boxed{}$

$3x = -2$

$x = \boxed{}$.

Thus, $\boxed{}$ is not in the domain of $f(x)$; all other real numbers are.

Domain $= \{x \,|\, x$ is a real number $and \, x \neq \boxed{}\}$

Answers

4. Domain: all real numbers; range: all real numbers   **5.** All real numbers

6. $\left\{x \,|\, x \text{ is a real number } and \, x \neq -\dfrac{2}{3}\right\}$

Guided Solution:

6. $0, -\dfrac{2}{3}, -\dfrac{2}{3}, -\dfrac{2}{3}$

EXAMPLE 4   Find the domain and the range of the function f whose graph is shown at left.

Since no endpoints are indicated, the graph extends indefinitely both horizontally and vertically. Thus the domain is the set of all real numbers. Likewise, the range is the set of all real numbers.

◀ Do Exercise 4.

When a function is given by an equation or a formula, the domain is understood to be the largest set of real numbers (inputs) for which function values (outputs) can be calculated. That is, the domain is the set of all possible allowable inputs into the formula. To find the domain, think, "What can we substitute?"

EXAMPLE 5   Find the domain: $f(x) = |x|$.

We ask, "What can we substitute?" Is there any number x for which we cannot calculate $|x|$? The answer is no. Thus the domain of f is the set of all real numbers.

EXAMPLE 6   Find the domain: $f(x) = \dfrac{3}{2x - 5}$.

We ask, "What can we substitute?" Is there any number x for which we cannot calculate $3/(2x - 5)$? Since $3/(2x - 5)$ cannot be calculated when the denominator $2x - 5$ is 0, we solve the following equation to find those real numbers that must be excluded from the domain of f:

$$
\begin{aligned}
2x - 5 &= 0 && \text{Setting the denominator equal to 0} \\
2x &= 5 && \text{Adding 5} \\
x &= \tfrac{5}{2}. && \text{Dividing by 2}
\end{aligned}
$$

Thus, $\tfrac{5}{2}$ is not in the domain, whereas all other real numbers are.
The domain of f is $\left\{x \,|\, x \text{ is a real number } and \, x \neq \tfrac{5}{2}\right\}$.

◀ Do Exercises 5 and 6.

Functions: A Review

The following is a review of the function concepts considered in Sections 16.1 and 16.2. Use the graph below to visualize the concepts.

Function Concepts

- Formula for f: $f(x) = x^2 - 7$
- For every input of f, there is exactly one output.
- When 1 is the input, -6 is the output.
- $f(1) = -6$
- $(1, -6)$ is on the graph.
- Domain $=$ The set of all inputs
  $=$ The set of all real numbers
- Range $=$ The set of all outputs
  $= \{y \,|\, y \geq -7\}$

Graph

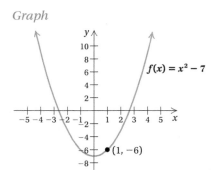

☑ Reading Check

Choose from the column on the right the domain of the function. Some choices may be used more than once; others not at all.

RC1. $f(x) = 5 - x$

RC2. $f(x) = \dfrac{-5}{5 - x}$

RC3. $f(x) = |5 - x|$

RC4. $f(x) = \dfrac{5}{|x - 5|}$

RC5. $f(x) = 5 - |x|$

RC6. $f(x) = \dfrac{x - 5}{x + 5}$

a) All real numbers

b) $\{x \mid x \text{ is a real number } and \ x \neq 5\}$

c) $\{x \mid x \text{ is a real number } and \ x \neq -5 \ and \ x \neq 5\}$

d) $\{x \mid x \text{ is a real number } and \ x \neq -5\}$

---

a   In Exercises 1–8, the graph is that of a function. Determine for each one **(a)** $f(1)$; **(b)** the domain; **(c)** all x-values such that $f(x) = 2$; and **(d)** the range. An open dot indicates that the point does not belong to the graph.

1.

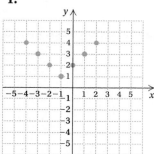

2.

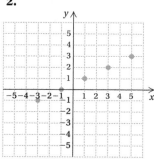

3.

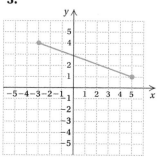

4.

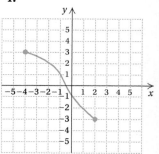

5.

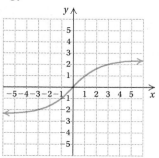

6.

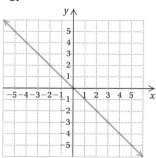

7.

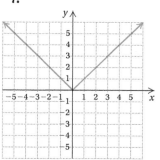

8.

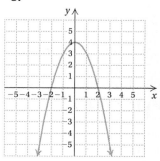

Find the domain.

9. $f(x) = \dfrac{2}{x + 3}$

10. $f(x) = \dfrac{7}{5 - x}$

11. $f(x) = 2x + 1$

12. $f(x) = 4 - 5x$

13. $f(x) = x^2 + 3$

14. $f(x) = x^2 - 2x + 3$

15. $f(x) = \dfrac{8}{5x - 14}$

16. $f(x) = \dfrac{x - 2}{3x + 4}$

17. $f(x) = |x| - 4$

18. $f(x) = |x - 4|$

19. $f(x) = \dfrac{x^2 - 3x}{|4x - 7|}$

20. $f(x) = \dfrac{4}{|2x - 3|}$

21. $g(x) = \dfrac{1}{x - 1}$

22. $g(x) = \dfrac{-11}{4 + x}$

23. $g(x) = x^2 - 2x + 1$

24. $g(x) = 8 - x^2$

25. $g(x) = x^3 - 1$

26. $g(x) = 4x^3 + 5x^2 - 2x$

27. $g(x) = \dfrac{7}{20 - 8x}$

28. $g(x) = \dfrac{2x - 3}{6x - 12}$

29. $g(x) = |x + 7|$

30. $g(x) = |x| + 1$

31. $g(x) = \dfrac{-2}{|4x + 5|}$

32. $g(x) = \dfrac{x^2 + 2x}{|10x - 20|}$

33. For the function f whose graph is shown below, find $f(-1), f(0),$ and $f(1)$.

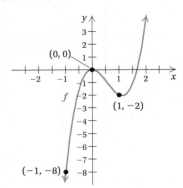

34. For the function g whose graph is shown below, find all the x-values for which $g(x) = 1$.

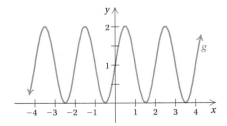

Skill Maintenance

Simplify.   [15.1c]

35. $\dfrac{a^2 - 1}{a + 1}$

36. $\dfrac{10y^2 + 10y - 20}{35y^2 + 210y - 245}$

37. $\dfrac{5x - 15}{x^2 - x - 6}$

Divide.   [13.8b]

38. $(t^2 + 3t - 28) \div (t + 7)$

39. $(w^2 + 4w + 5) \div (w + 3)$

Synthesis

40. Determine the range of each of the functions in Exercises 22, 23, 24, and 30.

41. Determine the range of each of the functions in Exercises 9, 14, 17, and 18.

Find the domain of each function.

42. $g(x) = \sqrt{2 - x}$

43. $f(x) = \sqrt[3]{x - 1}$

Concept Reinforcement

Determine whether each statement is true or false.

_____ **1.** Every function is a relation. [16.1a]

_____ **2.** It is possible for one input of a function to have two or more outputs. [16.1a]

_____ **3.** It is possible for all the inputs of a function to have the same output. [16.1a]

_____ **4.** If it is possible for a vertical line to cross a graph more than once, the graph is not the graph of a function. [16.1d]

_____ **5.** If the domain of a function is the set of real numbers, then the range is the set of real numbers. [16.2a]

Guided Solutions

GS Use the graph to complete the table of ordered pairs that name points on the graph. [16.1c]

6.

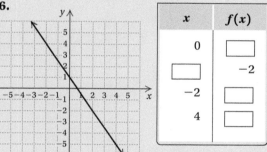

x	$f(x)$
0	☐
☐	-2
-2	☐
4	☐

7.

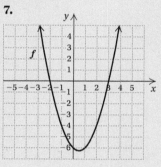

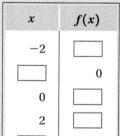

x	$f(x)$
-2	☐
☐	0
0	☐
2	☐
☐	-4
1	☐

Mixed Review

Determine whether the correspondence is a function. [16.1a]

8. *Domain   Range*

9. *Domain   Range*

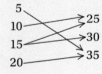

10. Find the domain and the range. [16.2a]

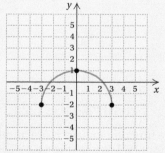

Find the function value. [16.1b]

11. $g(x) = 2 + x$; $g(-5)$

12. $f(x) = x - 7$; $f(0)$

13. $h(x) = 8$; $h\left(\dfrac{1}{2}\right)$

14. $f(x) = 3x^2 - x + 5$; $f(-1)$

15. $g(p) = p^4 - p^3$; $g(10)$

16. $f(t) = \dfrac{1}{2}t + 3$; $f(-6)$

Determine whether each of the following is the graph of a function. [16.1d]

17.
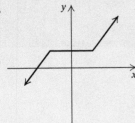

18.

19.
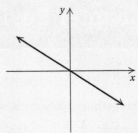

Find the domain. [16.2a]

20. $g(x) = \dfrac{3}{12 - 3x}$

21. $f(x) = x^2 - 10x + 3$

22. $h(x) = \dfrac{x - 2}{x + 2}$

23. $f(x) = |x - 4|$

Graph. [16.1c]

24. $g(x) = -\dfrac{2}{3}x - 2$
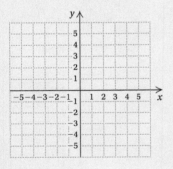

25. $f(x) = x - 1$
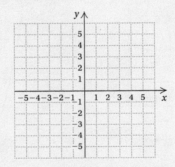

26. $h(x) = 2x + \dfrac{1}{2}$
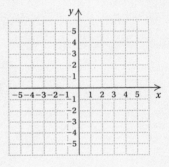

27. $g(x) = |x| - 3$
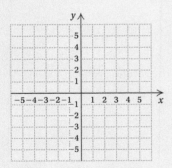

28. $f(x) = 1 + x^2$
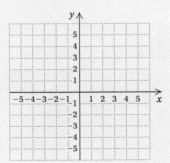

29. $f(x) = -\dfrac{1}{4}x$
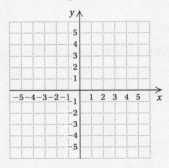

Understanding Through Discussion and Writing

30. Is it possible for a function to have more numbers as outputs than as inputs? Why or why not? [16.1a]

31. Without making a drawing, how can you tell that the graph of $f(x) = x - 30$ passes through three quadrants? [16.1c]

32. For a given function f, it is known that $f(2) = -3$. Give as many interpretations of this fact as you can. [16.1b], [16.2a]

33. Explain the difference between the domain and the range of a function. [16.2a]

Linear Functions: Graphs and Slope

16.3

We now turn our attention to functions whose graphs are straight lines. Such functions are called **linear** and can be written in the form $f(x) = mx + b$.

LINEAR FUNCTION

A **linear function** f is any function that can be described by $f(x) = mx + b$.

Compare the two equations $7y + 2x = 11$ and $y = 3x + 5$. Both are linear equations because their graphs are straight lines. Each can be expressed in an equivalent form that is a linear function.

The equation $y = 3x + 5$ can be expressed as $f(x) = mx + b$, where $m = 3$ and $b = 5$.

The equation $7y + 2x = 11$ also has an equivalent form $f(x) = mx + b$. To see this, we solve for y:

$$7y + 2x = 11$$
$$7y + 2x - 2x = -2x + 11 \qquad \text{Subtracting } 2x$$
$$7y = -2x + 11$$
$$\frac{7y}{7} = \frac{-2x + 11}{7} \qquad \text{Dividing by } 7$$
$$y = -\frac{2}{7}x + \frac{11}{7}. \qquad \text{Simplifying}$$

We now have an equivalent function in the form $f(x) = mx + b$:

$$f(x) = -\frac{2}{7}x + \frac{11}{7}, \qquad \text{where} \quad m = -\frac{2}{7} \quad \text{and} \quad b = \frac{11}{7}.$$

In this section, we consider the effects of the constants m and b on the graphs of linear functions.

OBJECTIVES

a Find the y-intercept of a line from the equation $y = mx + b$ or $f(x) = mx + b$.

b Given two points on a line, find the slope. Given a linear equation, derive the equivalent slope–intercept equation and determine the slope and the y-intercept.

c Solve applied problems involving slope.

SKILL TO REVIEW

Objective 10.4a: Subtract real numbers.

Subtract.
1. $11 - (-8)$
2. $-6 - (-6)$

Answers

Skill to Review:
1. 19   2. 0

1. Graph $y = 3x$ and $y = 3x - 6$ using the same set of axes. Then compare the graphs.

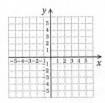

2. Graph $y = -2x$ and $y = -2x + 3$ using the same set of axes. Then compare the graphs.

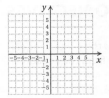

CALCULATOR CORNER

Exploring b We can use a graphing calculator to explore the effect of the constant b on the graph of a function of the form $f(x) = mx + b$. Graph $y_1 = x$ in the standard $[-10, 10, -10, 10]$ viewing window. Then graph $y_2 = x + 4$, followed by $y_3 = x - 3$, in the same viewing window.

EXERCISES:

1. Compare the graph of y_2 with the graph of y_1.

2. Compare the graph of y_3 with the graph of y_1.

Answers

1.

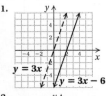

The graph of $y = 3x - 6$ is the graph of $y = 3x$ shifted down 6 units.

2.

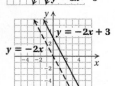

The graph of $y = -2x + 3$ is the graph of $y = -2x$ shifted up 3 units.

a   THE CONSTANT *b*: THE *y*-INTERCEPT

Let's first explore the effect of the constant b.

EXAMPLE 1   Graph $y = 2x$ and $y = 2x + 3$ using the same set of axes. Then compare the graphs.

We first make a table of solutions of both equations. Next, we plot these points. Drawing a red line for $y = 2x$ and a blue line for $y = 2x + 3$, we note that the graph of $y = 2x + 3$ is simply the graph of $y = 2x$ shifted, or *translated*, up 3 units. The lines are parallel.

x	y $y = 2x$	y $y = 2x + 3$
0	0	3
1	2	5
-1	-2	1
2	4	7
-2	-4	-1

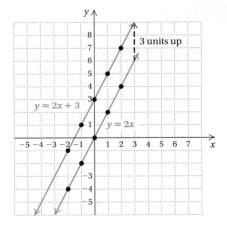

◀ Do Exercises 1 and 2.

EXAMPLE 2   Graph $f(x) = \frac{1}{3}x$ and $g(x) = \frac{1}{3}x - 2$ using the same set of axes. Then compare the graphs.

We first make a table of solutions of both equations. By choosing multiples of 3, we can avoid fractions.

x	$f(x)$ $f(x) = \frac{1}{3}x$	$g(x)$ $g(x) = \frac{1}{3}x - 2$
0	0	-2
3	1	-1
-3	-1	-3
6	2	0

We then plot these points. Drawing a red line for $f(x) = \frac{1}{3}x$ and a blue line for $g(x) = \frac{1}{3}x - 2$, we see that the graph of $g(x) = \frac{1}{3}x - 2$ is simply the graph of $f(x) = \frac{1}{3}x$ shifted, or translated, down 2 units. The lines are parallel.

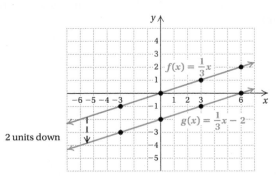

In Example 1, we saw that the graph of $y = 2x + 3$ is parallel to the graph of $y = 2x$ and that it passes through the point $(0, 3)$. Similarly, in Example 2, we saw that the graph of $y = \frac{1}{3}x - 2$ is parallel to the graph of $y = \frac{1}{3}x$ and that it passes through the point $(0, -2)$. In general, the graph of $y = mx + b$ is a line parallel to $y = mx$, passing through the point $(0, b)$. The point $(0, b)$ is called the **y-intercept** because it is the point at which the graph crosses the y-axis. Often it is convenient to refer to the number b as the y-intercept. The constant b has the effect of moving the graph of $y = mx$ up or down $|b|$ units to obtain the graph of $y = mx + b$.

Do Exercise 3. ▶

3. Graph $f(x) = \frac{1}{3}x$ and $g(x) = \frac{1}{3}x + 2$ using the same set of axes. Then compare the graphs.

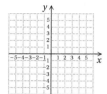

y-INTERCEPT

The y-intercept of the graph of $f(x) = mx + b$ is the point $(0, b)$ or, simply b.

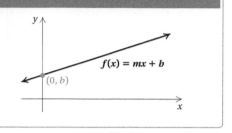

EXAMPLE 3 Find the y-intercept: $y = -5x + 4$.

$$y = -5x + 4 \qquad (0, 4) \text{ is the y-intercept.}$$

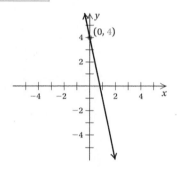

EXAMPLE 4 Find the y-intercept: $f(x) = 6.3x - 7.8$.

$$f(x) = 6.3x - 7.8 \qquad (0, -7.8) \text{ is the y-intercept.}$$

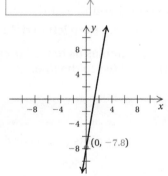

Find the y-intercept.

4. $y = 7x + 8$

5. $f(x) = -6x - \frac{2}{3}$

Do Exercises 4 and 5. ▶

Answers

3.
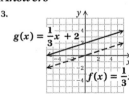
The graph of $g(x)$ is the graph of $f(x)$ shifted up 2 units.

$g(x) = \frac{1}{3}x + 2$

$f(x) = \frac{1}{3}x$

4. $(0, 8)$ 5. $\left(0, -\frac{2}{3}\right)$

b THE CONSTANT m: SLOPE

Look again at the graphs in Examples 1 and 2. Note that the slant of each red line seems to match the slant of each blue line. This leads us to believe that the number m in the equation $y = mx + b$ is related to the slant of the line. Let's consider some examples.

Graphs with $m < 0$:

$m = -1$

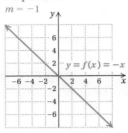

$m = -5$

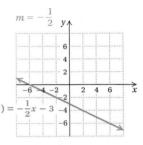

$m = -\dfrac{1}{2}$

Graphs with $m = 0$:

$m = 0$

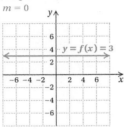

$m = 0$

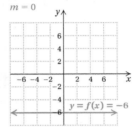

$m = 0$

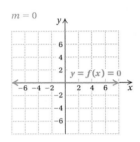

Graphs with $m > 0$:

$m = 1$

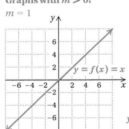

$m = 6$

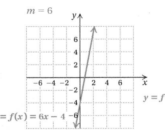

$m = \dfrac{1}{3}$

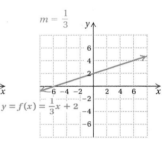

Note that

$m < 0 \rightarrow$ The graph slants down from left to right;

$m = 0 \rightarrow$ the graph is horizontal; and

$m > 0 \rightarrow$ the graph slants up from left to right.

The following definition enables us to visualize the slant and attach a number, a geometric ratio, or *slope*, to the line.

SLOPE

The **slope** of a line containing points (x_1, y_1) and (x_2, y_2) is given by

$$m = \frac{\text{rise}}{\text{run}}$$

$$= \frac{\text{change in } y}{\text{change in } x} = \frac{y_2 - y_1}{x_2 - x_1} = \frac{y_1 - y_2}{x_1 - x_2}.$$

Consider a line with two points marked P_1 and P_2, as follows. As we move from P_1 to P_2, the y-coordinate changes from 1 to 3 and the x-coordinate changes from 2 to 7. The change in y is $3 - 1$, or 2. The change in x is $7 - 2$, or 5.

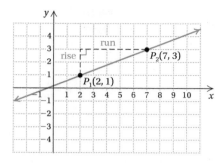

We call the change in y the **rise** and the change in x the **run**. The ratio rise/run is the same for any two points on a line. We call this ratio the **slope**. Slope describes the slant of a line. The slope of the line in the graph above is given by

$$\frac{\text{rise}}{\text{run}}, \quad \text{or} \quad \frac{\text{change in } y}{\text{change in } x}, \quad \text{or} \quad \frac{2}{5}.$$

Whenever x increases by 5 units, y increases by 2 units. Equivalently, whenever x increases by 1 unit, y increases by $\frac{2}{5}$ unit.

EXAMPLE 5 Graph the line containing the points $(-4, 3)$ and $(2, -5)$ and find the slope.

The graph is shown below. Going from $(-4, 3)$ to $(2, -5)$, we see that the change in y, or the rise, is $-5 - 3$, or -8. The change in x, or the run, is $2 - (-4)$, or 6.

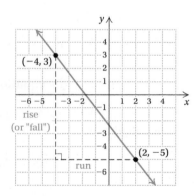

$$\text{Slope} = \frac{\text{rise}}{\text{run}} = \frac{\text{change in } y}{\text{change in } x}$$

$$= \frac{-5 - 3}{2 - (-4)}$$

$$= \frac{-8}{6} = -\frac{8}{6}, \text{ or } -\frac{4}{3}$$

The formula

$$m = \frac{y_2 - y_1}{x_2 - x_1} = \frac{y_1 - y_2}{x_1 - x_2}$$

tells us that we can subtract in two ways. We must remember, however, to subtract the x-coordinates in the same order that we subtract the y-coordinates.

Let's do Example 5 again:

$$\text{Slope} = \frac{\text{change in } y}{\text{change in } x} = \frac{3 - (-5)}{-4 - 2} = \frac{8}{-6} = -\frac{8}{6} = -\frac{4}{3}.$$

We see that both ways give the same value for the slope.

CALCULATOR CORNER

Visualizing Slope

EXERCISES: Use the window settings $[-6, 6, -4, 4]$, with Xscl $= 1$ and Yscl $= 1$.

1. Graph $y = x$, $y = 2x$, and $y = 5x$ in the same window. What do you think the graph of $y = 10x$ will look like?

2. Graph $y = x$, $y = 0.5x$, and $y = 0.1x$ in the same window. What do you think the graph of $y = 0.005x$ will look like?

3. Graph $y = -x$, $y = -2x$, and $y = -5x$ in the same window. What do you think the graph of $y = -10x$ will look like?

4. Graph $y = -x$, $y = -0.5x$, and $y = -0.1x$ in the same window. What do you think the graph of $y = -0.005x$ will look like?

Graph the line through the given points and find its slope.

6. $(-1, -1)$ and $(2, -4)$

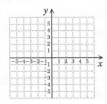

$$m = \frac{-4 - ()}{ - (-1)}$$

$$= \frac{}{3}$$

$$= \boxed{}$$

7. $(0, 2)$ and $(3, 1)$

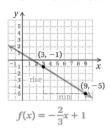

8. Find the slope of the line $f(x) = -\frac{2}{3}x + 1$. Use the points $(9, -5)$ and $(3, -1)$.

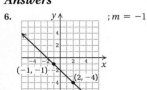

$$f(x) = -\frac{2}{3}x + 1$$

Answers

6. 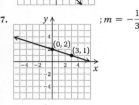 ; $m = -1$

7. ; $m = -\frac{1}{3}$

8. $m = -\frac{2}{3}$

Guided Solution:
6. $-1, 2, -3, -1$

The slope of a line tells how it slants. A line with positive slope slants up from left to right. The larger the positive number, the steeper the slant. A line with negative slope slants downward from left to right. The smaller the negative number, the steeper the line.

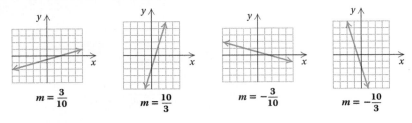

$m = \frac{3}{10}$ $m = \frac{10}{3}$ $m = -\frac{3}{10}$ $m = -\frac{10}{3}$

◀ **Do Exercises 6 and 7.**

How can we find the slope from a given equation? Let's consider the equation $y = 2x + 3$, which is in the form $y = mx + b$. We can find two points by choosing convenient values for x—say, 0 and 1—and substituting to find the corresponding y-values.

If $x = 0$, $y = 2 \cdot 0 + 3 = 3$.

If $x = 1$, $y = 2 \cdot 1 + 3 = 5$.

We find two points on the line to be

$(0, 3)$ and $(1, 5)$.

The slope of the line is found as follows, using the definition of slope:

$$m = \frac{\text{change in } y}{\text{change in } x}$$

$$= \frac{5 - 3}{1 - 0} = \frac{2}{1} = 2.$$

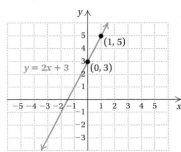

The slope is 2. Note that this is the coefficient of the x-term in the equation $y = 2x + 3$.

If we had chosen different points on the line—say, $(-2, -1)$ and $(4, 11)$—the slope would still be 2, as we see in the following calculation:

$$m = \frac{11 - (-1)}{4 - (-2)} = \frac{11 + 1}{4 + 2} = \frac{12}{6} = 2.$$

◀ **Do Exercise 8.**

We see that the slope of the line $y = mx + b$ is indeed the constant m, the coefficient of x.

SLOPE

The **slope** of the line $y = mx + b$ is m.

From a linear equation in the form $y = mx + b$, we can read the slope and the y-intercept of the graph directly.

SLOPE–INTERCEPT EQUATION

The equation $y = mx + b$ is called the **slope–intercept equation**. The slope is m and the y-intercept is $(0, b)$.

Note that any graph of an equation $y = mx + b$ passes the vertical-line test and thus represents a function.

EXAMPLE 6 Find the slope and the y-intercept of $y = 5x - 4$.

Since the equation is already in the form $y = mx + b$, we simply read the slope and the y-intercept from the equation:

$$y = 5x - 4.$$

The slope is 5.   The y-intercept is $(0, -4)$.

EXAMPLE 7 Find the slope and the y-intercept of $2x + 3y = 8$.

We first solve for y so we can easily read the slope and the y-intercept:

$2x + 3y = 8$

$\quad 3y = -2x + 8$ Subtracting $2x$

$\quad \dfrac{3y}{3} = \dfrac{-2x + 8}{3}$ Dividing by 3

$\quad\quad y = -\dfrac{2}{3}x + \dfrac{8}{3}.$ Finding the form $y = mx + b$

The slope is $-\frac{2}{3}$.   The y-intercept is $(0, \frac{8}{3})$.

Do Exercises 9 and 10. ▶

Find the slope and the y-intercept.

9. $f(x) = -8x + 23$

GS 10. $5x - 10y = 25$

First solve for y:

$5x - 10y = 25$

$\quad -10y = \boxed{} + 25$

$\quad\quad y = \dfrac{-5x + 25}{\boxed{}}$

$\quad\quad y = \boxed{}\, x - \dfrac{5}{\boxed{}}.$

Slope is $\boxed{}$; y-intercept is $(0, \boxed{})$.

C APPLICATIONS

Slope has many real-world applications. For example, numbers like 2%, 3%, and 6% are often used to represent the *grade* of a road, a measure of how steep a road on a hill or a mountain is. A 3% grade $\left(3\% = \frac{3}{100}\right)$ means that for every horizontal distance of 100 ft that the road runs, the road rises 3 ft, and a -3% grade means that for every horizontal distance of 100 ft, the road drops 3 ft. (Normally, the road-grade signs do not include negative signs, since it is obvious whether you are climbing or descending.)

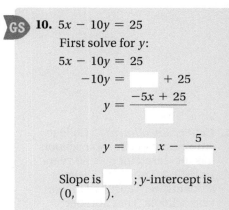

Road grade = $\frac{a}{b}$
(expressed as a percent)

An athlete might change the grade of a treadmill during a workout. An escape ramp on an airliner might have a slope of about −0.6.

Architects and carpenters use slope when designing and building stairs, ramps, or roof pitches. Another application occurs in hydrology. The strength or force of a river depends on how far the river falls vertically compared to how far it flows horizontally. Slope can also be considered as a **rate of change**.

EXAMPLE 8 *Student Debt.* The average educational debt per college student at his or her graduation has steadily increased. In 1993, the average debt was $14,500 (in 2011 dollars). By 2011, this amount had increased to $26,600. Find the rate of change in the average student debt with respect to time, in years.

Sources: Higher Education Research Institute, UCLA; Sallie Mae; NCES; FinAid; the College Board; McKinsey Global Institute; *Time*, October 29, 2012

11. *College Enrollment.* College enrollment in two-year schools has increased for over 40 years. In 1970, 2.3 million college students were enrolled in two-year schools. That number had increased to 7.8 million by 2011. Find the rate of change in enrollment in two-year schools with respect to time, in years.

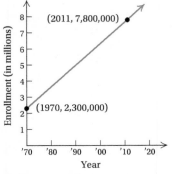

SOURCE: Higher Education Research Institute, UCLA; *TIME*, October 29, 2012

Student Educational Debt

The rate of change with respect to time, in years, is given by

$$\text{Rate of change} = \frac{\$26{,}600 - \$14{,}500}{2011 - 1993}$$

$$= \frac{\$12{,}100}{18 \text{ years}}$$

$$\approx \$672 \text{ per year.}$$

The average student debt at graduation is increasing at a rate of about $672 per year.

◀ Do Exercise 11.

Answer

11. The rate of change in enrollment in two-year colleges is about 134,146 per year.

EXAMPLE 9 *Volume of Mail.* The volume of first-class mail through the U.S. Postal Service has been decreasing since 2005. Find the rate of change of the volume of first-class mail with respect to time, in years.

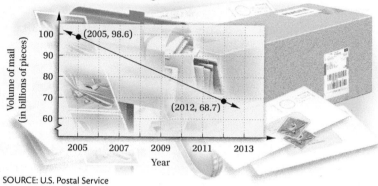

Volume of First-Class Mail Through U.S. Post Office

SOURCE: U.S. Postal Service

Since the graph is linear, we can use any pair of points to determine the rate of change:

$$\text{Rate of change} = \frac{68.7\,\text{billion} - 98.6\,\text{billion}}{2012 - 2005} = \frac{-29.9\,\text{billion}}{7\,\text{years}} \approx -4.27\,\text{billion per year.}$$

The volume of first-class mail through the U.S. Postal Service is decreasing at a rate of about 4.27 billion pieces per year.

12. *Newspaper Circulation.* Daily newspaper circulation has decreased in recent years. The following graph shows the circulation of daily newspapers, in millions, for three years. Find the rate of change in the circulation of daily newspapers per year.

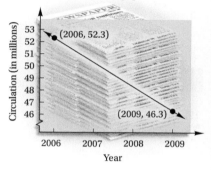

Circulation of Daily Newspapers

Do Exercise 12. ▶

Answer

12. The rate of change is −2 million papers per year.

☑ Reading Check

Choose from the column on the right the slope of each line.

RC1.
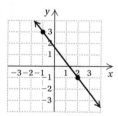

RC2.
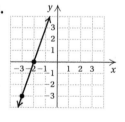

RC3.
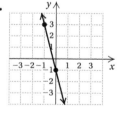

a) $-\dfrac{3}{4}$

b) 3

c) 0

d) -4

e) $\dfrac{3}{4}$

f) $-\dfrac{4}{3}$

RC4.
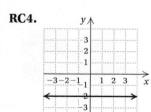

RC5.
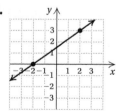

RC6.
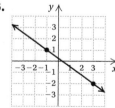

Find the slope and the *y*-intercept of each equation.

1. $y = 4x + 5$

2. $y = -5x + 10$

3. $f(x) = -2x - 6$

4. $g(x) = -5x + 7$

5. $y = -\frac{3}{8}x - \frac{1}{5}$

6. $y = \frac{15}{7}x + \frac{16}{5}$

7. $g(x) = 0.5x - 9$

8. $f(x) = -3.1x + 5$

9. $2x - 3y = 8$

10. $-8x - 7y = 24$

11. $9x = 3y + 6$

12. $9y + 36 - 4x = 0$

13. $3 - \frac{1}{4}y = 2x$

14. $5x = \frac{2}{3}y - 10$

15. $17y + 4x + 3 = 7 + 4x$

16. $3y - 2x = 5 + 9y - 2x$

Find the slope of each line.

17.
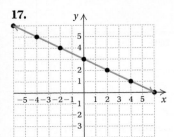

18.
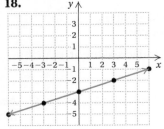

19.
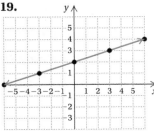

20.
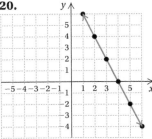

Find the slope of the line containing the given pair of points.

21. $(6, 9)$ and $(4, 5)$

22. $(8, 7)$ and $(2, -1)$

23. $(9, -4)$ and $(3, -8)$

24. $(17, -12)$ and $(-9, -15)$

25. $(-16.3, 12.4)$ and $(-5.2, 8.7)$

26. $(14.4, -7.8)$ and $(-12.5, -17.6)$

Find the slope (or rate of change).

27. Find the slope (or grade) of the treadmill.

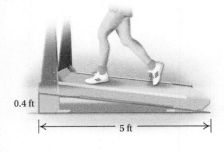

28. Find the slope (or head) of the river.

29. Find the slope (or pitch) of the roof.

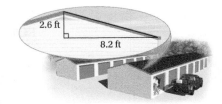

30. Public buildings regularly include steps with 7-in. risers and 11-in. treads. Find the grade of such a stairway.

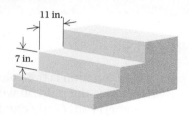

11 in.

7 in.

31. *Luxury Purchases.* Find the rate of change in luxury purchases in China with respect to time, in years.

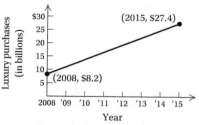

*Estimated values for 2010–2015
SOURCE: McKinsey Insights, China-Wealthy Consumer Studies (2008, 2010)

32. *People with Alzheimer's.* Find the rate of change in the number of people with Alzheimer's disease with respect to time, in years.

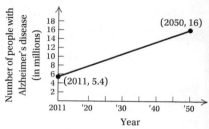

*Estimated values for 2012–2050
SOURCE: Alzheimer's Association

Find the rate of change.

33.

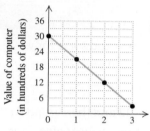

34.

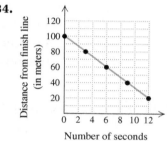

35.

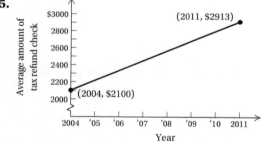

SOURCES: FDIC Consumer News Winter 2004/2005; *USA TODAY*, April 13, 2012; www.greenbaypressgazette.com, January 4, 2013

36.

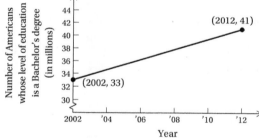

SOURCE: U.S. Census Bureau

Skill Maintenance

Simplify. [10.8c, d], [13.1b]

37. $3^2 - 24 \cdot 56 + 144 \div 12$

38. $9\{2x - 3[5x + 2(-3x + y^0 - 2)]\}$

39. $10\{2x + 3[5x - 2(-3x + y^1 - 2)]\}$

40. $5^4 \div 625 \div 5^2 \cdot 5^7 \div 5^3$

Solve. [11.6a]

41. One side of a square is 5 yd less than a side of an equilateral triangle. If the perimeter of the square is the same as the perimeter of the triangle, what is the length of a side of the square? of the triangle?

Factor. [14.6a]

42. $8 - 125x^3$

43. $c^6 - d^6$

44. $56x^3 - 7$

45. Divide: $(a^2 - 11a + 6) \div (a - 1)$. [13.8b]

16.4 More on Graphing Linear Equations

OBJECTIVES

a Graph linear equations using intercepts.

b Given a linear equation in slope–intercept form, use the slope and the y-intercept to graph the line.

c Graph linear equations of the form $x = a$ or $y = b$.

d Given the equations of two lines, determine whether their graphs are parallel or whether they are perpendicular.

SKILL TO REVIEW

Objective 12.1a: Plot points associated with ordered pairs of numbers.

1. Plot the following points: $A(0, 4)$, $B(0, -1)$, $C(0, 0)$, $D(3, 0)$, and $E(-\frac{7}{2}, 0)$.

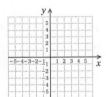

Answer

Skill to Review:

1.

a GRAPHING USING INTERCEPTS

The **x-intercept** of the graph of a linear equation or function is the point at which the graph crosses the x-axis. The **y-intercept** is the point at which the graph crosses the y-axis. We know from geometry that only one line can be drawn through two given points. Thus, if we know the intercepts, we can graph the line. To ensure that a computation error has not been made, it is a good idea to calculate a third point as a check.

Many equations of the type $Ax + By = C$ can be graphed conveniently using intercepts.

---

x- AND y-INTERCEPTS

A **y-intercept** is a point $(0, b)$. To find b, let $x = 0$ and solve for y.

An **x-intercept** is a point $(a, 0)$. To find a, let $y = 0$ and solve for x.

EXAMPLE 1 Find the intercepts of $3x + 2y = 12$ and then graph the line.

y-intercept: To find the y-intercept, we let $x = 0$ and solve for y:

$$3x + 2y = 12$$
$$3 \cdot 0 + 2y = 12 \qquad \text{Substituting 0 for } x$$
$$2y = 12$$
$$y = 6.$$

The y-intercept is $(0, 6)$.

x-intercept: To find the x-intercept, we let $y = 0$ and solve for x:

$$3x + 2y = 12$$
$$3x + 2 \cdot 0 = 12 \qquad \text{Substituting 0 for } y$$
$$3x = 12$$
$$x = 4.$$

The x-intercept is $(4, 0)$.

We plot these points and draw the line, using a third point as a check. We choose $x = 6$ and solve for y:

$$3(6) + 2y = 12$$
$$18 + 2y = 12$$
$$2y = -6$$
$$y = -3.$$

We plot $(6, -3)$ and note that it is on the line so the graph is probably correct.

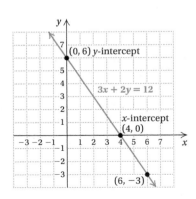

When both the x-intercept and the y-intercept are $(0, 0)$, as is the case with an equation such as $y = 2x$, whose graph passes through the origin, another point would have to be calculated and a third point used as a check.

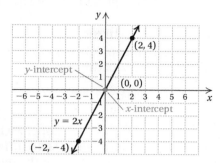

Do Exercise 1. ▶

 1. Find the intercepts of $4y - 12 = -6x$ and then graph the line.

To find the y-intercept, set $x = 0$ and solve for y:

$$4y - 12 = -6 \cdot \boxed{}$$
$$4y - 12 = \boxed{}$$
$$4y = \boxed{}$$
$$y = \boxed{}.$$

The y-intercept is $(0, \boxed{})$.

To find the x-intercept, set $y = 0$ and solve for x:

$$4 \cdot \boxed{} - 12 = -6x$$
$$\boxed{} = -6x$$
$$\boxed{} = x.$$

The x-intercept is $(\boxed{}, 0)$.

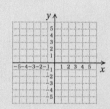

CALCULATOR CORNER

Viewing the Intercepts Knowing the intercepts of a linear equation helps us determine a good viewing window for the graph of the equation. For example, when we graph the equation $y = -x + 15$ in the standard window, we see only a small portion of the graph in the upper right-hand corner of the screen, as shown on the left below.

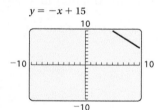

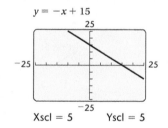

Using algebra, as we did in Example 1, we can find that the intercepts of the graph of this equation are $(0, 15)$ and $(15, 0)$. This tells us that, if we are to see a portion of the graph that includes the intercepts, both Xmax and Ymax should be greater than 15. We can try different window settings until we find one that suits us. One good choice, shown on the right above, is $[-25, 25, -25, 25]$, with Xscl = 5 and Yscl = 5.

EXERCISES: Find the intercepts of the equation algebraically. Then graph the equation on a graphing calculator, choosing window settings that allow the intercepts to be seen clearly. (Settings may vary.)

1. $y = -3.2x - 16$ **2.** $y - 4.25x = 85$

3. $6x + 5y = 90$ **4.** $5x - 6y = 30$

5. $8x + 3y = 9$ **6.** $y = 0.4x - 5$

7. $y = 1.2x - 12$ **8.** $4x - 5y = 2$

Answer

1.

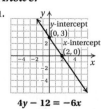

$$4y - 12 = -6x$$

Guided Solution:
1. 0, 0, 12, 3, 3; 0, −12, 2, 2

Graph using the slope and the y-intercept.

2. $y = \dfrac{3}{2}x + 1$

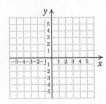

3. $f(x) = \dfrac{3}{4}x - 2$

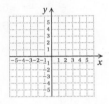

4. $g(x) = -\dfrac{3}{5}x + 5$

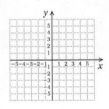

5. $y = -\dfrac{5}{3}x - 4$

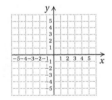

b # GRAPHING USING THE SLOPE AND THE y-INTERCEPT

We can also graph a line using its slope and y-intercept.

EXAMPLE 2 Graph: $y = -\dfrac{2}{3}x + 1$.

This equation is in slope–intercept form, $y = mx + b$. The y-intercept is $(0, 1)$. We plot $(0, 1)$. We can think of the slope $\left(m = -\dfrac{2}{3}\right)$ as $\dfrac{-2}{3}$.

$$m = \frac{\text{Rise}}{\text{Run}} = \frac{-2}{3} \qquad \begin{array}{l}\text{Move 2 units down.}\\ \text{Move 3 units right.}\end{array}$$

Starting at the y-intercept and using the slope, we find another point by moving 2 units down (since the numerator is *negative* and corresponds to the change in y) and 3 units to the right (since the denominator is *positive* and corresponds to the change in x). We get to a new point, $(3, -1)$. In a similar manner, we can move from the point $(3, -1)$ to find another point, $(6, -3)$.

We could also think of the slope $\left(m = -\dfrac{2}{3}\right)$ as $\dfrac{2}{-3}$.

$$m = \frac{\text{Rise}}{\text{Run}} = \frac{2}{-3} \qquad \begin{array}{l}\text{Move 2 units up.}\\ \text{Move 3 units left.}\end{array}$$

Then we can start again at $(0, 1)$, but this time we move 2 units up (since the numerator is *positive* and corresponds to the change in y) and 3 units to the left (since the denominator is *negative* and corresponds to the change in x). We get another point on the graph, $(-3, 3)$, and from it we can obtain $(-6, 5)$ and others in a similar manner. We plot the points and draw the line.

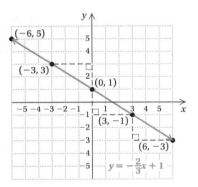

EXAMPLE 3 Graph: $f(x) = \dfrac{2}{5}x + 4$.

First, we plot the y-intercept, $(0, 4)$. We then consider the slope $\dfrac{2}{5}$. A slope of $\dfrac{2}{5}$ tells us that, for every 2 units that the graph rises, it runs 5 units horizontally in the positive direction, or to the right. Thus, starting at the y-intercept and using the slope, we find another point by moving 2 units up (since the numerator is *positive* and corresponds to the change in y) and 5 units to the right (since the denominator is *positive* and corresponds to the change in x). We get to a new point, $(5, 6)$.

Answers

2.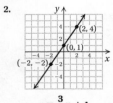
$y = \dfrac{3}{2}x + 1$

3.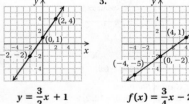
$f(x) = \dfrac{3}{4}x - 2$

4.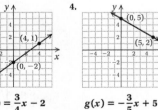
$g(x) = -\dfrac{3}{5}x + 5$

5.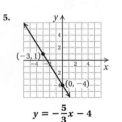
$y = -\dfrac{5}{3}x - 4$

We can also think of the slope $\frac{2}{5}$ as $\frac{-2}{-5}$. A slope of $\frac{-2}{-5}$ tells us that, for every 2 units that the graph drops, it runs 5 units horizontally in the negative direction, or to the left. We again start at the y-intercept, $(0, 4)$. We move 2 units down (since the numerator is *negative* and corresponds to the change in y) and 5 units to the left (since the denominator is *negative* and corresponds to the change in x). We get to another new point, $(-5, 2)$. We plot the points and draw the line.

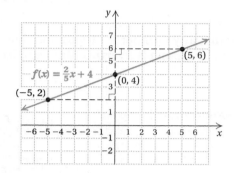

Do Exercises 2–5 on the preceding page. ▶

c HORIZONTAL LINES AND VERTICAL LINES

Some equations have graphs that are parallel to one of the axes. This happens when either A or B is 0 in $Ax + By = C$. These equations have a missing variable; that is, there is only one variable in the equation. In the following example, x is missing.

EXAMPLE 4 Graph: $y = 3$.

Since x is missing, any number for x will do. Thus all ordered pairs $(x, 3)$ are solutions. The graph is a **horizontal line** parallel to the x-axis.

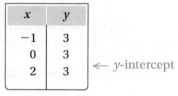

x	y
−1	3
0	3
2	3

← *y*-intercept

↑   ↑—— Regardless of x, y must be 3.
Choose *any* number for x.

What about the slope of a horizontal line? In Example 4, consider the points $(-1, 3)$ and $(2, 3)$, which are on the line $y = 3$. The change in y is $3 - 3$, or 0. The change in x is $-1 - 2$, or -3. Thus,

$$m = \frac{3 - 3}{-1 - 2} = \frac{0}{-3} = 0.$$

Any two points on a horizontal line have the same y-coordinate. Thus the change in y is always 0, so the slope is 0.

Do Exercises 6 and 7. ▶

Graph and determine the slope.

6. $f(x) = -4$

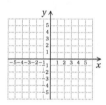

7. $y = 3.6$

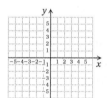

Answers

6. $m = 0$

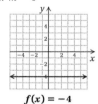

$f(x) = -4$

7. $m = 0$

$y = 3.6$

We can also determine the slope by noting that $y = 3$ can be written in slope–intercept form as $y = 0x + 3$, or $f(x) = 0x + 3$. From this equation, we read that the slope is 0. A function of this type is called a **constant function**. We can express it in the form $y = b$, or $f(x) = b$. Its graph is a horizontal line that crosses the y-axis at $(0, b)$.

In the following example, y is missing and the graph is parallel to the y-axis.

EXAMPLE 5   Graph: $x = -2$.

Since y is missing, any number for y will do. Thus all ordered pairs $(-2, y)$ are solutions. The graph is a **vertical line** parallel to the y-axis.

Graph.

8. $x = -5$

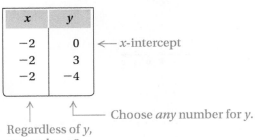

9. $8x - 5 = 19$ (*Hint:* Solve for x.)

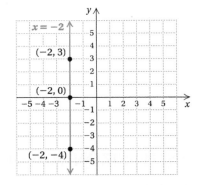

This graph is not the graph of a function because it fails the vertical-line test. The vertical line itself crosses the graph more than once.

◀ Do Exercises 8 and 9.

10. Determine, if possible, the slope of each line.

a) $x = -12$ b) $y = 6$
c) $2y + 7 = 11$ d) $x = 0$
e) $y = -\frac{3}{4}$
f) $10 - 5x = 15$

What about the slope of a vertical line? In Example 5, consider the points $(-2, 3)$ and $(-2, -4)$, which are on the line $x = -2$. The change in y is $3 - (-4)$, or 7. The change in x is $-2 - (-2)$, or 0. Thus,

$$m = \frac{3 - (-4)}{-2 - (-2)} = \frac{7}{0}. \quad \text{Not defined}$$

Since division by 0 is not defined, the slope of this line is not defined. Any two points on a vertical line have the same x-coordinate. Thus the change in x is always 0, so the slope of any vertical line is not defined.

The following summarizes the characteristics of horizontal lines and vertical lines and their equations.

HORIZONTAL LINE; VERTICAL LINE

The graph of $y = b$, or $f(x) = b$, is a **horizontal line** with y-intercept $(0, b)$. It is the graph of a constant function with slope 0.

The graph of $x = a$ is a **vertical line** with x-intercept $(a, 0)$. The slope is not defined. It is not the graph of a function.

◀ Do Exercise 10.

Answers

8.

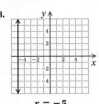

$x = -5$

9.

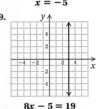

$8x - 5 = 19$

10. (a) Not defined; (b) 0; (c) 0;
(d) not defined; (e) 0; (f) not defined

We have graphed linear equations in several ways in this chapter. Although, in general, you can use any method that works best for you, we list some guidelines in the margin at right.

To graph a linear equation:

1. Is the equation of the type $x = a$ or $y = b$? If so, the graph will be a line parallel to an axis; $x = a$ is vertical and $y = b$ is horizontal.

2. If the line is of the type $y = mx$, both intercepts are the origin, $(0, 0)$. Plot $(0, 0)$ and one other point.

3. If the line is of the type $y = mx + b$, plot the y-intercept and one other point.

4. If the equation is of the form $Ax + By = C$, graph using intercepts. If the intercepts are too close together, choose another point farther from the origin.

5. In all cases, use a third point as a check.

d PARALLEL LINES AND PERPENDICULAR LINES

Parallel Lines

Parallel lines extend indefinitely without intersecting. If two lines are vertical, they are parallel. How can we tell whether nonvertical lines are parallel? We examine their slopes and y-intercepts.

> **PARALLEL LINES**
>
> Two nonvertical lines are **parallel** if they have the *same* slope and *different* y-intercepts.

EXAMPLE 6 Determine if the graphs of $y - 3x = 1$ and $3x + 2y = -2$ are parallel.

To determine if lines are parallel, we first find their slopes. To do this, we find the slope–intercept form of each equation by solving for y:

$$y - 3x = 1 \qquad\qquad 3x + 2y = -2$$
$$y = 3x + 1; \qquad\qquad 2y = -3x - 2$$
$$y = \tfrac{1}{2}(-3x - 2)$$
$$y = -\tfrac{3}{2}x - 1.$$

The slopes, 3 and $-\tfrac{3}{2}$, are different. Thus the lines are not parallel, as the graphs at right confirm.

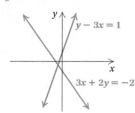

EXAMPLE 7 Determine if the graphs of $3x - y = -5$ and $y - 3x = -2$ are parallel.

We first find the slope–intercept form of each equation by solving for y:

$$3x - y = -5 \qquad\qquad y - 3x = -2$$
$$-y = -3x - 5 \qquad\qquad y = 3x - 2.$$
$$-1(-y) = -1(-3x - 5)$$
$$y = 3x + 5;$$

The slopes, 3, are the same. The y-intercepts, $(0, 5)$ and $(0, -2)$, are different. Thus the lines are parallel, as the graphs appear to confirm.

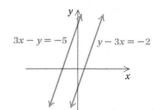

Do Exercises 11–13.

Determine whether the graphs of the given pair of lines are parallel.

 11. $x + 4 = y$,
$y - x = -3$
Write each equation in the form $y = mx + b$:
$x + 4 = y \rightarrow y = x + 4$;
$y - x = -3 \rightarrow y = \boxed{} - 3$.
The slope of each line is $\boxed{}$, and the y-intercepts, $(0, 4)$ and $(0, \boxed{})$, are different. Thus the lines $\underline{}$ parallel.
 are/are not

12. $y + 4 = 3x$,
$4x - y = -7$

13. $y = 4x + 5$,
$2y = 8x + 10$

Answers

11. Yes **12.** No **13.** No; they are the same line.

Guided Solution:
11. x, 1, -3, are

Perpendicular Lines

If one line is vertical and another is horizontal, they are perpendicular. For example, the lines $x = 5$ and $y = -3$ are perpendicular. Otherwise, how can we tell whether two lines are perpendicular?

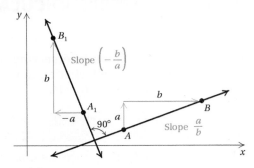

Consider a line $\overleftrightarrow{AB}$, as shown in the figure above, with slope a/b. Then think of rotating the line 90° to get a line $\overleftrightarrow{A_1 B_1}$ perpendicular to $\overleftrightarrow{AB}$. For the new line, the rise and the run are interchanged, but the run is now negative. Thus the slope of the new line is $-b/a$, which is the opposite of the reciprocal of the slope of the first line. Also note that when we multiply the slopes, we get

$$\frac{a}{b}\left(-\frac{b}{a}\right) = -1.$$

This is the condition under which lines will be perpendicular.

> ### PERPENDICULAR LINES
>
> Two lines are **perpendicular** if the product of their slopes is -1.
> (If one line has slope m, then the slope of a line perpendicular to it is $-1/m$. That is, to find the slope of a line perpendicular to a given line, we take the reciprocal of the given slope and change the sign.)
>
> Lines are also perpendicular if one of them is vertical ($x = a$) and one of them is horizontal ($y = b$).

EXAMPLE 8 Determine whether the graphs of $5y = 4x + 10$ and $4y = -5x + 4$ are perpendicular.

To determine whether the lines are perpendicular, we determine whether the product of their slopes is -1. We first find the slope-intercept form of each equation by solving for y.

We have

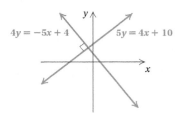

$$5y = 4x + 10 \qquad\qquad 4y = -5x + 4$$
$$y = \tfrac{1}{5}(4x + 10) \qquad\quad y = \tfrac{1}{4}(-5x + 4)$$
$$y = \tfrac{4}{5}x + 2; \qquad\qquad y = -\tfrac{5}{4}x + 1.$$

The slope of the first line is $\frac{4}{5}$, and the slope of the second line is $-\frac{5}{4}$. The product of the slopes is $\frac{4}{5} \cdot \left(-\frac{5}{4}\right) = -1$. Thus the lines are perpendicular.

◀ Do Exercises 14 and 15.

Determine whether the graphs of the given pair of lines are perpendicular.

14. $2y - x = 2,$
$\quad y + 2x = 4$ (GS)

Write each equation in the form $y = mx + b$:

$2y - x = 2 \rightarrow y = \boxed{} x + 1;$

$y + 2x = 4 \rightarrow y = \boxed{} x + 4.$

The slopes of these lines are $\boxed{}$ and -2. The product of the slopes $\frac{1}{2} \cdot (-2) = \boxed{}$.

Thus the lines $\underline{}$
$\qquad\qquad$ are/are not

perpendicular.

15. $3y = 2x + 15,$
$\quad 2y = 3x + 10$

Answers

14. Yes 15. No

Guided Solution:

14. $\frac{1}{2}, -2, \frac{1}{2}, -1$, are

Visualizing for Success

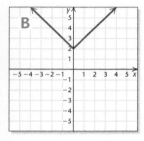

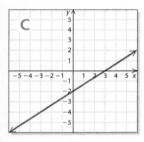

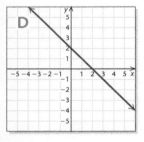

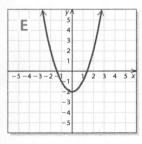

Match each equation with its graph.

1. $y = 2 - x$

2. $x - y = 2$

3. $x + 2y = 2$

4. $2x - 3y = 6$

5. $x = 2$

6. $y = 2$

7. $y = |x + 2|$

8. $y = |x| + 2$

9. $y = x^2 - 2$

10. $y = 2 - x^2$

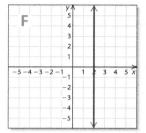

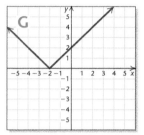

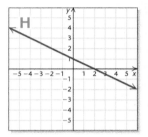

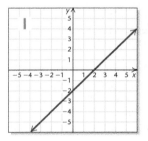

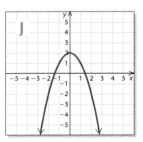

Answers on page A-35

Reading Check

Determine whether each statement is true or false.

RC1. The graphs of the lines $x = -4$ and $y = 5$ are perpendicular.

RC2. The y-intercept of $y = -2x + 7$ is $(0, -2)$.

RC3. Two lines are perpendicular if the product of their slopes is 1.

RC4. The x-intercept of $x = -\frac{2}{7}$ is $\left(-\frac{2}{7}, 0\right)$.

RC5. The slope of a horizontal line is 0.

RC6. Two nonvertical lines are parallel if they have the same slope and the same y-intercepts.

a   Find the intercepts and then graph the line.

1. $x - 2 = y$

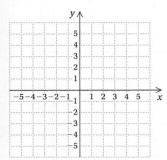

2. $x + 3 = y$

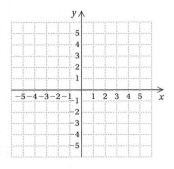

3. $x + 3y = 6$

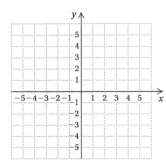

4. $x - 2y = 4$

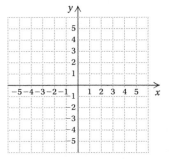

5. $2x + 3y = 6$

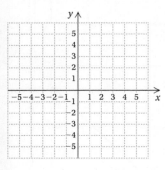

6. $5x - 2y = 10$

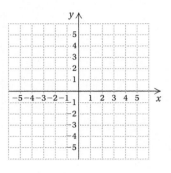

7. $f(x) = -2 - 2x$

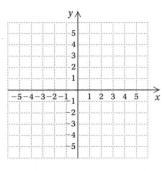

8. $g(x) = 5x - 5$

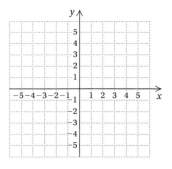

9. $5y = -15 + 3x$

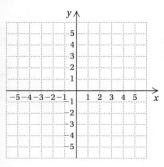

10. $5x - 10 = 5y$

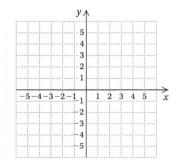

11. $2x - 3y = 6$

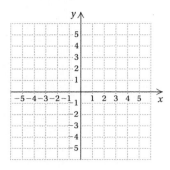

12. $4x + 5y = 20$

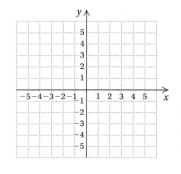

13. $2.8y - 3.5x = -9.8$

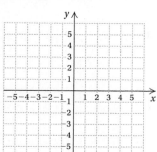

14. $10.8x - 22.68 = 4.2y$

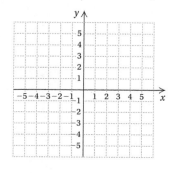

15. $5x + 2y = 7$

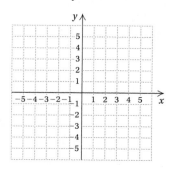

16. $3x - 4y = 10$

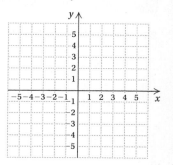

b Graph using the slope and the y-intercept.

17. $y = \dfrac{5}{2}x + 1$

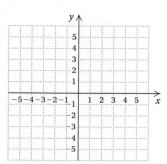

18. $y = \dfrac{2}{5}x - 4$

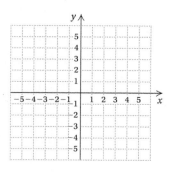

19. $f(x) = -\dfrac{5}{2}x - 4$

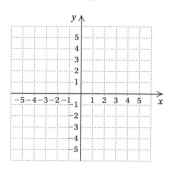

20. $f(x) = \dfrac{2}{5}x + 3$

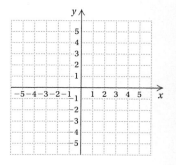

21. $x + 2y = 4$

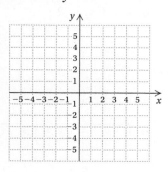

22. $x - 3y = 6$

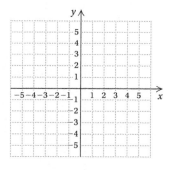

23. $4x - 3y = 12$

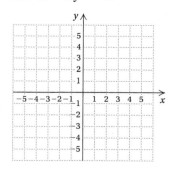

24. $2x + 6y = 12$

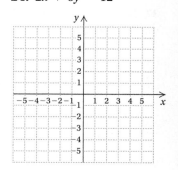

25. $f(x) = \dfrac{1}{3}x - 4$

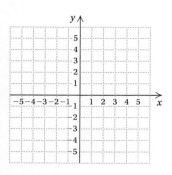

26. $g(x) = -0.25x + 2$

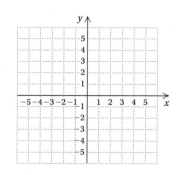

27. $5x + 4 \cdot f(x) = 4$
(*Hint*: Solve for $f(x)$.)

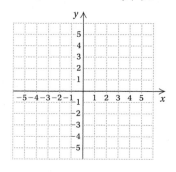

28. $3 \cdot f(x) = 4x + 6$

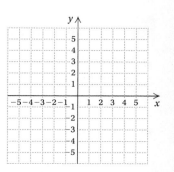

c   Graph and, if possible, determine the slope.

29. $x = 1$

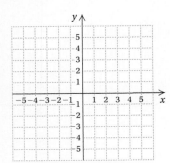

30. $x = -4$

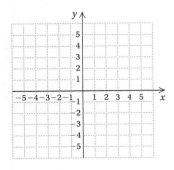

31. $y = -1$

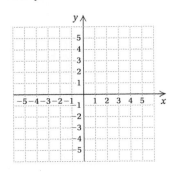

32. $y = \dfrac{3}{2}$

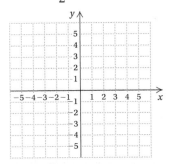

33. $f(x) = -6$

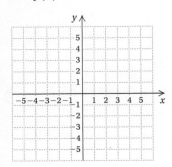

34. $f(x) = 2$

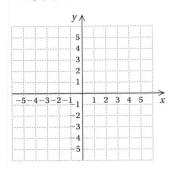

35. $y = 0$

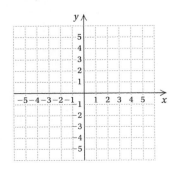

36. $x = 0$

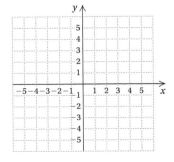

37. $2 \cdot f(x) + 5 = 0$

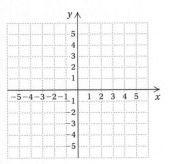

38. $4 \cdot g(x) + 3x = 12 + 3x$

39. $7 - 3x = 4 + 2x$

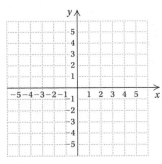

40. $3 - f(x) = 2$

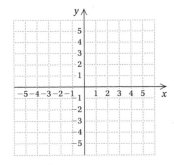

d   Determine whether the graphs of the given pair of lines are parallel.

41. $x + 6 = y,$
    $y - x = -2$

42. $2x - 7 = y,$
    $y - 2x = 8$

43. $y + 3 = 5x,$
    $3x - y = -2$

44. $y + 8 = -6x,$
    $-2x + y = 5$

45. $y = 3x + 9,$
    $2y = 6x - 2$

46. $y + 7x = -9,$
    $-3y = 21x + 7$

47. $12x = 3,$
    $-7x = 10$

48. $5y = -2,$
    $\frac{3}{4}x = 16$

Determine whether the graphs of the given pair of lines are perpendicular.

49. $y = 4x - 5$,
$4y = 8 - x$

50. $2x - 5y = -3$,
$2x + 5y = 4$

51. $x + 2y = 5$,
$2x + 4y = 8$

52. $y = -x + 7$,
$y = x + 3$

53. $2x - 3y = 7$,
$2y - 3x = 10$

54. $x = y$,
$y = -x$

55. $2x = 3$,
$-3y = 6$

56. $-5y = 10$,
$y = -\frac{4}{9}$

Skill Maintenance

Write in scientific notation.   [13.2c]

57. 53,000,000,000

58. 0.000047

59. 0.018

60. 99,902,000

Write in decimal notation.   [13.2c]

61. 2.13×10^{-5}

62. 9.01×10^{8}

63. 2×10^{4}

64. 8.5677×10^{-2}

Factor.   [10.7d]

65. $9x - 15y$

66. $12a + 21ab$

67. $21p - 7pq + 14p$

68. $64x - 128y + 256$

69. *Heaviest Pumpkin.*   In September 2012, Ron Wallace of Greene, Rhode Island, set a world record for the heaviest pumpkin. The previous record, set in 2010, was 1810.5 lb. The new record is 706.75 lb less than 1.5 times the record set in 2010. What was the record weight set in 2012?   [11.6a]

Source: www.huffingtonpost.com

70. Graph: $f(x) = -x^2 + 3x - 1$.   [16.1c]

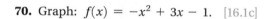
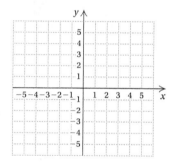

Synthesis

71. Find the value of a such that the graphs of $5y = ax + 5$ and $\frac{1}{4}y = \frac{1}{10}x - 1$ are parallel.

72. Find the value of k such that the graphs of $x + 7y = 70$ and $y + 3 = kx$ are perpendicular.

73. Write an equation of the line that has x-intercept $(-3, 0)$ and y-intercept $(0, \frac{2}{5})$.

74. Find the coordinates of the point of intersection of the graphs of the equations $x = -4$ and $y = 5$.

75. Write an equation for the x-axis. Is this equation a function?

76. Write an equation for the y-axis. Is this equation a function?

77. Find the value of m in $y = mx + 3$ so that the x-intercept of its graph will be $(4, 0)$.

78. Find the value of b in $2y = -7x + 3b$ so that the y-intercept of its graph will be $(0, -13)$.

OBJECTIVES

a Find an equation of a line when the slope and the y-intercept are given.

b Find an equation of a line when the slope and a point are given.

c Find an equation of a line when two points are given.

d Given a line and a point not on the given line, find an equation of the line parallel to the line and containing the point, and find an equation of the line perpendicular to the line and containing the point.

e Solve applied problems involving linear functions.

SKILL TO REVIEW

Objective 10.6b: Find the reciprocal of a real number.

Find the reciprocal of the number.

1. 3

2. $-\dfrac{4}{9}$

1. A line has slope 3.4 and y-intercept $(0, -8)$. Find an equation of the line.

Answers

Skill to Review:

1. $\dfrac{1}{3}$ **2.** $-\dfrac{9}{4}$

Margin Exercise:

1. $y = 3.4x - 8$

In this section, we will learn to find an equation of a line for which we have been given two pieces of information.

a FINDING AN EQUATION OF A LINE WHEN THE SLOPE AND THE y-INTERCEPT ARE GIVEN

If we know the slope and the y-intercept of a line, we can find an equation of the line using the slope–intercept equation $y = mx + b$.

EXAMPLE 1   A line has slope -0.7 and y-intercept $(0, 13)$. Find an equation of the line.

We use the slope–intercept equation and substitute -0.7 for m and 13 for b:

$$y = mx + b$$
$$y = -0.7x + 13.$$

◀ **Do Margin Exercise 1.**

b FINDING AN EQUATION OF A LINE WHEN THE SLOPE AND A POINT ARE GIVEN

Suppose we know the slope of a line and the coordinates of one point on the line. We can use the slope–intercept equation to find an equation of the line. Or, we can use the **point–slope equation**. We first develop a formula for such a line.

Suppose that a line of slope m passes through the point (x_1, y_1). For any other point (x, y) on this line, we must have

$$\frac{y - y_1}{x - x_1} = m.$$

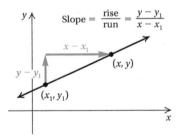

It is tempting to use this last equation as an equation of the line of slope m that passes through (x_1, y_1). The only problem with this form is that when x and y are replaced with x_1 and y_1, we have $\frac{0}{0} = m$, a false equation. To avoid this difficulty, we multiply by $x - x_1$ on both sides and simplify:

$$\frac{y - y_1}{x - x_1}(x - x_1) = m(x - x_1) \qquad \text{Multiplying by } x - x_1 \text{ on both sides}$$

$$y - y_1 = m(x - x_1). \qquad \text{Removing a factor of 1: } \frac{x - x_1}{x - x_1} = 1$$

This is the *point–slope* form of a linear equation.

POINT–SLOPE EQUATION

The **point–slope equation** of a line with slope m, passing through (x_1, y_1), is

$$y - y_1 = m(x - x_1).$$

If we know the slope of a line and a point on the line, we can find an equation of the line using either the point-slope equation,

$$y - y_1 = m(x - x_1),$$

or the slope-intercept equation,

$$y = mx + b.$$

EXAMPLE 2 Find an equation of the line with slope 5 and containing the point $\left(\frac{1}{2}, -1\right)$.

Using the Point-Slope Equation: We consider $\left(\frac{1}{2}, -1\right)$ to be (x_1, y_1) and 5 to be the slope m, and substitute:

$$
\begin{aligned}
y - y_1 &= m(x - x_1) && \text{Point-slope equation} \\
y - (-1) &= 5\left(x - \tfrac{1}{2}\right) && \text{Substituting} \\
y + 1 &= 5x - \tfrac{5}{2} && \text{Simplifying} \\
y &= 5x - \tfrac{5}{2} - 1 \\
y &= 5x - \tfrac{5}{2} - \tfrac{2}{2} \\
y &= 5x - \tfrac{7}{2}.
\end{aligned}
$$

Using the Slope-Intercept Equation: The point $\left(\frac{1}{2}, -1\right)$ is on the line, so it is a solution of the equation. Thus we can substitute $\frac{1}{2}$ for x and -1 for y in $y = mx + b$. We also substitute 5 for m, the slope. Then we solve for b:

$$
\begin{aligned}
y &= mx + b && \text{Slope-intercept equation} \\
-1 &= 5 \cdot \left(\tfrac{1}{2}\right) + b && \text{Substituting} \\
-1 &= \tfrac{5}{2} + b \\
-1 - \tfrac{5}{2} &= b \\
-\tfrac{2}{2} - \tfrac{5}{2} &= b \\
-\tfrac{7}{2} &= b. && \text{Solving for } b
\end{aligned}
$$

We then use the slope-intercept equation $y = mx + b$ again and substitute 5 for m and $-\frac{7}{2}$ for b:

$$y = 5x - \tfrac{7}{2}.$$

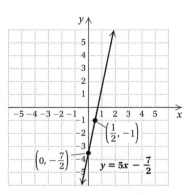

Do Exercises 2–5. ▶

Find an equation of the line with the given slope and containing the given point.

2. $m = -5,\ (-4, 2)$

3. $m = 3,\ (1, -2)$

4. $m = 8,\ (3, 5)$

5. $m = -\dfrac{2}{3},\ (1, 4)$

Answers

2. $y = -5x - 18$ **3.** $y = 3x - 5$

4. $y = 8x - 19$ **5.** $y = -\dfrac{2}{3}x + \dfrac{14}{3}$

C FINDING AN EQUATION OF A LINE WHEN TWO POINTS ARE GIVEN

We can also use the slope–intercept equation or the point–slope equation to find an equation of a line when two points are given.

EXAMPLE 3 Find an equation of the line containing the points $(2, 3)$ and $(-6, 1)$.

First, we find the slope:

$$m = \frac{3 - 1}{2 - (-6)} = \frac{2}{8}, \text{ or } \frac{1}{4}.$$

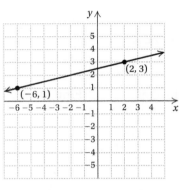

Now we have the slope and two points. We then proceed as we did in Example 2, using either point, and either the point–slope equation or the slope–intercept equation.

Using the Point–Slope Equation: We choose $(2, 3)$ and substitute 2 for x_1, 3 for y_1, and $\frac{1}{4}$ for m:

$$y - y_1 = m(x - x_1) \qquad \text{Point-slope equation}$$
$$y - 3 = \tfrac{1}{4}(x - 2) \qquad \text{Substituting}$$
$$y - 3 = \tfrac{1}{4}x - \tfrac{1}{2}$$
$$y = \tfrac{1}{4}x - \tfrac{1}{2} + 3$$
$$y = \tfrac{1}{4}x - \tfrac{1}{2} + \tfrac{6}{2}$$
$$y = \tfrac{1}{4}x + \tfrac{5}{2}.$$

Using the Slope–Intercept Equation: We choose $(2, 3)$ and substitute 2 for x, 3 for y, and $\frac{1}{4}$ for m:

$$y = mx + b \qquad \text{Slope-intercept equation}$$
$$3 = \tfrac{1}{4} \cdot 2 + b \qquad \text{Substituting}$$
$$3 = \tfrac{1}{2} + b$$
$$3 - \tfrac{1}{2} = \tfrac{1}{2} + b - \tfrac{1}{2}$$
$$\tfrac{6}{2} - \tfrac{1}{2} = b$$
$$\tfrac{5}{2} = b. \qquad \text{Solving for } b$$

Finally, we use the slope–intercept equation $y = mx + b$ again and substitute $\frac{1}{4}$ for m and $\frac{5}{2}$ for b:

$$y = \tfrac{1}{4}x + \tfrac{5}{2}.$$

◀ **Do Exercises 6 and 7.**

6. Find an equation of the line containing the points $(4, -3)$ and $(1, 2)$. **GS**

First, find the slope:

$$m = \frac{\boxed{} - (-3)}{1 - 4} = \frac{\boxed{}}{-3} = -\frac{5}{\boxed{}}.$$

Using the point-slope equation,
$$y - y_1 = m(x - x_1),$$

substitute 4 for x_1, -3 for y_1, and $-\dfrac{5}{3}$ for m:

$$y - (\boxed{}) = \boxed{}(x - \boxed{})$$
$$y + \boxed{} = -\frac{5}{3}x + \boxed{}$$
$$y = -\frac{5}{3}x + \frac{20}{3} - 3$$
$$y = -\frac{5}{3}x + \frac{20}{3} - \frac{\boxed{}}{3}$$
$$y = -\frac{5}{3}x + \frac{\boxed{}}{3}.$$

7. Find an equation of the line containing the points $(-3, -5)$ and $(-4, 12)$.

Answers

6. $y = -\dfrac{5}{3}x + \dfrac{11}{3}$ **7.** $y = -17x - 56$

Guided Solution:

6. $2, 5, 3; -3, -\dfrac{5}{3}, 4, 3, \dfrac{20}{3}, 9, 11$

1090 CHAPTER 16 Graphs, Functions, and Applications

d FINDING AN EQUATION OF A LINE PARALLEL OR PERPENDICULAR TO A GIVEN LINE THROUGH A POINT NOT ON THE LINE

We can also use the methods of Example 2 to find an equation of a line parallel or perpendicular to a given line and containing a point not on the line.

EXAMPLE 4 Find an equation of the line containing the point $(-1, 3)$ and parallel to the line $2x + y = 10$.

A line parallel to the given line $2x + y = 10$ must have the same slope as the given line. To find that slope, we first find the slope–intercept equation by solving for y:

$$2x + y = 10$$
$$y = -2x + 10.$$

Thus the line we want to find through $(-1, 3)$ must also have slope -2.

Using the Point–Slope Equation: We use the point $(-1, 3)$ and the slope -2, substituting -1 for x_1, 3 for y_1, and -2 for m:

$$y - y_1 = m(x - x_1)$$
$$y - 3 = -2(x - (-1)) \qquad \text{Substituting}$$
$$y - 3 = -2(x + 1) \qquad \text{Simplifying}$$
$$y - 3 = -2x - 2$$
$$y = -2x + 1.$$

Using the Slope–Intercept Equation: We substitute -1 for x, 3 for y, and -2 for m in $y = mx + b$. Then we solve for b:

$$y = mx + b$$
$$3 = -2(-1) + b \qquad \text{Substituting}$$
$$3 = 2 + b$$
$$1 = b. \qquad \text{Solving for } b$$

We then use the equation $y = mx + b$ again and substitute -2 for m and 1 for b:

$$y = -2x + 1.$$

The given line $2x + y = 10$, or $y = -2x + 10$, and the line $y = -2x + 1$ have the same slope but different y-intercepts. Thus their graphs are parallel.

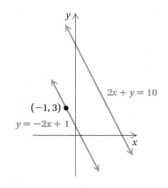

Do Exercise 8. ▶

GS **8.** Find an equation of the line containing the point $(2, -1)$ and parallel to the line $9x - 3y = 5$.

Find the slope of the given line:
$$9x - 3y = 5$$
$$-3y = \boxed{} + 5$$
$$y = \boxed{}\, x - \frac{5}{3}.$$

The slope is $\boxed{}$.

The line parallel to $9x - 3y = 5$ must have slope $\boxed{}$.

Using the slope–intercept equation,
$$y = mx + b,$$
substitute 3 for m, 2 for x, and -1 for y, and solve for b:
$$\boxed{} = 3 \cdot \boxed{} + b$$
$$-1 = 6 + b$$
$$\boxed{} = b.$$

Substitute 3 for m and -7 for b in $y = mx + b$:
$$y = 3x + (\boxed{})$$
$$y = 3x - 7.$$

Answer

8. $y = 3x - 7$
Guided Solution:
8. $-9x, 3, 3, 3; -1, 2, -7, -7$

EXAMPLE 5 Find an equation of the line containing the point $(2, -3)$ and perpendicular to the line $4y - x = 20$.

To find the slope of the given line, we first find its slope–intercept form by solving for y:

$$4y - x = 20$$
$$4y = x + 20$$
$$\frac{4y}{4} = \frac{x + 20}{4} \qquad \text{Dividing by 4}$$
$$y = \tfrac{1}{4}x + 5.$$

We know that the slope of the perpendicular line must be the opposite of the reciprocal of $\tfrac{1}{4}$. Thus the new line through $(2, -3)$ must have slope -4.

Using the Point–Slope Equation: We use the point $(2, -3)$ and the slope -4, substituting 2 for x_1, -3 for y_1, and -4 for m:

$$y - y_1 = m(x - x_1)$$
$$y - (-3) = -4(x - 2) \qquad \text{Substituting}$$
$$y + 3 = -4x + 8$$
$$y = -4x + 5.$$

Using the Slope–Intercept Equation: We now substitute 2 for x and -3 for y in $y = mx + b$. We also substitute -4 for m, the slope. Then we solve for b:

$$y = mx + b$$
$$-3 = -4(2) + b \qquad \text{Substituting}$$
$$-3 = -8 + b$$
$$5 = b. \qquad \text{Solving for } b$$

Finally, we use the equation $y = mx + b$ again and substitute -4 for m and 5 for b:

$$y = -4x + 5.$$

The product of the slopes of the lines $4y - x = 20$ and $y = -4x + 5$ is $\tfrac{1}{4} \cdot (-4) = -1$. Thus their graphs are perpendicular.

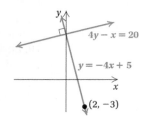

◀ **Do Exercise 9.**

---

9. Find an equation of the line containing the point $(5, 4)$ and perpendicular to the line $2x - 4y = 9$. **GS**

Find the slope of the given line:

$2x - 4y = 9$

$-4y = \boxed{} + 9$

$y = \boxed{}\, x - \dfrac{9}{4}.$

The slope is $\boxed{}$.

The slope of a line perpendicular to $2x - 4y = 9$ is the opposite of the reciprocal of $\tfrac{1}{2}$, or $\boxed{}$.

Using the point-slope equation,

$$y - y_1 = m(x - x_1),$$

substitute -2 for m, 5 for x_1, and 4 for y_1:

$y - \boxed{} = \boxed{}\,(x - 5)$

$y - 4 = -2x + \boxed{}$

$y = -2x + 14.$

e APPLICATIONS OF LINEAR FUNCTIONS

When the essential parts of a problem are described in mathematical language, we say that we have a **mathematical model**. We have already studied many kinds of mathematical models in this text—for example, the formulas in Section 2.4 and the functions in Section 16.1. Here we study linear functions as models.

EXAMPLE 6 *Cost of a Necklace.* Amelia's Beads offers a class in designing necklaces. For a necklace made of 6-mm beads, 4.23 beads per inch are needed. The cost of a necklace of 6-mm gemstone beads that sell for 40¢ each is $7 for the clasp and the crimps and approximately $1.70 per inch.

a) Formulate a linear function that models the total cost of a necklace $C(n)$, where n is the length of the necklace, in inches.

b) Graph the model.

c) Use the model to determine the cost of a 30-in. necklace.

a) The problem describes a situation in which cost per inch is charged in addition to the fixed cost of the clasp and the crimps. The total cost of a 16-in. necklace is

$$\$7 + \$1.70 \cdot 16 = \$34.20.$$

For a 17-in. necklace, the total cost is

$$\$7 + \$1.70 \cdot 17 = \$35.90.$$

These calculations lead us to generalize that for a necklace that is n inches long, the total cost is given by $C(n) = 7 + 1.7n$, where $n \geq 0$ since the length of the necklace cannot be negative. (Actually most necklaces are at least 14 in. long.) The notation $C(n)$ indicates that the cost C is a function of the length n.

b) Before we draw the graph, we rewrite the model in slope–intercept form:

$$C(n) = 1.7n + 7.$$

The y-intercept is $(0, 7)$ and the slope, or rate of change, is 1.70, or $\frac{\$17}{10}$, per inch. We first plot $(0, 7)$; from that point, we move 17 units up and 10 units to the right to the point $(10, 24)$. We then draw a line through these points. We also calculate a third value as a check:

$$C(20) = 1.7 \cdot 20 + 7 = 41.$$

The point $(20, 41)$ lines up with the other two points so the graph is correct.

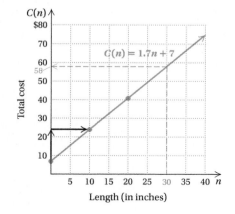

10. *Cost of a Service Call.* For a service call, Belmont Heating and Air Conditioning charges a $65 trip fee and $80 per hour for labor.

a) Formulate a linear function for the total cost of the service call $C(t)$, where t is the length of the call, in hours.

b) Graph the model.

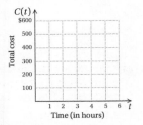

c) Use the model to determine the cost of a $2\frac{1}{2}$-hr service call.

c) To determine the total cost of a 30-in. necklace, we find $C(30)$:

$$C(30) = 1.7 \cdot 30 + 7 = 58.$$

From the graph, we see that the input 30 corresponds to the output 58. Thus we see that a 30-in. necklace costs $58.

◀ **Do Exercise 10.**

In the following example, we use two points and find an equation for the linear function through these points. Then we use the equation to estimate.

EXAMPLE 7 *Farmers' Markets.* The number of farmers' markets has increased steadily in recent years. The following table lists data regarding the correspondence between the year and the number of farmers' markets.

YEAR, x (in number of years since 2002)	NUMBER OF FARMERS' MARKETS, n
2002, 0	3137
2012, 10	7864

SOURCE: U.S. Department of Agriculture

a) Assuming a constant rate of change, use the two data points to find a linear function that fits the data.

b) Use the function to determine the number of farmers' markets in 2010.

c) In which year will the number of farmers' markets reach 12,000?

a) We let $x = $ the number of years since 2002 and $N = $ the number of farmers' markets. The table gives us two ordered pairs, $(0, 3137)$ and $(10, 7864)$. We use them to find a linear function that fits the data. First, we find the slope:

$$m = \frac{7864 - 3137}{10 - 0} = \frac{4727}{10} = 472.7.$$

Next, we find an equation $N = mx + b$ that fits the data. One of the data points, $(0, 3137)$, is the y-intercept. Thus we know b in the slope–intercept equation, $y = mx + b$. We use the equation $N = mx + b$ and substitute 472.7 for m and 3137 for b:

$$N = 472.7x + 3137.$$

Using function notation, we have

$$N(x) = 472.7x + 3137.$$

Answer

10. (a) $C(t) = 80t + 65$;

(b)

C(t)
$600
Total cost
500
400
300
200
100
1 2 3 4 5 6 t
Time (in hours)

(c) $265

b) To determine the number of farmers' markets in 2010, we substitute 8 for x (2010 is 8 years since 2002) in the function $N(x) = 472.7x + 3137$:

$$N(x) = 472.7x + 3137$$
$$N(8) = 472.7(8) + 3137 \qquad \text{Substituting}$$
$$= 3781.6 + 3137$$
$$= 6918.6 \approx 6919.$$

There were about 6919 farmers' markets in 2010.

c) To find the year in which the number of farmers' markets will reach 12,000, we substitute 12,000 for $N(x)$ and solve for x:

$$N(x) = 472.7x + 3137$$
$$12{,}000 = 472.7x + 3137 \qquad \text{Substituting}$$
$$8863 = 472.7x \qquad \text{Subtracting 3137}$$
$$19 \approx x. \qquad \text{Dividing by 472.7}$$

The number of farmers' markets will reach 12,000 about 19 years after 2002, or in 2021.

Do Exercise 11. ▶

11. *Hat Size as a Function of Head Circumference.* The following table lists data relating hat size to head circumference.

HEAD CIRCUMFERENCE, C (in inches)	HAT SIZE, H
21.2	$6\frac{3}{4}$
22	7

Source: Shushan's New Orleans

a) Assuming a constant rate of change, use the two data points to find a linear function that fits the data.

b) Use the function to determine the hat size of a person whose head has a circumference of 24.8 in.

c) Jerome's hat size is 8. What is the circumference of his head?

Answer

11. (a) $H(C) = \dfrac{5}{16}C + \dfrac{1}{8}$, or
$H(C) = 0.3125C + 0.125$;

(b) $7\dfrac{7}{8}$, or 7.875; **(c)** 25.2 in.

16.5	**Exercise Set**	For Extra Help MyMathLab®	MathXL® PRACTICE	WATCH	READ	REVIEW

✓ Reading Check

For the given equation, determine the slope of the line **(a)** parallel to the given line and **(b)** perpendicular to the given line.

RC1. $y = \dfrac{4}{11}x - 2$

RC2. $y = -5$

RC3. $2x - y = -4$

RC4. $y - \dfrac{4}{3} = -\dfrac{5}{6}x$

RC5. $x = 3$

RC6. $10x + 5y = 14$

---

a Find an equation of the line having the given slope and y-intercept.

1. Slope: -8; y-intercept: $(0, 4)$

2. Slope: 5; y-intercept: $(0, -3)$

3. Slope: 2.3; y-intercept: $(0, -1)$

4. Slope: -9.1; y-intercept: $(0, 2)$

Find a linear function $f(x) = mx + b$ whose graph has the given slope and y-intercept.

5. Slope: $-\frac{7}{3}$; y-intercept: $(0, -5)$

6. Slope: $\frac{4}{5}$; y-intercept: $(0, 28)$

7. Slope: $\frac{2}{3}$; y-intercept: $\left(0, \frac{5}{8}\right)$

8. Slope: $-\frac{7}{8}$; y-intercept: $\left(0, -\frac{7}{11}\right)$

b Find an equation of the line having the given slope and containing the given point.

9. $m = 5$, $(4, 3)$

10. $m = 4$, $(5, 2)$

11. $m = -3$, $(9, 6)$

12. $m = -2$, $(2, 8)$

13. $m = 1$, $(-1, -7)$

14. $m = 3$, $(-2, -2)$

15. $m = -2$, $(8, 0)$

16. $m = -3$, $(-2, 0)$

17. $m = 0$, $(0, -7)$

18. $m = 0$, $(0, 4)$

19. $m = \frac{2}{3}$, $(1, -2)$

20. $m = -\frac{4}{5}$, $(2, 3)$

c Find an equation of the line containing the given pair of points.

21. $(1, 4)$ and $(5, 6)$

22. $(2, 5)$ and $(4, 7)$

23. $(-3, -3)$ and $(2, 2)$

24. $(-1, -1)$ and $(9, 9)$

25. $(-4, 0)$ and $(0, 7)$

26. $(0, -5)$ and $(3, 0)$

27. $(-2, -3)$ and $(-4, -6)$

28. $(-4, -7)$ and $(-2, -1)$

29. $(0, 0)$ and $(6, 1)$

30. $(0, 0)$ and $(-4, 7)$

31. $\left(\frac{1}{4}, -\frac{1}{2}\right)$ and $\left(\frac{3}{4}, 6\right)$

32. $\left(\frac{2}{3}, \frac{3}{2}\right)$ and $\left(-3, \frac{5}{6}\right)$

d Write an equation of the line containing the given point and parallel to the given line.

33. $(3, 7)$; $x + 2y = 6$

34. $(0, 3)$; $2x - y = 7$

35. $(2, -1)$; $5x - 7y = 8$

36. $(-4, -5)$; $2x + y = -3$

37. $(-6, 2)$; $3x = 9y + 2$

38. $(-7, 0)$; $2y + 5x = 6$

Write an equation of the line containing the given point and perpendicular to the given line.

39. $(2, 5)$; $2x + y = 3$

40. $(4, 1)$; $x - 3y = 9$

41. $(3, -2)$; $3x + 4y = 5$

42. $(-3, -5)$; $5x - 2y = 4$

43. $(0, 9)$; $2x + 5y = 7$

44. $(-3, -4)$; $6y - 3x = 2$

 Solve.

45. *School Fund-Raiser.* A school club is raising funds by having a "Shred It Day," when residents of the community can bring in their sensitive documents to be shredded. The club is charging $10 for the first three paper bags full of documents and $5 for each additional bag.
   a) Formulate a linear function that models the total cost $C(x)$ of shredding x additional bags of documents.
   b) Graph the model.
   c) Use the model to determine the total cost of shredding 7 bags of documents.

46. *Fitness Club Costs.* A fitness club charges an initiation fee of $165 plus $24.95 per month.
   a) Formulate a linear function that models the total cost $C(t)$ of a club membership for t months.
   b) Graph the model.
   c) Use the model to determine the total cost of a 14-month membership.

47. *Value of a Lawn Mower.* A landscaping business purchased a ZTR commercial lawn mower for $9400. The value $V(t)$ of the mower depreciates (declines) at a rate of $85 per month.
   a) Formulate a linear function that models the value $V(t)$ of the mower after t months.
   b) Graph the model.
   c) Use the model to determine the value of the mower after 18 months.

48. *Value of a Computer.* True Tone Graphics bought a computer for $3800. The value $V(t)$ of the computer depreciates at a rate of $50 per month.
   a) Formulate a linear function that models the value $V(t)$ of the computer after t months.
   b) Graph the model.
   c) Use the model to determine the value of the computer after $10\frac{1}{2}$ months.

In Exercises 49–54, assume that a constant rate of change exists for each model formed.

49. *Organic-Food Sales.* The following table lists data regarding sales, in billions of dollars, of organic food in 2004 and in 2012.

YEAR, x (in number of years since 2004)	ORGANIC FOOD SALES (in billions)
2004, 0	$11
2012, 8	27

SOURCE: *Nutrition Business Journal*

   a) Use the two data points to find a linear function that fits the data. Let $x = $ the number of years since 2004 and $S(x) = $ the total sales, in billions of dollars, of organic food.
   b) Use the function of part (a) to estimate and predict the sales of organic food in 2008 and in 2017.

50. *Cost of Diabetes.* The following table lists data regarding the health-care and work-related costs of diabetes in 2007 and in 2012.

YEAR, x (in number of years since 2007)	COSTS OF DIABETES (in billions)
2007, 0	$174
2012, 5	245

SOURCE: American Diabetes Association

   a) Use the two data points to find a function that fits the data. Let $x = $ the number of years since 2007 and $D(x) = $ the costs, in billions of dollars, of diabetes.
   b) Use the function of part (a) to estimate the costs of diabetes in 2010 and in 2015.

51. Auto Dealers. At the close of 1995, there were 22,800 new-auto dealers in the United States. By the end of 2012, this number had dropped to 17,540. Let $D(x) = $ the number of new-auto dealerships and $x = $ the number of years since 1995.

Source: Urban Science Automotive Dealer Census

a) Find a linear function that fits the data.
b) Use the function of part (a) to estimate the number of new-auto dealerships in 2000.
c) At this rate of decrease, when will the number of new-auto dealerships be 15,500?

52. Records in the 400-Meter Run. In 1930, the record for the 400-m run was 46.8 sec. In 1970, it was 43.8 sec. Let $R(t) = $ the record in the 400-m run and $t = $ the number of years since 1930.

a) Find a linear function that fits the data.
b) Use the function of part (a) to estimate the record in 2003 and in 2006.
c) When will the record be 40 sec?

53. Life Expectancy in South Africa. In 2003, the life expectancy in South Africa was 46.56 years. In 2011, it was 49.33 years. Let $E(t) = $ life expectancy and $t = $ the number of years since 2003.

Source: CIA World Factbook 2003–2012

a) Find a linear function that fits the data.
b) Use the function of part (a) to estimate life expectancy in 2016.

54. Life Expectancy in Monaco. In 2003, the life expectancy in Monaco was 79.27 years. In 2011, it was 89.73 years. Let $E(t) = $ life expectancy and $t = $ the number of years since 2003.

Source: CIA World Factbook 2003–2012

a) Find a linear function that fits the data.
b) Use the function of part (a) to estimate life expectancy in 2016.

Skill Maintenance

Simplify. [15.1c]

55. $\dfrac{w - t}{t - w}$

56. $\dfrac{b^2 - 1}{b - 1}$

57. $\dfrac{3x^2 + 15x - 72}{6x^2 + 18x - 240}$

58. $\dfrac{4y + 32}{y^2 - y - 72}$

Find the slope, if it exists, of the line. [16.3b], [16.4c]

59. $2x - 7y = -10$

60. $y = -1$

61. $x = 42$

62. $4 - 6y = 12x$

Synthesis

63. Find k such that the line containing the points $(-3, k)$ and $(4, 8)$ is parallel to the line containing the points $(5, 3)$ and $(1, -6)$.

64. Find an equation of the line passing through the point $(4, 5)$ and perpendicular to the line passing through the points $(-1, 3)$ and $(2, 9)$.

Vocabulary Reinforcement

Complete each statement with the correct term from the column on the right. Some of the choices may be used more than once, and some may not be used at all.

x-intercept

y-intercept

at least one

exactly one

slope–intercept

point–slope

slope

function

relation

parallel

perpendicular

vertical

horizontal

domain

range

1. The graph of $x = a$ is a(n) _____ line with x-intercept $(a, 0)$. [16.4c]

2. The _____ equation of a line with slope m and passing through (x_1, y_1) is $y - y_1 = m(x - x_1)$. [16.5b]

3. A(n) _____ is a correspondence between a first set, called the _____, and a second set called the _____, such that each member of the _____ corresponds to _____ member of the _____. [16.1a]

4. The _____ of a line containing points (x_1, y_1) and (x_2, y_2) is given by $m =$ the change in y/the change in x, also described as rise/run. [16.3b]

5. Two lines are _____ if the product of their slopes is -1. [16.4d]

6. The equation $y = mx + b$ is called the _____ equation of a line with slope m and y-intercept $(0, b)$. [16.3b]

7. Lines are _____ if they have the same slope and different y-intercepts. [16.4d]

Concept Reinforcement

Determine whether each statement is true or false.

_____ 1. The slope of a vertical line is 0. [16.4c]

_____ 2. A line with slope 1 slants less steeply than a line with slope -5. [16.3b]

_____ 3. Parallel lines have the same slope and y-intercept. [16.4d]

Study Guide

Objective 16.1a   Determine whether a correspondence is a function.

Example   Determine whether each correspondence is a function.

The correspondence f is a function because each member of the domain is matched to *only one* member of the range. The correspondence g *is not* a function because one member of the domain, Q, is matched to more than one member of the range.

Practice Exercise

1. Determine whether the correspondence is a function.

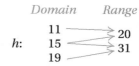

Objective 16.1b   Given a function described by an equation, find function values (outputs) for specified values (inputs).

Example   Find the indicated function value.
a) $f(0)$, for $f(x) = -x + 6$ **b)** $g(5)$, for $g(x) = -10$
c) $h(-1)$, for $h(x) = 4x^2 + x$

a) $f(x) = -x + 6$: $f(0) = -0 + 6 = 6$
b) $g(x) = -10$: $g(5) = -10$
c) $h(x) = 4x^2 + x$: $h(-1) = 4(-1)^2 + (-1) = 3$

Practice Exercise

2. Find $g(0)$, $g(-2)$, and $g(6)$ for $g(x) = \frac{1}{2}x - 2$.

Objective 16.1c   Draw the graph of a function.

Example   Graph: $f(x) = -\frac{2}{3}x + 2$.

    By choosing multiples of 3 for x, we can avoid fraction values for y. If $x = -3$, then $y = -\frac{2}{3} \cdot (-3) + 2$ $= 2 + 2 = 4$. We list three ordered pairs in a table, plot the points, draw the line, and label the graph.

x	$f(x)$
3	0
0	2
-3	4

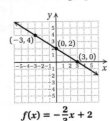

$$f(x) = -\frac{2}{3}x + 2$$

Practice Exercise

3. Graph: $f(x) = \frac{2}{5}x - 3$.

x	$f(x)$

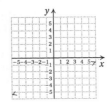

Objective 16.1d   Determine whether a graph is that of a function using the vertical-line test.

Example   Determine whether each of the following is the graph of a function.

a) 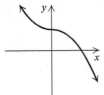   **b)**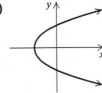

a) The graph is that of a function because no vertical line can cross the graph at more than one point.

b) The graph is not that of a function because a vertical line can cross the graph more than once.

Practice Exercise

4. Determine whether the graph is the graph of a function.

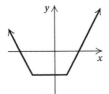

Objective 16.2a   Find the domain and the range of a function.

Example   For the function f whose graph is shown below, determine the domain and the range.

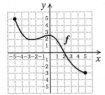 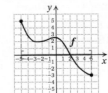 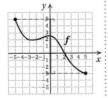

Domain: $\{x \mid -5 \le x \le 5\}$; range: $\{y \mid -3 \le y \le 5\}$

Practice Exercises

5. For the function g whose graph is shown below, determine the domain and the range.

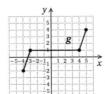

Example Find the domain of $g(x) = \dfrac{x + 1}{2x - 6}$.

Since $(x + 1)/(2x - 6)$ cannot be calculated when the denominator $2x - 6$ is 0, we solve $2x - 6 = 0$ to find the real numbers that must be excluded from the domain of g:

$$2x - 6 = 0$$
$$2x = 6$$
$$x = 3.$$

Thus, 3 is not in the domain. The domain of g is $\{x \mid x$ is a real number $and\ x \neq 3\}$.

6. Find the domain of
$$h(x) = \frac{x - 3}{3x + 9}.$$

Objective 16.3b Given two points on a line, find the slope. Given a linear equation, derive the equivalent slope–intercept equation and determine the slope and the y-intercept.

Example Find the slope of the line containing $(-5, 6)$ and $(-1, -4)$.

$$m = \frac{\text{change in } y}{\text{change in } x} = \frac{6 - (-4)}{-5 - (-1)} = \frac{6 + 4}{-5 + 1} = \frac{10}{-4} = -\frac{5}{2}$$

Example Find the slope and the y-intercept of
$$4x - 2y = 20.$$
We first solve for y:
$$4x - 2y = 20$$
$$-2y = -4x + 20 \qquad \text{Subtracting } 4x$$
$$y = 2x - 10. \qquad \text{Dividing by } -2$$
The slope is 2, and the y-intercept is $(0, -10)$.

Practice Exercises

7. Find the slope of the line containing $(2, -8)$ and $(-3, 2)$.

8. Find the slope and the y-intercept of
$$3x = -6y + 12.$$

Objective 16.4a Graph linear equations using intercepts.

Example Find the intercepts of $x - 2y = 6$ and then graph the line.

To find the y-intercept, we let $x = 0$ and solve for y:
$$0 - 2y = 6 \qquad \text{Substituting 0 for } x$$
$$-2y = 6$$
$$y = -3.$$
The y-intercept is $(0, -3)$.

To find the x-intercept, we let $y = 0$ and solve for x:
$$x - 2 \cdot 0 = 6 \qquad \text{Substituting 0 for } y$$
$$x = 6.$$
The x-intercept is $(6, 0)$.

We plot these points and draw the line, using a third point as a check. We let $x = -2$ and solve for y:
$$-2 - 2y = 6$$
$$-2y = 8$$
$$y = -4.$$
We plot $(-2, -4)$ and note that it is on the line. Thus the graph is correct.

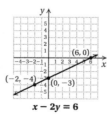

$x - 2y = 6$

Practice Exercise

9. Find the intercepts of $3y - 3 = x$ and then graph the line.

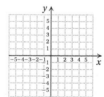

Objective 16.4b   Given a linear equation in slope–intercept form, use the slope and the y-intercept to graph the line.

Example   Graph using the slope and the y-intercept:

$$y = -\frac{3}{2}x + 5.$$

This equation is in slope–intercept form, $y = mx + b$. The y-intercept is $(0, 5)$. We plot $(0, 5)$. We can think of the slope $\left(m = -\frac{3}{2}\right)$ as $\frac{-3}{2}$.

Starting at the y-intercept, we use the slope to find another point on the graph. We move 3 units down and 2 units to the right. We get a new point: $(2, 2)$.

To get a third point for a check, we start at $(2, 2)$ and move 3 units down and 2 units to the right to the point $(4, -1)$. We plot the points and draw the line.

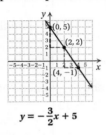

$$y = -\frac{3}{2}x + 5$$

Practice Exercise

10. Graph using the slope and the y-intercept:

$$y = \frac{1}{4}x - 3.$$

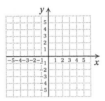

Objective 16.4c   Graph linear equations of the form $x = a$ or $y = b$.

Example   Graph: $y = -1$.

All ordered pairs $(x, -1)$ are solutions; y is -1 at each point. The graph is a horizontal line that intersects the y-axis at $(0, -1)$.

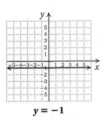

$$y = -1$$

Example   Graph: $x = 2$.

All ordered pairs $(2, y)$ are solutions; x is 2 at each point. The graph is a vertical line that intersects the x-axis at $(2, 0)$.

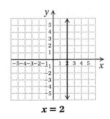

$$x = 2$$

Practice Exercises

11. Graph: $y = 3$.

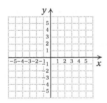

12. Graph: $x = -\dfrac{5}{2}$.

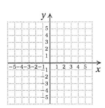

Objective 16.4d Given the equations of two lines, determine whether their graphs are parallel or whether they are perpendicular.

Example Determine whether the graphs of the given pair of lines are parallel, perpendicular, or neither.

a) $2y - x = 16$, $x + \frac{1}{2}y = 4$

b) $5x - 3 = 2y$, $2y + 12 = 5x$

a) Writing each equation in slope–intercept form, we have $y = \frac{1}{2}x + 8$ and $y = -2x + 8$. The slopes are $\frac{1}{2}$ and -2. The product of the slopes is -1: $\frac{1}{2} \cdot (-2) = -1$. The graphs are perpendicular.

b) Writing each equation in slope–intercept form, we have $y = \frac{5}{2}x - \frac{3}{2}$ and $y = \frac{5}{2}x - 6$. The slopes are the same, $\frac{5}{2}$, and the y-intercepts are different. The graphs are parallel.

Practice Exercises

Determine whether the graphs of the given pair of lines are parallel, perpendicular, or neither.

13. $-3x + 8y = -8$, $8y = 3x + 40$

14. $5x - 2y = -8$, $2x + 5y = 15$

Objective 16.5a Find an equation of a line when the slope and the y-intercept are given.

Example A line has slope 0.8 and y-intercept $(0, -17)$. Find an equation of the line.

We use the slope–intercept equation and substitute 0.8 for m and -17 for b:

$$y = mx + b \qquad \text{Slope–intercept equation}$$
$$y = 0.8x - 17.$$

Practice Exercise

15. A line has slope -8 and y-intercept $(0, 0.3)$. Find an equation of the line.

Objective 16.5b Find an equation of a line when the slope and a point are given.

Example Find an equation of the line with slope -2 and containing the point $\left(\frac{1}{3}, -1\right)$.

Using the *point–slope equation*, we substitute -2 for m, $\frac{1}{3}$ for x_1, and -1 for y_1:

$$y - (-1) = -2\left(x - \frac{1}{3}\right) \qquad \text{Using } y - y_1 = m(x - x_1)$$
$$y + 1 = -2x + \frac{2}{3}$$
$$y = -2x - \frac{1}{3}.$$

Using the *slope–intercept equation*, we substitute -2 for m, $\frac{1}{3}$ for x, and -1 for y, and then solve for b:

$$-1 = -2 \cdot \frac{1}{3} + b \qquad \text{Using } y = mx + b$$
$$-1 = -\frac{2}{3} + b$$
$$-\frac{1}{3} = b.$$

Then, substituting -2 for m and $-\frac{1}{3}$ for b in the slope–intercept equation $y = mx + b$, we have $y = -2x - \frac{1}{3}$.

Practice Exercise

16. Find an equation of the line with slope -4 and containing the point $\left(\frac{1}{2}, -3\right)$.

Objective 16.5c   Find an equation of a line when two points are given.

Example   Find an equation of the line containing the points $(-3, 9)$ and $(1, -2)$.

We first find the slope:

$$\frac{9 - (-2)}{-3 - 1} = \frac{11}{-4} = -\frac{11}{4}.$$

Using the slope–intercept equation and the point $(1, -2)$, we substitute $-\frac{11}{4}$ for m, 1 for x, and -2 for y, and then solve for b. We could also have used the point $(-3, 9)$.

$$y = mx + b$$
$$-2 = -\tfrac{11}{4} \cdot 1 + b$$
$$-\tfrac{8}{4} = -\tfrac{11}{4} + b$$
$$\tfrac{3}{4} = b$$

Then substituting $-\frac{11}{4}$ for m and $\frac{3}{4}$ for b in $y = mx + b$, we have $y = -\frac{11}{4}x + \frac{3}{4}$.

Practice Exercise

17. Find an equation of the line containing the points $(-2, 7)$ and $(4, -3)$.

Objective 16.5d   Given a line and a point not on the given line, find an equation of the line parallel to the line and containing the point, and find an equation of the line perpendicular to the line and containing the point.

Example   Write an equation of the line containing $(-1, 1)$ and parallel to $3y - 6x = 5$.

Solving $3y - 6x = 5$ for y, we get $y = 2x + \frac{5}{3}$. The slope of the given line is 2.

A line parallel to the given line must have the same slope, 2. We substitute 2 for m, -1 for x_1, and 1 for y_1 in the point–slope equation:

$$y - 1 = 2[x - (-1)] \qquad \text{Using } y - y_1 = m(x - x_1)$$
$$y - 1 = 2(x + 1)$$
$$y - 1 = 2x + 2$$
$$y = 2x + 3. \qquad \text{Line parallel to the given line and passing through } (-1, 1)$$

Example   Write an equation of the line containing the point $(2, -4)$ and perpendicular to $6x + 2y = 13$.

Solving $6x + 2y = 13$ for y, we get $y = -3x + \frac{13}{2}$. The slope of the given line is -3.

The slope of a line perpendicular to the given line is the opposite of the reciprocal of -3, or $\frac{1}{3}$. We substitute $\frac{1}{3}$ for m, 2 for x_1, and -4 for y_1 in the point–slope equation:

$$y - (-4) = \tfrac{1}{3}(x - 2) \qquad \text{Using } y - y_1 = m(x - x_1)$$
$$y + 4 = \tfrac{1}{3}x - \tfrac{2}{3}$$
$$y = \tfrac{1}{3}x - \tfrac{14}{3}. \qquad \text{Line perpendicular to the given line and passing through } (2, -4)$$

Practice Exercises

18. Write an equation of the line containing the point $(2, -5)$ and parallel to $4x - 3y = 6$.

19. Write an equation of the line containing $(2, -5)$ and perpendicular to $4x - 3y = 6$.

Review Exercises

Determine whether each correspondence is a function.   [16.1a]

1. 3 ⟶ a
5 ⟶ b
7 ⟶ c
9 ⟶ d
⟶ e

2. 1 ⟶ a
2 ⟶ b
3 ⟶ c
4 ⟶ d
5

Find the function values.   [16.1b]

3. $g(x) = -2x + 5$; $g(0)$ and $g(-1)$

4. $f(x) = 3x^2 - 2x + 7$; $f(0)$ and $f(-1)$

5. *Tuition Cost.* The function $C(t) = 309.2t + 3717.7$ can be used to approximate the average cost of tuition and fees for in-state students at public four-year colleges, where t is the number of years after 2000. Estimate the average cost of tuition and fees in 2010. That is, find $C(10)$.   [16.1b]

Source: U.S. National Center for Education Statistics

Graph.   [16.1c]

6. $f(x) = -3x + 2$

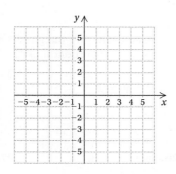

7. $g(x) = \frac{5}{2}x - 3$

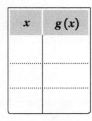
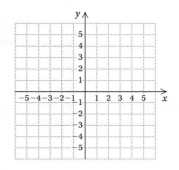

8. $f(x) = |x - 3|$

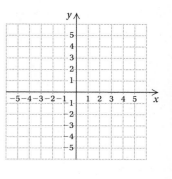

9. $h(x) = 3 - x^2$

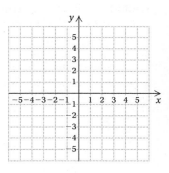

Determine whether each of the following is the graph of a function.   [16.1d]

10.
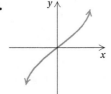

11.
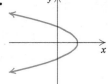

12. For the following graph of a function f, determine **(a)** $f(2)$; **(b)** the domain; **(c)** all x-values such that $f(x) = 2$; and **(d)** the range.   [16.2a]

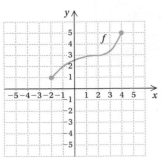

Find the domain.   [16.2a]

13. $f(x) = \dfrac{5}{x - 4}$ **14.** $g(x) = x - x^2$

Find the slope and the y-intercept.   [16.3a, b]

15. $y = -3x + 2$ **16.** $4y + 2x = 8$

17. Find the slope, if it exists, of the line containing the points $(13, 7)$ and $(10, -4)$.   [16.3b]

Find the intercepts. Then graph the equation.   [16.4a]

18. $2y + x = 4$

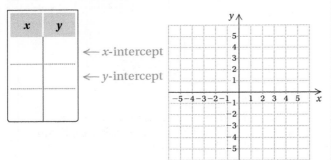

19. $2y = 6 - 3x$

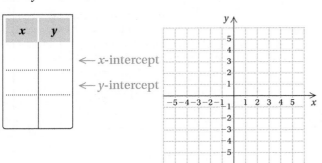

Graph using the slope and the y-intercept.   [16.4b]

20. $g(x) = -\frac{2}{3}x - 4$ **21.** $f(x) = \frac{5}{2}x + 3$

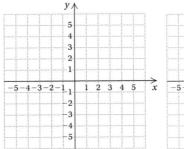   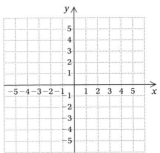

Graph.   [16.4c]

22. $x = -3$ **23.** $f(x) = 4$

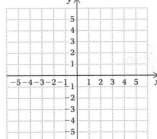   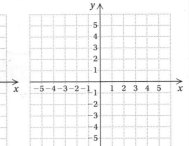

Determine whether the graphs of the given pair of lines are parallel or perpendicular.   [16.4d]

24. $y + 5 = -x$,   **25.** $3x - 5 = 7y$,
  $x - y = 2$ $7y - 3x = 7$

26. $4y + x = 3$,   **27.** $x = 4$,
  $2x + 8y = 5$ $y = -3$

28. Find a linear function $f(x) = mx + b$ whose graph has slope 4.7 and y-intercept $(0, -23)$.   [16.5a]

29. Find an equation of the line having slope -3 and containing the point $(3, -5)$.   [16.5b]

30. Find an equation of the line containing the points $(-2, 3)$ and $(-4, 6)$.   [16.5c]

31. Find an equation of the line containing the given point and parallel to the given line:

$$(14, -1); \quad 5x + 7y = 8. \quad [16.5d]$$

32. Find an equation of the line containing the given point and perpendicular to the given line:

$$(5, 2); \quad 3x + y = 5. \quad [16.5d]$$

33. *Records in the 400-Meter Run.*   The following table shows data regarding the Summer Olympics winning times in the men's 400-m run.   [16.5e]

YEAR	SUMMER OLYMPICS WINNING TIME IN MEN'S 400-M RUN (in seconds)
1972	44.66
2012	43.94

a) Use the two data points to find a linear function that fits the data. Let $x =$ the number of years since 1972 and $R(x) =$ the Summer Olympics winning time x years from 1972.

b) Use the function to estimate the winning time in the men's 400-m run in 2000 and in 2010.

34. What is the domain of $f(x) = \dfrac{x + 3}{x - 2}$?   [16.2a]

A. $\{x \mid x \geq -3\}$
B. $\{x \mid x \text{ is a real number } and\, x \neq -3 \, and\, x \neq 2\}$
C. $\{x \mid x \text{ is a real number } and\, x \neq 2\}$
D. $\{x \mid x > -3\}$

35. Find an equation of the line containing the point $(-2, 1)$ and perpendicular to $3y - \frac{1}{2}x = 0$.   [16.5d]

A. $6x + y = -11$ **B.** $y = -\dfrac{1}{6}x - 11$

C. $y = -2x - 3$ **D.** $2x + \dfrac{1}{3} = 0$

Synthesis

36. Homespun Jellies charges $2.49 for each jar of preserves. Shipping charges are $3.75 for handling, plus $0.60 per jar. Find a linear function for determining the cost of buying and shipping x jars of preserves.   [16.5e]

Understanding Through Discussion and Writing

1. Under what conditions will the x-intercept and the y-intercept of a line be the same? What would the equation for such a line look like?   [16.4a]

2. Explain the usefulness of the concept of slope when describing a line.   [16.3b, c], [16.4b], [16.5a, b, c, d]

3. A student makes a mistake when using a graphing calculator to draw $4x + 5y = 12$ and the following screen appears. Use algebra to show that a mistake has been made. What do you think the mistake was?   [16.3b]

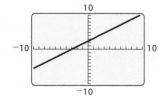

4. *Computer Repair.*   The cost $R(t)$, in dollars, of computer repair at PC Pros is given by

$$R(t) = 50t + 35,$$

where t is the number of hours that the repair requires. Determine m and b in this application and explain their meaning.   [16.5e]

5. Explain why the slope of a vertical line is not defined but the slope of a horizontal line is 0.   [16.4c]

6. A student makes a mistake when using a graphing calculator to draw $5x - 2y = 3$ and the following screen appears. Use algebra to show that a mistake has been made. What do you think the mistake was?   [16.3b]

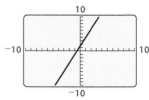

CHAPTER

16 **Test**

For Extra Help For step-by-step test solutions, access the Chapter Test Prep Videos in MyMathLab® or on YouTube (search "BittingerAlgebraFoundations" and click on "Channels").

Determine whether each correspondence is a function.

1.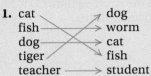
cat → dog
fish → worm
dog → cat
tiger → fish
teacher → student

2. Lake Placid → 1980
Oslo → 1976
Squaw Valley → 1960
Innsbruck → 1952
1932

Find the function values.

3. $f(x) = -3x - 4$; $f(0)$ and $f(-2)$

4. $g(x) = x^2 + 7$; $g(0)$ and $g(-1)$

5. $h(x) = -6$; $h(-4)$ and $h(-6)$

6. $f(x) = |x + 7|$; $f(-10)$ and $f(-7)$

Graph.

7. $h(x) = -2x - 5$

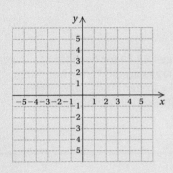

8. $f(x) = -\dfrac{3}{5}x$

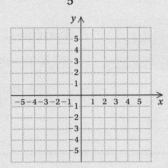

9. $g(x) = 2 - |x|$

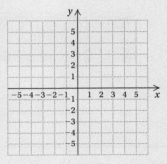

10. $f(x) = x^2 + 2x - 3$

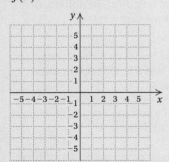

11. $y = f(x) = -3$

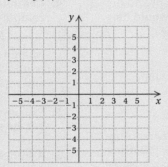

12. $2x = -4$

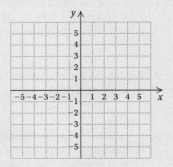

13. *Median Age of Cars.* The function

$$A(t) = 0.233t + 5.87$$

can be used to estimate the median age of cars in the United States t years after 1990. (This means, for example, that if the median age of cars is 3 years, then half the cars are older than 3 years and half are younger.)

Source: The Polk Co.

a) Find the median age of cars in 2005.

b) In what year was the median age of cars 7.734 years?

Determine whether each of the following is the graph of a function.

14.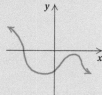

15.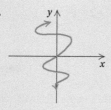

Find the domain.

16. $f(x) = \dfrac{8}{2x + 3}$

17. $g(x) = 5 - x^2$

18. For the following graph of function f, determine **(a)** $f(1)$; **(b)** the domain; **(c)** all x-values such that $f(x) = 2$; and **(d)** the range.

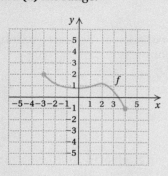

Find the slope and the y-intercept.

19. $f(x) = -\frac{3}{5}x + 12$

20. $-5y - 2x = 7$

Find the slope, if it exists, of the line containing the following points.

21. $(-2, -2)$ and $(6, 3)$

22. $(-3.1, 5.2)$ and $(-4.4, 5.2)$

23. Find the slope, or rate of change, of the graph at right.

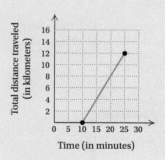

24. Find the intercepts. Then graph the equation.

$$2x + 3y = 6$$

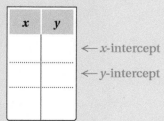

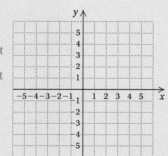

25. Graph using the slope and the y-intercept:

$$f(x) = -\frac{2}{3}x - 1.$$

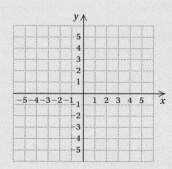

Determine whether the graphs of the given pair of lines are parallel or perpendicular.

26. $4y + 2 = 3x$,
$-3x + 4y = -12$

27. $y = -2x + 5$,
$2y - x = 6$

28. Find an equation of the line that has the given characteristics:

slope: -3; y-intercept: $(0, 4.8)$.

29. Find a linear function $f(x) = mx + b$ whose graph has the given slope and y-intercept:

slope: 5.2; y-intercept: $\left(0, -\frac{5}{8}\right)$.

30. Find an equation of the line having the given slope and containing the given point:

$m = -4$; $(1, -2)$.

31. Find an equation of the line containing the given pair of points:

$(4, -6)$ and $(-10, 15)$.

32. Find an equation of the line containing the given point and parallel to the given line:

$(4, -1)$; $x - 2y = 5$.

33. Find an equation of the line containing the given point and perpendicular to the given line:

$(2, 5)$; $x + 3y = 2$.

34. *Median Age of Men at First Marriage.* The following table lists data regarding the median age of men at first marriage in 1970 and in 2010.

YEAR	MEDIAN AGE OF MEN AT FIRST MARRIAGE
1970	23.2
2010	28.2

Source: U.S. Census Bureau

a) Use the two data points to find a linear function that fits the data. Let $x =$ the number of years since 1970 and $A =$ the median age at first marriage x years from 1970.

b) Use the function to estimate the median age of men at first marriage in 2008 and in 2015.

35. Find an equation of the line having slope -2 and containing the point $(3, 1)$.

A. $y - 1 = 2(x - 3)$ **B.** $y - 1 = -2(x - 3)$
C. $x - 1 = -2(y - 3)$ **D.** $x - 1 = 2(y - 3)$

Synthesis

36. Find k such that the line $3x + ky = 17$ is perpendicular to the line $8x - 5y = 26$.

37. Find a formula for a function f for which $f(-2) = 3$.

CHAPTER
17

Systems of Equations

STUDYING FOR SUCCESS   *Working with Others*

☐ Try being a tutor for a fellow student. You may find that you understand concepts better after explaining them to someone else.

☐ Consider forming a study group.

☐ Verbalize the math. Often simply talking about a concept with a classmate can help clarify it for you.

17.1 Systems of Equations in Two Variables

OBJECTIVE

a Solve a system of two linear equations or two functions by graphing and determine whether a system is consistent or inconsistent and whether the equations in a system are dependent or independent.

SKILL TO REVIEW

Objective 12.2a: Graph linear equations of the type $y = mx + b$ and $Ax + By = C$.

Graph.

1. $x + y = 3$

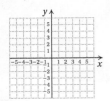

2. $y = x - 2$

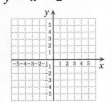

We can solve many applied problems more easily by translating them to two or more equations in two or more variables than by translating to a single equation. Let's look at such a problem.

School Enrollment.   In 2012, approximately 50 million children were enrolled in public elementary and secondary schools in the United States. There were 20 million more students enrolled in prekindergarten–grade 8 than there were in grades 9–12. How many were enrolled at each level?

Source: National Center for Education Statistics

To solve, we first let

$x =$ the number enrolled in prekindergarten–grade 8, and

$y =$ the number enrolled in grades 9–12,

where x and y are in millions of students. The problem gives us two statements that can be translated to equations.

First, we consider the total number enrolled:

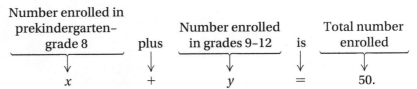

Number enrolled in prekindergarten–grade 8	plus	Number enrolled in grades 9–12	is	Total number enrolled
x	$+$	y	$=$	50.

The second statement of the problem compares the enrollment at the two levels:

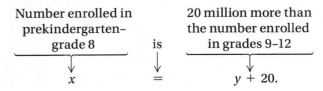

Number enrolled in prekindergarten–grade 8	is	20 million more than the number enrolled in grades 9–12
x	$=$	$y + 20.$

Answers

Answers to Skill Review Exercises 1 and 2 are on p. 1113.

We have now translated the problem to a pair of equations, or a **system of equations**:

$$x + y = 50,$$
$$x = y + 20.$$

A **solution** of a system of two equations in two variables is an ordered pair that makes *both* equations true. If we graph a system of equations, the point at which the graphs intersect will be a solution of *both* equations. To find the solution of the system above, we graph both equations, as shown here.

We see that the graphs intersect at the point $(35, 15)$—that is, $x = 35$ and $y = 15$. These numbers check in the statement of the original problem. This tells us that 35 million students were enrolled in prekindergarten–grade 8, and 15 million students were enrolled in grades 9–12.

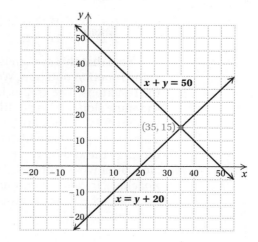

a SOLVING SYSTEMS OF EQUATIONS GRAPHICALLY

One Solution

EXAMPLE 1 Solve this system graphically:

$$y - x = 1,$$
$$y + x = 3.$$

We draw the graph of each equation and find the coordinates of the point of intersection.

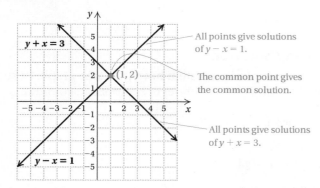

The point of intersection has coordinates that make *both* equations true. The solution seems to be the point $(1, 2)$. However, solving by graphing may give only approximate answers. Thus we check the pair $(1, 2)$ in both equations.

Check:

$y - x = 1$		$y + x = 3$	
$2 - 1 \; ? \; 1$		$2 + 1 \; ? \; 3$	
1	TRUE	3	TRUE

The solution is $(1, 2)$.

Do Exercises 1 and 2. ▶

Solve each system graphically.

1. $-2x + y = 1,$
   $3x + y = 1$

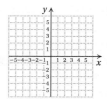

2. $y = \frac{1}{2}x,$
   $y = -\frac{1}{4}x + \frac{3}{2}$

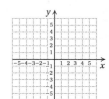

 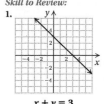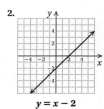

No Solution

Sometimes the equations in a system have graphs that are parallel lines.

EXAMPLE 2 Solve graphically:

$$f(x) = -3x + 5,$$
$$g(x) = -3x - 2.$$

Note that this system is written using function notation. We graph the functions. The graphs have the same slope, -3, and different y-intercepts, so they are parallel. There is no point at which they cross, so the system has no solution. No matter what point we try, it will *not* check in *both* equations. The solution set is thus the empty set, denoted $\varnothing$, or $\{\ \}$.

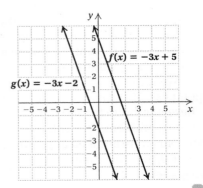

3. Solve graphically:

$$y + 2x = 3,$$
$$y + 2x = -4.$$

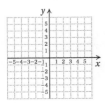

4. Classify each of the systems in Margin Exercises 1–3 as consistent or inconsistent.

The system in Margin Exercise 1 has a solution, so it is [   ] .

The system in Margin Exercise 2 has a solution, so it is [   ] .

The system in Margin Exercise 3 does not have a solution, so it is [   ] .

> ### CONSISTENT SYSTEMS AND INCONSISTENT SYSTEMS
>
> If a system of equations has at least one solution, then it is **consistent**.
> If a system of equations has no solution, then it is **inconsistent**.

The system in Example 1 is consistent. The system in Example 2 is inconsistent.

◀ Do Exercises 3 and 4.

Infinitely Many Solutions

Sometimes the equations in a system have the same graph. In such a case, the equations have an *infinite* number of solutions in common.

EXAMPLE 3 Solve graphically:

$$3y - 2x = 6,$$
$$-12y + 8x = -24.$$

We graph the equations and see that the graphs are the same. Thus any solution of one of the equations is a solution of the other. Each equation has an infinite number of solutions, two of which are shown on the graph.

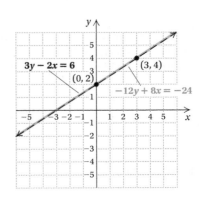

We check one such solution, $(0, 2)$, which is the y-intercept of each equation.

Check:

$$3y - 2x = 6$$
$$3(2) - 2(0) \overset{?}{=} 6$$
$$6 - 0 \quad$$
$$6 \quad | \quad \text{TRUE}$$

$$-12y + 8x = -24$$
$$-12(2) + 8(0) \overset{?}{=} -24$$
$$-24 + 0 \quad$$
$$-24 \quad | \quad \text{TRUE}$$

We leave it to the student to check that $(3, 4)$ is a solution of both equations. If $(0, 2)$ and $(3, 4)$ are solutions, then all points on the line containing them will be solutions. The system has an infinite number of solutions.

DEPENDENT EQUATIONS AND INDEPENDENT EQUATIONS

If for a system of two equations in two variables:

the graphs of the equations are the same line, then the equations are **dependent**.

the graphs of the equations are different lines, then the equations are **independent**.

When we graph a system of two equations, one of the following three things can happen.

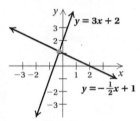

One solution.
Graphs intersect.
The system is consistent
and *the equations are*
independent.

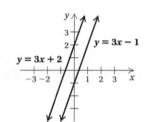

No solution.
Graphs are parallel.
The system is inconsistent
and *the equations are*
independent.

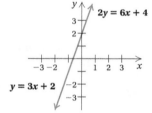

Infinitely many solutions.
Equations have the same
graph. *The system is* consistent
and *the equations are*
dependent.

Let's summarize what we know about the systems of equations shown in Examples 1–3.

	NUMBER OF SOLUTIONS	GRAPHS OF EQUATIONS
EXAMPLE 1	**1** System is consistent.	**Different** Equations are independent.
EXAMPLE 2	**0** System is inconsistent.	**Different** Equations are independent.
EXAMPLE 3	**Infinitely many** System is consistent.	**Same** Equations are dependent.

Do Exercises 5 and 6. ▶

5. Solve graphically:

$$2x - 5y = 10,$$
$$-6x + 15y = -30.$$

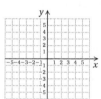

 6. Classify the equations in Margin Exercises 1, 2, 3, and 5 as dependent or independent.

In Margin Exercise 1, the graphs are different, so the equations are ____ .

In Margin Exercise 2, the graphs are different, so the equations are ____ .

In Margin Exercise 3, the graphs are different, so the equations are ____ .

In Margin Exercise 5, the graphs are the same, so the equations are ____ .

Answers

5. Infinitely many solutions
6. Independent: Margin Exercises 1, 2, and 3; dependent: Margin Exercise 5

Guided Solution:
6. Independent, independent, independent, dependent

Consider the equation $-2x + 13 = 4x - 17$. Let's solve it algebraically:

$$-2x + 13 = 4x - 17$$
$$13 = 6x - 17 \qquad \text{Adding } 2x$$
$$30 = 6x \qquad \text{Adding } 17$$
$$5 = x. \qquad \text{Dividing by } 6$$

We can also solve the equation graphically, as we see in the following two methods. Using method 1, we graph two functions. The solution of the original equation is the x-coordinate of the point of intersection. Using method 2, we graph one function. The solution of the original equation is the x-coordinate of the x-intercept of the graph.

Method 1: Solve $-2x + 13 = 4x - 17$ graphically.

We let $f(x) = -2x + 13$ and $g(x) = 4x - 17$. Graphing the system of equations, we get the graph shown below.

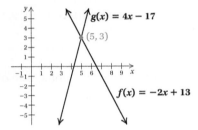

The point of intersection of the two graphs is $(5, 3)$. Note that the x-coordinate of this point is 5. This is the value of x for which $-2x + 13 = 4x - 17$, so it is the solution of the equation.

◀ **Do Exercises 7 and 8.**

Method 2: Solve $-2x + 13 = 4x - 17$ graphically.

Adding $-4x$ and 17 on both sides, we obtain an equation with 0 on one side: $-6x + 30 = 0$. This time we let $f(x) = -6x + 30$ and $g(x) = 0$. Since the graph of $g(x) = 0$, or $y = 0$, is the x-axis, we need only graph $f(x) = -6x + 30$ and see where it crosses the x-axis.

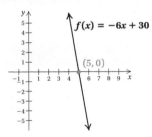

Note that the x-intercept of $f(x) = -6x + 30$ is $(5, 0)$, or just 5. This x-value is the solution of the equation $-2x + 13 = 4x - 17$.

◀ **Do Exercises 9 and 10.**

---

7. a) Solve $x + 1 = \frac{2}{3}x$ algebraically.

  b) Solve $x + 1 = \frac{2}{3}x$ graphically using method 1.

8. Solve $\frac{1}{2}x + 3 = 2$ graphically using method 1.

9. a) Solve $x + 1 = \frac{2}{3}x$ graphically using method 2.

  b) Compare your answers to Margin Exercises 7(a), 7(b), and 9(a).

10. Solve $\frac{1}{2}x + 3 = 2$ graphically using method 2.

Answers

7. (a) -3; **(b)** the same: -3 **8.** -2
9. (a) -3; **(b)** All are -3.   **10.** -2

Solving Systems of Equations   We can solve a system of two equations in two variables using a graphing calculator. Consider the system of equations in Example 1:

$$y - x = 1,$$
$$y + x = 3.$$

First, we solve the equations for y, obtaining $y = x + 1$ and $y = -x + 3$. Next, we enter $y_1 = x + 1$ and $y_2 = -x + 3$ on the equation-editor screen and graph the equations. We can use the standard viewing window, $[-10, 10, -10, 10]$.

We will use the **INTERSECT** feature to find the coordinates of the point of intersection of the lines. To access this feature, we press **2ND** **CALC** **5**. (CALC is the second operation associated with the **TRACE** key.) The query "First curve?" appears on the graph screen. The blinking cursor is positioned on the graph of y_1. We press **ENTER** to indicate that this is the first curve involved in the intersection. Next, the query "Second curve?" appears and the blinking cursor is positioned on the graph of y_2. We press **ENTER** to indicate that this is the second curve. Now the query "Guess?" appears. We use the ▷ and ◁ keys to move the cursor close to the point of intersection or we enter an x-value close to the first coordinate of the point of intersection. Then we press **ENTER**. The coordinates of the point of intersection of the graphs, $x = 1$, $y = 2$, appear at the bottom of the screen. Thus the solution of the system of equations is $(1, 2)$.

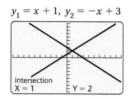

$y_1 = x + 1,\ y_2 = -x + 3$

Intersection
X = 1   Y = 2

EXERCISES: Use a graphing calculator to solve each system of equations.

1. $x + y = 5,$
   $y = x + 1$

2. $y = x + 3,$
   $2x - y = -7$

3. $x - y = -6,$
   $y = 2x + 7$

4. $x + 4y = -1,$
   $x - y = 4$

17.1 | Exercise Set

For Extra Help

MyMathLab®   MathXL®

PRACTICE   WATCH   READ   REVIEW

✓ Reading Check

Determine whether each statement is true or false.

RC1. Every system of equations has one solution.

RC2. A solution of a system of equations in two variables is an ordered pair.

RC3. Graphs of two lines may have one point, no points, or an infinite number of points in common.

RC4. If a system of two equations has only one solution, the system is consistent and the equations in the system are independent.

Solve each system of equations graphically. Then classify the system as consistent or inconsistent and the equations as dependent or independent. Complete the check for Exercises 1–4.

1. $x + y = 4,$
$x - y = 2$

Check: $x + y = 4$
_____
?
|

$x - y = 2$
_____
?
|

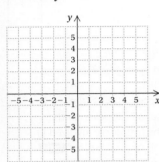

2. $x - y = 3,$
$x + y = 5$

Check: $x - y = 3$
_____
?
|

$x + y = 5$
_____
?
|

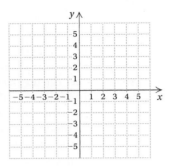

3. $2x - y = 4,$
$2x + 3y = -4$

Check: $2x - y = 4$
_____
?
|

$2x + 3y = -4$
_____
?
|

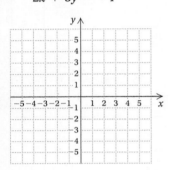

4. $3x + y = 5,$
$x - 2y = 4$

Check: $3x + y = 5$
_____
?
|

$x - 2y = 4$
_____
?
|

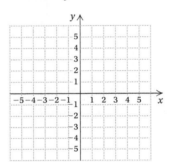

5. $2x + y = 6,$
$3x + 4y = 4$

6. $2y = 6 - x,$
$3x - 2y = 6$

7. $f(x) = x - 1,$
$g(x) = -2x + 5$

8. $f(x) = x + 1,$
$g(x) = \frac{2}{3}x$

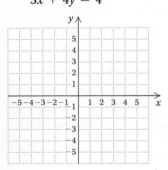
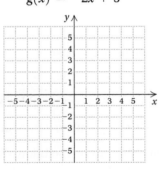
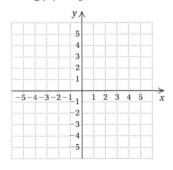

9. $2u + v = 3$,
$\quad 2u = v + 7$

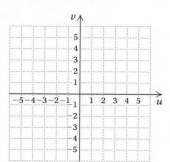

10. $2b + a = 11$,
$\quad a - b = 5$

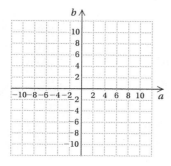

11. $f(x) = -\frac{1}{3}x - 1$,
$\quad g(x) = \frac{4}{3}x - 6$

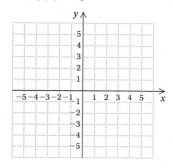

12. $f(x) = -\frac{1}{4}x + 1$,
$\quad g(x) = \frac{1}{2}x - 2$

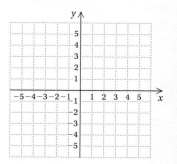

13. $6x - 2y = 2$,
$\quad 9x - 3y = 1$

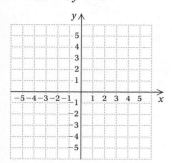

14. $y - x = 5$,
$\quad 2x - 2y = 10$

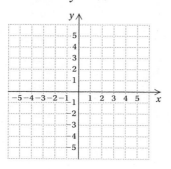

15. $2x - 3y = 6$,
$\quad 3y - 2x = -6$

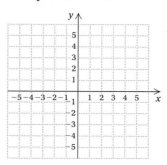

16. $y = 3 - x$,
$\quad 2x + 2y = 6$

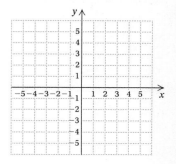

17. $x = 4$,
$\quad y = -5$

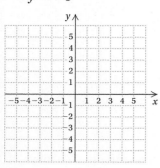

18. $x = -3$,
$\quad y = 2$

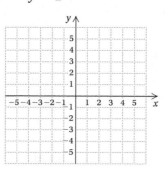

19. $y = -x - 1$,
$\quad 4x - 3y = 17$

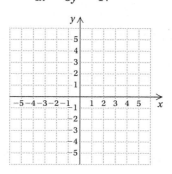

20. $a + 2b = -3$,
$\quad b - a = 6$

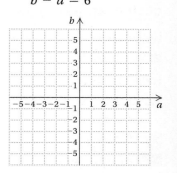

Matching. Each of Exercises 21–26 shows the graph of a system of equations and its solution. First, classify the system as consistent or inconsistent and the equations as dependent or independent. Then match it with one of the appropriate systems of equations (A)–(F), which follow.

21. Solution: $(3, 3)$

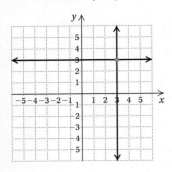

22. Solution: $(1, 1)$

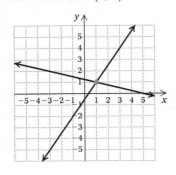

23. Solutions: Infinitely many

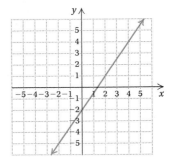

24. Solution: $(4, -3)$

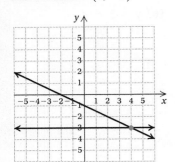

25. Solution: No solution

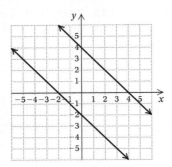

26. Solution: $(-1, 3)$

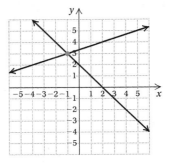

A. $3y - x = 10,$
$\quad x = -y + 2$

B. $9x - 6y = 12,$
$\quad y = \frac{3}{2}x - 2$

C. $2y - 3x = -1,$
$\quad x + 4y = 5$

D. $x + y = 4,$
$\quad y = -x - 2$

E. $\frac{1}{2}x + y = -1,$
$\quad y = -3$

F. $x = 3,$
$\quad y = 3$

Skill Maintenance

Write an equation of the line containing the given point and parallel to the given line. [16.5d]

27. $(-4, 2);\ 3x = 5y - 4$

28. $(-6, 0);\ 8y - 3x = 2$

Write an equation of the line containing the given point and perpendicular to the given line. [16.5d]

29. $(-4, 6);\ 2x = 3y - 12$

30. $(3, -10);\ 8y - 4 = -6x$

Synthesis

Use a graphing calculator to solve each system of equations. Round all answers to the nearest hundredth. You may need to solve for y first.

31. $2.18x + 7.81y = 13.78,$
$\quad 5.79x - 3.45y = 8.94$

32. $f(x) = 123.52x + 89.32,$
$\quad g(x) = -89.22x + 33.76$

Solve graphically.

33. $y = |x|,$
$\quad x + 4y = 15$

34. $x - y = 0,$
$\quad y = x^2$

Solving by Substitution

Consider this system of equations:

$$5x + 9y = 2,$$
$$4x - 9y = 10.$$

What is the solution? It is rather difficult to tell exactly by graphing. It would appear that fractions are involved. It turns out that the solution is

$$\left(\frac{4}{3}, -\frac{14}{27} \right).$$

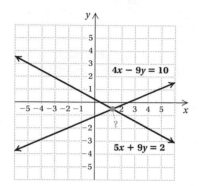

Solving by graphing, though useful in many applied situations, is not always fast or accurate in cases where solutions are not integers. We need techniques involving algebra to determine the solution exactly. Because these techniques use algebra, they are called **algebraic methods**.

a THE SUBSTITUTION METHOD

One nongraphical method for solving systems is known as the **substitution method**.

EXAMPLE 1 Solve this system:

$$x + y = 4, \qquad \textbf{(1)}$$
$$x = y + 1. \qquad \textbf{(2)}$$

Equation (2) says that x and $y + 1$ name the same number. Thus we can substitute $y + 1$ for x in equation (1):

$$x + y = 4 \qquad \text{Equation (1)}$$
$$(y + 1) + y = 4. \qquad \text{Substituting } y + 1 \text{ for } x$$

Since this equation has only one variable, we can solve for y using methods learned earlier:

$$(y + 1) + y = 4$$
$$2y + 1 = 4 \qquad \text{Removing parentheses and collecting like terms}$$
$$2y = 3 \qquad \text{Subtracting 1}$$
$$y = \tfrac{3}{2}. \qquad \text{Dividing by 2}$$

We return to the original pair of equations and substitute $\frac{3}{2}$ for y in *either* equation so that we can solve for x. Calculation will be easier if we choose equation (2) since it is already solved for x:

$$x = y + 1 \qquad \text{Equation (2)}$$
$$= \tfrac{3}{2} + 1 \qquad \text{Substituting } \tfrac{3}{2} \text{ for } y$$
$$= \tfrac{3}{2} + \tfrac{2}{2} = \tfrac{5}{2}.$$

We obtain the ordered pair $\left(\frac{5}{2}, \frac{3}{2} \right)$. Even though we solved for y *first*, it is still the *second* coordinate since x is before y alphabetically. We check to be sure that the ordered pair is a solution.

Answers

Skill to Review:
1. 1   **2.** −3

Solve by the substitution method.

1. $x + y = 6,$
   $y = x + 2$

2. $y = 7 - x,$
   $2x - y = 8$

   (*Caution*: Use parentheses when you substitute, being careful about removing them. Remember to solve for both variables.)

Solve by the substitution method.

3. $2y + x = 1,$
   $y - 2x = 8$

4. $8x - 5y = 12,$ **(1)**
   $x - y = 3$ **(2)**   **GS**

Solve for x in equation (2):

$x - y = 3$

$x = \boxed{} + 3.$ **(3)**

Substitute $y + 3$ for $\boxed{}$ in equation (1) and solve for $\boxed{}$:

$8x - 5y = 12$

$8(y + 3) - 5y = 12$

$8y + \boxed{} - 5y = 12$

$3y + 24 = 12$

$3y = -12$

$y = \boxed{}.$

Substitute -4 for y in equation (3) and solve for x:

$x = y + 3$

$ = -4 + 3$

$ = \boxed{}.$

The ordered pair checks in both equations. The solution is $(\boxed{}, \boxed{}).$

Check:
$$\begin{array}{c|c} x + y = 4 & x = y + 1 \\ \hline \frac{5}{2} + \frac{3}{2} \ ? \ 4 & \frac{5}{2} \ ? \ \frac{3}{2} + 1 \\ \frac{8}{2} & \frac{3}{2} + \frac{2}{2} \\ 4 \quad \text{TRUE} & \frac{5}{2} \quad \text{TRUE} \end{array}$$

Since $\left(\frac{5}{2}, \frac{3}{2}\right)$ checks, it is the solution. Even though exact fraction solutions are difficult to determine graphically, a graph can help us to visualize whether the solution is reasonable.

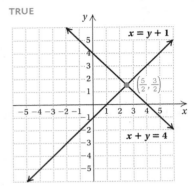

◀ **Do Exercises 1 and 2.**

Suppose neither equation of a pair has a variable alone on one side. We then solve one equation for one of the variables.

EXAMPLE 2   Solve this system:

$2x + y = 6,$ (1)

$3x + 4y = 4.$ (2)

First, we solve one equation for one variable. Since the coefficient of y is 1 in equation (1), it is the easier one to solve for y:

$y = 6 - 2x.$ (3)

Next, we substitute $6 - 2x$ for y in equation (2) and solve for x:

$3x + 4(6 - 2x) = 4$ Substituting $6 - 2x$ for y

···· **Caution!** ····

Remember to use parentheses when you substitute. Then remove them properly.

$3x + 24 - 8x = 4$ Multiplying to remove parentheses

$24 - 5x = 4$ Collecting like terms

$-5x = -20$ Subtracting 24

$x = 4.$ Dividing by -5

In order to find y, we return to either of the original equations, (1) or (2), or equation (3), which we solved for y. It is generally easier to use an equation like (3), where we have solved for the specific variable. We substitute 4 for x in equation (3) and solve for y:

$y = 6 - 2x = 6 - 2(4) = 6 - 8 = -2.$

We obtain the ordered pair $(4, -2)$.

Check:
$$\begin{array}{c|c} 2x + y = 6 & 3x + 4y = 4 \\ \hline 2(4) + (-2) \ ? \ 6 & 3(4) + 4(-2) \ ? \ 4 \\ 8 - 2 & 12 - 8 \\ 6 \quad \text{TRUE} & 4 \quad \text{TRUE} \end{array}$$

Since $(4, -2)$ checks, it is the solution.

◀ **Do Exercises 3 and 4.**

EXAMPLE 3 Solve this system of equations:

$$y = -3x + 5, \qquad (1)$$
$$y = -3x - 2. \qquad (2)$$

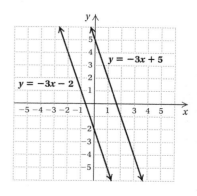

The graphs of the equations in the system are shown at right. Since the graphs are parallel, there is no solution. Let's try to solve this system algebraically using substitution. We substitute $-3x - 2$ for y in equation (1):

$$-3x - 2 = -3x + 5 \qquad \text{Substituting } -3x - 2 \text{ for } y$$
$$-2 = 5. \qquad \text{Adding } 3x$$

We have a false equation. The equation has no solution. This means that the system has **no solution**.

Do Exercise 5. ▶

5. Solve using substitution:

$$y + 2x = 3,$$
$$y + 2x = -4.$$

EXAMPLE 4 Solve this system of equations:

$$x = 2y - 1, \qquad (1)$$
$$4y - 2x = 2. \qquad (2)$$

The graphs of the equations in the system are shown at right. Since the graphs are the same, there is an infinite number of solutions.

Let's try to solve this system algebraically using substitution. We substitute $2y - 1$ for x in equation (2):

$$4y - 2(2y - 1) = 2 \qquad \text{Substituting } 2y - 1 \text{ for } x$$
$$4y - 4y + 2 = 2 \qquad \text{Removing parentheses}$$
$$2 = 2. \qquad \text{Simplifying; } 4y - 4y = 0$$

We have a true equation. Any value of y will make this equation true. This means that the system has **infinitely many solutions**.

Do Exercise 6. ▶

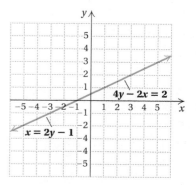

SPECIAL CASES

When solving a system of two linear equations in two variables:

1. If a false equation is obtained, then the system has no solution.
2. If a true equation is obtained, then the system has an infinite number of solutions.

6. Solve using substitution:

$$x - 3y = 1,$$
$$6y - 2x = -2.$$

b SOLVING APPLIED PROBLEMS INVOLVING TWO EQUATIONS

Many applied problems are easier to solve if we first translate to a system of two equations rather than to a single equation.

EXAMPLE 5 *Architecture.* The architects who designed the John Hancock Building in Chicago created a visually appealing building that slants on the sides. The ground floor is in the shape of a rectangle that is larger than the rectangle formed by the top floor. The ground floor has a perimeter of 860 ft. The length is 100 ft more than the width. Find the length and the width.

1. **Familiarize.** We first make a drawing and label it, using l for length and w for width. We recall or look up the formula for perimeter: $P = 2l + 2w$. This formula can be found at the back of this book.

2. **Translate.** We translate as follows:

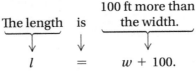

The perimeter is 860 ft.

$$2l + 2w = 860.$$

We can also write a second equation:

The length is 100 ft more than the width.

$$l = w + 100.$$

We now have a system of equations:

$$2l + 2w = 860, \quad (1)$$
$$l = w + 100. \quad (2)$$

3. **Solve.** We substitute $w + 100$ for l in equation (1) and solve for w:

$$2(w + 100) + 2w = 860 \qquad \text{Substituting in equation (1)}$$
$$2w + 200 + 2w = 860 \qquad \text{Multiplying to remove parentheses}$$
$$4w + 200 = 860 \qquad \text{Collecting like terms}$$
$$\left.\begin{array}{r} 4w = 660 \\ w = 165. \end{array}\right\} \text{Solving for } w$$

Next, we substitute 165 for w in equation (2) and solve for l:

$$l = 165 + 100 = 265.$$

4. **Check.** Consider the dimensions 265 ft and 165 ft. The length is 100 ft more than the width. The perimeter is $2(265 \text{ ft}) + 2(165 \text{ ft})$, or 860 ft. The dimensions 265 ft and 165 ft check in the original problem.

5. **State.** The length is 265 ft, and the width is 165 ft.

◀ Do Exercise 7.

7. *Architecture.* The top floor of the John Hancock Building is also in the shape of a rectangle, but its perimeter is 520 ft. The width is 60 ft less than the length. Find the length and the width.

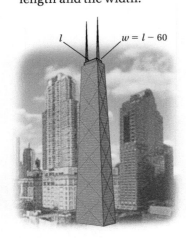

Answer

7. Length: 160 ft; width: 100 ft

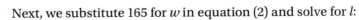

 17.2 | **Exercise Set**

For Extra Help **MyMathLab®** MathXL® PRACTICE WATCH READ REVIEW

✓ Reading Check

Determine whether each statement is true or false.

RC1. The substitution method is an algebraic method for solving systems of equations.

RC2. We can find solutions of systems of equations involving fractions using the substitution method.

RC3. When we are writing the solution of a system, the value that we found first is always the first number in the ordered pair.

RC4. When solving using substitution, if we obtain a false equation, then the system has many solutions.

a Solve each system of equations by the substitution method.

1. $y = 5 - 4x$,
$2x - 3y = 13$

2. $x = 8 - 4y$,
$3x + 5y = 3$

3. $2y + x = 9$,
$x = 3y - 3$

4. $9x - 2y = 3$,
$3x - 6 = y$

5. $3s - 4t = 14$,
$5s + \ t = 8$

6. $m - 2n = 3$,
$4m + \ n = 1$

7. $9x - 2y = -6$,
$7x + 8 = y$

8. $t = 4 - 2s$,
$t + 2s = 6$

9. $-5s + \ \ t = 11$,
$4s + 12t = 4$

10. $\ \ 5x + 6y = 14$,
$-3y + \ x = 7$

11. $2x - 3 = y$,
$y - 2x = 1$

12. $4p - 2w = 16$,
$5p + 7w = 1$

13. $3a - \ b = 7$,
$2a + 2b = 5$

14. $3x + y = 4$,
$12 - 3y = 9x$

15. $2x - 6y = 4$,
$3y + 2 = x$

16. $5x + 3y = 4$,
$x - 4y = 3$

17. $2x + 2y = 2$,
$3x - \ y = 1$

18. $\ \ 4x + 13y = 5$,
$-6x + \ \ y = 13$

b Solve.

19. *Archaeology.* The remains of an ancient ball court in Monte Alban, Mexico, include a rectangular playing alley with a perimeter of about 60 m. The length of the alley is five times the width. Find the length and the width of the playing alley.

20. *Soccer Field.* The perimeter of a soccer field is 340 m. The length exceeds the width by 50 m. Find the length and the width.

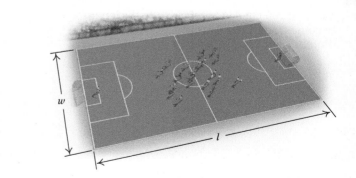

21. *Supplementary Angles.* **Supplementary angles** are angles whose sum is 180°. Two supplementary angles are such that one angle is 12° less than three times the other. Find the measures of the angles.

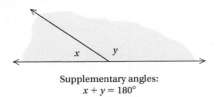

Supplementary angles:
$x + y = 180°$

22. *Complementary Angles.* **Complementary angles** are angles whose sum is 90°. Two complementary angles are such that one angle is 6° more than five times the other. Find the measures of the angles.

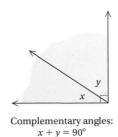

Complementary angles:
$x + y = 90°$

23. *Hockey Points.* At one time, hockey teams received two points when they won a game and one point when they tied. One season, a team won a championship with 60 points. They won 9 more games than they tied. How many wins and how many ties did the team have?

24. *Airplane Seating.* An airplane has a total of 152 seats. The number of coach-class seats is 5 more than six times the number of first-class seats. How many of each type of seat are there on the plane?

Skill Maintenance

25. Find the slope of the line $y = 1.3x - 7$. [16.3b]

26. Simplify: $-9(y + 7) - 6(y - 4)$. [10.8b]

27. Solve $A = \dfrac{pq}{7}$ for p. [11.4b]

28. Find the slope of the line containing the points $(-2, 3)$ and $(-5, -4)$. [16.3b]

Solve. [11.3c]

29. $-4x + 5(x - 7) = 8x - 6(x + 2)$

30. $-12(2x - 3) = 16(4x - 5)$

Synthesis

31. Two solutions of $y = mx + b$ are $(1, 2)$ and $(-3, 4)$. Find m and b.

32. Solve for x and y in terms of a and b:
$$5x + 2y = a,$$
$$x - y = b.$$

33. *Design.* A piece of posterboard has a perimeter of 156 in. If you cut 6 in. off the width, the length becomes four times the width. What are the dimensions of the original piece of posterboard?

$P = 156$ in.

34. *Nontoxic Scouring Powder.* A nontoxic scouring powder is made up of 4 parts baking soda and 1 part vinegar. How much of each ingredient is needed for a 16-oz mixture?

Solving by Elimination

a THE ELIMINATION METHOD

The **elimination method** for solving systems of equations makes use of the *addition principle* for equations. Some systems are much easier to solve using the elimination method rather than the substitution method.

EXAMPLE 1 Solve this system:

$$2x - 3y = 0, \qquad (1)$$
$$-4x + 3y = -1. \qquad (2)$$

The key to the advantage of the elimination method in this case is the $-3y$ in one equation and the $3y$ in the other. These terms are opposites. If we add them, these terms will add to 0, and in effect, the variable y will have been "eliminated."

We will use the addition principle for equations, adding the same number on both sides of the equation. According to equation (2), $-4x + 3y$ and -1 are the same number. Thus we can use a vertical form and add $-4x + 3y$ on the left side of equation (1) and -1 on the right side:

$$\begin{array}{ll} 2x - 3y = 0 & (1) \\ \underline{-4x + 3y = -1} & (2) \\ -2x + 0y = -1 & \text{Adding} \\ -2x + 0 = -1 \\ -2x = -1. \end{array}$$

We have eliminated the variable y, which is why we call this the *elimination method.** We now have an equation with just one variable, which we solve for x:

$$-2x = -1$$
$$x = \tfrac{1}{2}.$$

Next, we substitute $\tfrac{1}{2}$ for x in either equation and solve for y:

$$\begin{array}{ll} 2 \cdot \tfrac{1}{2} - 3y = 0 & \text{Substituting in equation (1)} \\ 1 - 3y = 0 \\ -3y = -1 & \text{Subtracting 1} \\ y = \tfrac{1}{3}. & \text{Dividing by } -3 \end{array}$$

We obtain the ordered pair $\left(\tfrac{1}{2}, \tfrac{1}{3}\right)$.

Check:

$$\begin{array}{c|c} 2x - 3y = 0 & -4x + 3y = -1 \\ \hline 2\left(\tfrac{1}{2}\right) - 3\left(\tfrac{1}{3}\right) \; ? \; 0 & -4\left(\tfrac{1}{2}\right) + 3\left(\tfrac{1}{3}\right) \; ? \; -1 \\ 1 - 1 & -2 + 1 \\ 0 \quad \text{TRUE} & -1 \quad \text{TRUE} \end{array}$$

Since $\left(\tfrac{1}{2}, \tfrac{1}{3}\right)$ checks, it is the solution. We can also see this in the graph shown at right.

Do Margin Exercises 1 and 2. ▶

_____

* This method is also called the *addition method.*

OBJECTIVES

a Solve systems of equations in two variables by the elimination method.

b Solve applied problems by solving systems of two equations using elimination.

SKILL TO REVIEW

Objective 11.3b: Solve equations by first clearing the equations of fractions or decimals.

Solve. Clear fractions or decimals first.

1. $4.2x - 10.4 = 45.4 - 5.1x$

2. $\tfrac{1}{4}x - \tfrac{2}{5} + \tfrac{1}{2}x = \tfrac{3}{5} + x$

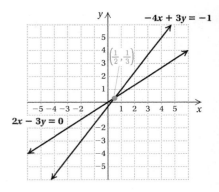

Solve by the elimination method.

1. $5x + 3y = 17,$
$\; -5x + 2y = 3$

2. $-3a + 2b = 0,$
$\; 3a - 4b = -1$

Answers

Skill to Review:
1. 6   2. -4

Margin Exercises:
1. $(1, 4)$ 2. $\left(\tfrac{1}{3}, \tfrac{1}{2}\right)$

3. Solve by the elimination method:

$$2y + 3x = 12, \qquad \textbf{(1)}$$
$$-4y + 5x = -2. \qquad \textbf{(2)}$$

Multiply by 2 on both sides of equation (1) and add:

$$\begin{array}{r} 4y + 6x = 24 \\ \underline{-4y + 5x = -2} \\ 0 + \boxed{} = \boxed{} \\ 11x = 22 \\ x = \boxed{}. \end{array}$$

Substitute $\boxed{}$ for x in equation (1) and solve for y:

$$2y + 3x = 12$$
$$2y + 3(\boxed{}) = 12$$
$$2y + 6 = 12$$
$$2y = \boxed{}$$
$$y = \boxed{}.$$

The ordered pair checks in both equations, so the solution is ($\boxed{}$, $\boxed{}$).

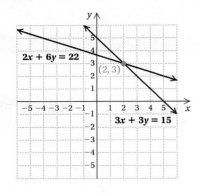

In order to eliminate a variable, we sometimes use the multiplication principle to multiply one or both of the equations by a particular number before adding.

EXAMPLE 2 Solve this system:

$$3x + 3y = 15, \qquad \textbf{(1)}$$
$$2x + 6y = 22. \qquad \textbf{(2)}$$

If we add directly, we will not eliminate a variable. However, note that if the $3y$ in equation (1) were $-6y$, we could eliminate y. Thus we multiply by -2 on both sides of equation (1) and add:

$$\begin{array}{rl} -6x - 6y = -30 & \text{Multiplying by } -2 \text{ on both sides of equation (1)} \\ \underline{2x + 6y = 22} & \text{Equation (2)} \\ -4x + 0 = -8 & \text{Adding} \\ -4x = -8 & \\ x = 2. & \text{Solving for } x \end{array}$$

Then

$$\begin{array}{rl} 2 \cdot 2 + 6y = 22 & \text{Substituting 2 for } x \text{ in equation (2)} \\ 4 + 6y = 22 \\ 6y = 18 \\ y = 3. \end{array} \Bigg\} \;\; \text{Solving for } y$$

We obtain $(2, 3)$, or $x = 2$, $y = 3$.

Check:

$$\begin{array}{c|c} 3x + 3y = 15 & 2x + 6y = 22 \\ \hline 3(2) + 3(3) \;?\; 15 & 2(2) + 6(3) \;?\; 22 \\ 6 + 9 & 4 + 18 \\ 15 \;\bigm|\; \text{TRUE} & 22 \;\bigm|\; \text{TRUE} \end{array}$$

Since $(2, 3)$ checks, it is the solution. We can also see this in the graph at left.

◀ **Do Exercise 3.**

Sometimes we must multiply twice in order to make two terms opposites.

EXAMPLE 3 Solve this system:

$$2x + 3y = 17, \qquad \textbf{(1)}$$
$$5x + 7y = 29. \qquad \textbf{(2)}$$

We must first multiply in order to make one pair of terms with the same variable opposites. We decide to do this with the x-terms in each equation. We multiply equation (1) by 5 and equation (2) by -2. Then we get $10x$ and $-10x$, which are opposites.

$$\begin{array}{llrl} \textit{From equation } (1): & 10x + 15y = & 85 & \text{Multiplying by 5} \\ \textit{From equation } (2): & \underline{-10x - 14y = -58} & & \text{Multiplying by } -2 \\ & 0 + y = & 27 & \text{Adding} \\ & y = & 27 & \text{Solving for } y \end{array}$$

Answer

3. $(2, 3)$

Guided Solution:

3. $11x$, 22, 2, 2, 2, 6, 3, 2, 3

Then

$$2x + 3 \cdot 27 = 17 \qquad \text{Substituting 27 for } y \text{ in equation (1)}$$

$$\left.\begin{array}{r} 2x + 81 = 17 \\ 2x = -64 \\ x = -32. \end{array}\right\} \quad \text{Solving for } x$$

We check the ordered pair $(-32, 27)$.

Check:

$$\begin{array}{c|c} 2x + 3y = 17 & \\ \hline 2(-32) + 3(27) \ ? \ 17 & \\ -64 + 81 & \\ 17 & \text{TRUE} \end{array}$$

$$\begin{array}{c|c} 5x + 7y = 29 & \\ \hline 5(-32) + 7(27) \ ? \ 29 & \\ -160 + 189 & \\ 29 & \text{TRUE} \end{array}$$

We obtain $(-32, 27)$, or $x = -32$, $y = 27$, as the solution.

Do Exercises 4 and 5. ▶

When solving a system of equations using the elimination method, it helps to first write the equations in the form $Ax + By = C$. When decimals or fractions occur, it also helps to *clear* before solving.

EXAMPLE 4 Solve this system:

$$0.2x + 0.3y = 1.7,$$
$$\tfrac{1}{7}x + \tfrac{1}{5}y = \tfrac{29}{35}.$$

We have

$$0.2x + 0.3y = 1.7, \xrightarrow{\substack{\text{Multiplying by 10} \\ \text{to clear decimals}}} 2x + 3y = 17,$$

$$\tfrac{1}{7}x + \tfrac{1}{5}y = \tfrac{29}{35} \xrightarrow{\substack{\text{Multiplying by 35} \\ \text{to clear fractions}}} 5x + 7y = 29.$$

We multiplied by 10 to clear the decimals. Multiplication by 35, the least common denominator, clears the fractions. The problem is now identical to Example 3. The solution is $(-32, 27)$, or $x = -32$, $y = 27$.

Do Exercises 6 and 7. ▶

To use the elimination method to solve systems of two equations:

1. Write both equations in the form $Ax + By = C$.
2. Clear any decimals or fractions.
3. Choose a variable to eliminate.
4. Make the chosen variable's terms opposites by multiplying one or both equations by appropriate numbers if necessary.
5. Eliminate a variable by adding the respective sides of the equations and then solve for the remaining variable.
6. Substitute in either of the original equations to find the value of the other variable.

Solve by the elimination method.

4. $4x + 5y = -8,$
   $7x + 9y = 11$

5. $4x - 5y = 38,$
   $7x - 8y = -22$

6. Clear the decimals. Then solve.
   $$0.02x + 0.03y = 0.01,$$
   $$0.3x - 0.1y = 0.7$$
   (*Hint*: Multiply the first equation by 100 and the second one by 10.)

7. Clear the fractions. Then solve.
   $$\frac{3}{5}x + \frac{2}{3}y = \frac{1}{3},$$
   $$\frac{3}{4}x - \frac{1}{3}y = \frac{1}{4}$$

Answers

4. $(-127, 100)$ 5. $(-138, -118)$
6. $2x + 3y = 1,$
   $3x - y = 7; (2, -1)$
7. $9x + 10y = 5,$
   $9x - 4y = 3; \left(\dfrac{25}{63}, \dfrac{1}{7}\right)$

Some systems have no solution. How do we recognize such systems if we are solving using elimination?

EXAMPLE 5 Solve this system:

$$y + 3x = 5, \qquad (1)$$
$$y + 3x = -2. \qquad (2)$$

If we find the slope–intercept equations for this system, we get

$$y = -3x + 5,$$
$$y = -3x - 2.$$

The graphs are parallel lines.
The system has no solution.

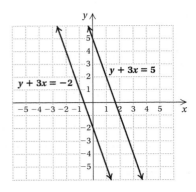

Let's attempt to solve the system by the elimination method:

$$y + 3x = 5 \qquad \text{Equation (1)}$$
$$\underline{-y - 3x = 2} \qquad \text{Multiplying equation (2) by } -1$$
$$0 = 7. \qquad \text{Adding, we obtain a false equation.}$$

The x-terms and the y-terms are eliminated and we have a *false* equation. If we obtain a false equation, such as $0 = 7$, when solving algebraically, we know that the system has **no solution**. The system is inconsistent, and the equations are independent.

◀ Do Exercise 8.

Some systems have infinitely many solutions. How can we recognize such a situation when we are solving systems using an algebraic method?

EXAMPLE 6 Solve this system:

$$3y - 2x = 6, \qquad (1)$$
$$-12y + 8x = -24. \qquad (2)$$

The graphs are the same line. The system has an infinite number of solutions.

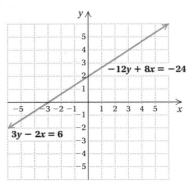

---

8. Solve by the elimination method:

$$y + 2x = 3,$$
$$y + 2x = -1.$$

Multiply the second equation by -1 and add:

$$y + 2x = 3$$
$$\underline{-y - 2x = 1}$$
$$0 = \boxed{}.$$

The equation is _____, so
$$\text{true/false}$$
the system has no solution.

Suppose we try to solve this system by the elimination method:

$$12y - 8x = 24 \qquad \text{Multiplying equation (1) by 4}$$
$$\underline{-12y + 8x = -24} \qquad \text{Equation (2)}$$
$$0 = 0. \qquad \text{Adding, we obtain a true equation.}$$

We have eliminated both variables, and what remains is a true equation, $0 = 0$. It can be expressed as $0 \cdot x + 0 \cdot y = 0$, and is true for all numbers x and y. If an ordered pair is a solution of one of the original equations, then it will be a solution of the other. The system has an infinite number of solutions. The system is consistent, and the equations are dependent.

SPECIAL CASES

When solving a system of two linear equations in two variables:

1. If a false equation is obtained, such as $0 = 7$, then the system has no solution. The system is *inconsistent*, and the equations are *independent*.

2. If a true equation is obtained, such as $0 = 0$, then the system has an infinite number of solutions. The system is *consistent*, and the equations are *dependent*.

Do Exercise 9. ▶

9. Solve by the elimination method:
$$2x - 5y = 10,$$
$$-6x + 15y = -30.$$

Comparing Methods

We can solve systems of equations graphically, or we can solve them algebraically using substitution or elimination. When deciding which method to use, consider the information in this table as well as directions from your instructor.

METHOD	STRENGTHS	WEAKNESSES
Graphical	Can "see" solutions.	Inexact when solutions involve numbers that are not integers. Solutions may not appear on the part of the graph drawn.
Substitution	Yields exact solutions. Convenient to use when a variable has a coefficient of 1.	Can introduce extensive computations with fractions. Cannot "see" solutions quickly.
Elimination	Yields exact solutions. Convenient to use when no variable has a coefficient of 1. The preferred method for systems of three or more equations in three or more variables. (See Section 17.5.)	Cannot "see" solutions quickly.

Answer

9. Infinitely many solutions

b SOLVING APPLIED PROBLEMS USING ELIMINATION

Let's now solve an applied problem using the elimination method.

EXAMPLE 7 *Stimulating the Hometown Economy.* To stimulate the economy in his town of Brewton, Alabama, in 2009, Danny Cottrell, co-owner of The Medical Center Pharmacy, gave each of his full-time employees $700 and each part-time employee $300. He asked that each person donate 15% to a charity of his or her choice and spend the rest locally. The money was paid in $2 bills, a rarely used currency, so that the business community could easily see how the money circulated. Cottrell gave away a total of $16,000 to his 24 employees. How many full-time employees and how many part-time employees were there?

Source: *The Press-Register,* March 4, 2009

1. **Familiarize.** We let f = the number of full-time employees and p = the number of part-time employees. Each full-time employee received $700, so a total of $700f$ was paid to them. Similarly, the part-time employees received a total of $300p$. Thus a total of $700f + 300p$ was given away.

2. **Translate.** We translate to two equations.

$$\underbrace{\text{Total amount given away}}_{700f + 300p} \quad \text{is} \quad \underset{16{,}000}{\$16{,}000.}$$

$$\underbrace{\text{Total number of employees}}_{f + p} \quad \text{is} \quad \underset{24}{24.}$$

We now have a system of equations:

$$700f + 300p = 16{,}000, \quad \textbf{(1)}$$
$$f + p = 24. \quad \textbf{(2)}$$

3. **Solve.** First, we multiply by -300 on both sides of equation (2) and add:

$700f + 300p = 16{,}000$	Equation (1)
$\underline{-300f - 300p = -7200}$	Multiplying by -300 on both sides of equation (2)
$400f \qquad\quad = 8800$	Adding
$f = 22.$	Solving for f

Next, we substitute 22 for f in equation (2) and solve for p:

$$22 + p = 24$$
$$p = 2.$$

4. **Check.** If there are 22 full-time employees and 2 part-time employees, there is a total of $22 + 2$, or 24, employees. The 22 full-time employees received a total of $700 \cdot 22$, or $15,400, and the 2 part-time employees received a total of $300 \cdot 2$, or $600. Then a total of $15,400 + $600, or $16,000, was given away. The numbers check in the original problem.

5. **State.** There were 22 full-time employees and 2 part-time employees.

◀ **Do Exercise 10.**

10. *Bonuses.* Monica gave each of the full-time employees in her small business a year-end bonus of $500 while each part-time employee received $250. She gave a total of $4000 in bonuses to her 10 employees. How many full-time employees and how many part-time employees did Monica have?

Answer

10. Full-time: 6; part-time: 4

✓ Reading Check

Choose from the column on the right the word that best completes each sentence. Words may be used more than once.

RC1. If a system of equations has a solution, then it is _____.

RC2. If a system of equations has no solution, then it is _____.

RC3. If a system of equations has infinitely many solutions, then it is _____.

RC4. If the graphs of the equations in a system of two equations in two variables are the same line, then the equations are _____.

RC5. If the graphs of the equations in a system of two equations in two variables are parallel, then the system is _____.

RC6. If the graphs of the equations in a system of two equations in two variables intersect at one point, then the equations are _____.

consistent

inconsistent

dependent

independent

a   Solve each system of equations using the elimination method.

1. $x + 3y = 7,$
$-x + 4y = 7$

2. $x + y = 9,$
$2x - y = -3$

3. $9x + 5y = 6,$
$2x - 5y = -17$

4. $2x - 3y = 18,$
$2x + 3y = -6$

5. $5x + 3y = -11,$
$3x - y = -1$

6. $2x + 3y = -9,$
$5x - 6y = -9$

7. $5r - 3s = 19,$
$2r - 6s = -2$

8. $2a + 3b = 11,$
$4a - 5b = -11$

9. $2x + 3y = 1,$
$4x + 6y = 2$

10. $3x - 2y = 1,$
$-6x + 4y = -2$

11. $5x - 9y = 7,$
$7y - 3x = -5$

12. $5x + 4y = 2,$
$2x - 8y = 4$

13. $3x + 2y = 24,$
$2x + 3y = 26$

14. $5x + 3y = 25,$
$3x + 4y = 26$

15. $2x - 4y = 5,$
$2x - 4y = 6$

16. $3x - 5y = -2,$
$5y - 3x = 7$

17. $2a + b = 12,$
$\quad a + 2b = -6$

18. $10x + y = 306,$
$\quad 10y + x = 90$

19. $\frac{1}{3}x + \frac{1}{5}y = 7,$
$\quad \frac{1}{6}x - \frac{2}{5}y = -4$

20. $\frac{2}{3}x + \frac{1}{7}y = -11,$
$\quad \frac{1}{7}x - \frac{1}{3}y = -10$

21. $\frac{1}{5}x + \frac{1}{2}y = 6,$
$\quad \frac{2}{5}x - \frac{3}{2}y = -8$

22. $\frac{2}{3}x + \frac{3}{5}y = -17,$
$\quad \frac{1}{2}x - \frac{1}{3}y = -1$

23. $\frac{1}{2}x - \frac{1}{3}y = -4,$
$\quad \frac{1}{4}x + \frac{5}{6}y = 4$

24. $\frac{4}{3}x + \frac{3}{2}y = 4,$
$\quad \frac{5}{6}x - \frac{1}{8}y = -6$

25. $0.3x - 0.2y = 4,$
$\quad 0.2x + 0.3y = 0.5$

26. $0.7x - 0.3y = 0.5,$
$\quad -0.4x + 0.7y = 1.3$

27. $0.05x + 0.25y = 22,$
$\quad 0.15x + 0.05y = 24$

28. $1.3x - 0.2y = 12,$
$\quad 0.4x + 17y = 89$

---

b Solve. Use the elimination method when solving the translated system.

29. *Finding Numbers.* The sum of two numbers is 63. The larger number minus the smaller number is 9. Find the numbers.

30. *Finding Numbers.* The sum of two numbers is 2. The larger number minus the smaller number is 20. Find the numbers.

31. *Finding Numbers.* The sum of two numbers is 3. Three times the larger number plus two times the smaller number is 24. Find the numbers.

32. *Finding Numbers.* The sum of two numbers is 9. Two times the larger number plus three times the smaller number is 2. Find the numbers.

33. *Complementary Angles.* Two angles are complementary. (**Complementary angles** are angles whose sum is 90°.) Their difference is 6°. Find the angles.

Complementary angles:
$x + y = 90°$

34. *Supplementary Angles.* Two angles are supplementary. (**Supplementary angles** are angles whose sum is 180°.) Their difference is 22°. Find the angles.

Supplementary angles:
$x + y = 180°$

35. *Basketball Scoring.* In their championship game, the Eastside Golden Eagles scored 60 points on a combination of two-point shots and three-point shots. If they made a total of 27 shots, how many of each kind of shot was made?

36. *Basketball Scoring.* Wilt Chamberlain once scored 100 points, setting a record for points scored in an NBA game. Chamberlain took only two-point shots and (one-point) foul shots and made a total of 64 shots. How many shots of each type did he make?

Copyright © 2015 Pearson Education, Inc.

37. Each course offered during the winter session at New Heights Community College is worth either 3 credits or 4 credits. The members of the Touring Concert Chorale took a total of 33 courses during the winter session, worth a total of 107 credits. How many of each type of class did the chorale members take?

38. Daphne's Lawn and Garden Center offered customers who bought a custom lawn-care package a free ornamental tree, either an Eastern Redbud or a Kousa Dogwood. The center's cost for each Eastern Redbud was $37, and its cost for each Kousa Dogwood was $45. A total of 18 customers took advantage of the offer. The center's total cost for the promotional items was $754. How many patrons chose each type of ornamental tree?

Skill Maintenance

Given the function $f(x) = 3x^2 - x + 1$, find each of the following function values. [16.1b]

39. $f(0)$ **40.** $f(-1)$ **41.** $f(-2)$ **42.** $f(2a)$

43. Find the domain of the function
$$f(x) = \frac{x - 5}{x + 7}. \quad [16.2a]$$

44. Find the domain and the range of the function
$$g(x) = 5 - x^2. \quad [16.2a]$$

45. Find an equation of the line with slope $-\frac{3}{5}$ and y-intercept $(0, -7)$. [16.5a]

46. Find an equation of the line containing the points $(-10, 2)$ and $(-2, 10)$. [16.5c]

Synthesis

47.  Use the INTERSECT feature to solve the following system of equations. You may need to first solve for y. Round answers to the nearest hundredth.
$$3.5x - 2.1y = 106.2,$$
$$4.1x + 16.7y = -106.28$$

48. Solve:
$$\frac{x + y}{2} - \frac{x - y}{5} = 1,$$
$$\frac{x - y}{2} + \frac{x + y}{6} = -2.$$

49. The solution of this system is $(-5, -1)$. Find A and B.
$$Ax - 7y = -3,$$
$$x - By = -1$$

50. Find an equation to pair with $6x + 7y = -4$ such that $(-3, 2)$ is a solution of the system.

51. The points $(0, -3)$ and $\left(-\frac{3}{2}, 6\right)$ are two of the solutions of the equation $px - qy = -1$. Find p and q.

52. Determine a and b for which $(-4, -3)$ will be a solution of the system
$$ax + by = -26,$$
$$bx - ay = 7.$$

Solving Applied Problems: Two Equations

OBJECTIVES

---

a Solve applied problems involving total value and mixture using systems of two equations.

b Solve applied problems involving motion using systems of two equations.

a TOTAL-VALUE PROBLEMS AND MIXTURE PROBLEMS

Systems of equations can be a useful tool in solving applied problems. Using systems often makes the *Translate* step easier than using a single equation. The first kind of problem we consider involves quantities of items purchased and the total value, or cost, of the items. We refer to this type of problem as a **total-value problem**.

EXAMPLE 1 *Lunch Orders.* In order to pick up lunch, Cathy collected $181.50 from her co-workers for a total of 21 salads and sandwiches. When she got to the deli, she forgot how many of each were ordered. If salads cost $7.50 and sandwiches cost $9.50, how many of each should she buy?

1. **Familiarize.** Let's begin by guessing that 5 salads were ordered. Since there was a total of 21 orders, this means that 16 sandwiches were ordered. The total cost of the order would then be

$$\underbrace{\$7.50(5)}_{\text{Cost of salads}} + \underbrace{\$9.50(16)}_{\text{Cost of sandwiches}} = \$37.50 + \$152 = \$189.50.$$

The guess is incorrect, but we can use the same process to translate the problem to a system of equations. We also note that our guess resulted in a total that was too high, so there were more salads and fewer sandwiches ordered than we guessed.

We let d = the number of salads and w = the number of sandwiches ordered. The cost of d salads is 7.50d$, and the cost of w sandwiches is 9.50w$. Organizing the information in a table can help us translate the information to a system of equations.

	SALADS	SANDWICHES	TOTAL	
NUMBER OF ORDERS	d	w	21	→ $d + w = 21$
COST PER ORDER	$7.50	$9.50		
TOTAL COST	7.50d$	9.50w$	$181.50	→ $7.50d + 9.50w = 181.50$

2. **Translate.** The first row of the table gives us one equation:

   $$d + w = 21.$$

   The last row of the table gives us a second equation:

   $$7.50d + 9.50w = 181.50.$$

Clearing decimals in the second equation gives us the following system of equations:

$$d + w = 21, \qquad (1)$$
$$75d + 95w = 1815. \qquad (2)$$

3. Solve. We use the elimination method to solve the system of equations. We eliminate d by multiplying by -75 on both sides of equation (1) and then adding the result to equation (2):

$-75d - 75w = -1575$	Multiplying equation (1) by -75
$\underline{75d + 95w = 1815}$	Equation (2)
$20w = 240$	Adding
$w = 12.$	Dividing by 20

Next, we substitute 12 for w in equation (1) and solve for d:

$d + w = 21$	Equation (1)
$d + 12 = 21$	Substituting 12 for w
$d = 9.$	Solving for d

We obtain $(9, 12)$, or $d = 9$, $w = 12$.

4. Check. We check in the original problem.

Total number of orders:	$d + w = 9 + 12 = 21$	
Cost of salads:	$\$7.50d = \$7.50(9)$	$= \$\ 67.50$
Cost of sandwiches:	$\$9.50w = \$9.50(12)$	$= \underline{\$114.00}$
	Total	$= \$181.50$

The numbers check.

5. State. Cathy should buy 9 salads and 12 sandwiches.

Do Exercise 1. ▶

The following problem, similar to Example 1, is called a **mixture problem**.

EXAMPLE 2 *Blending Spices.* Spice It Up sells ground turmeric for $1.35 per ounce and ground sumac for $1.85 per ounce. Ethan wants to make a 20-oz seasoning blend of the two spices that sells for $1.65 per ounce. How much of each should he use?

 1. Retail Sales of Sweatshirts. A campus bookstore sells college tee shirts. White tee shirts sell for $18.95 each and red ones sell for $19.50 each. If receipts for the sale of 30 tee shirts total $572.90, how many of each color did the shop sell?

Complete the following table, letting $w =$ the number of white tee shirts and $r =$ the number of red tee shirts.

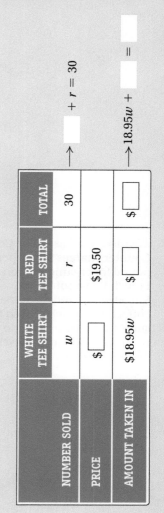

1. Familiarize. Suppose that Ethan uses 4 oz of sumac. Since he wants a total of 20 oz, he will need 16 oz of turmeric. We compare the value of the spices separately with the desired value of the blend:

Spices purchased separately: $\$1.85(4) + \$1.35(16)$, or $\$29.$
Blend: $\$1.65(20) = \$33$

Since these amounts are not the same, our guess is not correct, but these calculations help us to translate the problem.

We let $s =$ the number of ounces of sumac and $t =$ the number of ounces of turmeric. Then we organize the information in a table as follows.

	SUMAC	TURMERIC	BLEND	
NUMBER OF OUNCES	s	t	20	$\longrightarrow s + t = 20$
PRICE PER OUNCE	$\$1.85$	$\$1.35$	$\$1.65$	
VALUE OF SPICES	$\$1.85s$	$\$1.35t$	$\$1.65(20)$, or $\$33$	$\longrightarrow 1.85s + 1.35t = 33$

2. Translate. The total number of ounces in the blend is 20, so we have one equation:

$$s + t = 20.$$

The value of the sumac is $1.85s$, and the value of the turmeric is $1.35t$. These amounts are in dollars. Since the total value is to be $1.65(20)$, or 33, we have

$$1.85s + 1.35t = 33.$$

We can multiply by 100 on both sides of the second equation in order to clear the decimals. Thus we have translated to a system of equations:

$$s + t = 20, \qquad (1)$$
$$185s + 135t = 3300. \qquad (2)$$

3. Solve. We will solve this system using substitution, but elimination is also an appropriate method to use. When equation (1) is solved for t, we get $t = 20 - s$. We substitute $20 - s$ for t in equation (2) and solve:

$185s + 135(20 - s) = 3300$	Substituting
$185s + 2700 - 135s = 3300$	Using the distributive law
$50s = 600$	Subtracting 2700 and collecting like terms
$s = 12.$	

We have $s = 12$. Substituting 12 for s in the equation $t = 20 - s$, we obtain $t = 20 - 12$, or 8.

4. Check. We check in a manner similar to our guess in the *Familiarize* step.

Total number of ounces: $12 + 8 = 20$
Value of the blend: $1.85(12) + 1.35(8) = 33$

Thus the number of ounces of each spice checks.

5. State. Ethan should use 12 oz of sumac and 8 oz of turmeric.

◀ **Do Exercise 2.**

---

2. *Blending Coffees.* The Coffee Counter charges $18.00 per pound for organic Kenyan French Roast coffee and $16.00 per pound for Sumatran coffee. How much of each type should be used in order to make a 20-lb blend that sells for $16.70 per pound?

Answer

2. Kenyan: 7 lb; Sumatran: 13 lb

EXAMPLE 3 *Student Loans.* Jeron's student loans totaled $16,200. Part was a Perkins loan made at 5% interest and the rest was a Stafford loan made at 4% interest. After one year, Jeron's loans accumulated $715 in interest. What was the amount of each loan?

1. **Familiarize.** Listing the given information in a table will help. The columns in the table come from the formula for simple interest: $I = Prt$. We let $x =$ the number of dollars in the Perkins loan and $y =$ the number of dollars in the Stafford loan.

	PERKINS LOAN	STAFFORD LOAN	TOTAL
PRINCIPAL	x	y	$16,200
RATE OF INTEREST	5%	4%	
TIME	1 year	1 year	
INTEREST	0.05x	0.04y	$715

$\longrightarrow x + y = 16{,}200$

$\longrightarrow 0.05x + 0.04y = 715$

2. **Translate.** The total of the amounts of the loans is found in the first row of the table. This gives us one equation:

$$x + y = 16{,}200.$$

Look at the last row of the table. The interest totals $715. This gives us a second equation:

$$5\%x + 4\%y = 715, \quad \text{or} \quad 0.05x + 0.04y = 715.$$

After we multiply on both sides to clear the decimals, we have

$$5x + 4y = 71{,}500.$$

3. **Solve.** Using either elimination or substitution, we solve the resulting system:

$$x + y = 16{,}200,$$
$$5x + 4y = 71{,}500.$$

We find that $x = 6700$ and $y = 9500$.

4. **Check.** The sum is $6700 + $9500, or $16,200. The interest from $6700 at 5% for one year is 5%($6700), or $335. The interest from $9500 at 4% for one year is 4%($9500), or $380. The total amount of interest is $335 + $380, or $715. The numbers check in the problem.

5. **State.** The Perkins loan was for $6700, and the Stafford loan was for $9500.

Do Exercise 3. ▶

 3. *Client Investments.* Infinite Financial Services invested Jasmine's IRA contribution of $3700 for one year at simple interest, yielding $297. Part of the money is invested at 7% and the rest at 9%. How much was invested at each rate?

Do the *Familiarize* and *Translate* steps by completing the following table. Let $x =$ the number of dollars invested at 7% and $y =$ the number of dollars invested at 9%.

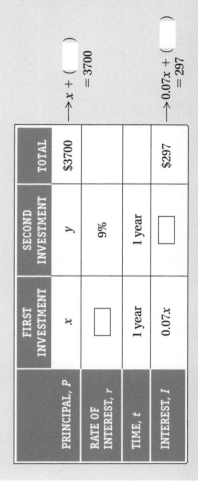

Answer

3. $1800 at 7%; $1900 at 9%

Guided Solution:
3.

	First Investment	Second Investment	Total
	x	y	$3700
	7%	9%	
	1 year	1 year	
	0.07x	0.09y	$297

$\longrightarrow x + y = 3700$

$\longrightarrow 0.07x + 0.09y = 297$

EXAMPLE 4 *Mixing Fertilizers.* Nature's Landscapes carries two kinds of fertilizer containing nitrogen and water. "Gently Green" is 5% nitrogen and "Sun Saver" is 15% nitrogen. Nature's Landscapes needs to combine the two types of solution in order to make 90 L of a solution that is 12% nitrogen. How much of each brand should be used?

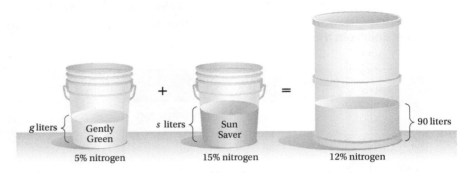

1. **Familiarize.** We first make a drawing and a guess to become familiar with the problem.

  We choose two numbers that total 90 L—say, 40 L of Gently Green and 50 L of Sun Saver—for the amounts of each fertilizer. Will the resulting mixture have the correct percentage of nitrogen?

  To find out, we multiply as follows:

  $5\%(40\text{ L}) = 2\text{ L of nitrogen}$ and $15\%(50\text{ L}) = 7.5\text{ L of nitrogen.}$

  Thus the total amount of nitrogen in the mixture is 2 L + 7.5 L, or 9.5 L. The final mixture of 90 L is supposed to be 12% nitrogen. Now

  $12\%(90\text{ L}) = 10.8\text{ L.}$

  Since 9.5 L and 10.8 L are not the same, our guess is incorrect. But these calculations help us to make the translation.

  We let g = the number of liters of Gently Green and s = the number of liters of Sun Saver in the mixture.

	GENTLY GREEN	SUN SAVER	MIXTURE	
NUMBER OF LITERS	g	s	90	$\longrightarrow g + s = 90$
PERCENT OF NITROGEN	5%	15%	12%	
AMOUNT OF NITROGEN	0.05g	0.15s	0.12 × 90, or 10.8 liters	$\longrightarrow 0.05g + 0.15s$ $= 10.8$

2. **Translate.** If we add g and s in the first row, we get 90, and this gives us one equation:

  $g + s = 90.$

  If we add the amounts of nitrogen listed in the third row, we get 10.8, and this gives us another equation:

  $5\%g + 15\%s = 10.8,$ or $0.05g + 0.15s = 10.8.$

After clearing the decimals, we have the following system:

$$g + s = 90, \qquad \textbf{(1)}$$
$$5g + 15s = 1080. \qquad \textbf{(2)}$$

3. Solve. We solve the system using elimination. We multiply equation (1) by -5 and add the result to equation (2):

$$
\begin{array}{rl}
-5g - 5s = -450 & \text{Multiplying equation (1) by } -5 \\
\underline{5g + 15s = 1080} & \text{Equation (2)} \\
10s = 630 & \text{Adding} \\
s = 63. & \text{Dividing by 10}
\end{array}
$$

Next, we substitute 63 for s in equation (1) and solve for g:

$$
\begin{array}{ll}
g + 63 = 90 & \text{Substituting in equation (1)} \\
 g = 27. & \text{Solving for } g
\end{array}
$$

We obtain $(27, 63)$, or $g = 27, s = 63$.

4. Check. Remember that g is the number of liters of Gently Green, with 5% nitrogen, and s is the number of liters of Sun Saver, with 15% nitrogen.

Total number of liters of mixture: $g + s = 27 + 63 = 90$ L

Amount of nitrogen: $5\%(27) + 15\%(63) = 1.35 + 9.45 = 10.8$ L

Percentage of nitrogen in mixture: $\dfrac{10.8}{90} = 0.12 = 12\%$

The numbers check in the original problem.

5. State. Nature's Landscapes should mix 27 L of Gently Green and 63 L of Sun Saver.

Do Exercise 4. ▶

b MOTION PROBLEMS

When a problem deals with speed, distance, and time, we can expect to use the following *motion formula*.

THE MOTION FORMULA

Distance = Rate (or speed) · Time
$$d = rt$$

TIPS FOR SOLVING MOTION PROBLEMS

1. Make a drawing using an arrow or arrows to represent distance and the direction of each object in motion.
2. Organize the information in a table or a chart.
3. Look for as many things as you can that are the same, so you can write equations.

4. Mixing Cleaning Solutions. King's Service Station uses two kinds of cleaning solution containing acid and water. "Attack" is 2% acid and "Blast" is 6% acid. They want to mix the two to get 60 qt of a solution that is 5% acid. How many quarts of each should they use?

Do the *Familiarize* and *Translate* steps by completing the following table. Let $a = $ the number of quarts of Attack and $b = $ the number of quarts of Blast.

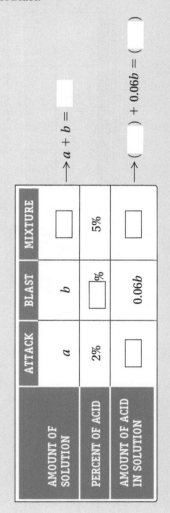

	ATTACK	BLAST	MIXTURE
AMOUNT OF SOLUTION	a	b	☐
PERCENT OF ACID	2%	☐%	5%
AMOUNT OF ACID IN SOLUTION	☐	$0.06b$	☐

Guided Solution:

4.

Attack	Blast	Mixture	
a	b	60	→ $a + b = 60$
2%	6%	5%	
$0.02a$	$0.06b$	0.05×60, or 3	→ $0.02a + 0.06b = 3$

5. Train Travel. A train leaves Barstow traveling east at 35 km/h. One hour later, a faster train leaves Barstow, also traveling east on a parallel track at 40 km/h. How far from Barstow will the faster train catch up with the slower one?

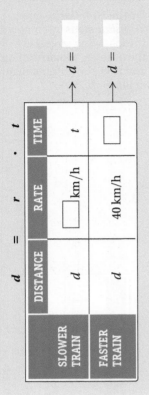

EXAMPLE 5 *Auto Travel.* Keri left Monday morning to drive to a seminar that began Monday evening. An hour after she had left the office, her assistant, Matt, realized that she had forgotten to take a large portfolio needed for a presentation. Knowing Keri would not answer her cell phone when driving, Matt left immediately with the portfolio to try to catch up with her. If Keri drove at a speed of 55 mph and Matt drove at a speed of 65 mph, how long did it take Matt to catch up with her? Assume that neither driver stopped to take a break.

1. **Familiarize.** We first make a drawing. From the drawing, we see that when Matt catches up with Keri, the distances from the office are the same. We let d = the distance, in miles. If we let t = the time, in hours, for Matt to catch Keri, then $t + 1$ = the time traveled by Keri at a slower speed.

Cars meet here

Matt's car
65 mph
t hours, d miles

Keri's car
55 mph
$t + 1$ hours, d miles

We organize the information in a table as follows.

d	$=$	r	$\cdot$	t
	DISTANCE	RATE	TIME	
KERI	d	55	$t + 1$	$\longrightarrow d = 55(t + 1)$
MATT	d	65	t	$\longrightarrow d = 65t$

2. **Translate.** Using $d = rt$ in each row of the table, we get an equation. Thus we have a system of equations:

$$d = 55(t + 1), \quad \text{(1)}$$
$$d = 65t. \quad \text{(2)}$$

3. **Solve.** We solve the system using the substitution method:

$65t = 55(t + 1)$ Substituting $65t$ for d in equation (1)

$65t = 55t + 55$ Multiplying to remove parentheses on the right

$\left.\begin{array}{l} 10t = 55 \\ t = 5.5. \end{array}\right\}$ Solving for t

Matt's time is 5.5 hr, which means that Keri's time is $5.5 + 1$, or 6.5 hr.

4. **Check.** At 65 mph, Matt will travel $65 \cdot 5.5$, or 357.5 mi, in 5.5 hr. At 55 mph, Keri will travel $55 \cdot 6.5$, or the same 357.5 mi, in 6.5 hr. The distances are the same, so the numbers check.

5. **State.** Matt will catch up with Keri in 5.5 hr.

◀ Do Exercise 5.

Answer

5. 280 km

Guided Solution:

5.

EXAMPLE 6 *Marine Travel.* A Coast-Guard patrol boat travels 4 hr on a trip downstream with a 6-mph current. The return trip against the same current takes 5 hr. Find the speed of the boat in still water.

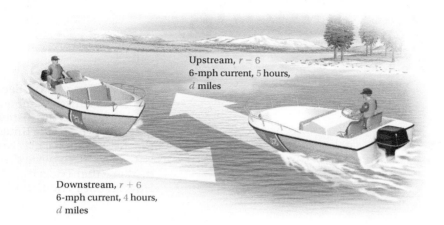

Upstream, $r - 6$
6-mph current, 5 hours,
d miles

Downstream, $r + 6$
6-mph current, 4 hours,
d miles

1. **Familiarize.** We first make a drawing. From the drawing, we see that the distances are the same. We let d = the distance, in miles, and r = the speed of the boat in still water, in miles per hour. Then, when the boat is traveling downstream, its speed is $r + 6$. (The current helps the boat along.) When it is traveling upstream, its speed is $r - 6$. (The current holds the boat back.) We can organize the information in a table. We use the formula $d = rt$.

$$d = r \cdot t$$

	DISTANCE	RATE	TIME
DOWNSTREAM	d	$r + 6$	4	$\longrightarrow d = (r + 6)4$
UPSTREAM	d	$r - 6$	5	$\longrightarrow d = (r - 6)5$

2. **Translate.** From each row of the table, we get an equation, $d = rt$:

$d = 4r + 24,$ **(1)**
$d = 5r - 30.$ **(2)**

3. **Solve.** We solve the system using the substitution method:

$4r + 24 = 5r - 30$ Substituting $4r + 24$ for d in equation (2)
$24 = r - 30$ ⎫
$54 = r.$ ⎭   Solving for r

4. **Check.** If $r = 54$, then $r + 6 = 60$; and $60 \cdot 4 = 240$ mi, the distance traveled downstream. If $r = 54$, then $r - 6 = 48$; and $48 \cdot 5 = 240$ mi, the distance traveled upstream. The distances are the same.

When checking your answer, always ask, "Have I found what the problem asked for?" We could solve for a certain variable but still have not answered the question of the original problem. For example, we might have found speed when the problem wanted distance. In this problem, we want the speed of the boat in still water, and that is r.

5. **State.** The speed in still water is 54 mph.

Do Exercise 6. ▶

 6. Air Travel. An airplane flew for 4 hr with a 20-mph tailwind. The return flight against the same wind took 5 hr. Find the speed of the plane in still air.

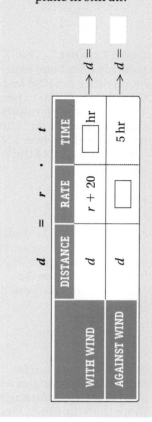

Answer
6. 180 mph

Guided Solution:
6.

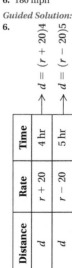

Translating for Success

1. **Office Expense.** The monthly telephone expense for an office is $1094 less than the janitorial expense. Three times the janitorial expense minus four times the telephone expense is $248. What is the total of the two expenses?

2. **Dimensions of a Triangle.** The sum of the base and the height of a triangle is 192 in. The height is twice the base. Find the base and the height.

3. **Supplementary Angles.** Two supplementary angles are such that twice one angle is 7° more than the other. Find the measures of the angles.

4. **SAT Scores.** The total of Megan's writing and math scores on the SAT was 1094. Her math score was 248 points higher than her writing score. What were her math and writing SAT scores?

5. **Sightseeing Boat.** A sightseeing boat travels 3 hr on a trip downstream with a 2.5-mph current. The return trip against the same current takes 3.5 hr. Find the speed of the boat in still water.

The goal of these matching questions is to practice step (2), Translate, of the five-step problem-solving process. Translate each word problem to a system of equations and select a correct translation from systems A–J.

A. $x = y + 248,$
$x + y = 1094$

B. $5x = 2y - 3,$
$y = \frac{2}{3}x + 5$

C. $y = \frac{1}{2}x,$
$2x + 2y = 192$

D. $2x = 7 + y,$
$x + y = 180$

E. $x + y = 192,$
$x = 2y$

F. $x + y = 180,$
$x = 2y + 7$

G. $x - 1094 = y,$
$3x - 4y = 248$

H. $3\%x + 2.5\%y = 97.50,$
$x + y = 2500$

I. $2x = 5 + \frac{2}{3}y,$
$3y = 15x - 4$

J. $x = (y + 2.5) \cdot 3,$
$3.5(y - 2.5) = x$

Answers on page A-38

6. **Running Distances.** Each day Tricia runs 5 mi more than two-thirds the distance that Chris runs. Five times the distance that Chris runs is 3 mi less than twice the distance that Tricia runs. How far does Tricia run daily?

7. **Dimensions of a Rectangle.** The perimeter of a rectangle is 192 in. The width is half the length. Find the length and the width.

8. **Mystery Numbers.** Teka asked her students to determine the two numbers that she placed in a sealed envelope. Twice the smaller number is 5 more than two-thirds the larger number. Three times the larger number is 4 less than fifteen times the smaller. Find the numbers.

9. **Supplementary Angles.** Two supplementary angles are such that one angle is 7° more than twice the other. Find the measures of the angles.

10. **Student Loans.** Brandt's student loans totaled $2500. Part was borrowed at 3% interest and the rest at 2.5%. After one year, Brandt had accumulated $97.50 in interest. What was the amount of each loan?

☑ Reading Check

Consider the following mixture problem and the table used to translate the problem.

Cherry Breeze is 30% fruit juice and Berry Choice is 15% fruit juice. How much of each should be used in order to make 10 gal of a drink that is 20% fruit juice?

Choose from the options below the expression that best fits each numbered space in the table.

2   10   15 $0.15y$

	CHERRY BREEZE	BERRY CHOICE	MIXTURE
GALLONS OF DRINK	x	y	**RC1.** ____
PERCENT OF FRUIT JUICE	30%	**RC2.** ____ %	20%
GALLONS OF FRUIT JUICE IN MIXTURE	$0.3x$	**RC3.** _____	**RC4.** ____

a   Solve.

1. *Entertainment.* For her personal-finance class, Laura was required to estimate her annual entertainment expenditures. She discovered that during the previous year, she spent $225.32 on a total of 68 e-books and game applications. If each book cost $3.99 and each game cost $1.99, how many books and how many games did she purchase?

2. *Flowers.* Kevin's Floral Emporium offers two types of sunflowers for sale by the stem. When in season, the small ones sell for $2.50 per stem, and the large ones sell for $3.95 per stem. One late summer weekend, Kevin sold a total of 118 stems for $376.20. How many of each size did he sell?

3. *Balloon Bouquets.* When the Southeast Cougars women's soccer team won the state championship, the parent boosters welcomed the team back to school with a balloon bouquet for each of the 18 players. The parents spent a total of $86.76 (excluding tax) on foil balloons that cost $1.99 each and latex school-color balloons that cost $0.12 each. Each player received 9 balloons, and all the balloon bouquets were identical. How many of each type of balloon did each bouquet include?

4. *Chocolate Assortments.* For a fundraiser, the Greenfield Merchants Association spent a total of $1872 on an assortment of chocolate truffles at $2.95 each and chocolate cream mints at $1.79 each. They then packaged 75 boxes to sell, each containing 12 pieces of candy. If the boxes were identical, how many of each kind of candy did each box contain?

5. *Furniture Polish.* A nontoxic furniture polish can be made by combining vinegar and olive oil. The amount of oil should be three times the amount of vinegar. How much of each ingredient is needed in order to make 30 oz of furniture polish?

6. *Nontoxic Floor Wax.* A nontoxic floor wax can be made by combining lemon juice and food-grade linseed oil. The amount of oil should be twice the amount of lemon juice. How much of each ingredient is needed in order to make 32 oz of floor wax? (The mix should be spread with a rag and buffed when dry.)

7. *Catering.* Stella's Catering is planning a wedding reception. The bride and groom would like to serve a nut mixture containing 25% peanuts. Stella has available mixtures that are either 40% or 10% peanuts. How much of each type should be mixed in order to get a 10-lb mixture that is 25% peanuts?

8. *Blending Granola.* Deep Thought Granola is 25% nuts and dried fruit. Oat Dream Granola is 10% nuts and dried fruit. How much of Deep Thought and how much of Oat Dream should be mixed in order to form a 20-lb batch of granola that is 19% nuts and dried fruit?

9. *Ink Remover.* Etch Clean Graphics uses one cleanser that is 25% acid and a second that is 50% acid. How many liters of each should be mixed in order to get 10 L of a solution that is 40% acid?

10. *Livestock Feed.* Soybean meal is 16% protein and corn meal is 9% protein. How many pounds of each should be mixed in order to get a 350-lb mixture that is 12% protein?

11. *Vegetable Seeds.* Tara's Web site, verdantveggies.com, specializes in the sale of rare or unusual vegetable seeds. Tara sells packets of sweet-pepper seeds for $2.85 each and packets of hot-pepper seeds for $4.29 each. She also offers a 16-packet mixed-pepper assortment combining packets of both types of seeds at $3.30 per packet. How many packets of each type of seed are in the assortment?

12. *Flower Bulbs.* Heritage Bulbs sells heirloom flower bulbs. Acuminata tulip bulbs cost $4.85 each, and Cafe Brun tulip bulbs cost $9.50 each. An assortment of 12 of these bulbs is priced at $7.95 per bulb. How many of each type of bulb are in the assortment?

Sweet peppers   Hot peppers   Assorted

13. *Student Loans.* Sarah's two student loans totaled $12,000. One of her loans was at 6% simple interest and the other at 3%. After one year, Sarah owed $585 in interest. What was the amount of each loan?

14. *Investments.* Ana and Johnny made two investments totaling $45,000. In one year, these investments yielded $2430 in simple interest. Part of the money was invested at 4% and the rest at 6%. How much was invested at each rate?

15. *Food Science.* The following bar graph shows the milk fat percentages in three dairy products. How many pounds each of whole milk and cream should be mixed in order to form 200 lb of milk for cream cheese?

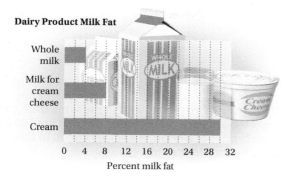

Dairy Product Milk Fat

Whole milk

Milk for cream cheese

Cream

0 4 8 12 16 20 24 28 32
Percent milk fat

16. *Automotive Maintenance.* Arctic Antifreeze is 18% alcohol and Frost No-More is 10% alcohol. How many liters of Arctic Antifreeze should be mixed with 7.5 L of Frost No-More in order to get a mixture that is 15% alcohol?

17. *Investments.* William opened two investment accounts for his daughter's college fund. The first year, these investments, which totaled $3200, yielded $155 in simple interest. Part of the money was invested at 5.5% and the rest at 4%. How much was invested at each rate?

18. *Student Loans.* Cole's two student loans totaled $31,000. One of his loans was at 2.8% simple interest and the other at 4.5%. After one year, Cole owed $1024.40 in interest. What was the amount of each loan?

19. *Making Change.* Juan goes to a bank and gets change for a $50 bill consisting of all $5 bills and $1 bills. There are 22 bills in all. How many of each kind are there?

20. *Making Change.* Christina makes a $9.25 purchase at a bookstore with a $20 bill. The store has no bills and gives her the change in quarters and dollar coins. There are 19 coins in all. How many of each kind are there?

b Solve.

21. *Train Travel.* A train leaves Danville Junction and travels north at a speed of 75 mph. Two hours later, a second train leaves on a parallel track and travels north at 125 mph. How far from the station will they meet?

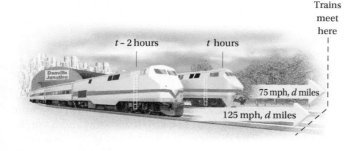

Trains meet here

$t - 2$ hours t hours

75 mph, d miles

125 mph, d miles

22. *Car Travel.* Max leaves Kansas City and drives east at a speed of 80 km/h. One hour later, Olivia leaves Kansas City traveling in the same direction as Max but at 96 km/h. Assuming neither driver stops for a break, how far from Kansas City will they be when Olivia catches up with Max?

23. *Canoeing.* Darren paddled for 4 hr with a 6-km/h current to reach a campsite. The return trip against the same current took 10 hr. Find the speed of Darren's canoe in still water.

24. *Boating.* Mia's motorboat took 3 hr to make a trip downstream with a 6-mph current. The return trip against the same current took 5 hr. Find the speed of the boat in still water.

25. *Air Travel.* Christie pilots her Cessna 150 plane for 270 mi against a headwind in 3 hr. The flight would take 1 hr and 48 min with a tailwind of the same speed. Find the headwind and the speed of the plane in still air.

26. *Air Travel.* Rod is a pilot for Crossland Airways. He computes his flight time against a headwind for a trip of 2900 mi at 5 hr. The flight would take 4 hr and 50 min if the headwind were half as great. Find the headwind and the plane's air speed in still air.

27. *Air Travel.* Two airplanes start at the same time and fly toward each other from points 1000 km apart at rates of 420 km/h and 330 km/h. After how many hours will they meet?

28. *Air Travel.* Two planes start at the same time and travel toward each other from cities that are 780 km apart at rates of 190 km/h and 200 km/h. In how many hours will they meet?

29. 🖩 *Point of No Return.* A plane flying the 3458-mi trip from New York City to London has a 50-mph tailwind. The flight's *point of no return* is the point at which the flight time required to return to New York is the same as the time required to continue to London. If the speed of the plane in still air is 360 mph, how far is New York from the point of no return?

30. 🖩 *Point of No Return.* A plane is flying the 2553-mi trip from Los Angeles to Honolulu into a 60-mph headwind. If the speed of the plane in still air is 310 mph, how far from Los Angeles is the plane's point of no return? (See Exercise 29.)

Skill Maintenance

Find the domain. [16.2a]

31. $g(x) = x^2 + 5x - 14$

32. $f(x) = \dfrac{2}{x - 14}$

Given the function $f(x) = 4x - 7$, find each of the following function values. [16.1b]

33. $f\left(\dfrac{3}{4}\right)$

34. $f(-2.5)$

35. $f(-3h)$

36. $f(1000)$

Synthesis

37. *Automotive Maintenance.* The radiator in Michelle's car contains 16 L of antifreeze and water. This mixture is 30% antifreeze. How much of this mixture should she drain and replace with pure antifreeze so that there will be a mixture of 50% antifreeze?

38. *Physical Exercise.* Natalie jogs and walks to school each day. She averages 4 km/h walking and 8 km/h jogging. The distance from home to school is 6 km and Natalie makes the trip in 1 hr. How far does she jog in a trip?

39. *Fuel Economy.* Ashlee's SUV gets 18 miles per gallon (mpg) in city driving and 24 mpg in highway driving. The SUV is driven 465 mi on 23 gal of gasoline. How many miles were driven in the city and how many were driven on the highway?

40. *Siblings.* Phil and Maria are siblings. Maria has twice as many brothers as she has sisters. Phil has the same number of brothers as he has sisters. How many girls and how many boys are in the family?

Mid-Chapter Review

Concept Reinforcement

Determine whether each statement is true or false.

_____ 1. If, when solving a system of two linear equations in two variables, a false equation is obtained, the system has infinitely many solutions. [17.2a], [17.3a]

_____ 2. Every system of equations has at least one solution. [17.1a]

_____ 3. If the graphs of two linear equations intersect, then the system is consistent. [17.1a]

_____ 4. The intersection of the graphs of the lines $x = a$ and $y = b$ is (a, b). [17.1a]

Guided Solutions

Fill in each box with the number, variable, or expression that creates a correct statement or solution.

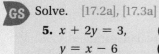 Solve. [17.2a], [17.3a]

5. $x + 2y = 3$,   **(1)**
    $y = x - 6$ **(2)**

$x + 2(\boxed{}) = 3$ Substituting for y in equation (1)

$x + \boxed{}x - \boxed{} = 3$ Removing parentheses

$\boxed{}x - 12 = 3$ Collecting like terms

$3x = \boxed{}$

$x = \boxed{}$

$y = \boxed{} - 6$ Substituting in equation (2)

$y = \boxed{}$ Subtracting

The solution is $(\boxed{}, \boxed{})$.

6. $3x - 2y = 5$,   **(1)**
    $2x + 4y = 14$ **(2)**

$\boxed{}x - \boxed{}y = \boxed{}$ Multiplying equation (1) by 2

$2x + 4y = 14$ Equation (2)

$\boxed{}x \qquad = \boxed{}$ Adding

$x = \boxed{}$

$2 \cdot \boxed{} + 4y = 14$ Substituting for x in equation (2)

$\boxed{} + 4y = 14$ Multiplying

$4y = \boxed{}$

$y = \boxed{}$

The solution is $(\boxed{}, \boxed{})$.

Mixed Review

Solve each system of equations graphically. Then classify the system as consistent or inconsistent and the equations as dependent or independent. [17.1a]

7. $y = x - 6$,
    $y = 4 - x$

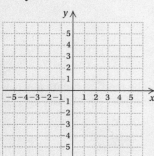

8. $x + y = 3$,
    $3x + y = 3$

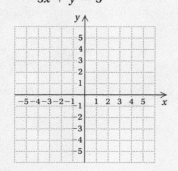

9. $y = 2x - 3$,
    $4x - 2y = 6$

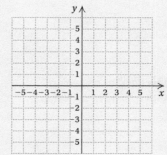

10. $x - y = 3$,
    $2y - 2x = 6$

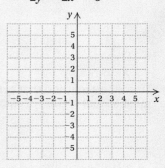

Solve using the substitution method. [17.2a]

11. $x = y + 2,$
  $2x - 3y = -2$

12. $y = x - 5,$
  $x - 2y = 8$

13. $4x + 3y = 3,$
  $y = x + 8$

14. $3x - 2y = 1,$
  $x = y + 1$

Solve using the elimination method. [17.3a]

15. $2x + y = 2,$
  $x - y = 4$

16. $x - 2y = 13,$
  $x + 2y = -3$

17. $3x - 4y = 5,$
  $5x - 2y = -1$

18. $3x + 2y = 11,$
  $2x + 3y = 9$

19. $x - 2y = 5,$
  $3x - 6y = 10$

20. $4x - 6y = 2,$
  $-2x + 3y = -1$

21. $\frac{1}{2}x + \frac{1}{3}y = 1,$
  $\frac{1}{5}x - \frac{3}{4}y = 11$

22. $0.2x + 0.3y = 0.6,$
  $0.1x - 0.2y = -2.5$

Solve.

23. *Garden Dimensions.* A landscape architect designs a garden with a perimeter of 44 ft. The width is 2 ft less than the length. Find the length and the width. [17.2b]

24. *Investments.* Sandy made two investments totaling $5000. Part of the money was invested at 2% and the rest at 3%. In one year, these investments earned $129 in simple interest. How much was invested at each rate? [17.4a]

25. *Mixing Solutions.* A lab technician wants to mix a solution that is 20% acid with a second solution that is 50% acid in order to get 84 L of a solution that is 30% acid. How many liters of each solution should be used? [17.4a]

26. *Boating.* Monica's motorboat took 5 hr to make a trip downstream with a 6-mph current. The return trip against the same current took 8 hr. Find the speed of the boat in still water. [17.4b]

Understanding Through Discussion and Writing

27. Explain how to find the solution of $\frac{3}{4}x + 2 = \frac{2}{5}x - 5$ in two ways graphically and in two ways algebraically. [17.1a], [17.2a], [17.3a]

28. Write a system of equations with the given solution. Answers may vary. [17.1a], [17.2a], [17.3a]

  a) $(4, -3)$ **b)** No solution
  c) Infinitely many solutions

29. Describe a method that could be used to create an inconsistent system of equations. [17.1a], [17.2a], [17.3a]

30. Describe a method that could be used to create a system of equations with dependent equations. [17.1a], [17.2a], [17.3a]

Systems of Equations in Three Variables

17.5

a SOLVING SYSTEMS IN THREE VARIABLES

A **linear equation in three variables** is an equation equivalent to one of the type $Ax + By + Cz = D$. A **solution** of a system of three equations in three variables is an ordered triple (x, y, z) that makes *all three* equations true.

The substitution method can be used to solve systems of three equations, but it is not efficient unless a variable has already been eliminated from one or more of the equations. Therefore, we will use only the elimination method.* The first step is to eliminate a variable and obtain a system of two equations in two variables.

EXAMPLE 1 Solve the following system of equations:

$$x + y + z = 4, \quad (1)$$
$$x - 2y - z = 1, \quad (2)$$
$$2x - y - 2z = -1. \quad (3)$$

a) We first use *any* two of the three equations to get an equation in two variables. In this case, let's use equations (1) and (2) and add to eliminate z:

$$\begin{array}{ll} x + y + z = 4 & (1) \\ \underline{x - 2y - z = 1} & (2) \\ 2x - y \phantom{{}+z} = 5. & (4) \quad \text{Adding to eliminate } z \end{array}$$

b) We use a *different* pair of equations and eliminate the **same variable** that we did in part (a). Let's use equations (1) and (3) and again eliminate z.

·············· **Caution!** ··············

A common error is to eliminate a different variable the second time.
···

$$\begin{array}{ll} x + y + z = 4, & (1) \\ 2x - y - 2z = -1; & (3) \end{array}$$

$$\begin{array}{ll} 2x + 2y + 2z = 8 & \text{Multiplying equation (1) by 2} \\ \underline{2x - y - 2z = -1} & (3) \\ 4x + y \phantom{{}- 2z} = 7 & (5) \quad \text{Adding to eliminate } z \end{array}$$

_____

*Other methods for solving systems of equations are considered in Appendixes B and C.

OBJECTIVE
·····································

a Solve systems of three equations in three variables.

·····································
SKILL TO REVIEW
·····································

Objective 17.3a: Solve systems of equations in two variables by the elimination method.

Solve.

1. $3x + y = 1,$
    $5x - y = 7$

2. $2x + 3y = 9,$
    $3x + 2y = 1$

Answers

Skill to Review:
1. $(1, -2)$ **2.** $(-3, 5)$

c) Now we solve the resulting system of equations, (4) and (5). That solution will give us two of the numbers. Note that we now have two equations in two variables. Had we eliminated two *different* variables in parts (a) and (b), this would not be the case.

$$
\begin{array}{ll}
2x - y = 5 & \textbf{(4)} \\
\underline{4x + y = 7} & \textbf{(5)} \\
6x = 12 & \text{Adding} \\
 x = 2 &
\end{array}
$$

We can use either equation (4) or (5) to find y. We choose equation (5):

$$
\begin{array}{ll}
4x + y = 7 & \textbf{(5)} \\
4(2) + y = 7 & \text{Substituting 2 for } x \\
8 + y = 7 & \\
y = -1. &
\end{array}
$$

d) We now have $x = 2$ and $y = -1$. To find the value for z, we use any of the original three equations, substitute, and solve for z. Let's use equation (1) and substitute our two numbers in it:

$$
\begin{array}{ll}
x + y + z = 4 & \textbf{(1)} \\
2 + (-1) + z = 4 & \text{Substituting 2 for } x \text{ and } -1 \text{ for } y \\
\left.\begin{array}{l} 1 + z = 4 \\ z = 3. \end{array}\right\} & \text{Solving for } z
\end{array}
$$

We have obtained the ordered triple $(2, -1, 3)$. To check, we substitute $(2, -1, 3)$ into each of the three equations using alphabetical order of the variables.

Check:

$$
\begin{array}{c}
x + y + z = 4 \\
\hline
2 + (-1) + 3 \;?\; 4 \\
4 \;\big|\; \quad \text{TRUE}
\end{array}
$$

$$
\begin{array}{c}
x - 2y - z = 1 \\
\hline
2 - 2(-1) - 3 \;?\; 1 \\
2 + 2 - 3 \;\big|\; \\
1 \;\big|\; \quad \text{TRUE}
\end{array}
$$

$$
\begin{array}{c}
2x - y - 2z = -1 \\
\hline
2(2) - (-1) - 2 \cdot 3 \;?\; -1 \\
4 + 1 - 6 \;\big|\; \\
-1 \;\big|\; \quad \text{TRUE}
\end{array}
$$

The triple $(2, -1, 3)$ checks and is the solution.

To use the elimination method to solve systems of three equations:

1. Write all equations in the standard form, $Ax + By + Cz = D$.
2. Clear any decimals or fractions.
3. Choose a variable to eliminate. Then use *any* two of the three equations to eliminate that variable, getting an equation in two variables.
4. Next, use a different pair of equations and get another equation in *the same two variables*. That is, eliminate the same variable that you did in step (3).
5. Solve the resulting system (pair) of equations. That will give two of the numbers.
6. Then use any of the original three equations to find the third number.

Do Exercise 1. ▶

1. Solve. Don't forget to check.
$$4x - y + z = 6,$$
$$-3x + 2y - z = -3,$$
$$2x + y + 2z = 3$$

EXAMPLE 2 Solve this system:

$$4x - 2y - 3z = 5, \qquad (1)$$
$$-8x - y + z = -5, \qquad (2)$$
$$2x + y + 2z = 5. \qquad (3)$$

a) The equations are in standard form and do not contain decimals or fractions.

b) We decide to eliminate the variable y since the y-terms are opposites in equations (2) and (3). We add:

$$-8x - y + z = -5 \qquad (2)$$
$$\underline{2x + y + 2z = \ \ 5} \qquad (3)$$
$$-6x \qquad + 3z = \ \ 0. \qquad (4) \qquad \text{Adding}$$

c) We use another pair of equations to get an equation in the same two variables, x and z. We use equations (1) and (3) and eliminate y:

$$4x - 2y - 3z = 5, \qquad (1)$$
$$2x + y + 2z = 5; \qquad (3)$$

$$4x - 2y - 3z = \ \ 5 \qquad (1)$$
$$\underline{4x + 2y + 4z = 10} \qquad \text{Multiplying equation (3) by 2}$$
$$8x \qquad + z = 15. \qquad (5) \qquad \text{Adding}$$

d) Next, we solve the resulting system of equations (4) and (5). That will give us two of the numbers:

$$-6x + 3z = 0, \qquad (4)$$
$$8x + z = 15. \qquad (5)$$

We multiply equation (5) by -3:

$$-6x + 3z = \qquad 0 \qquad (4)$$
$$\underline{-24x - 3z = -45} \qquad \text{Multiplying equation (5) by } -3$$
$$-30x \qquad = -45 \qquad \text{Adding}$$
$$x = \frac{-45}{-30} = \frac{3}{2}.$$

Answer

1. $(2, 1, -1)$

We now use equation (5) to find z:

$$8x + z = 15 \qquad \textbf{(5)}$$

$$8\left(\tfrac{3}{2}\right) + z = 15 \qquad \text{Substituting } \tfrac{3}{2} \text{ for } x$$

$$\left.\begin{array}{r} 12 + z = 15 \\ z = 3. \end{array}\right\} \quad \text{Solving for } z$$

e) Next, we use any of the original equations and substitute to find the third number, y. We choose equation (3) since the coefficient of y there is 1:

$$2x + y + 2z = 5 \qquad \textbf{(3)}$$

$$2\left(\tfrac{3}{2}\right) + y + 2(3) = 5 \qquad \text{Substituting } \tfrac{3}{2} \text{ for } x \text{ and } 3 \text{ for } z$$

$$\left.\begin{array}{r} 3 + y + 6 = 5 \\ y + 9 = 5 \\ y = -4. \end{array}\right\} \quad \text{Solving for } y$$

The solution is $\left(\tfrac{3}{2}, -4, 3\right)$. The check is as follows.

Check:

$$\begin{array}{c} \underline{4x - 2y - 3z = 5} \\ 4 \cdot \tfrac{3}{2} - 2(-4) - 3(3) \ ? \ 5 \\ 6 + 8 - 9 \ \Big| \\ 5 \ \Big| \qquad \text{TRUE} \end{array}$$

$$\begin{array}{c} \underline{-8x - y + z = -5} \\ -8 \cdot \tfrac{3}{2} - (-4) + 3 \ ? \ -5 \\ -12 + 4 + 3 \ \Big| \\ -5 \ \Big| \qquad \text{TRUE} \end{array}$$

$$\begin{array}{c} \underline{2x + y + 2z = 5} \\ 2 \cdot \tfrac{3}{2} + (-4) + 2(3) \ ? \ 5 \\ 3 - 4 + 6 \ \Big| \\ 5 \ \Big| \qquad \text{TRUE} \end{array}$$

2. Solve. Don't forget to check.

$$\begin{aligned} 2x + y - 4z &= 0, \\ x - y + 2z &= 5, \\ 3x + 2y + 2z &= 3 \end{aligned}$$

◀ **Do Exercise 2.**

In Example 3, two of the equations have a missing variable.

EXAMPLE 3 Solve this system:

$$\begin{aligned} x + y + z &= 180, \qquad \textbf{(1)} \\ x \quad - z &= -70, \qquad \textbf{(2)} \\ 2y - z &= 0. \qquad \textbf{(3)} \end{aligned}$$

We note that there is no y in equation (2). In order to have a system of two equations in the variables x and z, we need to find another equation without a y. We use equations (1) and (3) to eliminate y:

$$\begin{aligned} x + y + z &= 180, \qquad \textbf{(1)} \\ 2y - z &= 0; \qquad \textbf{(3)} \end{aligned}$$

$$\begin{array}{r} -2x - 2y - 2z = -360 \qquad \text{Multiplying equation (1) by } -2 \\ \underline{2y - z = \quad 0} \qquad \textbf{(3)} \\ -2x \quad - 3z = -360. \qquad \textbf{(4)} \qquad \text{Adding} \end{array}$$

Answer

2. $\left(2, -2, \tfrac{1}{2}\right)$

Next, we solve the resulting system of equations (2) and (4):

$$x - z = -70, \quad (2)$$
$$-2x - 3z = -360; \quad (4)$$

$$
\begin{array}{ll}
2x - 2z = -140 & \text{Multiplying equation (2) by 2} \\
-2x - 3z = -360 & (4) \\
\hline
-5z = -500 & \text{Adding} \\
z = 100.
\end{array}
$$

To find x, we substitute 100 for z in equation (2) and solve for x:

$$x - z = -70$$
$$x - 100 = -70$$
$$x = 30.$$

To find y, we substitute 100 for z in equation (3) and solve for y:

$$2y - z = 0$$
$$2y - 100 = 0$$
$$2y = 100$$
$$y = 50.$$

The triple $(30, 50, 100)$ is the solution. The check is left to the student.

Do Exercise 3. ▶

It is possible for a system of three equations to have no solution, that is, to be inconsistent. An example is the system

$$x + y + z = 14,$$
$$x + y + z = 11,$$
$$2x - 3y + 4z = -3.$$

Note the first two equations. It is not possible for a sum of three numbers to be both 14 and 11. Thus the system has no solution. We will not consider such systems here, nor will we consider systems with infinitely many solutions, which also exist.

 3. Solve. Don't forget to check.
$$x + y + z = 100, \quad (1)$$
$$x - y \phantom{{}+ z} = -10, \quad (2)$$
$$x \phantom{{}+ y} - z = -30 \quad (3)$$

Add equations (1) and (3):
$$
\begin{array}{ll}
x + y + z = 100 & (1) \\
x \phantom{{}+ y} - z = -30 & (3) \\
\hline
2x + y \phantom{{}- z} = \boxed{}. & (4)
\end{array}
$$

Add equations (2) and (4) and solve for x:
$$
\begin{array}{ll}
x - y = -10 & (2) \\
2x + y = 70 & (4) \\
\hline
3x \phantom{{}+ y} = \boxed{} \\
x = \boxed{}.
\end{array}
$$

Substitute 20 for x in equation (4) and solve for y:
$$2(20) + y = 70$$
$$y = \boxed{}.$$

Substitute 20 for x and 30 for y in equation (1) and solve for z:
$$20 + 30 + z = 100$$
$$z = \boxed{}.$$

The numbers check. The solution is $(20, 30, \boxed{})$.

Answer
3. $(20, 30, 50)$

Guided Solution:
3. 70, 60, 20, 30, 50, 50

17.5 **Exercise Set**

✓ Reading Check

Choose from the column on the right the option that is an example of each term. Choices may be used more than once.

RC1. A linear equation in three variables

RC2. A system of equations in three variables

RC3. A solution of a linear equation in three variables

RC4. A solution of a system of equations in three variables

a) $(4, -3, 0)$

b) $a + b - c = 1$

c) $a + 3b - c = 1,$
$ 2a + 3b - c = -1,$
$ a - 2b + 3c = 10$

1. $x + y + z = 2,$
$2x - y + 5z = -5,$
$-x + 2y + 2z = 1$

2. $2x - y - 4z = -12,$
$2x + y + z = 1,$
$x + 2y + 4z = 10$

3. $2x - y + z = 5,$
$6x + 3y - 2z = 10,$
$x - 2y + 3z = 5$

4. $x - y + z = 4,$
$3x + 2y + 3z = 7,$
$2x + 9y + 6z = 5$

5. $2x - 3y + z = 5,$
$x + 3y + 8z = 22,$
$3x - y + 2z = 12$

6. $6x - 4y + 5z = 31,$
$5x + 2y + 2z = 13,$
$x + y + z = 2$

7. $3a - 2b + 7c = 13,$
$a + 8b - 6c = -47,$
$7a - 9b - 9c = -3$

8. $x + y + z = 0,$
$2x + 3y + 2z = -3,$
$-x + 2y - 3z = -1$

9. $2x + 3y + z = 17,$
$x - 3y + 2z = -8,$
$5x - 2y + 3z = 5$

10. $2x + y - 3z = -4,$
$4x - 2y + z = 9,$
$3x + 5y - 2z = 5$

11. $2x + y + z = -2,$
$2x - y + 3z = 6,$
$3x - 5y + 4z = 7$

12. $2x + y + 2z = 11,$
$3x + 2y + 2z = 8,$
$x + 4y + 3z = 0$

13. $x - y + z = 4,$
$5x + 2y - 3z = 2,$
$3x - 7y + 4z = 8$

14. $2x + y + 2z = 3,$
$x + 6y + 3z = 4,$
$3x - 2y + z = 0$

15. $4x - y - z = 4,$
$2x + y + z = -1,$
$6x - 3y - 2z = 3$

16. $2r + s + t = 6,$
$3r - 2s - 5t = 7,$
$r + s - 3t = -10$

17. $a - 2b - 5c = -3,$
$3a + b - 2c = -1,$
$2a + 3b + c = 4$

18. $x + 4y - z = 5,$
$2x - y + 3z = -5,$
$4x + 3y + z = 5$

19. $2r + 3s + 12t = 4,$
$4r - 6s + 6t = 1,$
$r + s + t = 1$

20. $10x + 6y + z = 7,$
$5x - 9y - 2z = 3,$
$15x - 12y + 2z = -5$

21. $a + 2b + c = 1,$
$7a + 3b - c = -2,$
$a + 5b + 3c = 2$

22. $3p + 2r = 11,$
$q - 7r = 4,$
$p - 6q = 1$

23. $x + y + z = 57,$
$-2x + y = 3,$
$x - z = 6$

24. $4a + 9b = 8,$
$8a + 6c = -1,$
$6b + 6c = -1$

25. $r + s = 5,$
$3s + 2t = -1,$
$4r + t = 14$

26. $a - 5c = 17,$
$b + 2c = -1,$
$4a - b - 3c = 12$

27. $x + y + z = 105,$
$10y - z = 11,$
$2x - 3y = 7$

Skill Maintenance

Solve for the indicated letter.   [11.4b]

28. $F = 3ab,$ for a

29. $Q = 4(a + b),$ for a

30. $F = \frac{1}{2}t(c - d),$ for d

31. $F = \frac{1}{2}t(c - d),$ for c

32. $Ax + By = c,$ for y

33. $Ax - By = c,$ for y

Find the slope and the y-intercept.   [16.3b]

34. $y = -\frac{2}{3}x - \frac{5}{4}$

35. $y = 5 - 4x$

36. $2x - 5y = 10$

37. $7x - 6.4y = 20$

Synthesis

Solve.

38. $w + x - y + z = 0,$
$w - 2x - 2y - z = -5,$
$w - 3x - y + z = 4,$
$2w - x - y + 3z = 7$

39. $w + x + y + z = 2,$
$w + 2x + 2y + 4z = 1,$
$w - x + y + z = 6,$
$w - 3x - y + z = 2$

17.6

Solving Applied Problems: Three Equations

OBJECTIVE

a Solve applied problems using systems of three equations.

SKILL TO REVIEW

Objective 11.6a: Solve applied problems by translating to equations.

Solve.

1. The second angle of a triangle is twice as large as the first angle. The third angle is 5° larger than the second angle. Find the measures of the angles.

2. Giovanni invested his tax refund check in a fund paying 5% interest. After one year, he had earned $63 in interest. How much did he invest?

---

a USING SYSTEMS OF THREE EQUATIONS

Solving systems of three or more equations is important in many applications occurring in the natural and social sciences, business, and engineering.

EXAMPLE 1 *Jewelry Design.* Kim is designing a triangular-shaped pendant for a client of her custom jewelry business. The largest angle of the triangle is 70° greater than the smallest angle. The largest angle is twice as large as the remaining angle. Find the measure of each angle.

1. **Familiarize.** We first make a drawing. We let $x =$ the smallest angle, $z =$ the largest angle, and $y =$ the remaining angle.

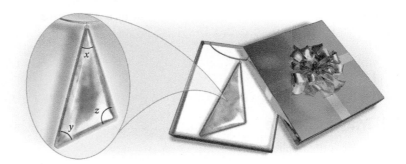

2. **Translate.** In order to translate the problem, we use the fact that the sum of the measures of the angles of a triangle is 180°:

$$x + y + z = 180.$$

There are two statements in the problem that we can translate directly.

The largest angle is 70° greater than the smallest angle.

$$z \quad = \quad 70 \quad + \quad x$$

The largest angle is twice as large as the remaining angle.

$$z \quad = \quad 2y$$

We now have a system of three equations:

$$x + y + z = 180, \qquad x + y + z = 180,$$
$$x + 70 = z, \qquad \text{or} \quad x \quad - z = -70,$$
$$2y = z; \qquad \qquad 2y - z = 0.$$

3. **Solve.** Solving the system, we find that the solution is $(30, 50, 100)$.

Answers

Skill to Review:
1. 35°, 70°, 75°   2. $1260

4. Check. The sum of the numbers is 180. The largest angle measures 100° and the smallest measures 30°, so the largest angle is 70° greater than the smallest. The largest angle is twice as large as 50°, the remaining angle. We have an answer to the problem.

5. State. The measures of the angles of the triangle are 30°, 50°, and 100°.

Do Exercise 1. ▷

1. *Triangle Measures.* One angle of a triangle is twice as large as a second angle. The remaining angle is 20° greater than the first angle. Find the measure of each angle.

EXAMPLE 2 *Super Bowl Commercials.* For commercials aired during Super Bowl XLVII, advertisers paid an average of $3.8 million to air a 30-sec commercial. Even at this rate, a number of commercials were longer than 30 sec. A total of 42 commercials ran for either 30 sec, 1 min, or $1\frac{1}{2}$ min. Together, these 42 commercials ran for 30 min. The number of 30-sec commercials was 10 more than the sum of the number of 1-min and $1\frac{1}{2}$-min commercials. How many of each length commercial aired during the Super Bowl?

Sources: businessinsider.com, kantarmediana.com

1. Familiarize. As we read the problem, we note that the price paid to air the commercials is not needed to solve the problem. We also note that the units of time are not all the same, so we convert 30 sec to $\frac{1}{2}$ min. We let $x =$ the number of $\frac{1}{2}$-min commercials, $y =$ the number of 1-min commercials, and $z =$ the number of $1\frac{1}{2}$-min commercials.

2. Translate. We can now translate three statements to equations.

A total of 42 commercials ran. →	$x + y + z = 42$
The commercials ran for 30 min. →	$\frac{1}{2}x + y + 1\frac{1}{2}z = 30$
The number of 30-sec commercials was 10 more than the sum of the number of 1-min and $1\frac{1}{2}$-min commercials. →	$x = y + z + 10$

3. Solve. We write the equations in standard form and convert the mixed numeral to fraction notation:

$$x + y + z = 42,$$
$$\tfrac{1}{2}x + y + \tfrac{3}{2}z = 30,$$
$$x - y - z = 10.$$

After clearing fractions, we have the system

$$x + y + z = 42, \quad (1)$$
$$x + 2y + 3z = 60, \quad (2)$$
$$x - y - z = 10. \quad (3)$$

This system is unusual, because we can eliminate *both* y and z by adding equations (1) and (3):

$$x + y + z = 42 \quad (1)$$
$$\underline{x - y - z = 10} \quad (3)$$
$$2x \qquad\quad = 52$$
$$x = 26.$$

2. Client Investments.
Client Investments. Kaufman Financial Corporation makes investments for corporate clients. One year, a client receives $1120 in simple interest from three investments that total $25,000. Part is invested at 3%, part at 4%, and part at 5%. There is $11,000 more invested at 5% than at 4%. How much was invested at each rate?

Let x = the amount invested at 3%, y = the amount invested at 4%, and z = the amount invested at 5%. Complete the following table to help in the translation.

	PRINCIPAL, P	RATE OF INTEREST, r	TIME, t	INTEREST, I
FIRST INVESTMENT	x	3%	1 year	0.03x
SECOND INVESTMENT	y	4%	1 year	0.04y
THIRD INVESTMENT	z	[ ]%	1 year	0.05z
TOTAL	[ ] $			[ ] $

We can now substitute 26 for x in equations (1) and (2) and solve for y and z.

Equation (1) becomes

$$26 + y + z = 42 \qquad \text{Substituting 26 for } x$$
$$y + z = 16. \qquad \text{Simplifying}$$

Equation (2) becomes

$$26 + 2y + 3z = 60 \qquad \text{Substituting 26 for } x$$
$$2y + 3z = 34. \qquad \text{Simplifying}$$

We then solve the following system for y and z:

$$y + z = 16, \qquad (4)$$
$$2y + 3z = 34. \qquad (5)$$

Multiplying equation (4) by -2 and adding, we have

$$-2y - 2z = -32$$
$$\underline{2y + 3z = 34}$$
$$z = 2.$$

Finally, we find y by substituting 2 for z in equation (4):

$$y + 2 = 16 \qquad \text{Substituting 2 for } z$$
$$y = 14.$$

We have $x = 26$, $y = 14$, and $z = 2$.

4. **Check.** We check our answers in each statement of the problem.
   - The total number of commercials is $26 + 14 + 2 = 42$.
   - The total time for the commercials is
     $$\tfrac{1}{2}(26) + (14) + 1\tfrac{1}{2}(2) = 13 + 14 + 3 = 30.$$
   - Ten more than the sum of the number of 1-min and $1\tfrac{1}{2}$-min commercials is $14 + 2 + 10 = 26$, which is the number of 30-sec commercials. The answer checks.

5. **State.** There were 26 30-sec commercials, 14 1-min commercials, and 2 $1\tfrac{1}{2}$-min commercials.

◀ Do Exercise 2.

✓ Reading Check

Match each statement with an appropriate translation from the column on the right.

RC1. The sum of three numbers is 60.

RC2. The first number minus the second number plus the third number is 60.

RC3. The first number is 60 more than the sum of the other two numbers.

RC4. The first number is 60 less than the sum of the other two numbers.

a) $x = y + z + 60$

b) $x = y + z - 60$

c) $x + y + z = 60$

d) $x - y + z = 60$

a Solve.

1. *Scholastic Aptitude Test.* More than 1.66 million members of the class of 2012 took the Scholastic Aptitude Test, making it the largest class of SAT takers in history. Students taking the SAT receive a critical reading score, a mathematics score, and a writing score. The average total score of the students from the class of 2012 was 1498. The average math score exceeded the average reading score by 18 points. The average math score was 470 points less than the sum of the average reading and writing scores. Find the average score on each part of the test.

Source: College Board

2. *Fat Content of Fast Food.* A meal at McDonald's consisting of a Big Mac, a medium order of fries, and a 21-oz vanilla milkshake contains 66 g of fat. The Big Mac has 11 more grams of fat than the milkshake. The total fat content of the fries and the shake exceeds that of the Big Mac by 8 g. Find the fat content of each food item.

Source: McDonald's

3. *Triangle Measures.* In triangle ABC, the measure of angle B is three times that of angle A. The measure of angle C is 20° more than that of angle A. Find the measure of each angle.

4. *Triangle Measures.* In triangle ABC, the measure of angle B is twice the measure of angle A. The measure of angle C is 80° more than that of angle A. Find the measure of each angle.

5. The sum of three numbers is 55. The difference of the largest and the smallest is 49, and the sum of the two smaller numbers is 13. Find the numbers.

6. *History.* Find the year in which the first U.S. transcontinental railroad was completed. The following are some facts about the number. The sum of the digits in the year is 24. The ones digit is 1 more than the hundreds digit. Both the tens and the ones digits are multiples of 3.

7. *Smoothies.* Jamba Juice sells fruit and veggie smoothies in three sizes: a 16-oz "Sixteen," a 22-oz "Original," and a 30-oz "Power." A Sixteen smoothie sells for $3.90, an Original smoothie for $4.90, and a Power smoothie for $5.70. One hot summer afternoon, Elliot sold 34 smoothies for a total of $163. In all, he sold 752 oz of smoothies. How many of each size did he sell?

Source: Jamba Juice

8. *Coffee.* A Starbucks® store on campus sells coffee in three sizes: a 12-oz tall, a 16-oz grande, and a 20-oz venti. A tall coffee sells for $1.75, a grande coffee for $1.95, and a venti coffee for $2.25. One morning, Brandie served 50 coffees for a total of $98.70. She made the coffee in 80-oz batches, and used exactly 10 of the batches during the morning. How many of each size did she sell?

Source: Starbucks®

9. *Cholesterol Levels.* Recent studies indicate that a child's intake of cholesterol should be no more than 300 mg per day. By eating 1 egg, 1 cupcake, and 1 slice of pizza, a child consumes 302 mg of cholesterol. If the child eats 2 cupcakes and 3 slices of pizza, he or she takes in 65 mg of cholesterol. By eating 2 eggs and 1 cupcake, a child consumes 567 mg of cholesterol. How much cholesterol is in each item?

10. *Book Sale.* Katie, Rachel, and Logan went together to a library book sale. Katie bought 22 children's books, 10 paperbacks, and 5 hardbacks for a total of $63.50. Rachel bought 12 paperbacks and 15 hardbacks for a total of $52.50. Logan bought 8 children's books and 6 hardbacks for a total of $29.00. How much did each type of book cost?

11. *Automobile Pricing.* A recent basic model of a particular automobile had a price of $14,685. The basic model with the added features of automatic transmission and power door locks was $16,070. The basic model with air conditioning (AC) and power door locks was $15,580. The basic model with AC and automatic transmission was $15,925. What was the individual cost of each of the three options?

12. *Computer Pricing.* Lindsay plans to buy a new desktop computer for gaming. The base price of the computer is $480. If she upgrades the processor and the memory, the price of the computer is $745. If she upgrades the memory and the graphics card, the price of the computer is $690. If she upgrades the processor and the graphics card, the price of the computer is $805. What is the price of each upgrade?

13. *Veterinary Expenditure.* The sum of the average amounts Americans spent, per animal, for veterinary expenses for dogs, cats, and birds in a recent year was $290. The average expenditure per dog exceeded the sum of the averages for cats and birds by $110. The amount spent per cat was nine times the amount spent per bird. Find the average amount spent on each type of animal.

Source: American Veterinary Medical Association

14. *Nutrition Facts.* A meal at Subway consisting of a 6-in. turkey breast sandwich, a bowl of minestrone soup, and a chocolate chip cookie contains 580 calories. The number of calories in the sandwich is 20 less than in the soup and the cookie together. The cookie has 120 calories more than the soup. Find the number of calories in each item.

Source: Subway

15. *Nutrition.* A dietician in a hospital prepares meals under the guidance of a physician. Suppose that for a particular patient a physician prescribes a meal to have 800 calories, 55 g of protein, and 220 mg of vitamin C. The dietician prepares a meal of roast beef, baked potato, and broccoli according to the data in the following table. How many servings of each food are needed in order to satisfy the doctor's orders?

16. *Nutrition.* Repeat Exercise 15 but replace the broccoli with asparagus, for which one 180-g serving contains 50 calories, 5 g of protein, and 44 mg of vitamin C. Which meal would you prefer eating?

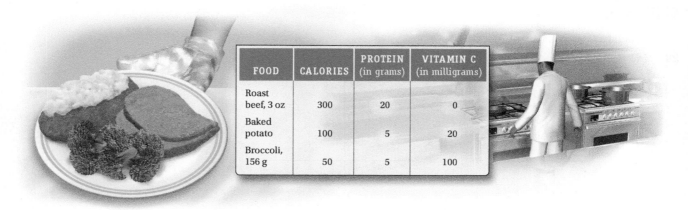

FOOD	CALORIES	PROTEIN (in grams)	VITAMIN C (in milligrams)
Roast beef, 3 oz	300	20	0
Baked potato	100	5	20
Broccoli, 156 g	50	5	100

17. *Investments.* A business class divided an imaginary investment of $80,000 among three mutual funds. The first fund grew by 2%, the second by 6%, and the third by 3%. Total earnings were $2250. The earnings from the first fund were $150 more than the earnings from the third. How much was invested in each fund?

18. *Student Loans.* Terrence owes $32,000 in student loans. The interest rate on his Perkins loan is 5%, the rate on his Stafford loan is 4%, and the rate on his bank loan is 7%. Interest for one year totaled $1500. The interest for one year from the Perkins loan is $220 more than the interest from the bank loan. What is the amount of each loan?

19. *Golf.* On an 18-hole golf course, there are par-3 holes, par-4 holes, and par-5 holes. A golfer who shoots par on every hole has a total of 70. There are twice as many par-4 holes as there are par-5 holes. How many of each type of hole are there on the golf course?

20. *Basketball Scoring.* The New York Knicks once scored a total of 92 points on a combination of 2-point field goals, 3-point field goals, and 1-point foul shots. Altogether, the Knicks made 50 baskets and 19 more 2-pointers than foul shots. How many shots of each kind were made?

21. *Lens Production.* When Sight-Rite's three polishing machines, A, B, and C, are all working, 5700 lenses can be polished in one week. When only A and B are working, 3400 lenses can be polished in one week. When only B and C are working, 4200 lenses can be polished in one week. How many lenses can be polished in a week by each machine alone?

22. *Telemarketing.* Steve, Teri, and Isaiah can process 740 telephone orders per day. Steve and Teri together can process 470 orders, while Teri and Isaiah together can process 520 orders per day. How many orders can each person process alone?

Skill Maintenance

Graph each function.   [16.1c]

23. $f(x) = 2x - 3$

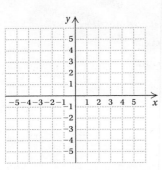

24. $g(x) = |x + 1|$

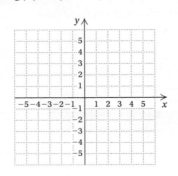

25. $h(x) = x^2 - 2$

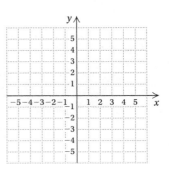

Determine whether each of the following is the graph of a function.   [16.1d]

26.

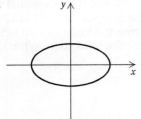

27.

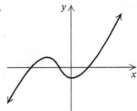

28.

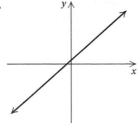

Synthesis

29. Find the sum of the angle measures at the tips of the star in this figure.

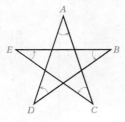

30. *Sharing Raffle Tickets.* Hal gives Tom as many raffle tickets as Tom has and Gary as many as Gary has. In like manner, Tom then gives Hal and Gary as many tickets as each then has. Similarly, Gary gives Hal and Tom as many tickets as each then has. If each finally has 40 tickets, with how many tickets does Tom begin?

31. *Digits.* Find a three-digit positive integer such that the sum of all three digits is 14, the tens digit is 2 more than the ones digit, and if the digits are reversed, the number is unchanged.

32. *Ages.* Tammy's age is the sum of the ages of Carmen and Dennis. Carmen's age is 2 more than the sum of the ages of Dennis and Mark. Dennis's age is four times Mark's age. The sum of all four ages is 42. How old is Tammy?

Vocabulary Reinforcement

Complete each statement with the correct term from the column on the right. Some of the choices may be used more than once and some may not be used at all.

algebraic
graphical
independent
dependent
consistent
inconsistent
pair
triple

1. A solution of a system of two equations in two variables is an ordered _____ that makes both equations true.   [17.1a]

2. A(n) _____ system of equations has at least one solution.   [17.1a]

3. The substitution method is a(n) _____ method for solving systems of equations.   [17.2a]

4. A solution of a system of three equations in three variables is an ordered _____ that makes all three equations true.   [17.5a]

5. If, for a system of two equations in two variables, the graphs of the equations are different lines, then the equations are _____.   [17.1a]

Concept Reinforcement

Determine whether each statement is true or false.

_____ **1.** A system of equations with infinitely many solutions is inconsistent.   [17.1a]

_____ **2.** It is not possible for the equations in an inconsistent system of two equations to be dependent.   [17.1a]

_____ **3.** When $(0, b)$ is a solution of each equation in a system of two equations, the graphs of the two equations have the same y-intercept.   [17.1a]

_____ **4.** The system of equations $x = 4$ and $y = -4$ is inconsistent.   [17.1a]

Study Guide

Objective 17.1a   Solve a system of two linear equations or two functions by graphing and determine whether a system is consistent or inconsistent and whether the equations in a system are dependent or independent.

Example   Solve this system of equations graphically. Then classify the system as consistent or inconsistent and the equations as dependent or independent.

$$x - y = 3,$$
$$y = 2x - 4$$

We graph the equations. The point of intersection appears to be $(1, -2)$. This checks in both equations, so it is the solution. The system has one solution, so it is consistent and the equations are independent.

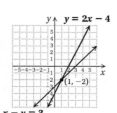

Practice Exercise

1. Solve this system of equations graphically. Then classify the system as consistent or inconsistent and the equations as dependent or independent.

$$x + 3y = 1,$$
$$x + \ y = 3$$

Objective 17.2a Solve systems of equations in two variables by the substitution method.

Example Solve the system
$$x - 2y = 1, \quad \textbf{(1)}$$
$$2x - 3y = 3. \quad \textbf{(2)}$$

We solve equation (1) for x, since the coefficient of x is 1 in that equation:
$$x - 2y = 1$$
$$x = 2y + 1. \quad \textbf{(3)}$$

Next, we substitute for x in equation (2) and solve for y:
$$2x - 3y = 3$$
$$2(2y + 1) - 3y = 3$$
$$4y + 2 - 3y = 3$$
$$y + 2 = 3$$
$$y = 1.$$

Then we substitute 1 for y in equation (1), (2), or (3) and find x. We choose equation (3) since it is already solved for x:
$$x = 2y + 1 = 2 \cdot 1 + 1 = 2 + 1 = 3$$

Check:

$x - 2y = 1$	$2x - 3y = 3$
$3 - 2 \cdot 1 \overset{?}{\mid} 1$	$2 \cdot 3 - 3 \cdot 1 \overset{?}{\mid} 3$
$3 - 2$	$6 - 3$
$1 \mid$ TRUE	$3 \mid$ TRUE

The ordered pair $(3, 1)$ checks in both equations, so it is the solution of the system of equations.

Practice Exercise

2. Solve the system
$$2x + y = 2,$$
$$3x + 2y = 5$$
using the substitution method.

Objective 17.3a Solve systems of equations in two variables by the elimination method.

Example Solve the system
$$2a + 3b = -1, \quad \textbf{(1)}$$
$$3a + 2b = 6. \quad \textbf{(2)}$$

We could eliminate either a or b. In this case, we decide to eliminate the a-terms. We multiply equation (1) by 3 and equation (2) by -2 and then add and solve for b:
$$6a + 9b = -3$$
$$\underline{-6a - 4b = -12}$$
$$5b = -15$$
$$b = -3.$$

Next, we substitute -3 for b in either of the original equations:
$$2a + 3b = -1 \quad \textbf{(1)}$$
$$2a + 3(-3) = -1$$
$$2a - 9 = -1$$
$$2a = 8$$
$$a = 4.$$

The ordered pair $(4, -3)$ checks in both equations, so it is a solution of the system of equations.

Practice Exercise

3. Solve the system
$$2x + 3y = 5,$$
$$3x + 4y = 6$$
using the elimination method.

Example   To start a small business, Michael took two loans totaling $18,000. One of the loans was at 7% interest and the other at 8%. After one year, Michael owed $1365 in interest. What was the amount of each loan?

1. **Familiarize.**   We let x and y represent the amounts of the two loans. Next, we organize the information in a table and use the simple interest formula, $I = Prt$.

	LOAN 1	LOAN 2	TOTAL
PRINCIPAL	x	y	$18,000
RATE OF INTEREST	7%	8%	
TIME	1 year	1 year	
INTEREST	7%x, or 0.07x	8%y, or 0.08y	$1365

2. **Translate.**   The total amount of the loans is found in the first row of the table. This gives us one equation:

$x + y = 18,000.$

From the last row of the table, we see that the interest totals $1365. This gives us a second equation:

$0.07x + 0.08y = 1365.$

3. **Solve.**   We solve the resulting system of equations:

$x + y = 18,000,$ **(1)**

$0.07x + 0.08y = 1365.$ **(2)**

We multiply by -0.07 on both sides of equation (1) and add:

$$-0.07x - 0.07y = -1260$$
$$\underline{0.07x + 0.08y = \quad 1365} \quad \textbf{(2)}$$
$$0.01y = \quad 105 \quad \text{Adding}$$
$$y = 10,500. \quad \text{Solving for } y$$

Then

$x + 10,500 = 18,000$ Substituting 10,500 for y in equation (1)

$x = 7500.$ Solving for x

We find that $x = 7500$ and $y = 10,500$.

4. **Check.**   The sum is $7500 + $10,500, or $18,000. The interest from $7500 at 7% for one year is 7%($7500), or $525. The interest from $10,500 at 8% for one year is 8%($10,500), or $840. The total amount of interest is $525 + $840, or $1365. The numbers check in the problem.

5. **State.**   Michael took loans of $7500 at 7% interest and $10,500 at 8% interest.

Practice Exercise

4. Jaretta made two investments totaling $23,000. In one year, these investments yielded $1237 in simple interest. Part of the money was invested at 6% and the rest at 5%. How much was invested at each rate?

Objective 17.5a Solve systems of three equations in three variables.

Example Solve:

$$x - y - z = -2, \quad \textbf{(1)}$$
$$2x + 3y + z = 3, \quad \textbf{(2)}$$
$$5x - 2y - 2z = -1. \quad \textbf{(3)}$$

The equations are in standard form and do not contain decimals or fractions. We choose to eliminate z since the z-terms in equations (1) and (2) are opposites.

First, we add these two equations:

$$x - y - z = -2$$
$$\underline{2x + 3y + z = 3}$$
$$3x + 2y \phantom{{}- z} = 1. \quad \textbf{(4)}$$

Next, we multiply equation (2) by 2 and add it to equation (3) to eliminate z from another pair of equations:

$$4x + 6y + 2z = 6$$
$$\underline{5x - 2y - 2z = -1}$$
$$9x + 4y \phantom{{}- 2z} = 5. \quad \textbf{(5)}$$

Now we solve the system consisting of equations (4) and (5). We multiply equation (4) by -2 and add:

$$-6x - 4y = -2$$
$$\underline{9x + 4y = 5}$$
$$3x \phantom{{}- 4y} = 3$$
$$x = 1.$$

Then we use either equation (4) or (5) to find y:

$$3x + 2y = 1 \quad \textbf{(4)}$$
$$3 \cdot 1 + 2y = 1$$
$$3 + 2y = 1$$
$$2y = -2$$
$$y = -1.$$

Finally, we use one of the original equations to find z:

$$2x + 3y + z = 3 \quad \textbf{(2)}$$
$$2 \cdot 1 + 3(-1) + z = 3$$
$$-1 + z = 3$$
$$z = 4.$$

Check:

$$\frac{x - y - z = -2}{1 - (-1) - 4 \; ? \; -2}$$
$$1 + 1 - 4 \quad$$
$$-2 \quad | \quad \text{TRUE}$$

$$\frac{2x + 3y + z = 3}{2 \cdot 1 + 3(-1) + 4 \; ? \; 3}$$
$$2 - 3 + 4 \quad$$
$$3 \quad | \quad \text{TRUE}$$

$$\frac{5x - 2y - 2z = -1}{5 \cdot 1 - 2(-1) - 2 \cdot 4 \; ? \; -1}$$
$$5 + 2 - 8 \quad$$
$$-1 \quad | \quad \text{TRUE}$$

The ordered triple $(1, -1, 4)$ checks in all three equations, so it is the solution of the system of equations.

Practice Exercise

5. Solve:

$$x - y + z = 9,$$
$$2x + y + 2z = 3,$$
$$4x + 2y - 3z = -1.$$

Review Exercises

Solve graphically. Then classify the system as consistent or inconsistent and the equations as dependent or independent.   [17.1a]

1. $4x - y = -9,$
$x - y = -3$

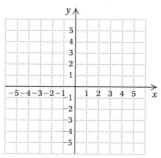

2. $15x + 10y = -20,$
$3x + 2y = -4$

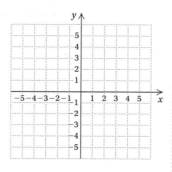

3. $y - 2x = 4,$
$y - 2x = 5$

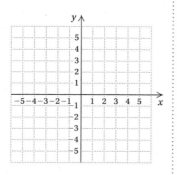

Solve using the substitution method.   [17.2a]

4. $2x - 3y = 5,$
$x = 4y + 5$

5. $y = x + 2,$
$y - x = 8$

6. $7x - 4y = 6,$
$y - 3x = -2$

Solve using the elimination method.   [17.3a]

7. $x + 3y = -3,$
$2x - 3y = 21$

8. $3x - 5y = -4,$
$5x - 3y = 4$

9. $\dfrac{1}{3}x + \dfrac{2}{9}y = 1,$
$\dfrac{3}{2}x + \dfrac{1}{2}y = 6$

10. $1.5x - 3 = -2y,$
$3x + 4y = 6$

11. *Air Travel.*   An airplane flew for 3 hr with a 30-mph tailwind. The return flight against the same wind took 4.5 hr. Find the speed of the plane in still air.   [17.4b]

12. *Retail Sales.*   Paint Town sold 45 paintbrushes, one kind at \$8.50 each and another at \$9.75 each. In all, \$398.75 was taken in for the brushes. How many of each kind were sold?   [17.4a]

13. *Orange Drink Mixtures.*   "Orange Thirst" is 15% orange juice and "Quencho" is 5% orange juice. How many liters of each should be combined in order to get 10 L of a mixture that is 10% orange juice?   [17.4a]

14. *Train Travel.*   A train leaves Watsonville at noon traveling north at 44 mph. One hour later, another train, going 52 mph, travels north on a parallel track. How many hours will the second train travel before it overtakes the first train?   [17.4b]

Solve. [17.5a]

15. $x + 2y + z = 10,$
$\quad 2x - y + z = 8,$
$\quad 3x + y + 4z = 2$

16. $3x + 2y + z = 1,$
$\quad 2x - y - 3z = 1,$
$\quad -x + 3y + 2z = 6$

17. $2x - 5y - 2z = -4,$
$\quad 7x + 2y - 5z = -6,$
$\quad -2x + 3y + 2z = 4$

18. $x + y + 2z = 1,$
$\quad x - y + z = 1,$
$\quad x + 2y + z = 2$

19. *Triangle Measure.*   In triangle ABC, the measure of angle A is four times the measure of angle C, and the measure of angle B is $45°$ more than the measure of angle C. What are the measures of the angles of the triangle? [17.6a]

20. *Popcorn.*   Paul paid a total of $49 for 1 bag of caramel nut crunch popcorn, 1 bag of plain popcorn, and 1 bag of mocha choco latte popcorn. The price of the caramel nut crunch popcorn was six times the price of the plain popcorn and $16 more than the mocha choco latte popcorn. What was the price of each type of popcorn? [17.6a]

21. Solve using the elimination method:
$$x - y = -9,$$
$$y - 2x = 9.$$
The first coordinate of the solution is which of the following? [17.3a]
A. 9
B. -9
C. 0
D. $\frac{9}{2}$

22. The sum of two numbers is -2. The sum of twice one number and the other is 4. One number is which of the following? [17.3b]
A. -6
B. 2
C. 6
D. 8

23. *Motorcycle Travel.*   Sally and Elliot travel on motorcycles toward each other from Chicago and Indianapolis, which are about 350 km apart, and they are biking at rates of 110 km/h and 90 km/h. They started at the same time. In how many hours will they meet? [17.4b]
A. 1.75 hr
B. 3.9 hr
C. 3.2 hr
D. 17.5 hr

Synthesis

24. Solve graphically: [16.1c], [17.1a]
$$y = x + 2,$$
$$y = x^2 + 2.$$

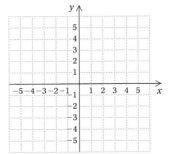

Understanding Through Discussion and Writing

1. Write a problem for a classmate to solve. Design the problem so the answer is "The florist sold 14 hanging baskets and 9 flats of petunias." [17.4a]

2. Exercise 21 in Exercise Set 17.6 can be solved mentally after a careful reading of the problem. Explain how this can be done. [17.6a]

3. *Ticket Revenue.*   A pops-concert audience of 100 people consists of adults, senior citizens, and children. The ticket prices are $10 each for adults, $3 each for senior citizens, and $0.50 each for children. The total amount of money taken in is $100. How many adults, senior citizens, and children are in attendance? Does there seem to be some information missing? Do some careful reasoning and explain. [17.6a]

CHAPTER

17   **Test**

For Extra Help   For step-by-step test solutions, access the Chapter Test Prep Videos in MyMathLab® or on YouTube (search "BittingerAlgebraFoundations" and click on "Channels").

Solve graphically. Then classify the system as consistent or inconsistent and the equations as dependent or independent.

1. $y = 3x + 7,$
$\quad 3x + 2y = -4$

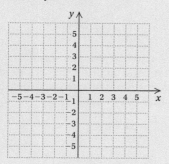

2. $y = 3x + 4,$
$\quad y = 3x - 2$

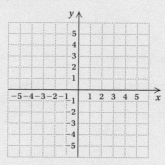

3. $y - 3x = 6,$
$\quad 6x - 2y = -12$

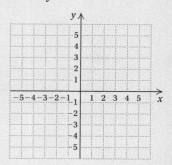

Solve using the substitution method.

4. $4x + 3y = -1,$
$\quad y = 2x - 7$

5. $x = 3y + 2,$
$\quad 2x - 6y = 4$

6. $x + 2y = 6,$
$\quad 2x + 3y = 7$

7. $t = 2 - r,$
$\quad 3r - 2t = 36$

Solve using the elimination method.

8. $2x + 5y = 3,$
$\quad -2x + 3y = 5$

9. $x + y = -2,$
$\quad 4x - 6y = -3$

10. $\dfrac{2}{3}x - \dfrac{4}{5}y = 1,$
$\quad \dfrac{1}{3}x - \dfrac{2}{5}y = 2$

11. $0.3a - 0.4b = 11,$
$\quad 0.7a + 1.2b = -17$

Solve.

12. *Tennis Court.* The perimeter of a standard tennis court used for playing doubles is 288 ft. The width of the court is 42 ft less than the length. Find the length and the width.

13. *Air Travel.* An airplane flew for 5 hr with a 20-km/h tailwind and returned in 7 hr against the same wind. Find the speed of the plane in still air.

14. *Chicken Dinners.* High Flyin' Wings charges $12 for a bucket of chicken wings and $7 for a chicken dinner. After filling 28 orders for buckets and dinners during a football game, the waiters had collected $281. How many buckets and how many dinners did they sell?

15. *Mixing Solutions.* A chemist has one solution that is 20% salt and a second solution that is 45% salt. How many liters of each should be used in order to get 20 L of a solution that is 30% salt?

16. Solve:
$$6x + 2y - 4z = 15,$$
$$-3x - 4y + 2z = -6,$$
$$4x - 6y + 3z = 8.$$

17. *Repair Rates.* An electrician, a carpenter, and a plumber are hired to work on a house. The electrician earns $21 per hour, the carpenter $19.50 per hour, and the plumber $24 per hour. The first day on the job, they worked a total of 21.5 hr and earned a total of $469.50. If the plumber worked 2 hr more than the carpenter did, how many hours did the electrician work?

18. A business class divided an imaginary $30,000 investment among three funds. The first fund grew 2%, the second grew 3%, and the third grew 5%. Total earnings were $990. The earnings from the third fund were $280 more than the earnings from the first. How much was invested at 5%?
   A. $9000   **B.** $10,000
   C. $11,000   **D.** $12,000

Synthesis

19. The graph of the function $f(x) = mx + b$ contains the points $(-1, 3)$ and $(-2, -4)$. Find m and b.

CHAPTER

18

More on Inequalities

18.1

Sets, Inequalities, and Interval Notation

OBJECTIVES

a Determine whether a given number is a solution of an inequality.

b Write interval notation for the solution set or the graph of an inequality.

c Solve an inequality using the addition principle and the multiplication principle and then graph the inequality.

d Solve applied problems by translating to inequalities.

SKILL TO REVIEW

Objective 11.7b: Graph an inequality on the number line.

Graph each inequality.

1. $x > -2$

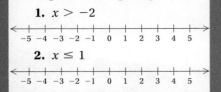

2. $x \leq 1$

---

a INEQUALITIES

We begin this chapter with a review of solving inequalities. Here we will write solution sets using both set-builder notation and *interval notation*.

> **INEQUALITY**
>
> An **inequality** is a sentence containing $<, >, \leq, \geq,$ or $\neq$.

Some examples of inequalities are

$$-2 < a, \quad x > 4, \quad x + 3 \leq 6,$$
$$6 - 7y \geq 10y - 4, \quad \text{and} \quad 5x \neq 10.$$

> **SOLUTION OF AN INEQUALITY**
>
> Any replacement or value for the variable that makes an inequality true is called a **solution** of the inequality. The set of all solutions is called the **solution set**. When all the solutions of an inequality have been found, we say that we have **solved** the inequality.

EXAMPLES   Determine whether the given number is a solution of the inequality.

1. $x + 3 < 6$; 5

We substitute 5 for x and get $5 + 3 < 6$, or $8 < 6$, a *false* sentence. Therefore, 5 is not a solution.

2. $2x - 3 > -3$; 1

We substitute 1 for x and get $2(1) - 3 > -3$, or $-1 > -3$, a *true* sentence. Therefore, 1 is a solution.

3. $4x - 1 \leq 3x + 2$; -3

We substitute -3 for x and get $4(-3) - 1 \leq 3(-3) + 2$, or $-13 \leq -7$, a *true* sentence. Therefore, -3 is a solution.

◀ Do Margin Exercises 1–3 on the following page.

Answers

Skill To Review:

1.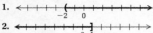

2.

b INEQUALITIES AND INTERVAL NOTATION

The **graph** of an inequality is a drawing that represents its solutions. An inequality in one variable can be graphed on the number line.

EXAMPLE 4 Graph $x < 4$ on the number line.

The solutions are all real numbers less than 4, so we shade all numbers less than 4 on the number line. To indicate that 4 is not a solution, we use a right parenthesis ") " at 4.

We can write the solution set for $x < 4$ using **set-builder notation**: $\{x \mid x < 4\}$. This is read "The set of all x such that x is less than 4."

Another way to write solutions of an inequality in one variable is to use **interval notation**. Interval notation uses parentheses ( ) and brackets [ ].

If a and b are real numbers such that $a < b$, we define the interval (a, b) as the set of all numbers between but not including a and b—that is, the set of all x for which $a < x < b$. Thus,

$$(a, b) = \{x \mid a < x < b\}.$$

The points a and b are the **endpoints** of the interval. The parentheses indicate that the endpoints are *not* included in the graph.

The interval $[a, b]$ is defined as the set of all numbers x for which $a \le x \le b$. Thus,

$$[a, b] = \{x \mid a \le x \le b\}.$$

The brackets indicate that the endpoints *are* included in the graph.*

The following intervals include one endpoint and exclude the other:

$$(a, b] = \{x \mid a < x \le b\}. \quad \text{The graph excludes } a \text{ and includes } b.$$

$$[a, b) = \{x \mid a \le x < b\}. \quad \text{The graph includes } a \text{ and excludes } b.$$

*Some books use the representations ──◇──◇── and ──●──●── instead of, respectively, ──(──)── and ──[──]──.

Determine whether the given number is a solution of the inequality.

1. $3 - x < 2$; 8

2. $3x + 2 > -1$; -2

3. $3x + 2 \le 4x - 3$; 5

.... Caution! ....

Do not confuse the *interval* (a, b) with the *ordered pair* (a, b), which denotes a point in the plane, as we saw in Chapter 3. The context in which the notation appears usually makes the meaning clear.

Answers
1. Yes   2. No   3. Yes

Some intervals extend without bound in one or both directions. We use the symbols ∞, read "infinity," and $-\infty$, read "negative infinity," to name these intervals. The notation (a, ∞) represents the set of all numbers greater than a—that is,

$$(a, \infty) = \{x \,|\, x > a\}.$$

Similarly, the notation $(-\infty, a)$ represents the set of all numbers less than a—that is,

$$(-\infty, a) = \{x \,|\, x < a\}.$$

The notations $[a, \infty)$ and $(-\infty, a]$ are used when we want to include the endpoint a. The interval $(-\infty, \infty)$ names the set of all real numbers.

$$(-\infty, \infty) = \{x \,|\, x \text{ is a real number}\}$$

Interval notation is summarized in the following table.

Intervals: Notation and Graphs

INTERVAL NOTATION	SET NOTATION	GRAPH
(a, b)	$\{x \,|\, a < x < b\}$	
$[a, b]$	$\{x \,|\, a \leq x \leq b\}$	
$[a, b)$	$\{x \,|\, a \leq x < b\}$	
$(a, b]$	$\{x \,|\, a < x \leq b\}$	
(a, ∞)	$\{x \,|\, x > a\}$	
$[a, \infty)$	$\{x \,|\, x \geq a\}$	
$(-\infty, b)$	$\{x \,|\, x < b\}$	
$(-\infty, b]$	$\{x \,|\, x \leq b\}$	
$(-\infty, \infty)$	$\{x \,|\, x \text{ is a real number}\}$	

.... Caution! ....

Whenever the symbol ∞ is included in interval notation, a right parenthesis ") " is used. Similarly, when $-\infty$ is included, a left parenthesis " (" is used.

....

EXAMPLES Write interval notation for the given set or graph.

5. $\{x \,|\, -4 < x < 5\} = (-4, 5)$

6. $\{x \,|\, x \geq -2\} = [-2, \infty)$

7. $\{x \,|\, 7 > x \geq 1\} = \{x \,|\, 1 \leq x < 7\} = [1, 7)$

EXAMPLES Write interval notation for the given graph.

8.

$$-6\ -5\ -4\ -3\ -2\ -1\ \ 0\ \ 1\ \ 2\ \ 3\ \ 4\ \ 5\ \ 6$$

$$(-2, 4]$$

9.

$$-6\ -5\ -4\ -3\ -2\ -1\ \ 0\ \ 1\ \ 2\ \ 3\ \ 4\ \ 5\ \ 6$$

$$(-\infty, -1)$$

Do Exercises 4–8.

Do Exercises 4–8.

c SOLVING INEQUALITIES

Two inequalities are **equivalent** if they have the same solution set. For example, the inequalities $x > 4$ and $4 < x$ are equivalent. Just as the addition principle for equations gives us equivalent equations, the addition principle for inequalities gives us equivalent inequalities.

THE ADDITION PRINCIPLE FOR INEQUALITIES

For any real numbers a, b, and c:

$a < b$ is equivalent to $a + c < b + c$;

$a > b$ is equivalent to $a + c > b + c$.

Similar statements hold for $\leq$ and $\geq$.

Since subtracting c is the same as adding $-c$, there is no need for a separate subtraction principle.

EXAMPLE 10 Solve and graph: $x + 5 > 1$.

We have

$$x + 5 > 1$$
$$x + 5 - 5 > 1 - 5 \qquad \text{Using the addition principle:}$$
$$\text{adding } -5 \text{ or subtracting } 5$$
$$x > -4.$$

We used the addition principle to show that the inequalities $x + 5 > 1$ and $x > -4$ are equivalent. The solution set is $\{x \mid x > -4\}$ and consists of an infinite number of solutions. We cannot possibly check them all. Instead, we can perform a partial check by substituting one member of the solution set (here we use -1) into the original inequality:

$$\frac{x + 5 > 1}{-1 + 5\ \ ?\ \ 1}$$
$$4\ \ \big|\ \ \text{TRUE}$$

Since $4 > 1$ is true, we have a partial check. The solution set is $\{x \mid x > -4\}$, or $(-4, \infty)$. The graph is as follows:

$$(-4, \infty)$$
$$-7\ -6\ -5\ -4\ -3\ -2\ -1\ \ 0\ \ 1\ \ 2\ \ 3\ \ 4\ \ 5\ \ 6\ \ 7$$

Do Exercises 9 and 10.

Do Exercises 9 and 10.

Write interval notation for the given set or graph.

4. $\{x \mid -4 \leq x < 5\}$

5. $\{x \mid x \leq -2\}$

6. $\{x \mid 6 \geq x > 2\}$

7.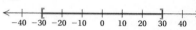

$$-40\ -30\ -20\ -10\ \ \ 0\ \ \ 10\ \ \ 20\ \ \ 30\ \ \ 40$$

8.

$$-40\ -30\ -20\ -10\ \ \ 0\ \ \ 10\ \ \ 20\ \ \ 30\ \ \ 40$$

Solve and graph.

9. $x + 6 > 9$

$$-5\ -4\ -3\ -2\ -1\ \ 0\ \ 1\ \ 2\ \ 3\ \ 4\ \ 5$$

10. $x + 4 \leq 7$

$$-5\ -4\ -3\ -2\ -1\ \ 0\ \ 1\ \ 2\ \ 3\ \ 4\ \ 5$$

Answers

4. $[-4, 5)$ **5.** $(-\infty, -2]$ **6.** $(2, 6]$
7. $[10, \infty)$ **8.** $[-30, 30]$
9. $\{x \mid x > 3\}$, or $(3, \infty)$;

$$0\ \ \ 3$$

10. $\{x \mid x \leq 3\}$, or $(-\infty, 3]$;

$$0\ \ \ 3$$

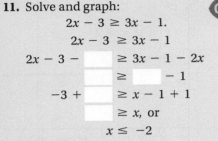

11. Solve and graph:
$$2x - 3 \geq 3x - 1.$$
$$2x - 3 \geq 3x - 1$$
$$2x - 3 - \boxed{} \geq 3x - 1 - 2x$$
$$\boxed{} \geq \boxed{} - 1$$
$$-3 + \boxed{} \geq x - 1 + 1$$
$$\boxed{} \geq x, \text{ or }$$
$$x \leq -2$$

The solution set is
$\{x \mid x \leq -2\}$, or $(-\infty, \boxed{}]$.

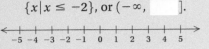

EXAMPLE 11 Solve and graph: $4x - 1 \geq 5x - 2$.

We have

$$4x - 1 \geq 5x - 2$$
$$4x - 1 + 2 \geq 5x - 2 + 2 \qquad \text{Adding 2}$$
$$4x + 1 \geq 5x \qquad \text{Simplifying}$$
$$4x + 1 - 4x \geq 5x - 4x \qquad \text{Subtracting } 4x$$
$$1 \geq x. \qquad \text{Simplifying}$$

The inequalities $1 \geq x$ and $x \leq 1$ have the same meaning and the same solutions. The solution set is $\{x \mid 1 \geq x\}$ or, more commonly, $\{x \mid x \leq 1\}$. Using interval notation, we write that the solution set is $(-\infty, 1]$. The graph is as follows:

◀ **Do Exercise 11.**

The multiplication principle for inequalities differs from the multiplication principle for equations. Consider the true inequality

$$-4 < 9.$$

If we multiply both numbers by 2, we get another true inequality:

$$-4(2) < 9(2), \quad \text{or} \quad -8 < 18. \qquad \text{True}$$

If we multiply both numbers by -3, we get a false inequality:

$$-4(-3) < 9(-3), \quad \text{or} \quad 12 < -27. \qquad \text{False}$$

However, if we now *reverse* the inequality symbol above, we get a true inequality:

$$12 > -27. \qquad \text{True}$$

THE MULTIPLICATION PRINCIPLE FOR INEQUALITIES

For any real numbers a and b, and any *positive* number c:

$$a < b \quad \text{is equivalent to} \quad ac < bc;$$
$$a > b \quad \text{is equivalent to} \quad ac > bc.$$

For any real numbers a and b, and any *negative* number c:

$$a < b \quad \text{is equivalent to} \quad ac > bc;$$
$$a > b \quad \text{is equivalent to} \quad ac < bc.$$

Similar statements hold for $\leq$ and $\geq$.

Since division by c is the same as multiplication by $1/c$, there is no need for a separate division principle.

The multiplication principle tells us that when we multiply or divide on both sides of an inequality by a negative number, we must reverse the inequality symbol to obtain an equivalent inequality.

Answer

11. $\{x \mid x \leq -2\}$, or $(-\infty, -2]$;

Guided Solution:
11. $2x, -3, x, 1, -2, -2$

EXAMPLE 12 Solve and graph: $3y < \frac{3}{4}$.

We have

$$3y < \frac{3}{4}$$

$$\frac{1}{3} \cdot 3y < \frac{1}{3} \cdot \frac{3}{4} \qquad \text{Multiplying by } \frac{1}{3}. \text{ Since } \frac{1}{3} > 0,$$
$$\text{the symbol stays the same.}$$

$$y < \frac{1}{4}. \qquad \text{Simplifying}$$

Any number less than $\frac{1}{4}$ is a solution. The solution set is $\{y | y < \frac{1}{4}\}$, or $(-\infty, \frac{1}{4})$. The graph is as follows:

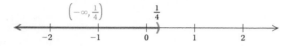

EXAMPLE 13 Solve and graph: $-5x \geq -80$.

We have

$$-5x \geq -80$$

$$\frac{-5x}{-5} \leq \frac{-80}{-5} \qquad \text{Dividing by } -5. \text{ Since } -5 < 0, \text{ the}$$
$$\text{inequality symbol must be reversed.}$$

$$x \leq 16.$$

The solution set is $\{x | x \leq 16\}$, or $(-\infty, 16]$. The graph is as follows:

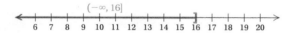

Do Exercises 12–14. ▶

We use the addition and multiplication principles together in solving inequalities in much the same way as in solving equations.

EXAMPLE 14 Solve: $16 - 7y \geq 10y - 4$.

We have

$$16 - 7y \geq 10y - 4$$
$$-16 + 16 - 7y \geq -16 + 10y - 4 \qquad \text{Adding } -16$$
$$-7y \geq 10y - 20 \qquad \text{Collecting like terms}$$
$$-10y + (-7y) \geq -10y + 10y - 20 \qquad \text{Adding } -10y$$
$$-17y \geq -20 \qquad \text{Collecting like terms}$$

$$\frac{-17y}{-17} \leq \frac{-20}{-17} \qquad \text{Dividing by } -17. \text{ The symbol}$$
$$\text{must be reversed.}$$

$$y \leq \frac{20}{17}. \qquad \text{Simplifying}$$

The solution set is $\{y | y \leq \frac{20}{17},\}$ or $(-\infty, \frac{20}{17}]$.

We can avoid multiplying or dividing by a negative number by using the addition principle in a different way. Let's rework Example 14 by adding $7y$ instead of $-10y$.

Solve and graph.

12. $5y \leq \frac{3}{2}$

<---+--+--+--+--+--+--+--+--+--+--->
-5 -4 -3 -2 -1 0 1 2 3 4 5

13. $-2y > 10$

<---+--+--+--+--+--+--+--+--+--+--->
-5 -4 -3 -2 -1 0 1 2 3 4 5

14. $-\frac{1}{3}x \leq -4$

<---+--+--+--+--+--+--+--+--->
-40 -30 -20 -10 0 10 20 30 40

Answers

12. $\{y | y \leq \frac{3}{10}\}$, or $(-\infty, \frac{3}{10}]$

<---+--+--+--+--+--+--+--+--+--->
0 $\frac{3}{10}$

13. $\{y | y < -5\}$, or $(-\infty, -5)$

<---+--+--+--+--+--+--+--+--->
-5 0

14. $\{x | x \geq 12\}$, or $[12, \infty)$

<---+--+--+--+--+--+--+--+--->
0 4 12

$$16 - 7y \geq 10y - 4$$

$$16 - 7y + 7y \geq 10y - 4 + 7y \qquad \text{Adding } 7y. \text{ This makes the coefficient of the } y\text{-term positive.}$$

$$16 \geq 17y - 4 \qquad \text{Collecting like terms}$$

$$16 + 4 \geq 17y - 4 + 4 \qquad \text{Adding } 4$$

$$20 \geq 17y \qquad \text{Collecting like terms}$$

$$\frac{20}{17} \geq \frac{17y}{17} \qquad \text{Dividing by 17. The symbol stays the same.}$$

$$\frac{20}{17} \geq y, \text{ or } y \leq \frac{20}{17}$$

Solve.

15. $6 - 5y \geq 7$ ⒼⓈ

$$6 - 5y \geq 7$$

$$6 - 5y - 6 \geq 7 - \boxed{}$$

$$\boxed{} \geq 1$$

$$\frac{-5y}{-5} \boxed{} \frac{1}{\boxed{}}$$

$$\boxed{} \leq -\frac{1}{5}$$

The solution set is

$$\left\{ y \mid y \leq -\frac{1}{5} \right\}, \text{ or } \left(\boxed{}, -\frac{1}{5} \right].$$

16. $3x + 5x < 4$

17. $17 - 5(y - 2) \leq$
$45y + 8(2y - 3) - 39y$

EXAMPLE 15 Solve: $-3(x + 8) - 5x > 4x - 9$.

$$-3(x + 8) - 5x > 4x - 9$$

$$-3x - 24 - 5x > 4x - 9 \qquad \text{Using the distributive law}$$

$$-24 - 8x > 4x - 9 \qquad \text{Collecting like terms}$$

$$-24 - 8x + 8x > 4x - 9 + 8x \qquad \text{Adding } 8x$$

$$-24 > 12x - 9 \qquad \text{Collecting like terms}$$

$$-24 + 9 > 12x - 9 + 9 \qquad \text{Adding } 9$$

$$-15 > 12x$$

Dividing by 12. The symbol stays the same.

$$\frac{-15}{12} > \frac{12x}{12}$$

$$-\frac{5}{4} > x.$$

The solution set is $\left\{ x \mid -\frac{5}{4} > x \right\}$, or $\left\{ x \mid x < -\frac{5}{4} \right\}$, or $\left(-\infty, -\frac{5}{4} \right)$.

◀ Do Exercises 15–17.

Ⓓ APPLICATIONS AND PROBLEM SOLVING

Many problem-solving and applied situations translate to inequalities. In addition to "is less than" and "is more than," other phrases are commonly used.

IMPORTANT WORDS	SAMPLE SENTENCE	TRANSLATION
is at least	Max is at least 5 years old.	$m \geq 5$
is at most	At most 6 people could fit in the elevator.	$n \leq 6$
cannot exceed	Total weight in the elevator cannot exceed 2000 pounds.	$w \leq 2000$
must exceed	The speed must exceed 15 mph.	$s > 15$
is between	Heather's income is between $23,000 and $35,000.	$23{,}000 < h < 35{,}000$
no more than	Bing weighs no more than 90 pounds.	$w \leq 90$
no less than	Saul would accept no less than $4000 for the piano.	$t \geq 4000$

Answers

15. $\left\{ y \mid y \leq -\frac{1}{5} \right\}$, or $\left(-\infty, -\frac{1}{5} \right]$

16. $\left\{ x \mid x < \frac{1}{2} \right\}$, or $\left(-\infty, \frac{1}{2} \right)$

17. $\left\{ y \mid y \geq \frac{17}{9} \right\}$, or $\left[\frac{17}{9}, \infty \right)$

Guided Solution:
15. $6, -5y, \leq, -5, y, -\infty$

The following phrases deserve special attention.

TRANSLATING "AT LEAST" AND "AT MOST"

A quantity x is **at least** some amount q: $x \geq q$.
 (If x is at least q, it cannot be less than q.)

A quantity x is **at most** some amount q: $x \leq q$.
 (If x is at most q, it cannot be more than q.)

Do Exercises 18–24. ▶

EXAMPLE 16 *Physical Therapists.* As a result of the aging population staying active longer than previous generations, the employment demand for physical therapists is expected to increase 39% from 2010 to 2020. The equation

$$P = 7745t + 198{,}600$$

can be used to estimate the number of licensed physical therapists in the work force, where t is the number of years since 2010. Determine the years for which the number of physical therapists will be more than 252,000.

Source: U.S. Department of Labor

1. **Familiarize.** We already have an equation. To become more familiar with it, we might make a substitution for t. Suppose that we want to know the number of physical therapists 8 years after 2010, or in 2018. We substitute 8 for t:

$$P = 7745(8) + 198{,}600 = 260{,}560.$$

We see that in 2018, the number of physical therapists will be more than 252,000. To find all the years in which the number of physical therapists exceeds 252,000, we could make other guesses less than 8, but it is more efficient to proceed to the next step.

2. **Translate.** The number of physical therapists is to be more than 252,000. Thus we have

$$P > 252{,}000.$$

We replace P with $7745t + 198{,}600$:

$$7745t + 198{,}600 > 252{,}000.$$

3. **Solve.** We solve the inequality:

$$7745t + 198{,}600 > 252{,}000$$
$$7745t > 53{,}400 \qquad \text{Subtracting } 198{,}600$$
$$t > 6.89. \qquad \text{Dividing by } 7745$$

4. **Check.** As a partial check, we can substitute a value for t that is greater than 6.89. We did that in the *Familiarize* step and found that the number of physical therapists was more than 252,000.

5. **State.** The number of physical therapists will be more than 252,000 for years more than 6.89 years after 2010, so we have $\{t \mid t > 6.89\}$.

Do Exercise 25. ▶

Translate.

18. Russell will pay at most $250 for that plane ticket.

19. Emma scored at least an 88 on her Spanish test.

20. The time of the test was between 50 min and 60 min.

21. The University of Southern Indiana is more than 25 mi away.

22. Sarah's weight is less than 110 lb.

23. That number is greater than −8.

24. The costs of production of that bar-code scanner cannot exceed $135,000.

25. *Physical Therapists.* Refer to Example 16. Determine, in terms of an inequality, the years for which the number of physical therapists is more than 254,200.

Answers

18. $t \leq 250$ 19. $s \geq 88$ 20. $50 < t < 60$
21. $d > 25$ 22. $w < 110$ 23. $n > -8$
24. $c \leq 135{,}000$ 25. More than 7.18 years after 2010, or $\{t \mid t > 7.18\}$

EXAMPLE 17 *Salary Plans.* On her new job, Rose can be paid in one of two ways: *Plan A* is a salary of $600 per month, plus a commission of 4% of sales; and *Plan B* is a salary of $800 per month, plus a commission of 6% of sales in excess of $10,000. For what amount of monthly sales is plan A better than plan B, if we assume that sales are always more than $10,000?

1. **Familiarize.** Listing the given information in a table will be helpful.

PLAN A: MONTHLY INCOME	PLAN B: MONTHLY INCOME
$600 salary 4% of sales *Total*: $600 + 4% of sales	$800 salary 6% of sales over $10,000 *Total*: $800 + 6% of sales over $10,000

Next, suppose that Rose had sales of $12,000 in one month. Which plan would be better? Under plan A, she would earn $600 plus 4% of $12,000, or

$$600 + 0.04(12{,}000) = \$1080.$$

Since with plan B commissions are paid only on sales in excess of $10,000, Rose would earn $800 plus 6% of ($12,000 − $10,000), or

$$800 + 0.06(12{,}000 - 10{,}000) = 800 + 0.06(2000) = \$920.$$

This shows that for monthly sales of $12,000, plan A is better. Similar calculations will show that for sales of $30,000 per month, plan B is better. To determine *all* values for which plan A pays more money, we must solve an inequality that is based on the calculations above.

2. **Translate.** We let $S =$ the amount of monthly sales. If we examine the calculations in the *Familiarize* step, we see that the monthly income from plan A is $600 + 0.04S$ and from plan B is $800 + 0.06(S − 10,000)$. Thus we want to find all values of S for which

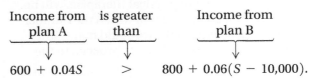

$$600 + 0.04S > 800 + 0.06(S - 10{,}000).$$

3. **Solve.** We solve the inequality:

$600 + 0.04S > 800 + 0.06(S - 10{,}000)$	
$600 + 0.04S > 800 + 0.06S - 600$	Using the distributive law
$600 + 0.04S > 200 + 0.06S$	Collecting like terms
$400 > 0.02S$	Subtracting 200 and $0.04S$
$20{,}000 > S$, or $S < 20{,}000$.	Dividing by 0.02

4. **Check.** For $S = 20{,}000$, the income from plan A is

$$600 + 4\% \cdot 20{,}000, \text{ or } \$1400.$$

The income from plan B is

$$800 + 6\% \cdot (20{,}000 - 10{,}000), \text{ or } \$1400.$$

This confirms that for sales of $20,000, Rose's pay is the same under either plan.

In the *Familiarize* step, we saw that for sales of $12,000, plan A pays more. Since 12,000 < 20,000, this is a partial check. Since we cannot check all possible values of S, we will stop here.

5. **State.** For monthly sales of less than $20,000, plan A is better.

◀ Do Exercise 26.

26. *Salary Plans.* A painter can be paid in one of two ways:

Plan A: $500 plus $4 per hour;

Plan B: Straight $9 per hour.

Suppose that the job takes n hours. For what values of n is plan A better for the painter?

Answer

26. For $\{n \mid n < 100\}$, plan A is better.

Translating for Success

1. *Consecutive Integers.* The sum of two consecutive even integers is 102. Find the integers.

2. *Salary Increase.* After Susanna earned a 5% raise, her new salary was $25,750. What was her former salary?

3. *Dimensions of a Rectangle.* The length of a rectangle is 6 in. more than the width. The perimeter of the rectangle is 102 in. Find the length and the width.

4. *Population.* The population of Doddville is decreasing at a rate of 5% per year. The current population is 25,750. What was the population the previous year?

5. *Reading Assignment.* Quinn has 6 days to complete a 150-page reading assignment. How many pages must he read the first day so that he has no more than 102 pages left to read on the 5 remaining days?

The goal of these matching questions is to practice step (2), Translate, of the five-step problem-solving process. Translate each word problem to an equation or an inequality and select a correct translation from A–O.

A. $0.05(25{,}750) = x$

B. $x + 2x = 102$

C. $2x + 2(x + 6) = 102$

D. $150 - x \leq 102$

E. $x - 0.05x = 25{,}750$

F. $x + (x + 2) = 102$

G. $x + (x + 6) > 102$

H. $x + 5x = 150$

I. $x + 0.05x = 25{,}750$

J. $x + (2x + 6) = 102$

K. $x + (x + 1) = 102$

L. $102 + x > 150$

M. $0.05x = 25{,}750$

N. $102 + 5x > 150$

O. $x + (x + 6) = 102$

Answer on page A-39

---

6. *Numerical Relationship.* One number is 6 more than twice another. The sum of the numbers is 102. Find the numbers.

7. *DVD Collections.* Together Ella and Ken have 102 DVDs. If Ken has 6 more DVDs than Ella, how many does each have?

8. *Sales Commissions.* Will earns a commission of 5% on his sales. One year he earned commissions totaling $25,750. What were his total sales for the year?

9. *Fencing.* Jess has 102 ft of fencing that he plans to use to enclose two dog runs. The perimeter of one run is to be twice the perimeter of the other. Into what lengths should the fencing be cut?

10. *Quiz Scores.* Lupe has a total of 102 points on the first 6 quizzes in her sociology class. How many total points must she earn on the 5 remaining quizzes in order to have more than 150 points for the semester?

✓ Reading Check

For each solution set expressed in set-builder notation, select from the column on the right the equivalent interval notation.

RC1. $\{x \mid a \leq x < b\}$

RC2. $\{x \mid x < b\}$

RC3. $\{x \mid x \text{ is a real number}\}$

RC4. $\{x \mid a < x < b\}$

RC5. $\{x \mid a \leq x \leq b\}$

RC6. $\{x \mid x \geq a\}$

a) (a, b)

b) $[a, b)$

c) $(-\infty, \infty)$

d) $[a, \infty)$

e) $(-\infty, b]$

f) $(a, b]$

g) $[a, b]$

h) $(-\infty, b)$

i) (a, ∞)

---

a   Determine whether the given numbers are solutions of the inequality.

1. $x - 2 \geq 6$; $-4, 0, 4, 8$

2. $3x + 5 \leq -10$; $-5, -10, 0, 27$

3. $t - 8 > 2t - 3$; $0, -8, -9, -3, -\frac{7}{8}$

4. $5y - 7 < 8 - y$; $2, -3, 0, 3, \frac{2}{3}$

---

b   Write interval notation for the given set or graph.

5. $\{x \mid x < 5\}$

6. $\{t \mid t \geq -5\}$

7. $\{x \mid -3 \leq x \leq 3\}$

8. $\{t \mid -10 < t \leq 10\}$

9. $\{x \mid -4 > x > -8\}$

10. $\{x \mid 13 > x \geq 5\}$

11.

12.

13.

14.

---

c   Solve and graph.

15. $x + 2 > 1$

16. $x + 8 > 4$

17. $y + 3 < 9$

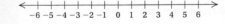

18. $y + 4 < 10$

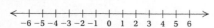

19. $a - 9 \leq -31$

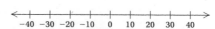

20. $a + 6 \leq -14$

21. $t + 13 \geq 9$

22. $x - 8 \leq 17$

23. $y - 8 > -14$

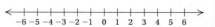

24. $y - 9 > -18$

25. $x - 11 \leq -2$

26. $y - 18 \leq -4$

27. $8x \geq 24$

28. $8t < -56$

29. $0.3x < -18$

30. $0.6x < 30$

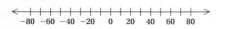

31. $\frac{2}{3}x > 2$

32. $\frac{3}{5}x > -3$

Solve.

33. $-9x \geq -8.1$

34. $-5y \leq 3.5$

35. $-\frac{3}{4}x \geq -\frac{5}{8}$

36. $-\frac{1}{8}y \leq -\frac{9}{8}$

37. $2x + 7 < 19$

38. $5y + 13 > 28$

39. $5y + 2y \leq -21$

40. $-9x + 3x \geq -24$

41. $2y - 7 < 5y - 9$

42. $8x - 9 < 3x - 11$

43. $0.4x + 5 \leq 1.2x - 4$

44. $0.2y + 1 > 2.4y - 10$

45. $5x - \frac{1}{12} \leq \frac{5}{12} + 4x$

46. $2x - 3 < \frac{13}{4}x + 10 - 1.25x$

47. $4(4y - 3) \geq 9(2y + 7)$

48. $2m + 5 \geq 16(m - 4)$

49. $3(2 - 5x) + 2x < 2(4 + 2x)$

50. $2(0.5 - 3y) + y > (4y - 0.2)8$

51. $5[3m - (m + 4)] > -2(m - 4)$

52. $[8x - 3(3x + 2)] - 5 \geq 3(x + 4) - 2x$

53. $3(r - 6) + 2 > 4(r + 2) - 21$

54. $5(t + 3) + 9 < 3(t - 2) + 6$

55. $19 - (2x + 3) \leq 2(x + 3) + x$

56. $13 - (2c + 2) \geq 2(c + 2) + 3c$

57. $\frac{1}{4}(8y + 4) - 17 < -\frac{1}{2}(4y - 8)$

58. $\frac{1}{3}(6x + 24) - 20 > -\frac{1}{4}(12x - 72)$

59. $2[4 - 2(3 - x)] - 1 \geq 4[2(4x - 3) + 7] - 25$

60. $5[3(7 - t) - 4(8 + 2t)] - 20 \leq -6[2(6 + 3t) - 4]$

61. $\frac{4}{5}(7x - 6) < 40$

62. $\frac{2}{3}(4x - 3) > 30$

63. $\frac{3}{4}(3 + 2x) + 1 \geq 13$

64. $\frac{7}{8}(5 - 4x) - 17 \geq 38$

65. $\frac{3}{4}\left(3x - \frac{1}{2}\right) - \frac{2}{3} < \frac{1}{3}$

66. $\frac{2}{3}\left(\frac{7}{8} - 4x\right) - \frac{5}{8} < \frac{3}{8}$

67. $0.7(3x + 6) \geq 1.1 - (x + 2)$

68. $0.9(2x + 8) < 20 - (x + 5)$

69. $a + (a - 3) \leq (a + 2) - (a + 1)$

70. $0.8 - 4(b - 1) > 0.2 + 3(4 - b)$

 Solve.

Body Mass Index.   *Body mass index I* can be used to de-termine whether an individual has a healthy weight for his or her height. An index in the range 18.5–24.9 indicates a normal weight. Body mass index is given by the formula, or model,

$$I = \frac{703W}{H^2},$$

where *W* is weight, in pounds, and *H* is height, in inches. Use this formula for Exercises 71 and 72.

Source: Centers for Disease Control and Prevention

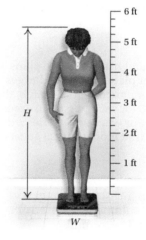

71. *Body Mass Index.*   Alexandra's height is 62 in. Determine, in terms of an inequality, those weights *W* that will keep her body mass index below 25.

72. *Body Mass Index.*   Josiah's height is 77 in. Determine, in terms of an inequality, those weights *W* that will keep his body mass index below 25.

73. *Grades.*   David is taking an economics course in which there will be 4 tests, each worth 100 points. He has scores of 89, 92, and 95 on the first three tests. He must make a total of at least 360 in order to get an A. What scores on the last test will give David an A?

74. *Grades.*   Elizabeth is taking a mathematics course in which there will be 5 tests, each worth 100 points. She has scores of 94, 90, and 89 on the first three tests. She must make a total of at least 450 in order to get an A. What scores on the fourth test will keep Elizabeth eligible for an A?

75. Insurance Claims. After a serious automobile accident, most insurance companies will replace the damaged car with a new one if repair costs exceed 80% of the N.A.D.A., or "blue-book," value of the car. Miguel's car recently sustained $9200 worth of damage but was not replaced. What was the blue-book value of his car?

76. Delivery Service. Jay's Express prices cross-town deliveries at $15 for the first 10 miles plus $1.25 for each additional mile. PDQ, Inc., prices its cross-town deliveries at $25 for the first 10 miles plus $0.75 for each additional mile. For what number of miles is PDQ less expensive?

77. Salary Plans. Imani can be paid in one of two ways:

Plan A: A salary of $400 per month plus a commission of 8% of gross sales;

Plan B: A salary of $610 per month, plus a commission of 5% of gross sales.

For what amount of gross sales should Imani select plan A?

78. Salary Plans. Aiden can be paid for his masonry work in one of two ways:

Plan A: $300 plus $9.00 per hour;

Plan B: Straight $12.50 per hour.

Suppose that the job takes n hours. For what values of n is plan B better for Aiden?

79. Prescription Coverage. Low Med offers two prescription-drug insurance plans. With plan 1, James would pay the first $150 of his prescription costs and 30% of all costs after that. With plan 2, James would pay the first $280 of costs, but only 10% of the rest. For what amount of prescription costs will plan 2 save James money? (Assume that his prescription costs exceed $280.)

80. Insurance Benefits. Bayside Insurance offers two plans. Under plan A, Giselle would pay the first $50 of her medical bills and 20% of all bills after that. Under plan B, Giselle would pay the first $250 of bills, but only 10% of the rest. For what amount of medical bills will plan B save Giselle money? (Assume that her bills will exceed $250.)

81. Wedding Costs. The Arnold Inn offers two plans for wedding parties. Under plan A, the inn charges $30 for each person in attendance. Under plan B, the inn charges $1300 plus $20 for each person in excess of the first 25 who attend. For what size parties will plan B cost less? (Assume that more than 25 guests will attend.)

82. Investing. Matthew is about to invest $20,000, part at 3% and the rest at 4%. What is the most that he can invest at 3% and still be guaranteed at least $650 in interest per year?

83. Renting Office Space. An investment group is renovating a commercial building and will rent offices to small businesses. The formula

$$R = 2(s + 70)$$

can be used to determine the monthly rent for an office with s square feet. All utilities are included in the monthly payment. For what square footage will the rent be less than $2100?

84. Temperatures of Solids. The formula

$$C = \tfrac{5}{9}(F - 32)$$

can be used to convert Fahrenheit temperatures F to Celsius temperatures C.

a) Gold is a solid at Celsius temperatures less than 1063°C. Find the Fahrenheit temperatures for which gold is a solid.

b) Silver is a solid at Celsius temperatures less than 960.8°C. Find the Fahrenheit temperatures for which silver is a solid.

85. *Tuition and Fees at Two-Year Colleges.* The equation

$$C = 82t + 1923$$

can be used to estimate the average cost of tuition and fees at two-year public institutions of higher education, where t is the number of years since 2005.

Source: National Center for Education Statistics, U.S. Department of Education

a) What was the average cost of tuition and fees in 2010? in 2014?

b) For what years will the cost of tuition and fees be more than $3000?

86. *Dewpoint Spread.* Pilots use the **dewpoint spread**, or the difference between the current temperature and the dewpoint (the temperature at which dew occurs), to estimate the height of the cloud cover. Each 3° of dewpoint spread corresponds to an increased height of cloud cover of 1000 ft. A plane, flying with limited instruments, must have a cloud cover higher than 3500 ft. What dewpoint spreads will allow the plane to fly?

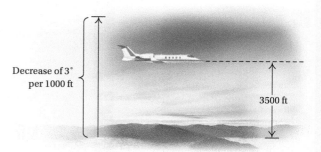

Decrease of 3° per 1000 ft

3500 ft

Skill Maintenance

Multiply. [13.6a]

87. $(3x - 4)(x + 8)$

88. $(r - 4s)(6r + s)$

89. $(2a - 5)(3a + 11)$

90. $(t + 2s)(t - 9s)$

Factor. [14.7a]

91. $4x^2 - 36x + 81$

92. $400y^2 - 16$

93. $27w^3 - 8$

94. $80 - 14x - 6x^2$

Find the domain. [16.2a]

95. $f(x) = \dfrac{-3}{x + 8}$

96. $f(x) = 3x - 5$

97. $f(x) = |x| - 4$

98. $f(x) = \dfrac{x + 7}{3x - 2}$

Synthesis

99. *Supply and Demand.* The supply S and demand D for a certain product are given by

$$S = 460 + 94p \quad \text{and} \quad D = 2000 - 60p.$$

a) Find those values of p for which supply exceeds demand.

b) Find those values of p for which supply is less than demand.

Determine whether each statement is true or false. If false, give a counterexample.

100. For any real numbers x and y, if $x < y$, then $x^2 < y^2$.

101. For any real numbers a, b, c, and d, if $a < b$ and $c < d$, then $a + c < b + d$.

102. Determine whether the inequalities

$$x < 3 \quad \text{and} \quad 0 \cdot x < 0 \cdot 3$$

are equivalent. Give reasons to support your answer.

Solve.

103. $x + 5 \leq 5 + x$

104. $x + 8 < 3 + x$

105. $x^2 + 1 > 0$

OBJECTIVES

a Find the intersection of two sets. Solve and graph conjunctions of inequalities.

b Find the union of two sets. Solve and graph disjunctions of inequalities.

c Solve applied problems involving conjunctions and disjunctions of inequalities.

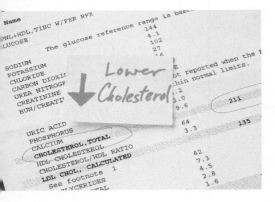

Cholesterol is a substance that is found in every cell of the human body. High levels of cholesterol can cause fatty deposits in the blood vessels that increase the risk of heart attack or stroke. A blood test can be used to measure *total cholesterol*. The following table shows the health risks associated with various cholesterol levels.

TOTAL CHOLESTEROL	RISK LEVEL
Less than 200	Normal
From 200 to 239	Borderline high
240 or higher	High

A total-cholesterol level T from 200 to 239 is considered border-line high. We can express this by the sentence

$$200 \leq T \quad and \quad T \leq 239$$

or more simply by

$$200 \leq T \leq 239.$$

This is an example of a *compound inequality*. **Compound inequalities** consist of two or more inequalities joined by the word *and* or the word *or*. We now "solve" such sentences—that is, we find the set of all solutions.

a INTERSECTIONS OF SETS AND CONJUNCTIONS OF INEQUALITIES

INTERSECTION

The **intersection** of two sets A and B is the set of all members that are common to A and B. We denote the intersection of sets A and B as

$$A \cap B.$$

The intersection of two sets is often illustrated as shown below.

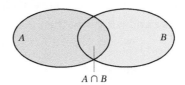

$A \cap B$

EXAMPLE 1 Find the intersection: $\{1, 2, 3, 4, 5\} \cap \{-2, -1, 0, 1, 2, 3\}$.

The numbers 1, 2, and 3 are common to the two sets, so the intersection is $\{1, 2, 3\}$.

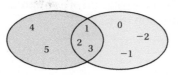

Do Exercises 1 and 2. ▶

1. Find the intersection:
   $\{0, 3, 5, 7\} \cap \{0, 1, 3, 11\}$.

2. Shade the intersection of sets A and B.

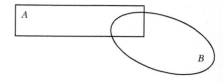

CONJUNCTION

When two or more sentences are joined by the word *and* to make a compound sentence, the new sentence is called a **conjunction** of the sentences.

The following is a conjunction of inequalities:

$$-2 < x \quad and \quad x < 1.$$

A number is a solution of a conjunction if it is a solution of *both* inequalities. For example, 0 is a solution of $-2 < x$ and $x < 1$ because $-2 < 0$ *and* $0 < 1$. Shown below is the graph of $-2 < x$, followed by the graph of $x < 1$, and then by the graph of the conjunction $-2 < x$ and $x < 1$. As the graphs demonstrate, *the solution set of a conjunction is the intersection of the solution sets of the individual inequalities.*

$\{x \mid -2 < x\}$

<---+---+---+---+---+---(---+---+---+---+---+---+---+---> $(-2, \infty)$
-7 -6 -5 -4 -3 -2 -1 0 1 2 3 4 5 6 7

$\{x \mid x < 1\}$

<---+---+---+---+---+---+---+---)---+---+---+---+---+---> $(-\infty, 1)$
-7 -6 -5 -4 -3 -2 -1 0 1 2 3 4 5 6 7

$\{x \mid -2 < x\} \cap \{x \mid x < 1\}$
$= \{x \mid -2 < x \text{ and } x < 1\}$

<---+---+---+---+---+---(---+---)---+---+---+---+---+---> $(-2, 1)$
-7 -6 -5 -4 -3 -2 -1 0 1 2 3 4 5 6 7

Because there are numbers that are both greater than -2 and less than 1, the conjunction $-2 < x$ and $x < 1$ can be abbreviated by $-2 < x < 1$. Thus the interval $(-2, 1)$ can be represented as $\{x \mid -2 < x < 1\}$, the set of all numbers that are *simultaneously* greater than -2 *and* less than 1. Note that, in general, for $a < b$,

$$a < x \quad and \quad x < b \quad \text{can be abbreviated} \quad a < x < b;$$
$$and \quad b > x \quad and \quad x > a \quad \text{can be abbreviated} \quad b > x > a.$$

.... Caution! ··································

"$a > x$ and $x < b$" cannot be abbreviated as "$a > x < b$".
....

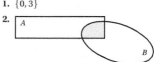

3. Graph and write interval notation:

$-1 < x$ *and* $x < 4$.

"AND"; "INTERSECTION"

The word **"and"** corresponds to **"intersection"** and to the symbol "∩". In order for a number to be a solution of a conjunction, it must make each part of the conjunction true.

◀ **Do Exercise 3.**

EXAMPLE 2 Solve and graph: $-1 \leq 2x + 5 < 13$.

This inequality is an abbreviation for the conjunction

$$-1 \leq 2x + 5 \quad and \quad 2x + 5 < 13.$$

The word *and* corresponds to set *intersection*, ∩. To solve the conjunction, we solve each of the two inequalities separately and then find the intersection of the solution sets:

$-1 \leq 2x + 5$	*and*	$2x + 5 < 13$	
$-6 \leq 2x$	*and*	$2x < 8$	Subtracting 5
$-3 \leq x$	*and*	$x < 4$.	Dividing by 2

We now abbreviate the result:

$$-3 \leq x < 4.$$

The solution set is $\{x \mid -3 \leq x < 4\}$, or, in interval notation, $[-3, 4)$. The graph is the intersection of the two separate solution sets.

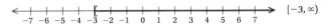

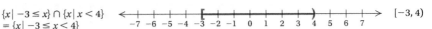

The steps above are generally combined as follows:

$-1 \leq 2x + 5 < 13$	$2x + 5$ appears in both inequalities.
$-6 \leq 2x < 8$	Subtracting 5
$-3 \leq x < 4$.	Dividing by 2

Such an approach saves some writing and will prove useful in Section 18.3.

◀ **Do Exercise 4.**

EXAMPLE 3 Solve and graph: $2x - 5 \geq -3$ *and* $5x + 2 \geq 17$.

We first solve each inequality separately:

$2x - 5 \geq -3$	*and*	$5x + 2 \geq 17$
$2x \geq 2$	*and*	$5x \geq 15$
$x \geq 1$	*and*	$x \geq 3$.

4. Solve and graph:

$-22 < 3x - 7 \leq 23$.

Answers

3. ; $(-1, 4)$

4. $\{x \mid -5 < x \leq 10\}$, or $(-5, 10]$;

Next, we find the intersection of the two separate solution sets:

$\{x \mid x \geq 1\}$ 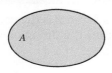 $[1, \infty)$

$\{x \mid x \geq 3\}$ [number line with bracket at 3 extending right] $[3, \infty)$

$\{x \mid x \geq 1\} \cap \{x \mid x \geq 3\}$
$= \{x \mid x \geq 3\}$ [number line with bracket at 3 extending right] $[3, \infty)$

The numbers common to both sets are those that are greater than or equal to 3. Thus the solution set is $\{x \mid x \geq 3\}$, or, in interval notation, $[3, \infty)$. You should check that any number in $[3, \infty)$ satisfies the conjunction whereas numbers outside $[3, \infty)$ do not.

Do Exercise 5. ▶

5. Solve and graph:

$3x + 4 < 10 \ and \ 2x - 7 < -13.$

[number line from -5 to 5]

EMPTY SET; DISJOINT SETS

Sometimes two sets have no elements in common. In such a case, we say that the intersection of the two sets is the **empty set**, denoted { } or $\varnothing$. Two sets with an empty intersection are said to be **disjoint**.

$A \cap B = \varnothing$ 

EXAMPLE 4 Solve and graph: $2x - 3 > 1 \ and \ 3x - 1 < 2.$

We solve each inequality separately:

$$2x - 3 > 1 \quad and \quad 3x - 1 < 2$$
$$2x > 4 \quad and \quad 3x < 3$$
$$x > 2 \quad and \quad x < 1.$$

The solution set is the intersection of the solution sets of the individual inequalities.

$\{x \mid x > 2\}$ [number line with parenthesis at 2 extending right] $(2, \infty)$

$\{x \mid x < 1\}$ [number line with parenthesis at 1 extending left] $(-\infty, 1)$

$\{x \mid x > 2\} \cap \{x \mid x < 1\}$
$= \{x \mid x > 2 \ and \ x < 1\}$
$= \varnothing$ [number line with no shading] $\varnothing$

Since no number is both greater than 2 and less than 1, the solution set is the empty set, $\varnothing$.

Do Exercise 6. ▶

6. Solve and graph:

$3x - 7 \leq -13 \ and \ 4x + 3 > 8.$

[number line from -6 to 6]

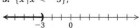

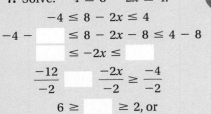

7. Solve: $-4 \le 8 - 2x \le 4$.   GS

$-4 \le 8 - 2x \le 4$

$-4 - \boxed{} \le 8 - 2x - 8 \le 4 - 8$

$\boxed{} \le -2x \le \boxed{}$

$\dfrac{-12}{-2} \boxed{} \dfrac{-2x}{-2} \ge \dfrac{-4}{-2}$

$6 \ge \boxed{} \ge 2$, or

$2 \boxed{} x \le 6$

The solution set is

$\{x \mid 2 \le x \le 6\}$, or $[\,\boxed{}, 6\,]$.

EXAMPLE 5   Solve: $3 \le 5 - 2x < 7$.

We have

$3 \le 5 - 2x < 7$

$3 - 5 \le 5 - 2x - 5 < 7 - 5$ Subtracting 5

$-2 \le -2x < 2$ Simplifying

$\dfrac{-2}{-2} \ge \dfrac{-2x}{-2} > \dfrac{2}{-2}$ Dividing by -2. The symbols must be reversed.

$1 \ge x > -1$.   Simplifying

The solution set is $\{x \mid 1 \ge x > -1\}$, or $\{x \mid -1 < x \le 1\}$, since the inequalities $1 \ge x > -1$ and $-1 < x \le 1$ are equivalent. The solution, in interval notation, is $(-1, 1\,]$.

◀ Do Exercise 7.

b UNIONS OF SETS AND DISJUNCTIONS OF INEQUALITIES

UNION

The **union** of two sets A and B is the collection of elements belonging to A and/or B. We denote the union of A and B by

$A \cup B$.

The union of two sets is often illustrated as shown below.

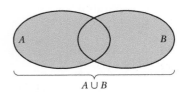

8. Find the union:

$\{0, 1, 3, 4\} \cup \{0, 1, 7, 9\}$.

9. Shade the union of sets A and B.

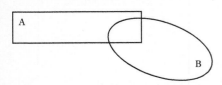

EXAMPLE 6   Find the union: $\{2, 3, 4\} \cup \{3, 5, 7\}$.

The numbers in either or both sets are 2, 3, 4, 5, and 7, so the union is $\{2, 3, 4, 5, 7\}$. We don't list the number 3 twice.

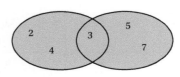

◀ Do Exercises 8 and 9.

Answers

7. $\{x \mid 2 \le x \le 6\}$, or $[2, 6]$ **8.** $\{0, 1, 3, 4, 7, 9\}$

9.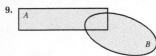

Guided Solution:

7. $8, -12, -4, \ge, x, \le, 2$

DISJUNCTION

When two or more sentences are joined by the word *or* to make a compound sentence, the new sentence is called a **disjunction** of the sentences.

The following is an example of a disjunction:

$$x < -3 \quad or \quad x > 3.$$

A number is a solution of a disjunction if it is a solution of at least one of the individual inequalities. For example, -7 is a solution of $x < -3$ or $x > 3$ because $-7 < -3$. Similarly, 5 is also a solution because $5 > 3$.

Shown below is the graph of $x < -3$, followed by the graph of $x > 3$, and then by the graph of the disjunction $x < -3$ or $x > 3$. As the graphs demonstrate, *the solution set of a disjunction is the union of the solution sets of the individual sentences.*

$\{x \mid x < -3\}$ [graph: number line -7 to 7, shading left of -3 with open circle] $(-\infty, -3)$

$\{x \mid x > 3\}$ [graph: number line -7 to 7, shading right of 3 with open circle] $(3, \infty)$

$\{x \mid x < -3\} \cup \{x \mid x > 3\}$
$= \{x \mid x < -3 \text{ or } x > 3\}$ [graph: number line -7 to 7, shading left of -3 and right of 3] $(-\infty, -3)$ $\cup (3, \infty)$

The solution set of

$$x < -3 \quad or \quad x > 3$$

is written $\{x \mid x < -3 \text{ or } x > 3\}$, or, in interval notation, $(-\infty, -3) \cup (3, \infty)$. This cannot be written in a more condensed form.

"OR"; "UNION"

The word "**or**" corresponds to "**union**" and the symbol "$\cup$". In order for a number to be in the solution set of a disjunction, it must be in *at least one* of the solution sets of the individual sentences.

Do Exercise 10. ▶

EXAMPLE 7 Solve and graph: $7 + 2x < -1$ or $13 - 5x \le 3$.

We solve each inequality separately, retaining the word *or*:

$$7 + 2x < -1 \quad or \quad 13 - 5x \le 3$$
$$2x < -8 \quad or \quad -5x \le -10$$

             Dividing by -5. The symbol must be reversed.

$$x < -4 \quad or \quad x \ge 2.$$

To find the solution set of the disjunction, we consider the individual graphs. We graph $x < -4$ and then $x \ge 2$. Then we take the union of the graphs.

$\{x \mid x < -4\}$ [graph: number line -7 to 7, shading left of -4 with open circle] $(-\infty, -4)$

$\{x \mid x \ge 2\}$ [graph: number line -7 to 7, shading right of 2 with closed bracket] $[2, \infty)$

$\{x \mid x < -4 \text{ or } x \ge 2\}$ [graph: number line -7 to 7, shading left of -4 and right of 2] $(-\infty, -4)$ $\cup [2, \infty)$

The solution set is written $\{x \mid x < -4 \text{ or } x \ge 2\}$, or, in interval notation, $(-\infty, -4) \cup [2, \infty)$.

10. Graph and write interval notation:

$$x \le -2 \text{ or } x > 4.$$

[graph: number line -5 to 5]

Answer

10.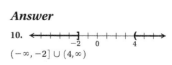
$(-\infty, -2] \cup (4, \infty)$

Solve and graph.

11. $x - 4 < -3 \, or \, x - 3 \geq 3$

<---+++++++++++++++++++++++--->
  −10 −8 −6 −4 −2   0   2   4   6   8   10

12. $-2x + 4 \leq -3 \, or \, x + 5 < 3$

<---+--+--+--+--+--+--+--+--+--+--+--+--->
 −6 −5 −4 −3 −2 −1   0   1   2   3   4   5   6

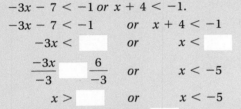

13. Solve:

$-3x - 7 < -1 \, or \, x + 4 < -1.$

$-3x - 7 < -1 \quad or \quad x + 4 < -1$

$-3x < \boxed{} \quad or \quad x < \boxed{}$

$\dfrac{-3x}{-3} \; \dfrac{6}{-3} \quad or \quad x < -5$

$x > \boxed{} \quad or \quad x < -5$

The solution set is $\{x \mid x < \boxed{} \, or \, x > -2\}$, or, in interval notation,

$(-\infty, -5) \cup (-2, \boxed{}).$

14. Solve and graph:

$5x - 7 \leq 13 \, or \, 2x - 1 \geq -7.$

<---+--+--+--+--+--+--+--+--+--+--+--+--->
 −6 −5 −4 −3 −2 −1   0   1   2   3   4   5   6

Answers

11. $\{x \mid x < 1 \, or \, x \geq 6\}$, or $(-\infty, 1) \cup [6, \infty)$;

<---+++++)+++++[++--->
     0   1   6

12. $\left\{x \mid x < -2 \, or \, x \geq \dfrac{7}{2}\right\}$, or $(-\infty, -2) \cup \left[\dfrac{7}{2}, \infty\right)$;

              $\frac{7}{2}$
<---++++)++++[++--->
   −2   0

13. $\{x \mid x < -5 \, or \, x > -2\}$, or $(-\infty, -5) \cup (-2, \infty)$

14. All real numbers;

<---+--+--+--+--+--+--+--+--+--+--+--->
             0

Guided Solution:
13. 6, −5, >, −2, −5, ∞

·········· **Caution!** ··········

A compound inequality like

$$x < -4 \quad or \quad x \geq 2,$$

as in Example 7, *cannot* be expressed as $2 \leq x < -4$ because to do so would be to say that x is *simultaneously* less than −4 and greater than or equal to 2. No number is both less than −4 *and* greater than or equal to 2, but many are less than −4 *or* greater than or equal to 2.

◀ **Do Exercises 11 and 12.**

EXAMPLE 8   Solve: $-2x - 5 < -2 \, or \, x - 3 < -10.$

We solve the individual inequalities separately, retaining the word *or*:

$$-2x - 5 < -2 \quad or \quad x - 3 < -10$$
$$-2x < 3 \quad or \quad x < -7$$

Reversing the symbol ⟶

$$x > -\tfrac{3}{2} \quad or \quad x < -7.$$

Keep the word "or."

The solution set is written $\left\{x \mid x < -7 \, or \, x > -\tfrac{3}{2}\right\}$, or, in interval notation, $(-\infty, -7) \cup \left(-\tfrac{3}{2}, \infty\right)$.

◀ **Do Exercise 13.**

EXAMPLE 9   Solve: $3x - 11 < 4 \, or \, 4x + 9 \geq 1.$

We solve the individual inequalities separately, retaining the word *or*:

$$3x - 11 < 4 \quad or \quad 4x + 9 \geq 1$$
$$3x < 15 \quad or \quad 4x \geq -8$$
$$x < 5 \quad or \quad x \geq -2.$$

To find the solution set, we first look at the individual graphs.

$\{x \mid x < 5\}$

<---+--+--+--+--+--+--+--+--+--+--+--+--+--)--+--+--->
 −7 −6 −5 −4 −3 −2 −1   0   1   2   3   4   5   6   7
     $(-\infty, 5)$

$\{x \mid x \geq -2\}$

<---+--+--+--+--[--+--+--+--+--+--+--+--+--+--+--->
 −7 −6 −5 −4 −3 −2 −1   0   1   2   3   4   5   6   7
     $[-2, \infty)$

$\{x \mid x < 5\} \cup \{x \mid x \geq -2\}$
$= \{x \mid x < 5 \, or \, x \geq -2\}$
$= \{x \mid x \text{ is a real number}\}$

<---+--+--+--+--+--+--+--+--+--+--+--+--+--+--+--->
 −7 −6 −5 −4 −3 −2 −1   0   1   2   3   4   5   6   7

$(-\infty, \infty)$
= The set of all real numbers

Since any number is either less than 5 or greater than or equal to −2, the two sets fill the entire number line. Thus the solution set is the set of all real numbers, $(-\infty, \infty)$.

◀ **Do Exercise 14.**

C APPLICATIONS AND PROBLEM SOLVING

EXAMPLE 10   *Renting Office Space.*   The equation

$$R = 2(s + 70)$$

can be used to determine the monthly rent for an office in a renovated commercial building. All utilities are included in the monthly payment. A florist shop has a monthly rental budget between $1720 and $2560. What square footage can be rented and remain within budget?

1. **Familiarize.**   We have an equation for calculating the monthly rent. Thus we can substitute a value into the formula. For 720 ft^2, the rent is found as follows:

$$R = 2(720 + 70) = 2 \cdot 790 = \$1580.$$

   This familiarizes us with the equation and also tells us that the number of square feet that we are looking for can be larger than 720 since $1580 is less than $1720.

2. **Translate.**   We want the monthly rent to be between $1720 and $2560, so we need to find those values of s for which $1720 < R < 2560$. Substituting $2(s + 70)$ for R, we have

$$1720 < 2(s + 70) < 2560.$$

3. **Solve.**   We solve the inequality:

$$1720 < 2(s + 70) < 2560$$
$$\frac{1720}{2} < \frac{2(s + 70)}{2} < \frac{2560}{2} \qquad \text{Dividing by 2}$$
$$860 < s + 70 < 1280$$
$$790 < s < 1210. \qquad \text{Subtracting 70}$$

4. **Check.**   We substitute some values as we did in the *Familiarize* step.

5. **State.**   Square footage between 790 ft^2 and 1210 ft^2 can be rented for a budget between $1720 and $2560 per month.

Do Exercise 15. ▶

15. *Renting Office Space.*   Refer to Example 10. What square footage can be rented for a budget between $2000 and $3200?

Answer

15. Between 930 ft^2 and 1530 ft^2

| 18.2 | Exercise Set |

✓ Reading Check

Determine whether each statement is true or false.

RC1. A compound inequality like $x < 5$ *and* $x > -2$ can be expressed as $-2 < x < 5$.

RC2. A compound inequality like $x \geq 5$ *and* $x < -2$ can be expressed as $5 \leq x < -2$.

RC3. The solution set of $x < -4$ *and* $x > 4$ can be written as $\varnothing$.

RC4. The solution set of $x \leq 3$ *and* $x \geq 0$ can be written as [0, 3].

a , b   Find the intersection or union.

1. $\{9, 10, 11\} \cap \{9, 11, 13\}$

2. $\{1, 5, 10, 15\} \cap \{5, 15, 20\}$

3. $\{a, b, c, d\} \cap \{b, f, g\}$

4. $\{m, n, o, p\} \cap \{m, o, p\}$

5. $\{9, 10, 11\} \cup \{9, 11, 13\}$

6. $\{1, 5, 10, 15\} \cup \{5, 15, 20\}$

7. $\{a, b, c, d\} \cup \{b, f, g\}$

8. $\{m, n, o, p\} \cup \{m, o, p\}$

9. $\{2, 5, 7, 9\} \cap \{1, 3, 4\}$

10. $\{a, e, i, o, u\} \cap \{m, q, w, s, t\}$

11. $\{3, 5, 7\} \cup \varnothing$

12. $\{3, 5, 7\} \cap \varnothing$

a   Graph and write interval notation.

13. $-4 < a \text{ and } a \leq 1$

14. $-\frac{5}{2} \leq m \text{ and } m < \frac{3}{2}$

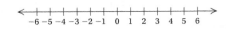

15. $1 < x < 6$

16. $-3 \leq y \leq 4$

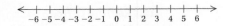

Solve and graph.

17. $-10 \leq 3x + 2 \text{ and } 3x + 2 < 17$

18. $-11 < 4x - 3 \text{ and } 4x - 3 \leq 13$

19. $3x + 7 \geq 4 \text{ and } 2x - 5 \geq -1$

20. $4x - 7 < 1 \text{ and } 7 - 3x > -8$

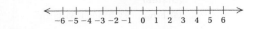

21. $4 - 3x \geq 10 \text{ and } 5x - 2 > 13$

22. $5 - 7x > 19 \text{ and } 2 - 3x < -4$

Solve.

23. $-4 < x + 4 < 10$ **24.** $-6 < x + 6 \leq 8$

25. $6 > -x \geq -2$ **26.** $3 > -x \geq -5$

27. $2 < x + 3 \leq 9$

28. $-6 \leq x + 1 < 9$

29. $1 < 3y + 4 \leq 19$

30. $5 \leq 8x + 5 \leq 21$

31. $-10 \leq 3x - 5 \leq -1$

32. $-6 \leq 2x - 3 < 6$

33. $-18 \leq -2x - 7 < 0$

34. $4 > -3m - 7 \geq 2$

35. $-\dfrac{1}{2} < \dfrac{1}{4}x - 3 \leq \dfrac{1}{2}$

36. $-\dfrac{2}{3} \leq 4 - \dfrac{1}{4}x < \dfrac{2}{3}$

37. $-4 \leq \dfrac{7 - 3x}{5} \leq 4$

38. $-3 < \dfrac{2x - 5}{4} < 8$

b   Graph and write interval notation.

39. $x < -2 \, or \, x > 1$

40. $x < -4 \, or \, x > 0$

41. $x \leq -3 \, or \, x > 1$

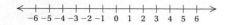

42. $x \leq -1 \, or \, x > 3$

Solve and graph.

43. $x + 3 < -2 \, or \, x + 3 > 2$

44. $x - 2 < -1 \, or \, x - 2 > 3$

45. $2x - 8 \leq -3 \, or \, x - 1 \geq 3$

46. $x - 5 \leq -4 \, or \, 2x - 7 \geq 3$

47. $7x + 4 \geq -17 \, or \, 6x + 5 \geq -7$

48. $4x - 4 < -8 \, or \, 4x - 4 < 12$

Solve.

49. $7 > -4x + 5 \, or \, 10 \leq -4x + 5$

50. $6 > 2x - 1 \, or \, -4 \leq 2x - 1$

51. $3x - 7 > -10 \text{ or } 5x + 2 \leq 22$

52. $3x + 2 < 2 \text{ or } 4 - 2x < 14$

53. $-2x - 2 < -6 \text{ or } -2x - 2 > 6$

54. $-3m - 7 < -5 \text{ or } -3m - 7 > 5$

55. $\frac{2}{3}x - 14 < -\frac{5}{6} \text{ or } \frac{2}{3}x - 14 > \frac{5}{6}$

56. $\frac{1}{4} - 3x \leq -3.7 \text{ or } \frac{1}{4} - 5x \geq 4.8$

57. $\frac{2x - 5}{6} \leq -3 \text{ or } \frac{2x - 5}{6} \geq 4$

58. $\frac{7 - 3x}{5} < -4 \text{ or } \frac{7 - 3x}{5} > 4$

 Solve.

59. *Pressure at Sea Depth.* The equation

$$P = 1 + \frac{d}{33}$$

gives the pressure P, in atmospheres (atm), at a depth of d feet in the sea. For what depths d is the pressure at least 1 atm and at most 7 atm?

60. *Temperatures of Liquids.* The formula

$$C = \tfrac{5}{9}(F - 32)$$

can be used to convert Fahrenheit temperatures F to Celsius temperatures C.

a) Gold is a liquid for Celsius temperatures C such that $1063° \leq C < 2660°$. Find such an inequality for the corresponding Fahrenheit temperatures.

b) Silver is a liquid for Celsius temperatures C such that $960.8° \leq C < 2180°$. Find such an inequality for the corresponding Fahrenheit temperatures.

61. *Aerobic Exercise.* In order to achieve maximum results from aerobic exercise, one should maintain one's heart rate at a certain level. A 30-year-old woman with a resting heart rate of 60 beats per minute should keep her heart rate between 138 and 162 beats per minute while exercising. She checks her pulse for 10 sec while exercising. What should the number of beats be?

62. *Minimizing Tolls.* A $6.00 toll is charged to cross the bridge from mainland Florida to Sanibel Island. A six-month pass, costing $50.00, reduces the toll to $2.00. A one-year pass, costing $400, allows for free crossings. How many crossings per year does it take, on average, for the two six-month passes to be the most economical choice? Assume a constant number of trips per month.

Source: leewayinfo.com

63. *Body Mass Index.* Refer to Exercises 71 and 72 in Exercise Set 18.1. Alexandra's height is 62 in. What weights *W* will allow Alexandra to keep her body mass index *I* in the 18.5–24.9 range?

64. *Body Mass Index.* Refer to Exercises 71 and 72 in Exercise Set 18.1. Josiah's height is 77 in. What weights *W* will allow Josiah to keep his body mass index in the 18.5–24.9 range?

65. *Young's Rule in Medicine.* Young's rule for determining the amount of a medicine dosage for a child is given by

$$c = \frac{ad}{a + 12},$$

where *a* is the child's age and *d* is the usual adult dosage, in milligrams. (*Warning!* Do not apply this formula without checking with a physician!) An 8-year-old child needs medication. What adult dosage can be used if a child's dosage must stay between 100 mg and 200 mg?

Source: Olsen, June L., et al., *Medical Dosage Calculations*, 6th ed. Reading, MA: Addison Wesley Longman, p. A-31.

66. *Young's Rule in Medicine.* Refer to Exercise 65. The dosage of a medication for a 5-year-old child must stay between 50 mg and 100 mg. Find the equivalent adult dosage.

Skill Maintenance

Solve.   [17.2a], [17.3a]

67. $3x - 2y = -7$,
$2x + 5y = 8$

68. $4x - 7y = 23$,
$x + 6y = -33$

69. $x + y = 0$,
$x - y = 8$

Find an equation of the line containing the given pair of points.   [16.5c]

70. $(2, 7), (3, -4)$

71. $(0, 7), (2, -1)$

72. $(4, -2), (-2, 4)$

Multiply.   [13.6a]

73. $(2a - b)(3a + 5b)$

74. $(5y + 6)(5y + 1)$

75. $(7x - 8)(3x - 5)$

76. $(13x - 2y)(x + 3y)$

Synthesis

Solve.

77. $x - 10 < 5x + 6 \leq x + 10$

78. $4m - 8 > 6m + 5 \ or \ 5m - 8 < -2$

79. $-\frac{2}{15} \leq \frac{2}{3}x - \frac{2}{5} \leq \frac{2}{15}$

80. $2[5(3 - y) - 2(y - 2)] > y + 4$

81. $x + 4 < 2x - 6 \leq x + 12$

82. $2x + 3 \leq x - 6 \ or \ 3x - 2 \leq 4x + 5$

Determine whether each sentence is true or false for all real numbers *a*, *b*, and *c*.

83. If $-b < -a$, then $a < b$.

84. If $a \leq c$ and $c \leq b$, then $b \geq a$.

85. If $a < c$ and $b < c$, then $a < b$.

86. If $-a < c$ and $-c > b$, then $a > b$.

Mid-Chapter Review

Concept Reinforcement

Determine whether each statement is true or false.

_____ **1.** The inequalities $x - 5 > 2$ and $x > 7$ are equivalent. [18.1c]

_____ **2.** If a is at most c, then it cannot be less than c. [18.1d]

_____ **3.** Sets A and B where $A = \{x \mid x < 2\}$ and $B = \{x \mid x \geq 2\}$ are disjoint sets. [18.2a]

_____ **4.** The union of two sets A and B is the collection of elements belonging to A and/or B. [18.2b]

Guided Solutions

 Fill in each blank with the number, variable, or symbol that creates a correct solution.

5. Solve: $8 - 5x \leq x + 20$. [18.1c]

$$8 - 5x \leq x + 20$$
$$-5x \leq x + \boxed{}$$
$$\boxed{} \, x \leq 12$$
$$x \; \boxed{} \; -2$$

6. Solve: $-17 < 3 - x < 36$. [18.2a]

$$-17 < 3 - x < 36$$
$$\boxed{} < -x < 33$$
$$20 \; \boxed{} \; x \; \boxed{} \; -33$$

Mixed Review

Match each graph of a solution with the correct set-builder notation or interval notation from selections A–H. [18.1b], [18.2a, b]

7. ⟨—+—+—(—+—+—+—+—+—+—+—⟩
 −5 −4 −3 −2 −1 0 1 2 3 4 5

8. ⟨—+—+—[—+—+—+—+—]—+—+—⟩
 −5 −4 −3 −2 −1 0 1 2 3 4 5

9. ⟨—+—+—+—+—+—+—+—+—(—+—+—⟩
 −5 −4 −3 −2 −1 0 1 2 3 4 5

10. ⟨—+—+—)—+—+—+—+—(—+—+—⟩
 −5 −4 −3 −2 −1 0 1 2 3 4 5

11. ⟨—+—+—[—+—+—+—+—)—+—+—⟩
 −5 −4 −3 −2 −1 0 1 2 3 4 5

12. ⟨—+—+—]—+—+—+—+—+—+—+—⟩
 −5 −4 −3 −2 −1 0 1 2 3 4 5

A. $(-\infty, -3]$

B. $\{x \mid -3 \leq x \leq 3\}$

C. $\{x \mid x \leq -3 \text{ or } x > 3\}$

D. $[-3, \infty)$

E. $(-\infty, -3) \cup (3, \infty)$

F. $\{x \mid -3 \leq x < 3\}$

G. $\{x \mid x > -3\}$

H. $(3, \infty)$

Find the intersection or union.   [18.2a, b]

13. $\{-1, 0, 10, 21, 40\} \cap \{-10, 0, 10\}$

14. $\{e, f, g, h\} \cup \{b, d, e\}$

15. $\left\{\dfrac{1}{4}, \dfrac{3}{8}\right\} \cup \varnothing$

16. $\{3, 6, 9, 12, 15\} \cap \{-12, -6, 7, 8\}$

Solve. Express the answer in both set-builder notation and interval notation.

17. $y - 8 \le -10$ [18.1c]

18. $-\dfrac{5}{11}x \ge -\dfrac{20}{11}$ [18.1c]

19. $x - 6 < -15 \text{ or } x + 2 > 3$
[18.2b]

20. $-6 \le x - 9 < 15$ [18.2a]

21. $x + 6 < -4 \text{ or } x + 8 > 9$
[18.2b]

22. $4(3t - 4) > 2(6 - t)$ [18.1c]

23. $3x - 2 \ge -11 \text{ or } 5x + 3 \ge -7$
[18.2b]

24. $0.1y + 3 < 5.6y - 2$ [18.1c]

25. $-6 < \dfrac{2x - 1}{3} < 8$ [18.2a]

26. $20 - (2x - 9) \le 3(x - 2) + x$
[18.1c]

27. $-\dfrac{1}{2} < 8 - \dfrac{1}{2}x < 6$ [18.2a]

28. $2x - 7 > -18 \text{ or } 3x - 7 \le 40$
[18.2b]

Solve.   [18.1d]

29. Jada is taking a chemistry course in which there will be 5 exams, each worth 100 points. She has scores of 85, 96, 88, and 95 on the first four exams. She must make a total of at least 450 points in order to get an A. What scores on the last exam will give Jada an A?

30. Michael is about to invest $12,500, part at 4.5% and the rest at 5%. What is the most he can invest at 4.5% and still be guaranteed at least $610 in interest per year?

Understanding Through Discussion and Writing

31. Explain in your own words why the inequality symbol must be reversed when both sides of an inequality are multiplied or divided by a negative number.   [18.1c]

32. Find the error or errors in each of the following steps: [18.1c]

$$7 - 9x + 6x < -9(x + 2) + 10x$$
$$7 - 9x + 6x < -9x + 2 + 10x \quad \textbf{(1)}$$
$$7 + 6x > 2 + 10x \quad \textbf{(2)}$$
$$-4x > 8 \quad \textbf{(3)}$$
$$x > -2. \quad \textbf{(4)}$$

33. Explain why the conjunction $3 < x \text{ and } x < 5$ is equivalent to $3 < x < 5$, but the disjunction $3 < x \text{ or } x < 5$ is not.   [18.2a, b]

STUDYING FOR SUCCESS *Your Textbook as a Resource*

☐ Study any drawings. Note the details in any sketches or graphs that accompany the explanations.

☐ Note the careful use of color to indicate substitutions and to highlight steps in a multistep problem.

18.3

Absolute-Value Equations and Inequalities

OBJECTIVES

a Simplify expressions containing absolute-value symbols.

b Find the distance between two points on the number line.

c Solve equations with absolute-value expressions.

d Solve equations with two absolute-value expressions.

e Solve inequalities with absolute-value expressions.

SKILL TO REVIEW

Objective 10.2e: Find the absolute value of a real number.

Find each absolute value.

1. $|-4|$ **2.** $|3.5|$

a PROPERTIES OF ABSOLUTE VALUE

We can think of the **absolute value** of a number as the number's distance from zero on the number line.

ABSOLUTE VALUE

The **absolute value** of x, denoted $|x|$, is defined as follows:
$$x \geq 0 \longrightarrow |x| = x; \qquad x < 0 \longrightarrow |x| = -x.$$

This definition tells us that, when x is nonnegative, the absolute value of x is x and, when x is negative, the absolute value of x is the opposite of x. For example, $|3| = 3$ and $|-3| = -(-3) = 3$. We see that absolute value is never negative.

Some simple properties of absolute value allow us to manipulate or simplify algebraic expressions.

PROPERTIES OF ABSOLUTE VALUE

a) $|ab| = |a| \cdot |b|$, for any real numbers a and b.
(The absolute value of a product is the product of the absolute values.)

b) $\left|\dfrac{a}{b}\right| = \dfrac{|a|}{|b|}$, for any real numbers a and b and $b \neq 0$.
(The absolute value of a quotient is the quotient of the absolute values.)

c) $|-a| = |a|$, for any real number a.
(The absolute value of the opposite of a number is the same as the absolute value of the number.)

EXAMPLES Simplify, leaving as little as possible inside the absolute-value signs.

1. $|5x| = |5| \cdot |x| = 5|x|$
2. $|-3y| = |-3| \cdot |y| = 3|y|$

Answers
Skill To Review:
1. 4 **2.** 3.5

EXAMPLES Simplify, leaving as little as possible inside the absolute-value signs.

3. $|7x^2| = |7| \cdot |x^2| = 7|x^2| = 7x^2$ Since x^2 is never negative for any number x

4. $\left|\dfrac{6x}{-3x^2}\right| = \left|\dfrac{-2}{x}\right| = \dfrac{|-2|}{|x|} = \dfrac{2}{|x|}$

Do Exercises 1–5. ▶

Simplify, leaving as little as possible inside the absolute-value signs.

1. $|7x|$ **2.** $|x^8|$

3. $|5a^2b|$ **4.** $\left|\dfrac{7a}{b^2}\right|$

5. $|-9x|$

b DISTANCE ON THE NUMBER LINE

The number line below shows that the distance between -3 and 2 is 5.

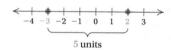

Another way to find the distance between two numbers on the number line is to determine the absolute value of the difference, as follows:

$$|-3 - 2| = |-5| = 5, \quad \text{or} \quad |2 - (-3)| = |5| = 5.$$

Note that the order in which we subtract does not matter because we are taking the absolute value after we have subtracted.

> **DISTANCE AND ABSOLUTE VALUE**
>
> For any real numbers a and b, the **distance** between them is $|a - b|$.

We should note that the distance is also $|b - a|$, because $a - b$ and $b - a$ are opposites and hence have the same absolute value.

EXAMPLE 5 Find the distance between -8 and -92 on the number line.

$$|-8 - (-92)| = |84| = 84, \quad \text{or} \quad |-92 - (-8)| = |-84| = 84$$

EXAMPLE 6 Find the distance between x and 0 on the number line.

$$|x - 0| = |x|$$

Do Exercises 6–8. ▶

Find the distance between the points.

GS **6.** $-6, -35$

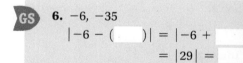
$$|-6 - (\quad)| = |-6 + \quad|$$
$$= |29| = \boxed{}$$

7. $19, 14$

8. $0, p$

c EQUATIONS WITH ABSOLUTE VALUE

EXAMPLE 7 Solve: $|x| = 4$. Then graph on the number line.

Note that $|x| = |x - 0|$, so that $|x - 0|$ is the distance from x to 0. Thus solutions of the equation $|x| = 4$, or $|x - 0| = 4$ are those numbers x whose distance from 0 is 4. Those numbers are -4 and 4. The solution set is $\{-4, 4\}$. The graph consists of just two points, as shown.

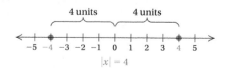

Answers

1. $7|x|$ 2. x^8 3. $5a^2|b|$ 4. $\dfrac{7|a|}{b^2}$
5. $9|x|$ 6. 29 7. 5 8. $|p|$
Guided Solution:
6. $-35, 35, 29$

EXAMPLE 8 Solve: $|x| = 0$.

The only number whose absolute value is 0 is 0 itself. Thus the solution is 0. The solution set is $\{0\}$.

EXAMPLE 9 Solve: $|x| = -7$.

The absolute value of a number is always nonnegative. Thus there is no number whose absolute value is -7; consequently, the equation has no solution. The solution set is $\varnothing$.

Examples 7–9 lead us to the following principle for solving linear equations with absolute value.

THE ABSOLUTE VALUE PRINCIPLE

For any positive number p and any algebraic expression X:

a) The solution of $|X| = p$ is those numbers that satisfy $X = -p$ *or* $X = p$.

b) The equation $|X| = 0$ is equivalent to the equation $X = 0$.

c) The equation $|X| = -p$ has no solution.

9. Solve: $|x| = 6$. Then graph on the number line.

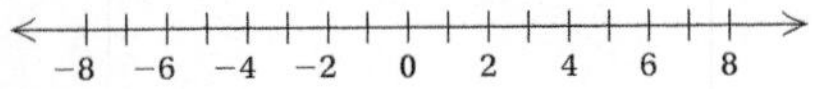

10. Solve: $|x| = -6$.

11. Solve: $|p| = 0$.

Do Exercises 9–11.

We can use the absolute-value principle with the addition and multiplication principles to solve equations with absolute value.

EXAMPLE 10 Solve: $2|x| + 5 = 9$.

We first use the addition and multiplication principles to get $|x|$ by itself. Then we use the absolute-value principle.

$$2|x| + 5 = 9$$
$$2|x| = 4 \quad \text{Subtracting 5}$$
$$|x| = 2 \quad \text{Dividing by 2}$$
$$x = -2 \quad or \quad x = 2 \quad \text{Using the absolute-value principle}$$

The solutions are -2 and 2. The solution set is $\{-2, 2\}$.

Solve.

12. $|3x| = 6$

13. $4|x| + 10 = 27$

14. $3|x| - 2 = 10$

Do Exercises 12–14.

EXAMPLE 11 Solve: $|x - 2| = 3$.

We can consider solving this equation in two different ways.

Method 1: This allows us to see the meaning of the solutions graphically. The solution set consists of those numbers that are 3 units from 2 on the number line.

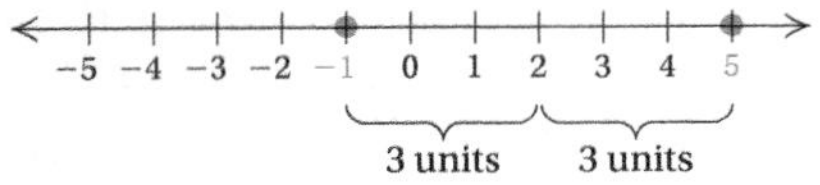

The solutions of $|x - 2| = 3$ are -1 and 5. The solution set is $\{-1, 5\}$.

Answers

9. $\{6, -6\}$;

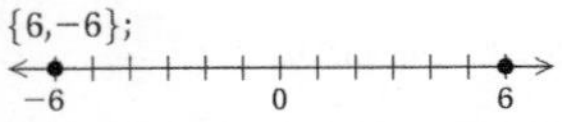

10. $\varnothing$ **11.** $\{0\}$ **12.** $\{-2, 2\}$ **13.** $\left\{-\frac{17}{4}, \frac{17}{4}\right\}$ **14.** $\{-4, 4\}$

Method 2: This method is more efficient. We use the absolute-value principle, replacing X with $x - 2$ and p with 3. Then we solve each equation separately.

$$|X| = p$$
$$|x - 2| = 3$$
$$x - 2 = -3 \quad or \quad x - 2 = 3 \qquad \text{Absolute-value principle}$$
$$x = -1 \quad or \quad x = 5$$

The solutions are -1 and 5. The solution set is $\{-1, 5\}$.

Do Exercise 15. ▶

15. Solve: $|x - 4| = 1$. Use two methods as in Example 11.

EXAMPLE 12 Solve: $|2x + 5| = 13$.

We use the absolute-value principle, replacing X with $2x + 5$ and p with 13:

$$|X| = p$$
$$|2x + 5| = 13$$
$$2x + 5 = -13 \quad or \quad 2x + 5 = 13 \qquad \text{Absolute-value principle}$$
$$2x = -18 \quad or \quad 2x = 8$$
$$x = -9 \quad or \quad x = 4.$$

The solutions are -9 and 4. The solution set is $\{-9, 4\}$.

Do Exercise 16. ▶

GS **16.** Solve: $|3x - 4| = 17$.
$$|3x - 4| = 17$$
$$3x - 4 = -17 \quad or \quad 3x - 4 = \boxed{}$$
$$3x = \boxed{} \quad or \quad 3x = 21$$
$$x = \boxed{} \quad or \quad x = \boxed{}$$
The solution set is $\{\boxed{}, 7\}$.

EXAMPLE 13 Solve: $|4 - 7x| = -8$.

Since absolute value is always nonnegative, this equation has no solution. The solution set is $\varnothing$.

Do Exercise 17. ▶

17. Solve: $|6 + 2x| = -3$.

d EQUATIONS WITH TWO ABSOLUTE-VALUE EXPRESSIONS

Sometimes equations have two absolute-value expressions. Consider $|a| = |b|$. This means that a and b are the same distance from 0. If a and b are the same distance from 0, then either they are the same number or they are opposites.

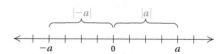

EXAMPLE 14 Solve: $|2x - 3| = |x + 5|$.

Either $2x - 3 = x + 5$ or $2x - 3 = -(x + 5)$. We solve each equation:

$$2x - 3 = x + 5 \quad or \quad 2x - 3 = -(x + 5)$$
$$x - 3 = 5 \quad or \quad 2x - 3 = -x - 5$$
$$x = 8 \quad or \quad 3x - 3 = -5$$
$$x = 8 \quad or \quad 3x = -2$$
$$x = 8 \quad or \quad x = -\tfrac{2}{3}.$$

The solutions are 8 and $-\tfrac{2}{3}$. The solution set is $\left\{8, -\tfrac{2}{3}\right\}$.

Answers

15. $\{3, 5\}$ **16.** $\left\{-\dfrac{13}{3}, 7\right\}$ **17.** $\varnothing$

Guided Solution:

16. $17, -13, -\dfrac{13}{3}, 7, -\dfrac{13}{3}$

EXAMPLE 15 Solve: $|x + 8| = |x - 5|$.

$$x + 8 = x - 5 \quad or \quad x + 8 = -(x - 5)$$
$$8 = -5 \quad or \quad x + 8 = -x + 5$$
$$8 = -5 \quad or \quad 2x = -3$$
$$8 = -5 \quad or \quad x = -\tfrac{3}{2}$$

The first equation has no solution. The solution of the second equation is $-\tfrac{3}{2}$. The solution set is $\left\{-\tfrac{3}{2}\right\}$.

◀ **Do Exercises 18 and 19.**

Solve.

18. $|5x - 3| = |x + 4|$

19. $|x - 3| = |x + 10|$

e INEQUALITIES WITH ABSOLUTE VALUE

We can extend our methods for solving equations with absolute value to those for solving inequalities with absolute value.

EXAMPLE 16 Solve: $|x| = 4$. Then graph on the number line.

From Example 7, we know that the solutions are -4 and 4. The solution set is $\{-4, 4\}$. The graph consists of just two points, as shown here.

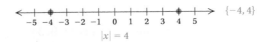

◀ **Do Exercise 20.**

20. Solve: $|x| = 5$. Then graph on the number line.

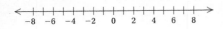

EXAMPLE 17 Solve: $|x| < 4$. Then graph.

The solutions of $|x| < 4$ are the solutions of $|x - 0| < 4$ and are those numbers x whose distance from 0 is less than 4. We can check by substituting or by looking at the number line that numbers like $-3, -2, -1, -\tfrac{1}{2}, -\tfrac{1}{4}, 0, \tfrac{1}{4}, \tfrac{1}{2}$, 1, 2, and 3 are all solutions. In fact, the solutions are all the real numbers x between -4 and 4. The solution set is $\{x | -4 < x < 4\}$ or, in interval notation, $(-4, 4)$. The graph is as follows.

◀ **Do Exercise 21.**

21. Solve: $|x| < 5$. Then graph.

EXAMPLE 18 Solve: $|x| \geq 4$. Then graph.

The solutions of $|x| \geq 4$ are solutions of $|x - 0| \geq 4$ and are those numbers whose distance from 0 is greater than or equal to 4—in other words, those numbers x such that $x \leq -4$ *or* $x \geq 4$. The solution set is $\{x | x \leq -4 \ or \ x \geq 4\}$, or $(-\infty, -4] \cup [4, \infty)$. The graph is as follows.

◀ **Do Exercise 22.**

22. Solve: $|x| \geq 5$. Then graph

Examples 16–18 illustrate three cases of solving equations and inequalities with absolute value. The following is a general principle for solving equations and inequalities with absolute value.

Answers

18. $\left\{\tfrac{7}{4}, -\tfrac{1}{6}\right\}$ **19.** $\left\{-\tfrac{7}{2}\right\}$

20. $\{-5, 5\}$;

21. $\{x | -5 < x < 5\}$, or $(-5, 5)$;

22. $\{x | x \leq -5 \ or \ x \geq 5\}$, or $(-\infty, -5] \cup [5, \infty)$;

SOLUTIONS OF ABSOLUTE-VALUE EQUATIONS AND INEQUALITIES

For any positive number p and any algebraic expression X:

a) The solutions of $|X| = p$ are those numbers that satisfy
$X = -p$ or $X = p$.

As an example, replacing X with $5x - 1$ and p with 8, we see that the solutions of $|5x - 1| = 8$ are those numbers x for which

$$5x - 1 = -8 \quad or \quad 5x - 1 = 8$$
$$5x = -7 \quad or \quad 5x = 9$$
$$x = -\tfrac{7}{5} \quad or \quad x = \tfrac{9}{5}.$$

The solution set is $\left\{-\tfrac{7}{5}, \tfrac{9}{5}\right\}$.

b) The solutions of $|X| < p$ are those numbers that satisfy
$-p < X < p$.

As an example, replacing X with $6x + 7$ and p with 5, we see that the solutions of $|6x + 7| < 5$ are those numbers x for which

$$-5 < 6x + 7 < 5$$
$$-12 < 6x < -2$$
$$-2 < x < -\tfrac{1}{3}.$$

The solution set is $\left\{x \mid -2 < x < -\tfrac{1}{3}\right\}$, or $\left(-2, -\tfrac{1}{3}\right)$.

c) The solutions of $|X| > p$ are those numbers that satisfy
$X < -p$ or $X > p$.

As an example, replacing X with $2x - 9$ and p with 4, we see that the solutions of $|2x - 9| > 4$ are those numbers x for which

$$2x - 9 < -4 \quad or \quad 2x - 9 > 4$$
$$2x < 5 \quad or \quad 2x > 13$$
$$x < \tfrac{5}{2} \quad or \quad x > \tfrac{13}{2}.$$

The solution set is $\left\{x \mid x < \tfrac{5}{2} \text{ or } x > \tfrac{13}{2}\right\}$, or $\left(-\infty, \tfrac{5}{2}\right) \cup \left(\tfrac{13}{2}, \infty\right)$.

EXAMPLE 19 Solve: $|3x - 2| < 4$. Then graph.

We use part (b). In this case, X is $3x - 2$ and p is 4:

$$|X| < p$$
$$|3x - 2| < 4 \qquad \text{Replacing } X \text{ with } 3x - 2 \text{ and } p \text{ with } 4$$
$$-4 < 3x - 2 < 4$$
$$-2 < 3x < 6$$
$$-\tfrac{2}{3} < x < 2.$$

The solution set is $\left\{ x \mid -\tfrac{2}{3} < x < 2 \right\}$, or $\left(-\tfrac{2}{3}, 2 \right)$. The graph is as follows.

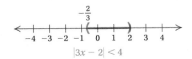

$|3x - 2| < 4$

Solve. Then graph.

23. $|2x - 3| < 7$

24. $|7 - 3x| \leq 4$

$$\boxed{} \leq 7 - 3x \leq 4$$
$$-11 \leq -3x \leq -3$$
$$\frac{-11}{-3} \quad \frac{-3x}{-3} \quad \frac{-3}{-3}$$
$$\frac{11}{3} \geq \boxed{} \geq 1$$

The solution set is
$\left\{ x \mid \boxed{} \leq x \leq \tfrac{11}{3} \right\}$, or $\left[1, \tfrac{11}{3} \right]$.

25. $|3x + 2| \geq 5$

$$3x + 2 \leq \boxed{} \quad \text{or} \quad 3x + 2 \geq 5$$
$$3x \leq -7 \quad \text{or} \quad 3x \geq \boxed{}$$
$$x \leq \boxed{} \quad \text{or} \quad x \geq 1$$

The solution set is
$\left\{ x \mid x \leq -\tfrac{7}{3} \, or \, x \geq \boxed{} \right\}$, or
$\left(\boxed{}, -\tfrac{7}{3} \right] \cup \left[\boxed{}, \infty \right)$.

EXAMPLE 20 Solve: $|8 - 4x| \leq 5$. Then graph.

We use part (b). In this case, X is $8 - 4x$ and p is 5:

$$|X| \leq p$$
$$|8 - 4x| \leq 5 \qquad \text{Replacing } X \text{ with } 8 - 4x \text{ and } p \text{ with } 5$$
$$-5 \leq 8 - 4x \leq 5$$
$$-13 \leq -4x \leq -3$$
$$\tfrac{13}{4} \geq x \geq \tfrac{3}{4}. \qquad \text{Dividing by } -4 \text{ and reversing the inequality symbols}$$

The solution set is $\left\{ x \mid \tfrac{13}{4} \geq x \geq \tfrac{3}{4} \right\}$, or $\left\{ x \mid \tfrac{3}{4} \leq x \leq \tfrac{13}{4} \right\}$, or $\left[\tfrac{3}{4}, \tfrac{13}{4} \right]$.

$|8 - 4x| \leq 5$

EXAMPLE 21 Solve: $|4x + 2| \geq 6$. Then graph.

We use part (c). In this case, X is $4x + 2$ and p is 6:

$$|X| \geq p$$
$$|4x + 2| \geq 6 \qquad \text{Replacing } X \text{ with } 4x + 2 \text{ and } p \text{ with } 6$$
$$4x + 2 \leq -6 \quad or \quad 4x + 2 \geq 6$$
$$4x \leq -8 \quad or \quad 4x \geq 4$$
$$x \leq -2 \quad or \quad x \geq 1.$$

The solution set is $\{ x \mid x \leq -2 \, or \, x \geq 1 \}$, or $(-\infty, -2] \cup [1, \infty)$.

$|4x + 2| \geq 6$

◀ **Do Exercises 23–25.**

Answers

23. $\{x \mid -2 < x < 5\}$, or $(-2, 5)$;

24. $\left\{ x \mid 1 \leq x \leq \tfrac{11}{3} \right\}$, or $\left[1, \tfrac{11}{3} \right]$;

25. $\left\{ x \mid x \leq -\tfrac{7}{3} \, or \, x \geq 1 \right\}$, or
$\left(-\infty, -\tfrac{7}{3} \right] \cup [1, \infty)$;

Guided Solutions:
24. $-4, \geq, \geq, x, 1$

25. $-5, 3, -\tfrac{7}{3}, 1, -\infty, 1$

✓ Reading Check

Solve the inequality and then select the correct graph of the solution from the column on the right.

RC1. $|x| > 3$

RC2. $|x| \geq 3$

RC3. $|x| < 3$

RC4. $|x| = 3$

RC5. $|x| \leq 3$

RC6. $|x| > -3$

a)

b)

c)

d)

e)

f)

---

a Simplify, leaving as little as possible inside absolute-value signs.

1. $|9x|$

2. $|26x|$

3. $|2x^2|$

4. $|8x^2|$

5. $|-2x^2|$

6. $|-20x^2|$

7. $|-6y|$

8. $|-17y|$

9. $\left| \dfrac{-2}{x} \right|$

10. $\left| \dfrac{y}{3} \right|$

11. $\left| \dfrac{x^2}{-y} \right|$

12. $\left| \dfrac{x^4}{-y} \right|$

13. $\left| \dfrac{-8x^2}{2x} \right|$

14. $\left| \dfrac{-9y^2}{3y} \right|$

15. $\left| \dfrac{4y^3}{-12y} \right|$

16. $\left| \dfrac{5x^3}{-25x} \right|$

---

b Find the distance between the points on the number line.

17. $-8, -46$

18. $-7, -32$

19. $36, 17$

20. $52, 18$

21. $-3.9, 2.4$

22. $-1.8, -3.7$

23. $-5, 0$

24. $\dfrac{2}{3}, -\dfrac{5}{6}$

c Solve.

25. $|x| = 3$

26. $|x| = 5$

27. $|x| = -3$

28. $|x| = -9$

29. $|q| = 0$

30. $|y| = 7.4$

31. $|x - 3| = 12$

32. $|3x - 2| = 6$

33. $|2x - 3| = 4$

34. $|5x + 2| = 3$

35. $|4x - 9| = 14$

36. $|9y - 2| = 17$

37. $|x| + 7 = 18$

38. $|x| - 2 = 6.3$

39. $574 = 283 + |t|$

40. $-562 = -2000 + |x|$

41. $|5x| = 40$

42. $|2y| = 18$

43. $|3x| - 4 = 17$

44. $|6x| + 8 = 32$

45. $7|w| - 3 = 11$

46. $5|x| + 10 = 26$

47. $\left|\dfrac{2x - 1}{3}\right| = 5$

48. $\left|\dfrac{4 - 5x}{6}\right| = 7$

49. $|m + 5| + 9 = 16$

50. $|t - 7| - 5 = 4$

51. $10 - |2x - 1| = 4$

52. $2|2x - 7| + 11 = 25$

53. $|3x - 4| = -2$

54. $|x - 6| = -8$

55. $\left|\frac{5}{9} + 3x\right| = \frac{1}{6}$

56. $\left|\frac{2}{3} - 4x\right| = \frac{4}{5}$

d Solve.

57. $|3x + 4| = |x - 7|$

58. $|2x - 8| = |x + 3|$

59. $|x + 3| = |x - 6|$

60. $|x - 15| = |x + 8|$

61. $|2a + 4| = |3a - 1|$

62. $|5p + 7| = |4p + 3|$

63. $|y - 3| = |3 - y|$ **64.** $|m - 7| = |7 - m|$ **65.** $|5 - p| = |p + 8|$

66. $|8 - q| = |q + 19|$ **67.** $\left|\dfrac{2x - 3}{6}\right| = \left|\dfrac{4 - 5x}{8}\right|$ **68.** $\left|\dfrac{6 - 8x}{5}\right| = \left|\dfrac{7 + 3x}{2}\right|$

69. $\left|\frac{1}{2}x - 5\right| = \left|\frac{1}{4}x + 3\right|$ **70.** $\left|2 - \frac{2}{3}x\right| = \left|4 + \frac{7}{8}x\right|$

e   Solve.

71. $|x| < 3$ **72.** $|x| \le 5$ **73.** $|x| \ge 2$ **74.** $|y| > 12$

75. $|x - 1| < 1$ **76.** $|x + 4| \le 9$ **77.** $5|x + 4| \le 10$ **78.** $2|x - 2| > 6$

79. $|2x - 3| \le 4$ **80.** $|5x + 2| \le 3$ **81.** $|2y - 7| > 10$ **82.** $|3y - 4| > 8$

83. $|4x - 9| \ge 14$ **84.** $|9y - 2| \ge 17$ **85.** $|y - 3| < 12$ **86.** $|p - 2| < 6$

87. $|2x + 3| \le 4$ **88.** $|5x + 2| \le 13$ **89.** $|4 - 3y| > 8$ **90.** $|7 - 2y| > 5$

91. $|9 - 4x| \ge 14$ **92.** $|2 - 9p| \ge 17$ **93.** $|3 - 4x| < 21$ **94.** $|-5 - 7x| \le 30$

95. $\left|\dfrac{1}{2} + 3x\right| \geq 12$ **96.** $\left|\dfrac{1}{4}y - 6\right| > 24$ **97.** $\left|\dfrac{x - 7}{3}\right| < 4$ **98.** $\left|\dfrac{x + 5}{4}\right| \leq 2$

99. $\left|\dfrac{2 - 5x}{4}\right| \geq \dfrac{2}{3}$ **100.** $\left|\dfrac{1 + 3x}{5}\right| > \dfrac{7}{8}$ **101.** $|m + 5| + 9 \leq 16$ **102.** $|t - 7| + 3 \geq 4$

103. $7 - |3 - 2x| \geq 5$ **104.** $16 \leq |2x - 3| + 9$ **105.** $\left|\dfrac{2x - 1}{3}\right| \leq 1$ **106.** $\left|\dfrac{3x - 2}{5}\right| \geq 1$

Skill Maintenance

Find the slope, if it exists, of each line.   [12.4b]

107. $x - 11y = 22$ **108.** $x = 10$ **109.** $10x = 5y - 3$ **110.** $y = -\dfrac{2}{5}$

Factor.   [14.6a]   Divide and simplify.   [15.2b]

111. $27w^3 - 1000$ **112.** $8 + 125t^3$ **113.** $\dfrac{w - z}{3w} \div \dfrac{w^2 - z^2}{9w^3}$ **114.** $\dfrac{t}{15} \div \dfrac{t}{25}$

Synthesis

115. *Motion of a Spring.* A weighted spring is bouncing up and down so that its distance d above the ground satisfies the inequality $|d - 6\text{ ft}| \leq \frac{1}{2}\text{ ft}$. Find all possible distances d.

116. *Container Sizes.* A container company is manufacturing rectangular boxes of various sizes. The length of any box must exceed the width by at least 3 in., but the perimeter cannot exceed 24 in. What widths are possible?

$$l \geq w + 3,$$
$$2l + 2w \leq 24$$

Solve.

117. $|x + 5| > x$ **118.** $1 - \left|\frac{1}{4}x + 8\right| = \frac{3}{4}$ **119.** $|7x - 2| = x + 4$

120. $|x - 1| = x - 1$ **121.** $|x - 6| \leq -8$ **122.** $|3x - 4| > -2$

Find an equivalent inequality with absolute value.

123. $-3 < x < 3$ **124.** $-5 \leq y \leq 5$ **125.** $x \leq -6 \text{ or } x \geq 6$

126. $-5 < x < 1$ **127.** $x < -8 \text{ or } x > 2$

Systems of Inequalities in Two Variables

A **graph** of an inequality is a drawing that represents its solutions. An inequality in one variable can be graphed on the number line. (See Section 18.1.) An inequality in two variables can be graphed on a coordinate plane.

A **linear inequality** is one that we can get from a related linear equation by changing the equals symbol to an inequality symbol. The graph of a linear inequality is a region on one side of a line. This region is called a **half-plane**. The graph sometimes includes the graph of the related line at the boundary of the half-plane.

a SOLUTIONS OF INEQUALITIES IN TWO VARIABLES

The solutions of an inequality in two variables are ordered pairs.

EXAMPLES Determine whether the ordered pair is a solution of the inequality $5x - 4y > 13$.

1. $(-3, 2)$

We have

$$\begin{array}{c} 5x - 4y > 13 \\ \hline 5(-3) - 4 \cdot 2 \ ? \ 13 \qquad \text{We use alphabetical order to replace } x \\ -15 - 8 \qquad \text{with } -3 \text{ and } y \text{ with } 2. \\ -23 \qquad \text{FALSE} \end{array}$$

Since $-23 > 13$ is false, $(-3, 2)$ is not a solution.

2. $(4, -3)$

We have

$$\begin{array}{c} 5x - 4y > 13 \\ \hline 5(4) - 4(-3) \ ? \ 13 \qquad \text{Replacing } x \text{ with } 4 \text{ and } y \text{ with } -3 \\ 20 + 12 \\ 32 \qquad \text{TRUE} \end{array}$$

Since $32 > 13$ is true, $(4, -3)$ is a solution.

Do Margin Exercises 1 and 2 on the following page. ▶

b GRAPHING INEQUALITIES IN TWO VARIABLES

Let's visualize the results of Examples 1 and 2. The equation $5x - 4y = 13$ is represented by the dashed line in the graphs on the following page. The solutions of the inequality $5x - 4y > 13$ are shaded below that dashed line. As shown in the graph on the left, the pair $(-3, 2)$ is not a solution of the inequality $5x - 4y > 13$ and is not in the shaded region.

OBJECTIVES

a Determine whether an ordered pair of numbers is a solution of an inequality in two variables.

b Graph linear inequalities in two variables.

c Graph systems of linear inequalities and find coordinates of any vertices.

SKILL TO REVIEW

Objective 16.4a: Graph linear equations using intercepts.

Find the intercepts. Then graph the equation.

1. $3x - 2y = 6$

2. $2x + y = 4$

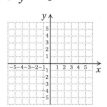

Answers

Skill to Review:

1.
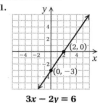
$3x - 2y = 6$

2.

$2x + y = 4$

Answers to Margin Exercises 1 and 2 are on p. 1216.

---

1. Determine whether $(1, -4)$ is a solution of $4x - 5y < 12$.

$$\frac{4x - 5y < 12}{\overset{?}{\underset{|}{}}}$$

2. Determine whether $(4, -3)$ is a solution of $3y - 2x \le 6$.

$$\frac{3y - 2x \le 6}{\overset{?}{\underset{|}{}}}$$

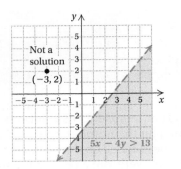 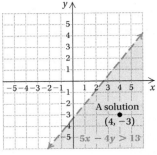

The pair $(4, -3)$ is a solution of the inequality $5x - 4y > 13$ and is in the shaded region. See the graph on the right above.

We now consider how to graph inequalities.

EXAMPLE 3 Graph: $y < x$.

We first graph the line $y = x$. Every solution of $y = x$ is an ordered pair like $(3, 3)$, where the first and second coordinates are the same. The graph of $y = x$ is shown on the left below. We draw it dashed because these points are *not* solutions of $y < x$.

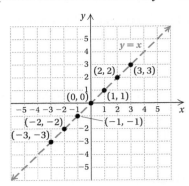 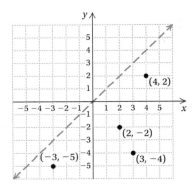

Now look at the graph on the right above. Several ordered pairs are plotted on the half-plane below $y = x$. Each is a solution of $y < x$. We can check the pair $(4, 2)$ as follows:

$$\frac{y < x}{2 \overset{?}{\underset{|}{}} 4} \quad \text{TRUE}$$

It turns out that any point on the same side of $y = x$ as $(4, 2)$ is also a solution. Thus, *if you know that one point in a half-plane is a solution of an inequality, then all points in that half-plane are solutions.* In this text, we will usually indicate this by color shading. We shade the half-plane below $y = x$.

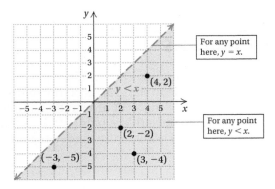

EXAMPLE 4 Graph: $8x + 3y \geq 24$.

First, we sketch the line $8x + 3y = 24$. Points on the line $8x + 3y = 24$ are also in the graph of $8x + 3y \geq 24$ because the inequality is greater than or equal to. We draw the line solid. This indicates that all points on the line are solutions. The rest of the solutions are in the half-plane either to the left or to the right of the line. To determine which, we select a point that is not on the line and determine whether it is a solution of $8x + 3y \geq 24$. We try $(-3, 4)$ as a test point:

$$
\begin{array}{c}
\underline{8x + 3y \geq 24} \\
8(-3) + 3(4) \ ? \ 24 \\
-24 + 12 \\
-12 \quad \text{FALSE}
\end{array}
$$

We see that $-12 \geq 24$ is *false*. Since $(-3, 4)$ is not a solution, none of the points in the half-plane containing $(-3, 4)$ is a solution. Thus the points in the opposite half-plane are solutions. We shade that half-plane and obtain the graph shown below.

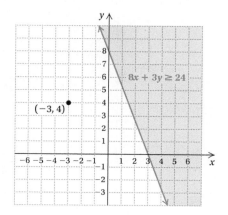

To graph an inequality in two variables:

1. Replace the inequality symbol with an equals sign and graph this related equation. This separates points that represent solutions from those that do not.

2. If the inequality symbol is $<$ or $>$, draw the line dashed. If the inequality symbol is $\leq$ or $\geq$, draw the line solid.

3. The graph consists of a half-plane that is either above or below or to the left or to the right of the line and, if the line is solid, the line as well. To determine which half-plane to shade, choose a point not on the line as a test point. If the line does not go through the origin, $(0, 0)$ is an easy point to use. Substitute to determine whether that point is a solution. If so, shade the half-plane containing that point. If not, shade the opposite half-plane.

EXAMPLE 5 Graph: $6x - 2y < 12$.

1. We first graph the related equation $6x - 2y = 12$.

2. Since the inequality uses the symbol $<$, points on the line are not solutions of the inequality, so we draw a dashed line.

3. To determine which half-plane to shade, we consider a test point *not* on the line. We try $(0, 0)$ and substitute:

$$\frac{6x - 2y < 12}{\begin{array}{c|c} 6(0) - 2(0) \; ? \; 12 & \\ 0 - 0 & \\ 0 & \text{TRUE} \end{array}}$$

Since the inequality $0 < 12$ is *true*, the point $(0, 0)$ is a solution; each point in the half-plane containing $(0, 0)$ is also a solution. Thus each point in the opposite half-plane is *not* a solution. The graph is shown below.

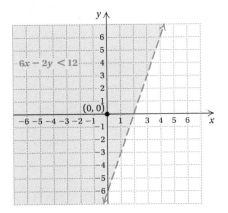

Graph.

3. $6x - 3y < 18$

4. $4x + 3y \geq 12$

◀ **Do Exercises 3 and 4.**

EXAMPLE 6 Graph $x > -3$ on a plane.

There is a missing variable in this inequality. If we graph the inequality on the number line, its graph is as follows:

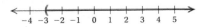

However, we can also write this inequality as $x + 0y > -3$ and consider graphing it in the plane. We are, in effect, determining which ordered pairs have x-values greater than -3. We use the same technique that we have used with the other examples. We first graph the related equation $x = -3$ in the plane. We draw the boundary with a dashed line. The rest of the graph is a half-plane to the right or to the left of the line $x = -3$. To determine which, we consider a test point, $(2, 5)$:

$$\frac{x + 0y > -3}{\begin{array}{c|c} 2 + 0(5) \; ? \; -3 & \\ 2 & \text{TRUE} \end{array}}$$

Answers

3.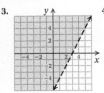
$6x - 3y < 18$

4.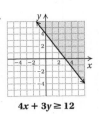
$4x + 3y \geq 12$

Since $(2, 5)$ is a solution, all the points in the half-plane containing $(2, 5)$ are solutions. We shade that half-plane.

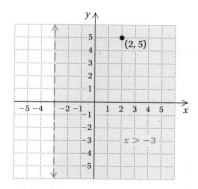

EXAMPLE 7 Graph $y \leq 4$ on a plane.

We first graph $y = 4$ using a solid line. We then use $(-2, 5)$ as a test point and substitute in $0x + y \leq 4$:

$$\begin{array}{c|c} 0x + y \leq 4 \\ \hline 0(-2) + 5 \stackrel{?}{} 4 \\ 0 + 5 \\ 5 & \text{FALSE} \end{array}$$

We see that $(-2, 5)$ is *not* a solution, so all the points in the half-plane containing $(-2, 5)$ are not solutions. Thus each point in the opposite half-plane is a solution.

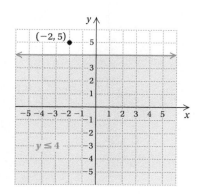

Do Exercises 5 and 6. ▶

c SYSTEMS OF LINEAR INEQUALITIES

The following is an example of a system of two linear inequalities in two variables:

$$x + y \leq 4,$$
$$x - y < 4.$$

A **solution** of a system of linear inequalities is an ordered pair that is a solution of *both* inequalities. We now graph solutions of systems of linear inequalities. To do so, we graph each inequality and determine where the graphs overlap, or intersect. That will be the region in which the ordered pairs are solutions of both inequalities.

Graph on a plane.

5. $x < 3$

6. $y \geq -4$

Answers

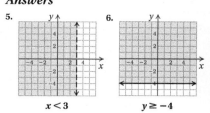

EXAMPLE 8 Graph the solutions of the system

$$x + y \leq 4,$$
$$x - y < 4.$$

We graph $x + y \leq 4$ by first graphing the equation $x + y = 4$ using a solid red line. We consider $(0, 0)$ as a test point and find that it is a solution, so we shade all points on that side of the line using red shading. (See the graph on the left below.) The arrows near the ends of the line also indicate the half-plane, or region, that contains the solutions.

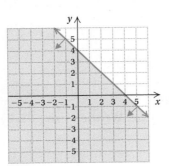 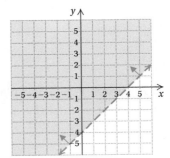

Next, we graph $x - y < 4$. We begin by graphing the equation $x - y = 4$ using a dashed blue line and consider $(0, 0)$ as a test point. Again, $(0, 0)$ is a solution so we shade that side of the line using blue shading. (See the graph on the right above.) The solution set of the system is the region that is shaded both red and blue and part of the line $x + y = 4$. (See the graph below.)

7. Graph:

$$x + y \geq 1,$$
$$y - x \geq 2.$$

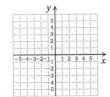

◀ **Do Exercise 7.**

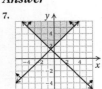

EXAMPLE 9 Graph: $-2 < x \leq 5$.

This is actually a system of inequalities:

$$-2 < x,$$
$$x \leq 5.$$

We graph the equation $-2 = x$ and see that the graph of the first inequality is the half-plane to the right of the line $-2 = x$. (See the graph on the left below.)

Next, we graph the second inequality, starting with the line $x = 5$, and find that its graph is the line and also the half-plane to the left of it. (See the graph on the right below.)

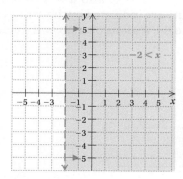 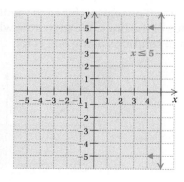

We shade the intersection of these graphs.

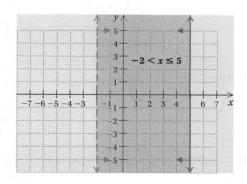

Do Exercise 8. ▶

8. Graph: $-3 \leq y < 4$.

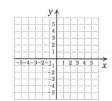

A system of inequalities may have a graph that consists of a polygon and its interior. In *linear programming*, which is a topic rich in application that you may study in a later course, it is important to be able to find the vertices of such a polygon.

EXAMPLE 10 Graph the following system of inequalities. Find the coordinates of any vertices formed.

$$6x - 2y \leq 12, \quad \text{(1)}$$
$$y - 3 \leq 0, \quad \text{(2)}$$
$$x + y \geq 0 \quad \text{(3)}$$

We graph the lines $6x - 2y = 12, y - 3 = 0,$ and $x + y = 0$ using solid lines. The regions for each inequality are indicated by the arrows near the ends of the lines. We then note where the regions overlap and shade the region of solutions using one color.

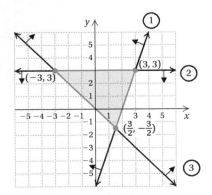

To find the vertices, we solve three different systems of equations. The system of equations from inequalities (1) and (2) is

$$6x - 2y = 12, \quad \text{(1)}$$
$$y - 3 = 0. \quad \text{(2)}$$

Solving, we obtain the vertex $(3, 3)$.

The system of equations from inequalities (1) and (3) is

$$6x - 2y = 12, \quad \text{(1)}$$
$$x + y = 0. \quad \text{(3)}$$

Solving, we obtain the vertex $\left(\frac{3}{2}, -\frac{3}{2}\right)$.

The system of equations from inequalities (2) and (3) is

$$y - 3 = 0, \quad \text{(2)}$$
$$x + y = 0. \quad \text{(3)}$$

Solving, we obtain the vertex $(-3, 3)$.

◀ Do Exercise 9.

9. Graph the system of inequalities. Find the coordinates of any vertices formed.

$$5x + 6y \leq 30,$$
$$0 \leq y \leq 3,$$
$$0 \leq x \leq 4$$

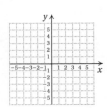

Answer

9.

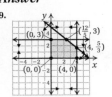

EXAMPLE 11   Graph the following system of inequalities. Find the coordinates of any vertices formed.

$$x + y \leq 16, \quad (1)$$
$$3x + 6y \leq 60, \quad (2)$$
$$x \geq 0, \quad (3)$$
$$y \geq 0 \quad (4)$$

We graph each inequality using solid lines. The regions for each inequality are indicated by the arrows near the ends of the lines. We then note where the regions overlap and shade the region of solutions using one color.

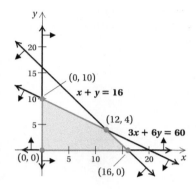

To find the vertices, we solve four different systems of equations. The system of equations from inequalities (1) and (2) is

$$x + y = 16, \quad (1)$$
$$3x + 6y = 60. \quad (2)$$

Solving, we obtain the vertex $(12, 4)$.

The system of equations from inequalities (1) and (4) is

$$x + y = 16, \quad (1)$$
$$y = 0. \quad (4)$$

Solving, we obtain the vertex $(16, 0)$.

The system of equations from inequalities (3) and (4) is

$$x = 0, \quad (3)$$
$$y = 0. \quad (4)$$

The vertex is $(0, 0)$.

The system of equations from inequalities (2) and (3) is

$$3x + 6y = 60, \quad (2)$$
$$x = 0. \quad (3)$$

Solving, we obtain the vertex $(0, 10)$.

Do Exercise 10. ▶

10. Graph the system of inequalities. Find the coordinates of any vertices formed.

$$2x + 4y \leq 8,$$
$$x + y \leq 3,$$
$$x \geq 0,$$
$$y \geq 0$$

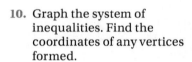

Answer

10.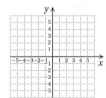

Visualizing for Success

A
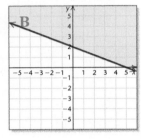

B
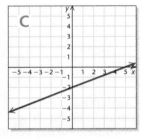

C
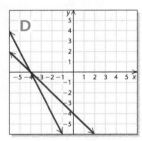

D
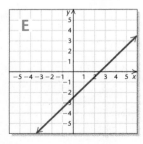

E
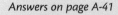

F
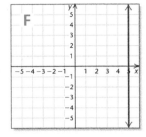

Match the equation, inequality, system of equations, or system of inequalities with its graph.

1. $x + y = -4$,
   $2x + y = -8$

2. $2x + 5y \geq 10$

3. $2x - 2y = 5$

4. $2x - 5y = 10$

5. $-2y < 8$

6. $5x - 2y = 10$

7. $2x = 10$

8. $5x + 2y < 10$,
   $2x - 5y > 10$

9. $5x \geq -10$

10. $y - 2x < 8$

G
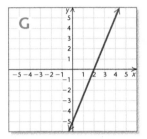

H
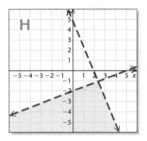

I

J

Answers on page A-41

☑ Reading Check

Choose from the selections below the inequality that matches the graph.

(a) $x - 3y > 3$ **(b)** $x - 3y < 3$ **(c)** $y > 3$ **(d)** $x - 3y \geq 3$ **(e)** $x < 3$ **(f)** $x - 3y \leq 3$

RC1.
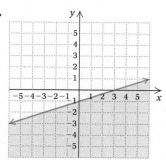

RC2.
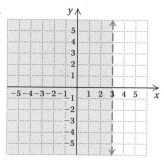

RC3.
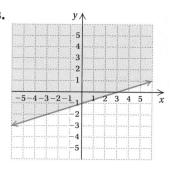

RC4.
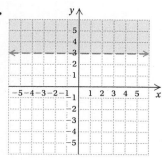

RC5.
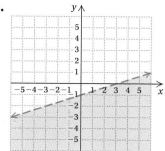

RC6.
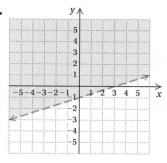

a   Determine whether the given ordered pair is a solution of the given inequality.

1. $(-3, 3)$; $3x + y < -5$ **2.** $(6, -8)$; $4x + 3y \geq 0$ **3.** $(5, 9)$; $2x - y > -1$ **4.** $(5, -2)$; $6y - x > 2$

b   Graph each inequality on a plane.

5. $y > 2x$
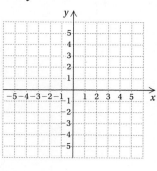

6. $y < 3x$
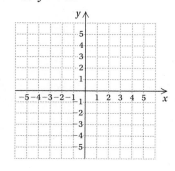

7. $y < x + 1$
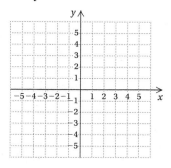

8. $y \leq x - 3$
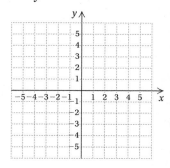

9. $y > x - 2$

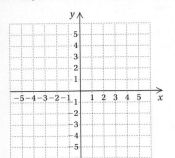

10. $y \geq x + 4$

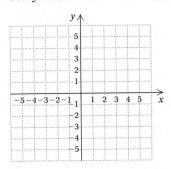

11. $x + y < 4$

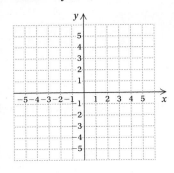

12. $x - y \geq 3$

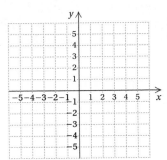

13. $3x + 4y \leq 12$

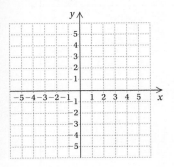

14. $2x + 3y < 6$

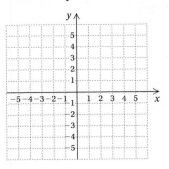

15. $2y - 3x > 6$

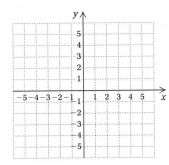

16. $2y - x \leq 4$

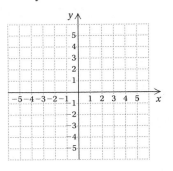

17. $3x - 2 \leq 5x + y$

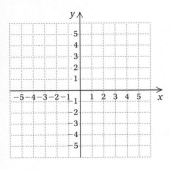

18. $2x - 2y \geq 8 + 2y$

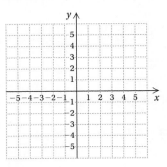

19. $x < 5$

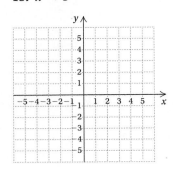

20. $y \geq -2$

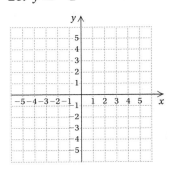

21. $y > 2$

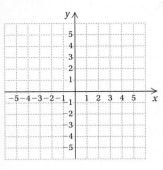

22. $x \leq -4$

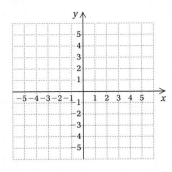

23. $2x + 3y \leq 6$

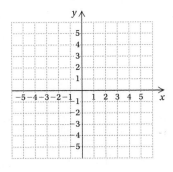

24. $7x + 2y \geq 21$

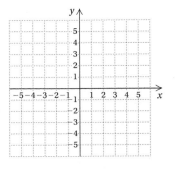

Matching. Each of Exercises 25–30 shows the graph of an inequality. Match the graph with one of the appropriate inequalities (A)–(F) that follow.

25.

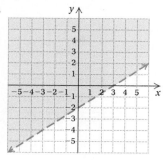

26.

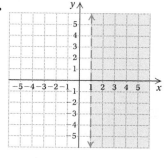

27.

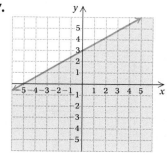

28.

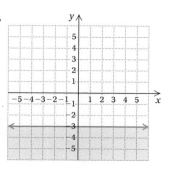

29.

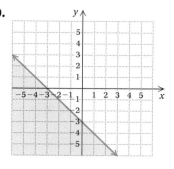

30.

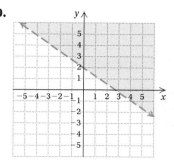

A. $4y > 8 - 3x$ **B.** $3x \geq 5y - 15$ **C.** $y + x \leq -3$ **D.** $x > 1$ **E.** $y \leq -3$ **F.** $2x - 3y < 6$

C   Graph each system of inequalities. Find the coordinates of any vertices formed.

31. $y \geq x,$
$\quad\ y \leq -x + 2$

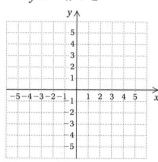

32. $y \geq x,$
$\quad\ y \leq -x + 4$

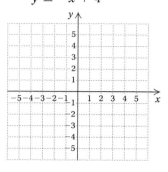

33. $y \geq x,$
$\quad\ y \leq -x + 1$

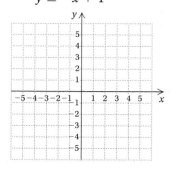

34. $y \leq x,$
$\quad\ y \geq -x + 3$

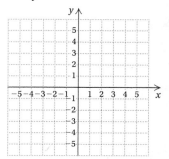

35. $x \leq 3,$
$\quad\ y \geq -3x + 2$

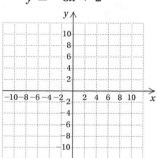

36. $x \geq -2,$
$\quad\ y \leq -2x + 3$

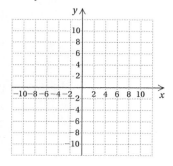

37. $x + y \leq 1,$
$\quad\ x - y \leq 2$

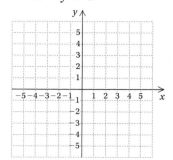

38. $x + y \leq 3,$
$\quad\ x - y \leq 4$

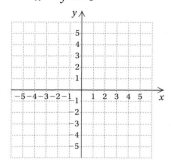

39. $y \leq 2x + 1$,
$y \geq -2x + 1$,
$x \leq 2$

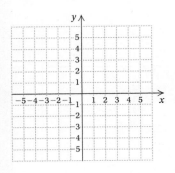

40. $x - y \leq 2$,
$x + 2y \geq 8$,
$y \leq 4$

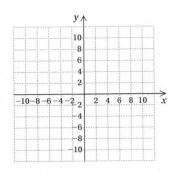

41. $x + 2y \leq 12$,
$2x + y \leq 12$,
$x \geq 0$,
$y \geq 0$

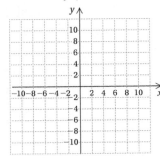

42. $y - x \geq 1$,
$y - x \leq 3$,
$2 \leq x \leq 5$

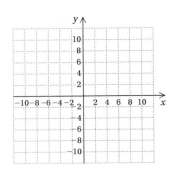

Skill Maintenance

Solve.   [11.3b, c]

43. $5(3x - 4) = -2(x + 5)$

44. $4(3x + 4) = 2 - x$

45. $2(x - 1) + 3(x - 2) - 4(x - 5) = 10$

46. $10x - 8(3x - 7) = 2(4x - 1)$

47. $5x + 7x = -144$

48. $0.5x - 2.34 + 2.4x = 7.8x - 9$

Given the function $f(x) = |2 - x|$, find each of the following function values.   [16.1b]

49. $f(0)$

50. $f(-1)$

51. $f(1)$

52. $f(10)$

53. $f(-2)$

54. $f(2a)$

55. $f(-4)$

56. $f(1.8)$

Synthesis

57. *Luggage Size.*   Unless an additional fee is paid, most major airlines will not check any luggage for which the sum of the item's length, width, and height exceeds 62 in. The U.S. Postal Service will ship a package only if the sum of the package's length and girth (distance around its midsection) does not exceed 130 in. A television network is ordering several 30-in.-long cases that will be both mailed and checked as luggage. Using w and h for width and height (in inches), respectively, write and graph an inequality that represents all acceptable combinations of width and height.

Source: U.S. Postal Service

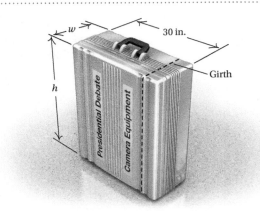

Vocabulary Reinforcement

Complete each statement with the correct term from the column on the right. Some of the choices may not be used and some may be used more than once.

1. A(n) _____ is a sentence containing $<, \leq, >, \geq,$ or $\neq$. [18.1a]

2. Using _____ notation, we write the solution set for $x < 7$ as $\{x \mid x < 7\}$. [18.1b]

3. Using _____ notation, we write the solution set of $-5 \leq y < 16$ as $[-5, 16)$. [18.1b]

4. The _____ of two sets A and B is the set of all members that are common to A and B. [18.2a]

5. When two or more sentences are joined by the word *and* to make a compound sentence, the new sentence is called a(n) _____ of the sentences. [18.2a]

6. When two sets have no elements in common, the intersection of the two sets is the _____. [18.2a]

7. Two sets with an empty intersection are said to be _____. [18.2a]

8. The _____ of two sets A and B is the collection of elements belonging to A and/or B. [18.2b]

9. When two or more sentences are joined by the word *or* to make a compound sentence, the new sentence is called a(n) _____ of the sentences. [18.2b]

10. A quantity is _____ some amount q: $x \leq q$. [18.1d]

11. A quantity is _____ some amount q: $x \geq q$. [18.1d]

12. For any real numbers a and b, the _____ between them is $|a - b|$. [18.3b]

union

set-builder

empty set

absolute value

disjunction

at least

at most

inequality

intersection

distance

interval

disjoint sets

compound

conjunction

Concept Reinforcement

Determine whether each statement is true or false.

_____ 1. If one point in a half-plane is a solution of a linear inequality, then all points in that half-plane are solutions. [18.4b]

_____ 2. Every system of linear inequalities has at least one solution. [18.4c]

_____ 3. For any real numbers a, b, and c, $c \neq 0$, $a \leq b$ is equivalent to $ac \leq bc$. [18.1c]

_____ 4. The inequalities $x < 2$ and $x \leq 1$ are equivalent. [18.1c]

_____ 5. If x is negative, $|x| = -x$. [18.3a]

_____ 6. $|x|$ is always positive. [18.3a]

_____ 7. $|a - b| = |b - a|$. [18.3b]

Study Guide

Objective 18.1a Determine whether a given number is a solution of an inequality.

Example Determine whether -3 and 1 are solutions of the inequality $4 - x \geq 2 - 5x$.

We substitute -3 for x and get

$$4 - (-3) \geq 2 - 5(-3), \text{ or } 7 \geq 17,$$

a *false* sentence. Therefore, -3 is not a solution.
We substitute 1 for x and get

$$4 - 1 \geq 2 - 5 \cdot 1, \text{ or } 3 \geq -3,$$

a *true* sentence. Therefore, 1 is a solution.

Practice Exercise

1. Determine whether -2 and 5 are solutions of the inequality $8 - 3x \leq 3x + 6$.

Objective 18.1b Write interval notation for the solution set of an inequality.

Example Write interval notation for the solution set.

a) $\{x \mid x \leq -12\} = (-\infty, -12]$
b) $\{r \mid r > -1\} = (-1, \infty)$
c) $\{y \mid -8 \leq y < 9\} = [-8, 9)$
d) $\{x \mid 0 \geq x \geq -6\} = [-6, 0]$
e) $\{c \mid -25 < c \leq 25\} = (-25, 25]$

Practice Exercise

2. Write interval notation for the solution set.

a) $\{t \mid t < -8\}$
b) $\{x \mid -7 \leq x < 10\}$
c) $\{b \mid b \geq 3\}$

Objective 18.1c Solve an inequality using the addition principle and the multiplication principle and then graph the inequality.

Example Solve and graph: $6x - 7 \leq 3x + 2$.

$$6x - 7 \leq 3x + 2$$
$$3x - 7 \leq 2 \qquad \text{Subtracting } 3x$$
$$3x \leq 9 \qquad \text{Adding } 7$$
$$x \leq 3 \qquad \text{Dividing by } 3$$

The solution set is $\{x \mid x \leq 3\}$, or $(-\infty, 3]$. We graph the solution set.

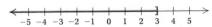

Practice Exercise

3. Solve and graph: $5y + 5 < 2y - 1$.

Objective 18.2a Find the intersection of two sets. Solve and graph conjunctions of inequalities.

Example Solve and graph: $-5 < 2x - 3 \leq 3$.

$$-5 < 2x - 3 \leq 3$$
$$-2 < 2x \leq 6 \qquad \text{Adding } 3$$
$$-1 < x \leq 3 \qquad \text{Dividing by } 2$$

The solution set is $\{x \mid -1 < x \leq 3\}$, or $(-1, 3]$. We graph the solution set.

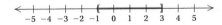

Practice Exercise

4. Solve and graph: $-4 \leq 5z + 6 < 11$.

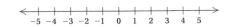

Objective 18.2b Find the union of two sets. Solve and graph disjunctions of inequalities.

Example Solve and graph:

$$2x + 1 \le -5 \text{ or } 3x + 1 > 7.$$
$$2x + 1 \le -5 \quad \text{or} \quad 3x + 1 > 7$$
$$2x \le -6 \quad \text{or} \qquad 3x > 6$$
$$x \le -3 \quad \text{or} \qquad x > 2$$

The solution set is $\{x \mid x \le -3 \text{ or } x > 2\}$, or $(-\infty, -3] \cup (2, \infty)$. We graph the solution set.

Practice Exercise

5. Solve and graph: $z + 4 < 3 \text{ or } 4z + 1 \ge 5$.

Objective 18.3c Solve equations with absolute-value expressions.

Example Solve: $|y - 2| = 1$.

$$y - 2 = -1 \quad \text{or} \quad y - 2 = 1$$
$$y = 1 \qquad \text{or} \qquad y = 3$$

The solution set is $\{1, 3\}$.

Practice Exercise

6. Solve: $|5x - 1| = 9$.

Objective 18.3d Solve equations with two absolute-value expressions.

Example Solve: $|4x - 4| = |2x + 8|$.

$$4x - 4 = 2x + 8 \quad \text{or} \quad 4x - 4 = -(2x + 8)$$
$$2x - 4 = 8 \qquad \text{or} \quad 4x - 4 = -2x - 8$$
$$2x = 12 \qquad \text{or} \quad 6x - 4 = -8$$
$$x = 6 \qquad \text{or} \qquad 6x = -4$$
$$x = 6 \qquad \text{or} \qquad x = -\frac{2}{3}$$

The solution set is $\left\{ 6, -\dfrac{2}{3} \right\}$.

Practice Exercise

7. Solve: $|z + 4| = |3z - 2|$.

Objective 18.3e Solve inequalities with absolute-value expressions.

Example Solve: **(a)** $|5x + 3| < 2$; **(b)** $|x + 3| \ge 1$.

a) $|5x + 3| < 2$
$$-2 < 5x + 3 < 2$$
$$-5 < 5x < -1$$
$$-1 < x < -\frac{1}{5}$$

The solution set is $\left\{ x \mid -1 < x < -\dfrac{1}{5} \right\}$, or $\left(-1, -\dfrac{1}{5} \right)$.

b) $|x + 3| \ge 1$
$$x + 3 \le -1 \quad \text{or} \quad x + 3 \ge 1$$
$$x \le -4 \quad \text{or} \qquad x \ge -2$$

The solution set is $\{x \mid x \le -4 \text{ or } x \ge -2\}$, or $(-\infty, -4] \cup [-2, \infty)$.

Practice Exercise

8. Solve: **(a)** $|2x + 3| < 5$; **(b)** $|3x + 2| \ge 8$.

Objective 18.4b Graph linear inequalities in two variables.

Example Graph: $2x + y \le 4$.

First, we graph the line $2x + y = 4$. The intercepts are $(0, 4)$ and $(2, 0)$. We draw the line solid since the inequality symbol is $\le$. Next, we choose a test point not on the line and determine whether it is a solution of the inequality. We choose $(0, 0)$, since it is usually an easy point to use.

$$\frac{2x + y \le 4}{2 \cdot 0 + 0 \; ? \; 4}$$
$$0 \quad | \quad \text{TRUE}$$

Since $(0, 0)$ is a solution, we shade the half-plane that contains $(0, 0)$.

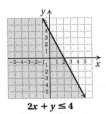

$2x + y \le 4$

Practice Exercise

9. Graph: $3x - 2y > 6$.

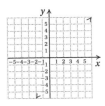

Objective 18.4c Graph systems of linear inequalities and find coordinates of any vertices.

Example Graph this system of inequalities and find the coordinates of any vertices formed:

$$x - 2y \ge -2, \quad \textbf{(1)}$$
$$3x - y \le 4, \quad \textbf{(2)}$$
$$y \ge -1. \quad \textbf{(3)}$$

We graph the related equations using solid lines. Then we indicate the region for each inequality by arrows near the ends of the line. Next, we shade the region of overlap.

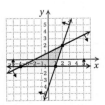

To find the vertices, we solve three different systems of related equations. From (1) and (2), we solve

$$x - 2y = -2,$$
$$3x - y = 4$$

to find the vertex $(2, 2)$. From (1) and (3), we solve

$$x - 2y = -2,$$
$$y = -1$$

to find the vertex $(-4, -1)$. From (2) and (3), we solve

$$3x - y = 4,$$
$$y = -1$$

to find the vertex $(1, -1)$.

Practice Exercise

10. Graph this system of inequalities and find the coordinates of any vertices found:

$$x - 2y \le 4,$$
$$x + y \le 4,$$
$$x - 1 \ge 0.$$

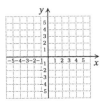

Review Exercises

1. Determine whether -3 and 7 are solutions of the inequality $2(1 - x) \le 3x + 15$.   [18.1a]

Write interval notation for the given set or graph. [18.1b]

2. $\{x \mid -8 \le x < 9\}$

3.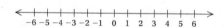

Solve and graph. Write interval notation for the solution set.   [18.1c]

4. $x - 2 \le -4$

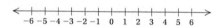

5. $x + 5 > 6$

Solve.   [18.1c]

6. $a + 7 \le -14$

7. $y - 5 \ge -12$

8. $4y > -16$

9. $-0.3y < 9$

10. $-6x - 5 < 13$

11. $4y + 3 \le -6y - 9$

12. $-\frac{1}{2}x - \frac{1}{4} > \frac{1}{2} - \frac{1}{4}x$

13. $0.3y - 8 < 2.6y + 15$

14. $-2(x - 5) \ge 6(x + 7) - 12$

15. *Moving Costs.*   Metro Movers charges $85 plus $40 per hour to move households across town. Champion Moving charges $60 per hour for cross-town moves. For what lengths of time is Champion more expensive?   [18.1d]

16. *Investments.*   Joe plans to invest $30,000, part at 3% and part at 4%, for one year. What is the most that can be invested at 3% in order to make at least $1100 interest in one year?   [18.1d]

Graph and write interval notation.   [18.2a,b]

17. $-2 \le x < 5$

18. $x \le -2 \text{ or } x > 5$

19. Find the intersection:
$$\{1, 2, 5, 6, 9\} \cap \{1, 3, 5, 9\}.   \text{[18.2a]}$$

20. Find the union:
$$\{1, 2, 5, 6, 9\} \cup \{1, 3, 5, 9\}.   \text{[18.2b]}$$

Solve.   [18.2a, b]

21. $2x - 5 < -7 \text{ and } 3x + 8 \ge 14$

22. $-4 < x + 3 \le 5$

23. $-15 < -4x - 5 < 0$

24. $3x < -9 \text{ or } -5x < -5$

25. $2x + 5 < -17 \text{ or } -4x + 10 \le 34$

26. $2x + 7 \le -5 \text{ or } x + 7 \ge 15$

Simplify.   [18.3a]

27. $\left| -\dfrac{3}{x} \right|$

28. $\left| \dfrac{2x}{y^2} \right|$

29. $\left| \dfrac{12y}{-3y^2} \right|$

30. Find the distance between -23 and 39.   [18.3b]

Solve.   [18.3c, d]

31. $|x| = 6$

32. $|x - 2| = 7$

33. $|2x + 5| = |x - 9|$

34. $|5x + 6| = -8$

Solve.   [18.3e]

35. $|2x + 5| < 12$

36. $|x| \ge 3.5$

37. $|3x - 4| \ge 15$

38. $|x| < 0$

Graph.　[18.4b]

39. $2x + 3y < 12$

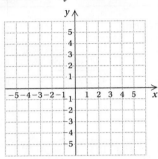

40. $y \leq 0$

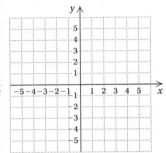

41. $x + y \geq 1$

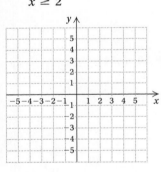

Graph. Find the coordinates of any vertices formed.　[18.4c]

42. $y \geq -3,$
　　$x \geq 2$

43. $x + 3y \geq -1,$
　　$x + 3y \leq 4$

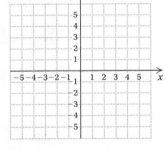

44. $x - y \leq 3,$
　　$x + y \geq -1,$
　　$y \leq 2$

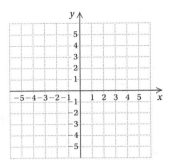

45. Solve: $-2(x + 3) - 1 < -2(2 - x)$.　[18.1c]

　A. $\left(-\infty, -\dfrac{3}{4}\right)$　　　　**B.** $(-6, \infty)$

　C. $(6, \infty)$　　　　**D.** $\left(-\dfrac{3}{4}, \infty\right)$

46. Solve: $\left|-\dfrac{1}{3}x - 10\right| \geq 30$.　[18.3e]

　A. $(-\infty, -100] \cup [80, \infty)$
　B. $(-\infty, -60] \cup [120, \infty)$
　C. $(-\infty, -120] \cup [60, \infty)$
　D. $(-\infty, -80] \cup [100, \infty)$

Synthesis

47. Solve: $|2x + 5| \leq |x + 3|$.　[18.3d, e]

Understanding Through Discussion and Writing

1. Describe the circumstances under which, for intervals, $[a, b] \cup [c, d] = [a, d]$.　[18.2b]

2. Explain in your own words why the solutions of the inequality $|x + 5| \leq 2$ can be interpreted as "all those numbers x whose distance from -5 is at most 2 units."　[18.3e]

3. When graphing linear inequalities, Ron always shades above the line when he sees a $\geq$ symbol. Is this wise? Why or why not?　[18.4a]

4. Explain in your own words why the interval $[6, \infty)$ is only part of the solution set of $|x| \geq 6$.　[18.3e]

CHAPTER

18   Test

For Extra Help   For step-by-step test solutions, access the Chapter Test Prep Videos in MyMathLab® or on You Tube (search "BittingerAlgebraFoundations" and click on "Channels").

Write interval notation for the given set or graph.

1. $\{x \mid -3 < x \le 2\}$

2.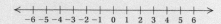

Solve and graph. Write interval notation for the solution set.

3. $x - 2 \le 4$

4. $-4y - 3 \ge 5$

Solve.

5. $x - 4 \ge 6$

6. $-0.6y < 30$

7. $3a - 5 \le -2a + 6$

8. $-5y - 1 > -9y + 3$

9. $4(5 - x) < 2x + 5$

10. $-8(2x + 3) + 6(4 - 5x) \ge 2(1 - 7x) - 4(4 + 6x)$

Solve.

11. *Moving Costs.* Mitchell Moving Company charges $105 plus $30 per hour to move households across town. Quick-Pak Moving charges $80 per hour for cross-town moves. For what lengths of time is Quick-Pak more expensive?

12. *Pressure at Sea Depth.* The equation
$$P = 1 + \frac{d}{33}$$
gives the pressure P, in atmospheres (atm), at a depth of d feet in the sea. For what depths d is the pressure at least 2 atm and at most 8 atm?

Graph and write interval notation.

13. $-3 \le x \le 4$

14. $x < -3 \text{ or } x > 4$

Solve.

15. $5 - 2x \le 1 \text{ and } 3x + 2 \ge 14$

16. $-3 < x - 2 < 4$

17. $-11 \le -5x - 2 < 0$

18. $-3x > 12 \text{ or } 4x > -10$

19. $x - 7 \le -5 \text{ or } x - 7 \ge -10$

20. $3x - 2 < 7 \text{ or } x - 2 > 4$

Simplify.

21. $\left| \dfrac{7}{x} \right|$

22. $\left| \dfrac{-6x^2}{3x} \right|$

23. Find the distance between 4.8 and -3.6.

24. Find the intersection:
$$\{1, 3, 5, 7, 9\} \cap \{3, 5, 11, 13\}.$$

25. Find the union:
$$\{1, 3, 5, 7, 9\} \cup \{3, 5, 11, 13\}.$$

Solve.

26. $|x| = 9$

27. $|x - 3| = 9$

28. $|x + 10| = |x - 12|$

29. $|2 - 5x| = -10$

30. $|4x - 1| < 4.5$

31. $|x| > 3$

32. $\left| \dfrac{6 - x}{7} \right| \leq 15$

33. $|-5x - 3| \geq 10$

Graph. Find the coordinates of any vertices formed.

34. $x - 6y < -6$

35. $x + y \geq 3,$
$x - y \geq 5$

36. $2y - x \geq -4,$
$2y + 3x \leq -6,$
$y \leq 0,$
$x \leq 0$

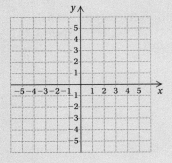

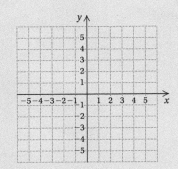

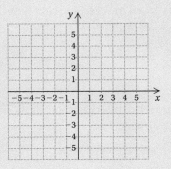

37. Solve: $\left| \dfrac{1}{2}x - 2 \right| \geq 2.2.$

   A. $[-0.4, 8.4]$
   C. $(-\infty, 8.4]$

   B. $(-\infty, -0.4] \cup [8.4, \infty)$
   D. $[-0.4, \infty)$

Synthesis

Solve.

38. $|3x - 4| \leq -3$

39. $7x < 8 - 3x < 6 + 7x$

Answers

CHAPTER 10

Exercise Set 10.1, p. 568

RC1. Division **RC2.** Multiplication **RC3.** Multiplication
RC4. Division
1. 32 min; 69 min; 81 min **3.** 260 mi **5.** 576 in² **7.** 1935 m²
9. 56 **11.** 8 **13.** 1 **15.** 6 **17.** 2 **19.** $b + 7$, or $7 + b$

21. $c - 12$ **23.** $a + b$, or $b + a$ **25.** $x \div y$, or $\dfrac{x}{y}$, or x/y

27. $x + w$, or $w + x$ **29.** $n - m$ **31.** $2z$ **33.** $3m$, or $m \cdot 3$
35. $4a + 6$, or $6 + 4a$ **37.** $xy - 8$ **39.** $2t - 5$ **41.** $3n + 11$,
or $11 + 3n$ **43.** $4x + 3y$, or $3y + 4x$ **45.** $s + 0.05s$
47. $65t$ miles **49.** $\$50 - x$ **51.** $\$12.50n$ **53.** $\frac{1}{4}$ **55.** 0

Calculator Corner, p. 575

1. 8.717797887 **2.** 17.80449381 **3.** 67.08203932
4. 35.4807407 **5.** 3.141592654 **6.** 91.10618695
7. 530.9291585 **8.** 138.8663978

Calculator Corner, p. 576

1. -0.75 **2.** -0.45 **3.** -0.125 **4.** -1.8 **5.** -0.675
6. -0.6875 **7.** -3.5 **8.** -0.76

Calculator Corner, p. 578

1. 5 **2.** 17 **3.** 0 **4.** 6.48 **5.** 12.7 **6.** 0.9 **7.** $\frac{5}{7}$ **8.** $\frac{4}{3}$

Exercise Set 10.2, p. 579

RC1. H **RC2.** E **RC3.** J **RC4.** D **RC5.** B **RC6.** G
RC7. True **RC8.** True **RC9.** False **RC10.** False
1. $24; -2$ **3.** $7{,}200{,}000{,}000{,}000; -460$ **5.** $1454; -55$

7. **9.**

11. **13.** -0.875 **15.** $0.8\overline{3}$
17. $-1.1\overline{6}$ **19.** $0.\overline{6}$ **21.** 0.1 **23.** -0.5 **25.** 0.16 **27.** $>$
29. $<$ **31.** $<$ **33.** $<$ **35.** $>$ **37.** $>$ **39.** $<$ **41.** $>$
43. $<$ **45.** $<$ **47.** $x < -6$ **49.** $y \geq -10$ **51.** False
53. True **55.** True **57.** False **59.** 3 **61.** 11 **63.** $\frac{2}{3}$
65. 0 **67.** 2.65 **69.** 3 **70.** 11 **71.** 3 **72.** 1
73. $-\frac{2}{3}, -\frac{2}{5}, -\frac{1}{3}, -\frac{2}{7}, -\frac{1}{7}, \frac{1}{3}, \frac{2}{5}, \frac{2}{3}, \frac{9}{8}$ **75.** $\frac{1}{1}$ **77.** $\frac{50}{9}$

Exercise Set 10.3, p. 587

RC1. Right; right **RC2.** Left; left **RC3.** Right; left
RC4. Left; right
1. -7 **3.** -6 **5.** 0 **7.** -8 **9.** -7 **11.** -27 **13.** 0
15. -42 **17.** 0 **19.** 0 **21.** 3 **23.** -9 **25.** 7 **27.** 0
29. 35 **31.** -3.8 **33.** -8.1 **35.** $-\frac{1}{5}$ **37.** $-\frac{7}{9}$ **39.** $-\frac{3}{8}$
41. $-\frac{19}{24}$ **43.** $\frac{1}{24}$ **45.** $\frac{8}{15}$ **47.** $\frac{16}{45}$ **49.** 37 **51.** 50 **53.** -24
55. 26.9 **57.** -8 **59.** $\frac{13}{8}$ **61.** -43 **63.** $\frac{4}{3}$ **65.** 24 **67.** $\frac{3}{8}$
69. 13,796 ft **71.** $-3°$F **73.** Profit of $\$4300$ **75.** He owes $\$85$.
77. 39 **78.** $y > -3$ **79.** $-0.08\overline{3}$ **80.** 0.625 **81.** 0 **82.** 21.4
83. All positive numbers **85.** B

Exercise Set 10.4, p. 593

RC1. (c) **RC2.** (b) **RC3.** (d) **RC4.** (a)
1. -7 **3.** -6 **5.** 0 **7.** -4 **9.** -7 **11.** -6 **13.** 0
15. 14 **17.** 11 **19.** -14 **21.** 5 **23.** -1 **25.** 18 **27.** -3
29. -21 **31.** 5 **33.** -8 **35.** 12 **37.** -23 **39.** -68
41. -73 **43.** 116 **45.** 0 **47.** -1 **49.** $\frac{1}{12}$ **51.** $-\frac{17}{12}$ **53.** $\frac{1}{8}$
55. 19.9 **57.** -8.6 **59.** -0.01 **61.** -193 **63.** 500
65. -2.8 **67.** -3.53 **69.** $-\frac{1}{2}$ **71.** $\frac{6}{7}$ **73.** $-\frac{41}{30}$ **75.** $-\frac{2}{15}$
77. $-\frac{1}{48}$ **79.** $-\frac{43}{60}$ **81.** 37 **83.** -62 **85.** -139 **87.** 6

89. 108.5 **91.** $\frac{1}{4}$ **93.** 30,383 ft **95.** $\$347.94$ **97.** 3780 m
99. 381 ft **101.** $1130°$F **103.** $y + 7$, or $7 + y$ **104.** $t - 41$
105. $a - h$ **106.** $6c$, or $c \cdot 6$ **107.** $r + s$, or $s + r$ **108.** $y - x$
109. False; $3 - 0 \neq 0 - 3$ **111.** True **113.** True

Mid-Chapter Review: Chapter 10, p. 597

1. True **2.** False **3.** True **4.** False **5.** $-x = -(-4) = 4$;
$-(-x) = -(-(-4)) = -(4) = -4$
6. $5 - 13 = 5 + (-13) = -8$ **7.** $-6 - 7 = -6 + (-7) = -13$
8. 4 **9.** 11 **10.** $3y$ **11.** $n - 5$ **12.** $450; -79$

13. **14.** -0.8 **15.** $2.\overline{3}$ **16.** $<$

17. $>$ **18.** False **19.** True **20.** $5 > y$ **21.** $t \leq -3$
22. 15.6 **23.** 18 **24.** 0 **25.** $\frac{12}{5}$ **26.** 5.6 **27.** $-\frac{7}{4}$ **28.** 0
29. 49 **30.** 19 **31.** 2.3 **32.** -2 **33.** $-\frac{1}{8}$ **34.** 0 **35.** -17
36. $-\frac{11}{24}$ **37.** -8.1 **38.** -9 **39.** -2 **40.** -10.4 **41.** 16
42. $\frac{7}{20}$ **43.** -12 **44.** -4 **45.** $-\frac{4}{3}$ **46.** -1.8 **47.** 13
48. 9 **49.** -23 **50.** 75 **51.** 14 **52.** $33°$C **53.** $\$54.80$
54. Answers may vary. Three examples are $\frac{6}{13}$, -23.8, and $\frac{43}{5}$. These
are rational numbers because they can be named in the form $\dfrac{a}{b}$,
where a and b are integers and b is not 0. They are not integers,
however, because they are neither whole numbers nor the opposites
of whole numbers. **55.** Answers may vary. Three examples are
π, $-\sqrt{7}$, and $0.31311311131111\ldots$. Irrational numbers cannot be
written as the quotient of two integers. Real numbers that are not
rational are irrational. Decimal notation for rational numbers either
terminates or repeats. Decimal notation for irrational numbers
neither terminates nor repeats. **56.** Answers may vary. If we think
of the addition on the number line, we start at 0, move to the left to a
negative number, and then move to the left again. This always brings
us to a point on the negative portion of the number line.
57. Yes; consider $m - (-n)$, where both m and n are positive.
Then $m - (-n) = m + n$. Now $m + n$, the sum of two positive
numbers, is positive.

Exercise Set 10.5, p. 602

RC1. Negative **RC2.** Positive **RC3.** Positive
RC4. Negative **RC5.** -9 **RC6.** 9 **RC7.** $-\frac{1}{4}$ **RC8.** $-\frac{1}{4}$
1. -8 **3.** -24 **5.** -72 **7.** 16 **9.** 42 **11.** -120 **13.** -238
15. 1200 **17.** 98 **19.** -72 **21.** -12.4 **23.** 30 **25.** 21.7
27. $-\frac{2}{3}$ **29.** $\frac{1}{12}$ **31.** -17.01 **33.** 420 **35.** $\frac{2}{7}$ **37.** -60
39. 150 **41.** 50.4 **43.** $\frac{10}{189}$ **45.** -960 **47.** 17.64 **49.** $-\frac{5}{784}$
51. 0 **53.** -720 **55.** $-30{,}240$ **57.** 1 **59.** 16, -16; 16, -16
61. $\frac{4}{25}, -\frac{4}{25}; \frac{4}{25}, -\frac{4}{25}$ **63.** $-9, -9; -9, -9$ **65.** 441, -147; 441, -147
67. 20; 20 **69.** -2; 2 **71.** $-24°$C **73.** -20 lb **75.** $\$12.71$
77. -32 m **79.** $38°$F **81.** 2 **82.** $\frac{4}{15}$ **83.** $-\frac{1}{3}$ **84.** -4.3
85. 44 **86.** $-\frac{1}{12}$ **87.** True **88.** False **89.** False
90. False **91.** A **93.** Largest quotient: $10 \div \frac{1}{5} = 50$; smallest
quotient: $-5 \div \frac{1}{5} = -25$

Calculator Corner, p. 611

1. -4 **2.** -2 **3.** -32 **4.** 1.4 **5.** 2.7 **6.** -9.5
7. -0.8 **8.** 14.44

Exercise Set 10.6, p. 611

RC1. Opposites **RC2.** 1 **RC3.** 0 **RC4.** Reciprocals
1. -8 **3.** -14 **5.** -3 **7.** 3 **9.** -8 **11.** 2 **13.** -12
15. -8 **17.** Not defined **19.** 0 **21.** $\frac{7}{15}$ **23.** $-\frac{13}{47}$ **25.** $\frac{1}{13}$
27. $-\frac{1}{32}$ **29.** -7.1 **31.** 9 **33.** $4y$ **35.** $\dfrac{3b}{2a}$ **37.** $4 \cdot \left(\frac{1}{17}\right)$
39. $8 \cdot \left(-\frac{1}{13}\right)$ **41.** $13.9 \cdot \left(-\frac{1}{1.5}\right)$ **43.** $\frac{2}{3} \cdot \left(-\frac{5}{4}\right)$ **45.** $x \cdot y$
47. $(3x + 4)\left(\frac{1}{5}\right)$ **49.** $-\frac{9}{8}$ **51.** $\frac{5}{3}$ **53.** $\frac{9}{14}$ **55.** $\frac{9}{64}$ **57.** $-\frac{5}{4}$
59. $-\frac{27}{5}$ **61.** $\frac{11}{13}$ **63.** -2 **65.** -16.2 **67.** -2.5 **69.** -1.25

71. Not defined   73. Percent increase is about 44%.
75. Percent decrease is about −21%.   77. $-\frac{1}{4}$ 78. 5
79. −42   80. −48   81. 8.5   82. $-\frac{1}{8}$ 83. $-0.0\overline{9}$ 84. $0.91\overline{6}$
85. 3.75   86. $-3.\overline{3}$ 87. $\frac{1}{-10.5}$; −10.5, the reciprocal of the reciprocal is the original number.   89. Negative   91. Positive
93. Negative

Exercise Set 10.7, p. 623

RC1. (g)   RC2. (h)   RC3. (f)   RC4. (e)   RC5. (d)

RC6. (a)   RC7. (b)   1. $\frac{3y}{5y}$ 3. $\frac{10x}{15x}$ 5. $\frac{2x}{x^2}$ 7. $-\frac{3}{2}$ 9. $-\frac{7}{6}$
11. $\frac{4s}{3}$ 13. $8+y$ 15. nm 17. $xy+9$, or $9+yx$, or $yx+9$
19. $c+ab$, or $ba+c$, or $c+ba$ 21. $(a+b)+2$ 23. $8(xy)$
25. $a+(b+3)$ 27. $(3a)b$ 29. $2+(b+a),(2+a)+b$, $(b+2)+a$; answers may vary   31. $(5+w)+v,(v+5)+w$, $(w+v)+5$; answers may vary   33. $(3x)y, y(x\cdot3), 3(yx)$; answers may vary   35. $a(7b), b(7a), (7b)a$; answers may vary
37. $2b+10$ 39. $7+7t$ 41. $30x+12$ 43. $7x+28+42y$
45. $7x-21$ 47. $-3x+21$ 49. $\frac{2}{3}b-4$ 51. $7.3x-14.6$
53. $-\frac{3}{5}x+\frac{3}{5}y-6$ 55. $45x+54y-72$ 57. $-4x+12y+8z$
59. $-3.72x+9.92y-3.41$ 61. $4x,3z$ 63. $7x,8y,-9z$
65. $2(x+2)$ 67. $5(6+y)$ 69. $7(2x+3y)$ 71. $7(2t-1)$
73. $8(x-3)$ 75. $6(3a-4b)$ 77. $-4(y-8)$, or $4(-y+8)$
79. $5(x+2+3y)$ 81. $8(2m-4n+1)$ 83. $4(3a+b-6)$
85. $2(4x+5y-11)$ 87. $a(x-1)$ 89. $a(x-y-z)$
91. $-6(3x-2y-1)$, or $6(-3x+2y+1)$ 93. $\frac{1}{3}(2x-5y+1)$
95. $6(6x-y+3z)$ 97. $19a$ 99. $9a$ 101. $8x+9z$
103. $7x+15y^2$ 105. $-19a+88$ 107. $4t+6y-4$ 109. b
111. $\frac{13}{4}y$ 113. $8x$ 115. $5n$ 117. $-16y$ 119. $17a-12b-1$
121. $4x+2y$ 123. $7x+y$ 125. $0.8x+0.5y$ 127. $\frac{35}{6}a+\frac{3}{2}b-42$
129. 38   130. −4.9   131. 4   132. 10   133. $-\frac{4}{49}$ 134. −106
135. 1   136. −34   137. 180   138. $\frac{4}{13}$ 139. True   140. False
141. True   142. True   143. Not equivalent;
$3\cdot2+5 \neq 3\cdot5+2$ 145. Equivalent; commutative law of
addition   147. $q(1+r+rs+rst)$

Calculator Corner, p. 632

1. −16   2. 9   3. 117,649   4. −1,419,857   5. −117,649
6. −1,419,857   7. −4   8. −2

Exercise Set 10.8, p. 633

RC1. Multiplication   RC2. Addition   RC3. Subtraction
RC4. Division   RC5. Division   RC6. Multiplication
1. $-2x-7$ 3. $-8+x$ 5. $-4a+3b-7c$
7. $-6x+8y-5$ 9. $-3x+5y+6$ 11. $8x+6y+43$
13. $5x-3$ 15. $-3a+9$ 17. $5x-6$ 19. $-19x+2y$
21. $9y-25z$ 23. $-7x+10y$ 25. $37a-23b+35c$
27. 7   29. −40   31. 19   33. $12x+30$ 35. $3x+30$
37. $9x-18$ 39. $-4x-64$ 41. −7   43. −1   45. −16
47. −334   49. 14   51. 1880   53. 12   55. 8   57. −86
59. 37   61. −1   63. −10   65. −67   67. −7988   69. −3000
71. 60   73. 1   75. 10   77. $-\frac{13}{45}$ 79. $-\frac{23}{18}$ 81. −122
83. 18   84. 35   85. 0.4   86. $\frac{15}{2}$ 87. $-\frac{1}{9}$ 88. $\frac{3}{7}$
89. −25   90. −35   91. 25   92. 35   93. $-2x-f$
95. (a) 52; 52; 28.130169; (b) −24; −24; −108.307025   97. −6

Summary and Review: Chapter 10, p. 637

Vocabulary Reinforcement

1. Integers   2. Additive inverses   3. Commutative law
4. Identity property of 1   5. Associative law
6. Multiplicative inverses   7. Identity property of 0

Concept Reinforcement

1. True   2. True   3. False   4. False

Study Guide

1. 14   2. <   3. $\frac{5}{4}$ 4. −8.5   5. −2   6. 56   7. −8
8. $\frac{9}{20}$ 9. $\frac{5}{3}$ 10. $5x+15y-20z$ 11. $9(3x+y-4z)$
12. $5a-2b$ 13. $4a-4b$ 14. −2

Review Exercises

1. 4   2. $19\%x$, or $0.19x$ 3. $620,-125$ 4. 38   5. 126
6. [number line: −2.5]   7. [number line: $\frac{8}{9}$]
8. <   9. >   10. >   11. <   12. $x>-3$ 13. True
14. False   15. −3.8   16. $\frac{3}{4}$ 17. $\frac{8}{3}$ 18. $-\frac{1}{7}$ 19. 34
20. 5   21. −3   22. −4   23. −5   24. 1   25. $-\frac{7}{5}$
26. −7.9   27. 54   28. −9.18   29. $-\frac{2}{7}$ 30. −210   31. −7
32. −3   33. $\frac{3}{4}$ 34. 24.8   35. −2   36. 2   37. −2
38. 8-yd gain   39. −$360   40. $4.64   41. $18.95
42. $15x-35$ 43. $-8x+10$ 44. $4x+15$ 45. $-24+48x$
46. $2(x-7)$ 47. $-6(x-1)$, or $6(-x+1)$ 48. $5(x+2)$
49. $-3(x-4y+4)$, or $3(-x+4y-4)$ 50. $7a-3b$
51. $-2x+5y$ 52. $5x-y$ 53. $-a+8b$ 54. $-3a+9$
55. $-2b+21$ 56. 6   57. $12y-34$ 58. $5x+24$
59. $-15x+25$ 60. D   61. B   62. $-\frac{5}{8}$ 63. −2.1
64. 1000   65. $4a+2b$

Understanding Through Discussion and Writing

1. The sum of each pair of opposites such as −50 and 50, −49 and 49, and so on, is 0. The sum of these sums and the remaining integer, 0, is 0.   2. The product of an even number of negative numbers is positive, and the product of an odd number of negative numbers is negative. Now $(-7)^8$ is the product of 8 factors of −7 so it is positive, and $(-7)^{11}$ is the product of 11 factors of −7 so it is negative.
3. Consider $\frac{a}{b}=q$, where a and b are both negative numbers. Then $q\cdot b=a$, so q must be a positive number in order for the product to be negative.   4. Consider $\frac{a}{b}=q$, where a is a negative number and b is a positive number. Then $q\cdot b=a$, so q must be a negative number in order for the product to be negative.   5. We use the distributive law when we collect like terms even though we might not always write this step.   6. Jake expects the calculator to multiply 2 and 3 first and then divide 18 by that product. This procedure does not follow the rules for order of operations.

Test: Chapter 10, p. 643

1. [10.1a] 6   2. [10.1b] $x-9$ 3. [10.2d] >   4. [10.2d] <
5. [10.2d] >   6. [10.2d] $-2>x$ 7. [10.2d] True   8. [10.2e] 7
9. [10.2e] $\frac{9}{4}$ 10. [10.2e] 2.7   11. [10.3b] $-\frac{2}{3}$ 12. [10.3b] 1.4
13. [10.6b] $-\frac{1}{2}$ 14. [10.6b] $\frac{7}{4}$ 15. [10.3b] 8   16. [10.4a] 7.8
17. [10.3a] −8   18. [10.3a] $\frac{7}{40}$ 19. [10.4a] 10   20. [10.4a] −2.5
21. [10.4a] $\frac{7}{8}$ 22. [10.5a] −48   23. [10.5a] $\frac{3}{16}$ 24. [10.6a] −9
25. [10.6c] $\frac{3}{4}$ 26. [10.6c] −9.728   27. [10.2e], [10.8d] −173
28. [10.8d] −5   29. [10.3c], [10.4b] Up 15 points   30. [10.4b] 2244 m
31. [10.5b] 16,080   32. [10.6d] −0.75°C each minute
33. [10.7c] $18-3x$ 34. [10.7c] $-5y+5$ 35. [10.7d] $2(6-11x)$
36. [10.7d] $7(x+3+2y)$ 37. [10.4a] 12   38. [10.8b] $2x+7$
39. [10.8b] $9a-12b-7$ 40. [10.8c] $68y-8$ 41. [10.8d] −4
42. [10.8d] 448   43. [10.2d] B   44. [10.2e], [10.8d] 15
45. [10.8c] $4a$ 46. [10.7e] $4x+4y$

CHAPTER 11

Exercise Set 11.1, p. 650

RC1. (f) **RC2.** (c) **RC3.** (e) **RC4.** (a)
1. Yes **3.** No **5.** No **7.** Yes **9.** Yes **11.** No **13.** 4
15. -20 **17.** -14 **19.** -18 **21.** 15 **23.** -14 **25.** 2
27. 20 **29.** -6 **31.** $6\frac{1}{2}$ **33.** 19.9 **35.** $\frac{7}{3}$ **37.** $-\frac{7}{4}$ **39.** $\frac{41}{24}$
41. $-\frac{1}{20}$ **43.** 5.1 **45.** 12.4 **47.** -5 **49.** $1\frac{5}{6}$ **51.** $-\frac{10}{21}$
53. $-\frac{3}{2}$ **54.** -5.2 **55.** $-\frac{1}{24}$ **56.** 172.72 **57.** $\$83 - x$
58. $65t$ miles **59.** $-\frac{26}{15}$ **61.** -10 **63.** All real numbers

Exercise Set 11.2, p. 655

RC1. (f) **RC2.** (d) **RC3.** (a) **RC4.** (b)
1. 6 **3.** 9 **5.** 12 **7.** -40 **9.** 1 **11.** -7 **13.** -6
15. 6 **17.** -63 **19.** -48 **21.** 36 **23.** -9 **25.** -21
27. $-\frac{3}{5}$ **29.** $-\frac{3}{2}$ **31.** $\frac{9}{2}$ **33.** 7 **35.** -7 **37.** 8 **39.** 15.9
41. -50 **43.** -14 **45.** $7x$ **46.** $-x + 5$ **47.** $8x + 11$
48. $-32y$ **49.** $x - 4$ **50.** $-5x - 23$ **51.** $-10y - 42$
52. $-22a + 4$ **53.** $8r$ miles **54.** $\frac{1}{2}b \cdot 10\ \text{m}^2$, or $5b\ \text{m}^2$
55. -8655 **57.** No solution **59.** No solution **61.** $\dfrac{b}{3a}$
63. $\dfrac{4b}{a}$

Calculator Corner, p. 661

1. Left to the student

Exercise Set 11.3, p. 665

RC1. (d) **RC2.** (a) **RC3.** (c) **RC4.** (e) **RC5.** (b)
1. 5 **3.** 8 **5.** 10 **7.** 14 **9.** -8 **11.** -8 **13.** -7
15. 12 **17.** 6 **19.** 4 **21.** 6 **23.** -3 **25.** 1 **27.** 6
29. -20 **31.** 7 **33.** 2 **35.** 5 **37.** 2 **39.** 10 **41.** 4
43. 0 **45.** -1 **47.** $-\frac{4}{3}$ **49.** $\frac{2}{5}$ **51.** -2 **53.** -4 **55.** $\frac{4}{5}$
57. $-\frac{28}{27}$ **59.** 6 **61.** 2 **63.** No solution **65.** All real numbers
67. 6 **69.** 8 **71.** 1 **73.** 17 **75.** $-\frac{5}{3}$ **77.** All real numbers
79. No solution **81.** -3 **83.** 2 **85.** $\frac{4}{7}$ **87.** No solution
89. All real numbers **91.** $-\frac{51}{31}$ **93.** -6.5 **94.** -75.14
95. $7(x - 3 - 2y)$ **96.** $8(y - 11x + 1)$ **97.** -160
98. $-17x + 18$ **99.** $91x - 242$ **100.** 0.25 **101.** $-\frac{5}{32}$ **103.** $\frac{52}{45}$

Exercise Set 11.4, p. 673

RC1. (b) **RC2.** (c) **RC3.** (d) **RC4.** (a)
1. $14\frac{1}{3}$ meters per cycle **3.** 10.5 calories per ounce
5. (a) 337.5 mi; **(b)** $t = \dfrac{d}{r}$ **7. (a)** 1423 students; **(b)** $n = 15f$
9. $x = \dfrac{y}{5}$ **11.** $c = \dfrac{a}{b}$ **13.** $m = n - 11$ **15.** $x = y + \dfrac{3}{5}$
17. $x = y - 13$ **19.** $x = y - b$ **21.** $x = 5 - y$
23. $x = a - y$ **25.** $y = \dfrac{5x}{8}$, or $\dfrac{5}{8}x$. **27.** $x = \dfrac{By}{A}$
29. $t = \dfrac{W - b}{m}$ **31.** $x = \dfrac{y - c}{b}$ **33.** $h = \dfrac{A}{b}$
35. $w = \dfrac{P - 2l}{2}$, or $\dfrac{1}{2}P - l$ **37.** $a = 2A - b$
39. $b = 3A - a - c$ **41.** $t = \dfrac{A - b}{a}$ **43.** $x = \dfrac{c - By}{A}$
45. $a = \dfrac{F}{m}$ **47.** $c^2 = \dfrac{E}{m}$ **49.** $t = \dfrac{3k}{v}$ **51.** 7
52. $-21a + 12b$ **53.** -13.2 **54.** $-\frac{3}{2}$ **55.** $-35\frac{1}{2}$ **56.** $-\frac{1}{6}$
57. -9.325 **58.** $3\frac{3}{4}$ **59.** $\frac{11}{8}$ **60.** -1 **61.** -3 **62.** $\frac{9}{7}$
63. $b = \dfrac{Ha - 2}{H}$, or $a - \dfrac{2}{H}$; $a = \dfrac{2 + Hb}{H}$, or $\dfrac{2}{H} + b$
65. A quadruples. **67.** A increases by $2h$ units.

Mid-Chapter Review: Chapter 11, p. 677

1. False **2.** True **3.** True **4.** False
5. $x + 5 = -3$
$\quad x + 5 - 5 = -3 - 5$
$\quad x + 0 = -8$
$\quad x = -8$

6. $-6x = 42$
$\quad \dfrac{-6x}{-6} = \dfrac{42}{-6}$
$\quad 1 \cdot x = -7$
$\quad\quad x = -7$

7. $5y + z = t$
$\quad 5y + z - z = t - z$
$\quad 5y = t - z$
$\quad \dfrac{5y}{5} = \dfrac{t - z}{5}$
$\quad y = \dfrac{t - z}{5}$

8. 6 **9.** -12 **10.** 7 **11.** -10 **12.** 20 **13.** 5 **14.** $\frac{3}{4}$
15. -1.4 **16.** 6 **17.** -17 **18.** -9 **19.** 17 **20.** 21
21. 18 **22.** -15 **23.** $-\frac{3}{2}$ **24.** 1 **25.** -3 **26.** $\frac{3}{2}$ **27.** -1
28. 3 **29.** -7 **30.** 4 **31.** 2 **32.** $\frac{9}{8}$ **33.** $-\frac{21}{5}$ **34.** 9
35. -2 **36.** 0 **37.** All real numbers **38.** No solution
39. $-\frac{13}{2}$ **40.** All real numbers **41.** $b = \dfrac{A}{4}$ **42.** $x = y + 1.5$
43. $m = s - n$ **44.** $t = \dfrac{9w}{4}$ **45.** $t = \dfrac{B + c}{a}$
46. $y = 2M - x - z$ **47.** Equivalent expressions have the same value for all possible replacements for the variable(s). Equivalent equations have the same solution(s). **48.** The equations are not equivalent because they do not have the same solutions. Although 5 is a solution of both equations, -5 is a solution of $x^2 = 25$ but not of $x = 5$. **49.** For an equation $x + a = b$, add the opposite of a (or subtract a) on both sides of the equation. **50.** The student probably added $\frac{1}{3}$ on both sides of the equation rather than adding $-\frac{1}{3}$ (or subtracting $\frac{1}{3}$) on both sides. The correct solution is -2. **51.** For an equation $ax = b$, multiply by $1/a$ (or divide by a) on both sides of the equation. **52.** Answers may vary. A walker who knows how far and how long she walks each day wants to know her average speed each day.

Exercise Set 11.5, p. 683

RC1. (d) **RC2.** (b) **RC3.** (e) **RC4.** (a) **RC5.** (f) **RC6.** (c)
1. 20% **3.** 150 **5.** 546 **7.** 24% **9.** 2.5 **11.** 5% **13.** 25%
15. 84 **17.** 24% **19.** 16% **21.** $46\frac{2}{3}$ **23.** 0.8 **25.** 5 **27.** 40
29. 811 million **31.** 5274 million **33.** 1764 million
35. $968 million **37.** $221 **39.** 21% **41. (a)** 12.5%;
(b) $13.50 **43. (a)** $31; **(b)** $35.65 **45.** About 85,821 acres
47. About 82.4% **49.** About 53.1% decrease **51.** About 53.4%
53. About 19.2% decrease **55.** $12 + 3q$ **56.** $5x - 21$ **57.** $\dfrac{15w}{8}$
58. $-\frac{3}{2}$ **59.** 44 **60.** $x + 8$ **61.** 6 ft 7 in.

Translating for Success, p. 698

1. B **2.** H **3.** G **4.** N **5.** J **6.** C **7.** L **8.** E
9. F **10.** D

Exercise Set 11.6, p. 699

RC1. Familiarize **RC2.** Translate **RC3.** Solve
RC4. Check **RC5.** State
1. 1522 Medals of Honor **3.** 180 in.; 60 in. **5.** 21.8 million
7. 4.37 mi **9.** 1204 and 1205 **11.** 41, 42, 43 **13.** 61, 63, 65
15. 36 in. $\times$ 110 in. **17.** $63 **19.** $24.95 **21.** 11 visits
23. 28°, 84°, 68° **25.** 33°, 38°, 109° **27.** $350 **29.** $852.94
31. 18 mi **33.** $38.60 **35.** 89 and 96 **37.** -12 **39.** $-\frac{47}{40}$
40. $-\frac{17}{40}$ **41.** $-\frac{3}{10}$ **42.** $-\frac{32}{15}$ **43.** -10 **44.** 1.6 **45.** 409.6
46. -9.6 **47.** -41.6 **48.** 0.1 **49.** $yz + 12, zy + 12$, or $12 + zy$ **50.** $c + (4 + d)$ **51.** 120 apples **53.** About 0.65 in.

Exercise Set 11.7, p. 712

RC1. Not equivalent **RC2.** Not equivalent **RC3.** Equivalent
RC4. Equivalent
1. (a) Yes; **(b)** yes; **(c)** no; **(d)** yes; **(e)** yes **3. (a)** No;
(b) no; **(c)** no; **(d)** yes; **(e)** no
5. $x > 4$ **7.** $t < -3$
9. $m \geq -1$ **11.** $-3 < x \leq 4$
13. $0 < x < 3$ **15.** $\{x \mid x > -5\}$
17. $\{x \mid x \leq -18\}$; **19.** $\{y \mid y > -5\}$

21. $\{x \mid x > 2\}$ **23.** $\{x \mid x \leq -3\}$ **25.** $\{x \mid x < 4\}$
27. $\{t \mid t > 14\}$ **29.** $\{y \mid y \leq \frac{1}{4}\}$ **31.** $\{x \mid x > \frac{7}{12}\}$
33. $\{x \mid x < 7\}$; **35.** $\{x \mid x < 3\}$;

37. $\{y \mid y \geq -\frac{2}{5}\}$ **39.** $\{x \mid x \geq -6\}$ **41.** $\{y \mid y \leq 4\}$
43. $\{x \mid x > \frac{17}{3}\}$ **45.** $\{y \mid y < -\frac{1}{14}\}$ **47.** $\{x \mid x \leq \frac{3}{10}\}$
49. $\{x \mid x < 8\}$ **51.** $\{x \mid x \leq 6\}$ **53.** $\{x \mid x < -3\}$
55. $\{x \mid x > -3\}$ **57.** $\{x \mid x \leq 7\}$ **59.** $\{x \mid x > -10\}$
61. $\{y \mid y < 2\}$ **63.** $\{y \mid y \geq 3\}$ **65.** $\{y \mid y > -2\}$
67. $\{x \mid x > -4\}$ **69.** $\{x \mid x \leq 9\}$ **71.** $\{y \mid y \leq -3\}$
73. $\{y \mid y < 6\}$ **75.** $\{m \mid m \geq 6\}$ **77.** $\{t \mid t < -\frac{5}{3}\}$
79. $\{r \mid r > -3\}$ **81.** $\{x \mid x \geq -\frac{57}{34}\}$ **83.** $\{x \mid x > -2\}$
85. $-\frac{5}{8}$ **86.** -1.11 **87.** -9.4 **88.** $-\frac{7}{8}$ **89.** 140 **90.** 41
91. $-2x - 23$ **92.** $37x - 1$ **93. (a)** Yes; **(b)** yes; **(c)** no;
(d) no; **(e)** no; **(f)** yes; **(g)** yes **95.** No solution

Exercise Set 11.8, p. 719

RC1. $r \leq q$ **RC2.** $q \leq r$ **RC3.** $r < q$ **RC4.** $q \leq r$
RC5. $r < q$ **RC6.** $r \leq q$
1. $n \geq 7$ **3.** $w > 2 \text{ kg}$ **5.** $90 \text{ mph} < s < 110 \text{ mph}$
7. $w \leq 20 \text{ hr}$ **9.** $c \geq \$3.20$ **11.** $x > 8$ **13.** $y \leq -4$
15. $n \geq 1300$ **17.** $W \leq 500 \text{ L}$ **19.** $3x + 2 < 13$
21. $\{x \mid x \geq 84\}$ **23.** $\{C \mid C < 1063°\}$ **25.** $\{Y \mid Y \geq 1935\}$
27. 15 or fewer copies **29.** $\{L \mid L \geq 5 \text{ in.}\}$ **31.** 5 min or more
33. 2 courses **35.** 4 servings or more **37.** Lengths greater than
or equal to 92 ft; lengths less than or equal to 92 ft **39.** Lengths less
than 21.5 cm **41.** The blue-book value is greater than or equal to
$10,625. **43.** It has at least 16 g of fat. **45.** Heights greater than or
equal to 4 ft **47.** Dates at least 6 weeks after July 1 **49.** 21 calls
or more **51.** 40 **52.** -22 **53.** 12 **54.** 6 **55.** 1250
60. 83.3%, or $83\frac{1}{3}$% **61.** Temperatures between $-15°$C and $-9\frac{4}{9}°$C
63. They contain at least 7.5 g of fat per serving.

Summary and Review: Chapter 11, p. 724

Vocabulary Reinforcement
1. Solution **2.** Addition principle **3.** Multiplication principle
4. Inequality **5.** Equivalent

Concept Reinforcement
1. True **2.** True **3.** False **4.** True

Study Guide
1. -12 **2.** All real numbers **3.** No solution **4.** $b = \dfrac{2A}{h}$
5. $x > 1$ **6.** $x \leq -1$
7. $\{y \mid y > -4\}$

Review Exercises

1. -22 **2.** 1 **3.** 25 **4.** 9.99 **5.** $\frac{1}{4}$ **6.** 7 **7.** -192
8. $-\frac{7}{3}$ **9.** $-\frac{15}{64}$ **10.** -8 **11.** 4 **12.** -5 **13.** $-\frac{1}{3}$
14. 3 **15.** 4 **16.** 16 **17.** All real numbers **18.** 6
19. -3 **20.** 28 **21.** 4 **22.** No solution **23.** Yes **24.** No
25. Yes **26.** $\{y \mid y \geq -\frac{1}{2}\}$ **27.** $\{x \mid x \geq 7\}$ **28.** $\{y \mid y > 2\}$
29. $\{y \mid y \leq -4\}$ **30.** $\{x \mid x < -11\}$ **31.** $\{y \mid y > -7\}$
32. $\{x \mid x > -\frac{9}{11}\}$ **33.** $\{x \mid x \geq -\frac{1}{12}\}$
34. $x < 3$ **35.** $-2 < x \leq 5$
36. $y > 0$ **37.** $d = \dfrac{C}{\pi}$ **38.** $B = \dfrac{3V}{h}$
39. $a = 2A - b$ **40.** $x = \dfrac{y - b}{m}$ **41.** Length: 365 mi; width:
275 mi **42.** 345, 346 **43.** $2117 **44.** 27 subscriptions
45. $35°, 85°, 60°$ **46.** 15 **47.** 18.75% **48.** 600
49. About 87.1% **50.** About 28.2% decrease **51.** $220
52. $53,400 **53.** $138.95 **54.** 86 **55.** $\{w \mid w > 17 \text{ cm}\}$
56. C **57.** A **58.** $23, -23$ **59.** $20, -20$ **60.** $a = \dfrac{y - 3}{2 - b}$

Understanding Through Discussion and Writing

1. The end result is the same either way. If s is the original salary, the
new salary after a 5% raise followed by an 8% raise is $1.08(1.05s)$. If
the raises occur the other way around, the new salary is $1.05(1.08s)$.
By the commutative and associative laws of multiplication, we see
that these are equal. However, it would be better to receive the 8%
raise first, because this increase yields a higher salary initially than a
5% raise. **2.** No; Erin paid 75% of the original price and was offered
credit for 125% of this amount, not to be used on sale items. Now,
125% of 75% is 93.75%, so Erin would have a credit of 93.75% of the
original price. Since this credit can be applied only to nonsale items,
she has less purchasing power than if the amount she paid were
refunded and she could spend it on sale items. **3.** The inequalities
are equivalent by the multiplication principle for inequalities. If we
multiply on both sides of one inequality by -1, the other inequality
results. **4.** For any pair of numbers, their relative position on
the number line is reversed when both are multiplied by the same
negative number. For example, -3 is to the left of 5 on the number
line $(-3 < 5)$, but 12 is to the right of -20 $(-3(-4) > 5(-4))$.
5. Answers may vary. Fran is more than 3 years older than Todd.
6. Let n represent "a number." Then "five more than a number"
translates to the *expression* $n + 5$, or $5 + n$, and "five is more than a
number" translates to the *inequality* $5 > n$.

Test: Chapter 11, p. 729

1. [11.1b] 8 **2.** [11.1b] 26 **3.** [11.2a] -6 **4.** [11.2a] 49
5. [11.3b] -12 **6.** [11.3a] 2 **7.** [11.3a] -8 **8.** [11.1b] $-\frac{7}{20}$
9. [11.3c] 7 **10.** [11.3c] $\frac{5}{3}$ **11.** [11.3b] $\frac{5}{2}$ **12.** [11.3c] No solution
13. [11.3c] All real numbers **14.** [11.7c] $\{x \mid x \leq -4\}$
15. [11.7c] $\{x \mid x > -13\}$ **16.** [11.7d] $\{x \mid x \leq 5\}$
17. [11.7d] $\{y \mid y \leq -13\}$ **18.** [11.7d] $\{y \mid y \geq 8\}$
19. [11.7d] $\{x \mid x \leq -\frac{1}{20}\}$ **20.** [11.7e] $\{x \mid x < -6\}$
21. [11.7e] $\{x \mid x \leq -1\}$
22. [11.7b] **23.** [11.7b, e]
$y \leq 9$ $x < 1$
24. [11.7b] $-2 \leq x \leq 2$ **25.** [11.5a] 18
26. [11.5a] 16.5% **27.** [11.5a] 40,000 **28.** [11.5a] About 60.4%
29. [11.6a] Width: 7 cm; length: 11 cm **30.** [11.5a] About $230,556
31. [11.6a] 2509, 2510, 2511 **32.** [11.6a] $880 **33.** [11.6a] 3 m, 5 m
34. [11.8b] $\{l \mid l \geq 174 \text{ yd}\}$ **35.** [11.8b] $\{b \mid b \leq \$105\}$

36. [11.8b] $\{c \mid c \le 119{,}531\}$ **37.** [11.4b] $r = \dfrac{A}{2\pi h}$

38. [11.4b] $x = \dfrac{y - b}{8}$ **39.** [11.5a] D

40. [11.4b] $d = \dfrac{1 - ca}{-c}$, or $\dfrac{ca - 1}{c}$ **41.** [10.2e], [11.3a] 15, -15

42. [11.6a] 60 tickets

CHAPTER 12

Exercise Set 12.1, p. 736

RC1. True   **RC2.** False   **RC3.** False   **RC4.** True

1.
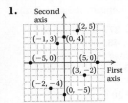

3. II   **5.** IV   **7.** III
9. On an axis, not in a quadrant
11. II   **13.** IV   **15.** I
17. Positive   **19.** II   **21.** I, IV
23. I, III

25. A: $(3, 3)$; B: $(0, -4)$; C: $(-5, 0)$; D: $(-1, -1)$; E: $(2, 0)$
27. No   **29.** No   **31.** Yes
33. $y = x - 5$

$$\dfrac{-1\ ?\ 4 - 5}{\quad\ \big|\ -1\quad} \text{TRUE}$$

$y = x - 5$

$$\dfrac{-4\ ?\ 1 - 5}{\quad\ \big|\ -4\quad} \text{TRUE}$$

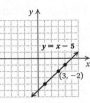

35. $y = \frac{1}{2}x + 3$

$$\dfrac{5\ ?\ \frac{1}{2} \cdot 4 + 3}{\quad\ \big|\ 2 + 3\quad}$$
$$\quad\ \big|\ 5\qquad \text{TRUE}$$

$y = \frac{1}{2}x + 3$

$$\dfrac{2\ ?\ \frac{1}{2}(-2) + 3}{\quad\ \big|\ -1 + 3\quad}$$
$$\quad\ \big|\ 2\qquad \text{TRUE}$$

37.

$$4x - 2y = 10$$
$$\dfrac{4 \cdot 0 - 2(-5)\ ?\ 10}{\quad\ 0 + 10\ \big|}$$
$$\quad\ 10\ \big|\qquad \text{TRUE}$$

$$4x - 2y = 10$$
$$\dfrac{4 \cdot 4 - 2 \cdot 3\ ?\ 10}{\quad\ 16 - 6\ \big|}$$
$$\quad\ 10\ \big|\qquad \text{TRUE}$$

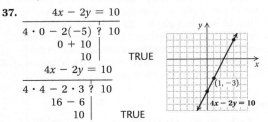

39. 8   **40.** $\frac{7}{4}$ **41.** All real numbers   **42.** No solution
43. $\frac{17}{10}$ **44.** $\frac{1}{3}$
45.
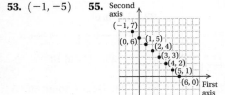
46.

47. $3.57   **48.** $48.60   **49.** 15%   **50.** $-\frac{3}{20}$ **51.** 0   **52.** -455
53. $(-1, -5)$ **55.**

57. 26 linear units

Calculator Corner, p. 739

1. Left to the student

Calculator Corner, p. 745

1. $y = -5x + 3$
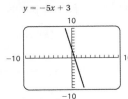

2. $y = 4x - 5$
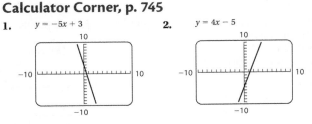

3. $y = \frac{4}{5}x + 2$
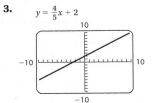

4. $y = -\frac{3}{5}x - 1$
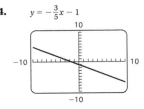

Exercise Set 12.2, p. 745

RC1. (c)   **RC2.** (b)   **RC3.** (d)   **RC4.** (a)

1.

x	y
-2	-1
-1	0
0	1
1	2
2	3

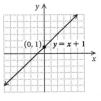

3.

x	y
-2	-2
-1	-1
0	0
1	1
2	2

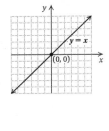

5.

x	y
-2	-1
0	0
4	2

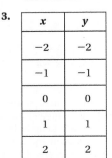

7.
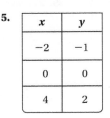

9.
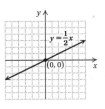

11.
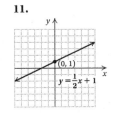

13.

15.

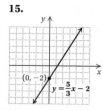

17.

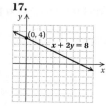

19.

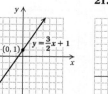

21.

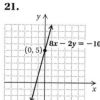

23.

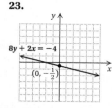

25. (a) 2010: \$68.38 billion; 2014: \$119.7 billion; 2015: \$132.53 billion

(b)

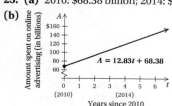

; about \$94 billion;

(c) 7 years after 2010, or in 2017

27. (a) 2000: 6.898 million; 2007: 6.023 million; 2012: 5.398 million

(b)

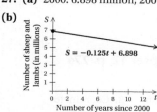

; about 5.6 million sheep and lambs; **(c)** 16 years after 2000, or in 2016

29. 12   **30.** 4.89   **31.** 0   **32.** $\frac{4}{5}$ **33.** $\frac{43}{2}$ **34.** -54
35. -10 **36.** 4   **37.** 16.6 million books   **38.** 157 concerts
39. $-\frac{3}{25}$ **40.** 3   **41.** $-\frac{15}{16}$ **42.** -420 **43.** $\frac{32}{7}$ **44.** -9

Calculator Corner, p. 752

1. y-intercept: $(0, -15)$;
   x-intercept: $(-2, 0)$;
   $y = -7.5x - 15$

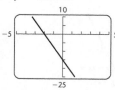

Xscl = 1   Yscl = 5

2. y-intercept: $(0, 43)$;
   x-intercept: $(-20, 0)$;
   $y = 2.15x + 43$

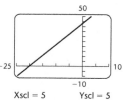

Xscl = 5   Yscl = 5

3. y-intercept: $(0, -30)$;
   x-intercept: $(25, 0)$;
   $y = (6x - 150)/5$

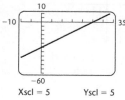

Xscl = 5   Yscl = 5

4. y-intercept: $(0, -4)$;
   x-intercept: $(20, 0)$;
   $y = 0.2x - 4$

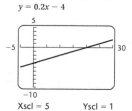

Xscl = 5   Yscl = 1

5. y-intercept: $(0, -15)$;
   x-intercept: $(10, 0)$;
   $y = 1.5x - 15$

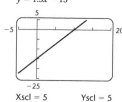

Xscl = 5   Yscl = 5

6. y-intercept: $\left(0, -\frac{1}{2}\right)$;
   x-intercept: $\left(\frac{2}{5}, 0\right)$;
   $y = (5x - 2)/4$

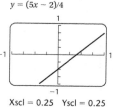

Xscl = 0.25   Yscl = 0.25

Visualizing for Success, p. 755

1. E   **2.** C   **3.** G   **4.** A   **5.** I   **6.** D   **7.** F   **8.** J
9. B   **10.** H

Exercise Set 12.3, p. 756

RC1. Horizontal; y-intercept   **RC2.** x-axis   **RC3.** $y = 0$
RC4. $(0,0)$ **RC5.** $x = 0$ **RC6.** Vertical; x-intercept
1. (a) $(0, 5)$; **(b)** $(2, 0)$ **3. (a)** $(0, -4)$; **(b)** $(3, 0)$
5. (a) $(0, 3)$; **(b)** $(5, 0)$ **7. (a)** $(0, -14)$; **(b)** $(4, 0)$
9. (a) $\left(0, \frac{10}{3}\right)$; **(b)** $\left(-\frac{5}{2}, 0\right)$ **11. (a)** $\left(0, -\frac{1}{3}\right)$; **(b)** $\left(\frac{1}{2}, 0\right)$

13.

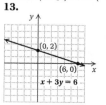

15.

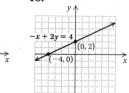

17.

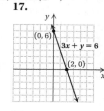

19.

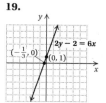

21.

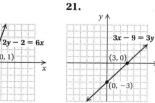

23.

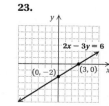

25.

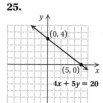

27.

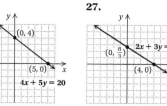

29.

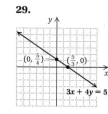

31.

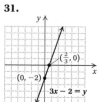

33.

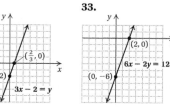

35.

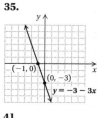

37.

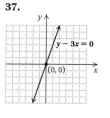

39.

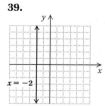

41.

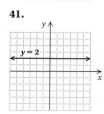

43.

45.

47.

49.

51.

53.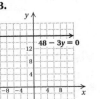

55. $y = -1$ **57.** $x = 4$ **59.** $\{x \mid x < 1\}$ **60.** $\{x \mid x \geq 2\}$
61. $\{x \mid x \leq 7\}$ **62.** $\{x \mid x > 1\}$ **63.** $y = -4$ **65.** $k = 12$

Mid-Chapter Review: Chapter 12, p. 761

1. True   **2.** False   **3.** True   **4.** False
5. (a) The y-intercept is $(0, -3)$. (b) The x-intercept is $(-3, 0)$.
6. (a) The x-intercept is $(c, 0)$. (b) The y-intercept is $(0, d)$.
7. A: $(-1, 0)$; B: $(2, 5)$; C: $(-5, -4)$; D: $(6, -2)$; E: $(-4, 2)$
8. F: $(0, -3)$; G: $(5, 0)$; H: $(1, 5)$; I: $(-4, 4)$; J: $(3, -6)$
9. No   **10.** Yes   **11.** x-intercept: $(-6, 0)$; y-intercept: $(0, 9)$
12. x-intercept: $\left(\frac{1}{2}, 0\right)$; y-intercept: $\left(0, -\frac{1}{20}\right)$
13. x-intercept: $(40, 0)$; y-intercept: $(0, -2)$
14. x-intercept: $(-42, 0)$; y-intercept: $(0, 105)$

15.

16.

17.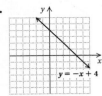

18.

19. D   **20.** C   **21.** B   **22.** E   **23.** A
24. No; an equation $x = a$, $a \neq 0$, does not have a y-intercept.
25. Most would probably say that the second equation would be easier to graph because it has been solved for y. This makes it more efficient to find the y-value that corresponds to a given x-value.
26. $A = 0$. If the line is horizontal, then the equation is of the form $y = $ a constant. Thus, Ax must be 0 and, hence, $A = 0$.
27. Any ordered pair $(7, y)$ is a solution of $x = 7$. Thus all points on the graph are 7 units to the right of the y-axis, so they lie on a vertical line.

Calculator Corner, p. 766

1. This line will pass through the origin and slant up from left to right. This line will be steeper than $y = 10x$.   **2.** This line will pass through the origin and slant up from left to right. This line will be less steep than $y = \frac{5}{32}x$.   **3.** This line will pass through the origin and slant down from left to right. This line will be steeper than $y = -10x$.   **4.** This line will pass through the origin and slant down from left to right. This line will be less steep than $y = -\frac{5}{32}x$.

Exercise Set 12.4, p. 769

RC1. (d)   **RC2.** (f)   **RC3.** (b)   **RC4.** (e)
RC5. (c)   **RC6.** (a)
1. $-\frac{3}{7}$ **3.** $\frac{2}{3}$ **5.** $\frac{3}{4}$ **7.** 0
9. $-\frac{4}{5}$;

11. 3;

13. $-\frac{2}{3}$;

15. $\frac{7}{8}$;

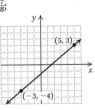

17. $\frac{2}{3}$ **19.** Not defined   **21.** $-\frac{5}{13}$ **23.** 0   **25.** -10
27. 3.78   **29.** 3   **31.** $-\frac{1}{5}$ **33.** $-\frac{3}{2}$ **35.** Not defined
37. -1 **39.** 3   **41.** $\frac{5}{4}$ **43.** 0   **45.** $\frac{4}{3}$ **47.** $-\frac{21}{8}$ **49.** $\frac{12}{41}$
51. $\frac{28}{129}$ **53.** 3.0%; yes   **55.** About 1.24 million children per year
57. About $-14,100$ people per year   **59.** 30,600,000 lb per year
61. $\frac{15}{2}$ **62.** -12 **63.** $-\frac{2}{3}p$ **64.** $5t - 1$
65. $y = -x + 5$ **67.** $y = x + 2$

Summary and Review: Chapter 12, p. 774

Vocabulary Reinforcement

1. Not defined   **2.** Horizontal; $(0, b)$ **3.** Coordinates
4. y-intercept   **5.** Vertical; $(a, 0)$ **6.** 0

Concept Reinforcement

1. True   **2.** False   **3.** True   **4.** True   **5.** False   **6.** False

Study Guide

1. F: $(2, 4)$; G: $(-2, 0)$; H: $(-3, -5)$

2.

3.

4.

5.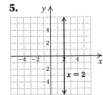

6. m is not defined.   **7.** $\frac{3}{2}$ **8.** 0   **9.** m is not defined.
10. -2 **11.** 0   **12.** About 27,400 people per year

Review Exercises

1.–3.

(2, 5)
(−4, −2)   (0, −3)

4. $(-5, -1)$ **5.** $(-2, 5)$ **6.** $(3, 0)$
7. IV   **8.** III   **9.** I   **10.** No
11. Yes

12.

$$2x - y = 3$$
$$\frac{2 \cdot 0 - (-3) \; ? \; 3}{}$$
$$0 + 3 \;|\;$$
$$3 \;|\; \text{TRUE}$$

$$2x - y = 3$$
$$\frac{2 \cdot 2 - 1 \; ? \; 3}{}$$
$$4 - 1 \;|\;$$
$$3 \;|\; \text{TRUE}$$

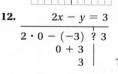

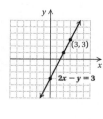

(3, 3)
$2x - y = 3$

13.

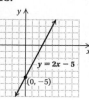

$y = 2x - 5$
(0, −5)

14.

$y = -\frac{3}{4}x$
(0, 0)

15.

(0, 4)
$y = -x + 4$

16.

(0, 3)
$y = 3 - 4x$

17. (a) $14\frac{1}{2}$ ft³, 16 ft³, $20\frac{1}{2}$ ft³, 28 ft³;
(b)

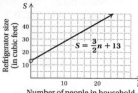

; 19 ft³;
(c) 6 residents

$S = \frac{3}{2}n + 13$

18.

(6, 0)
(0, −3)
$x - 2y = 6$

19.

(2, 0)
$5x - 2y = 10$
(0, −5)

20.

$y = 3$

21.

$5x - 4 = 0$

22. $\frac{1}{3}$ **23.** $-\frac{1}{3}$
24. $\frac{3}{5}$;

(5, 4)
(−5, −2)

25. -1;

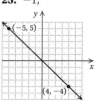

(−5, 5)
(4, −4)

---

26. $-\frac{5}{8}$ **27.** $\frac{1}{2}$ **28.** Not defined   **29.** 0   **30.** $\frac{1}{10}$ **31.** $\frac{3}{2}$
32. (a) 2.4 driveways per hour; **(b)** 25 minutes per driveway
33. 7%   **34.** $2.4 billion per year   **35.** D   **36.** C
37. 45 square units; 28 linear units   **38. (a)** 239.583 ft per minute;
(b) about 0.004 min per foot

Understanding Through Discussion and Writing

1. With slope $\frac{5}{3}$, for each horizontal change of 3 units, there is a vertical change of 5 units. With slope $\frac{4}{3}$, for each horizontal change of 3 units, there is a vertical change of 4 units. Since $5 > 4$, the line with slope $\frac{5}{3}$ has a steeper slant.   **2.** No; the equation $y = b, b \neq 0$, does not have an x-intercept.   **3.** The y-intercept is the point at which the graph crosses the y-axis. Since a point on the y-axis is neither left nor right of the origin, the first or x-coordinate of the point is 0.   **4.** Any ordered pair $(x, -2)$ is a solution of $y = -2$. All points on the graph are 2 units below the x-axis, so they lie on a horizontal line.

Test: Chapter 12, p. 780

1. [12.1a] II   **2.** [12.1a] III   **3.** [12.1b] $(-5, 1)$ **4.** [12.1b] $(0, -4)$
5. [12.1c]

$$y - 2x = 5$$
$$\frac{-3 - 2(-4) \; ? \; 5}{}$$
$$-3 + 8 \;|\;$$
$$5 \;|\; \text{TRUE}$$

$$y - 2x = 5$$
$$\frac{3 - 2(-1) \; ? \; 5}{}$$
$$3 + 2 \;|\;$$
$$5 \;|\; \text{TRUE}$$

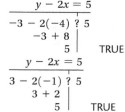

(−2, 1)
$y - 2x = 5$

6. [12.2a]

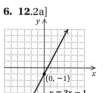

(0, −1)
$y = 2x - 1$

7. [12.2a]

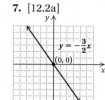

$y = -\frac{3}{2}x$
(0, 0)

8. [12.3a]

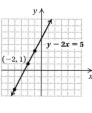

$2x - 4y = -8$
(0, 2)
(−4, 0)

9. [12.3a]

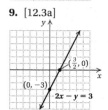

$(\frac{3}{2}, 0)$
(0, −3)
$2x - y = 3$

10. [12.3b]

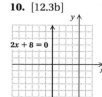

$2x + 8 = 0$

11. [12.3b]

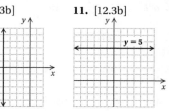

$y = 5$

12. [12.4a] -2
13. [12.4a] $\frac{3}{8}$;

(5, 4)
(−3, 1)

14. [12.4b] $\frac{2}{5}$ **15.** [12.4b] Not defined   **16.** [12.4b] 0
17. [12.4b] -11 **18.** [12.4c] $-\frac{1}{20}$, or -0.05 **19.** [12.2b] **(a)**
2007: $8593; 2009: $9805; 2012: $11,623
(b)

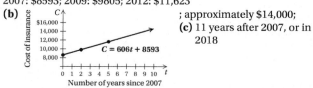

; approximately $14,000;
(c) 11 years after 2007, or in 2018

$C = 606t + 8593$

20. [12.4c] **(a)** 14.5 floors per minute; **(b)** $4\frac{4}{29}$ seconds per floor
21. [12.4c] 87.5 mph   **22.** [12.3a], [12.4b] B   **23.** [12.3b] $y = 3$
24. [12.1a] 25 square units; 20 linear units

CHAPTER 13

Exercise Set 13.1, p. 790

RC1. (c) **RC2.** (a) **RC3.** (e) **RC4.** (c) **RC5.** (d)
RC6. (c)
1. $3 \cdot 3 \cdot 3 \cdot 3$ **3.** $(-1.1)(-1.1)(-1.1)(-1.1)(-1.1)$
5. $\left(\frac{2}{3}\right)\left(\frac{2}{3}\right)\left(\frac{2}{3}\right)\left(\frac{2}{3}\right)$ **7.** $(7p)(7p)$ **9.** $8 \cdot k \cdot k \cdot k$
11. $-6 \cdot y \cdot y \cdot y \cdot y$ **13.** 1 **15.** b **17.** 1 **19.** -7.03
21. 1 **23.** ab **25.** a **27.** 27 **29.** 19 **31.** -81 **33.** 256
35. 93 **37.** 136 **39.** 10; 4 **41.** 3629.84 ft^2 **43.** $\frac{1}{3^2} = \frac{1}{9}$
45. $\frac{1}{10^3} = \frac{1}{1000}$ **47.** $\frac{1}{a^3}$ **49.** $8^2 = 64$ **51.** y^4 **53.** $\frac{5}{z^4}$ **55.** $\frac{x}{y^2}$
57. 4^{-3} **59.** x^{-3} **61.** a^{-5} **63.** 2^7 **65.** 9^{38} **67.** x^5 **69.** x^{17}
71. $(3y)^{12}$ **73.** $(7y)^{17}$ **75.** 3^3 **77.** 1 **79.** $\frac{1}{x^{13}}$ **81.** $\frac{1}{a^{10}}$
83. $x^6 y^{15}$ **85.** $s^3 t^7$ **87.** 7^3 **89.** y^8 **91.** $\frac{1}{16^6}$ **93.** $\frac{1}{m^6}$
95. $\frac{1}{(8x)^4}$ **97.** x^2 **99.** $\frac{1}{z^4}$ **101.** x^3 **103.** 1 **105.** $a^3 b^2$
107. $5^2 = 25; 5^{-2} = \frac{1}{25}; \left(\frac{1}{5}\right)^2 = \frac{1}{25}; \left(\frac{1}{5}\right)^{-2} = 25; -5^2 = -25;$
$(-5)^2 = 25; -\left(\frac{1}{5}\right)^2 = -\frac{1}{25}; \left(-\frac{1}{5}\right)^{-2} = 25$
109. 8 in.; 4 in. **110.** $51°, 27°, 102°$ **111.** 45%; 37.5%; 17.5%
112. Lengths less than 2.5 ft **113.** $\frac{7}{4}$ **114.** 2 **115.** $\frac{23}{14}$
116. $\frac{11}{10}$ **117.** No **119.** No **121.** y^{5x} **123.** a^{4t} **125.** 1
127. $>$ **129.** $<$ **131.** $-\frac{1}{10,000}$ **133.** No; for example,
$(3 + 4)^2 = 49$, but $3^2 + 4^2 = 25$.

Calculator Corner, p. 798

1. 1.3545×10^{-4} **2.** 3.2×10^5 **3.** 3×10^{-6} **4.** 8×10^{-26}

Exercise Set 13.2, p. 800

RC1. Multiply **RC2.** nth **RC3.** Right **RC4.** Positive
1. 2^6 **3.** $\frac{1}{5^6}$ **5.** x^{12} **7.** $\frac{1}{a^{18}}$ **9.** t^{18} **11.** $\frac{1}{t^{12}}$ **13.** x^8
15. $a^3 b^3$ **17.** $\frac{1}{a^3 b^3}$ **19.** $\frac{1}{m^3 n^6}$ **21.** $16x^6$ **23.** $\frac{9}{x^8}$
25. $\frac{1}{x^{12} y^{15}}$ **27.** $x^{24} y^8$ **29.** $\frac{a^{10}}{b^{35}}$ **31.** $\frac{25t^6}{r^8}$ **33.** $\frac{b^{21}}{a^{15} c^6}$
35. $\frac{9x^6}{y^{16} z^6}$ **37.** $\frac{16x^6}{y^4}$ **39.** $a^{12} b^8$ **41.** $\frac{y^6}{4}$ **43.** $\frac{a^8}{b^{12}}$ **45.** $\frac{8}{y^6}$
47. $49x^6$ **49.** $\frac{x^6 y^3}{z^3}$ **51.** $\frac{c^2 d^6}{a^4 b^2}$ **53.** 2.8×10^{10} **55.** 9.07×10^{17}
57. 3.04×10^{-6} **59.** 1.8×10^{-8} **61.** 10^{11} **63.** 4.19854×10^8
65. 6.8×10^{-7} **67.** 87,400,000 **69.** 0.00000005704
71. 10,000,000 **73.** 0.00001 **75.** 6×10^9 **77.** 3.38×10^4
79. 8.1477×10^{-13} **81.** 2.5×10^{13} **83.** 5.0×10^{-4}
85. 3.0×10^{-21} **87.** Approximately 1.325×10^{14} ft^3
89. 1×10^{22} **91.** The mass of Jupiter is 3.18×10^2 times the
mass of Earth. **93.** 1.25×10^9 bytes **95.** 3.1×10^2 sheets
97. 4.375×10^2 days
99. 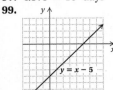 **100.**

Exercise Set 13.3, p. 813

RC1. (b) **RC2.** (f) **RC3.** (c) **RC4.** (d) **RC5.** (a)
RC6. (e)
1. $-18; 7$ **3.** $19; 14$ **5.** $-12; -7$ **7.** $\frac{13}{3}; 5$ **9.** $9; 1$
11. $56; -2$ **13.** 1112 ft **15.** 55 oranges **17.** $-4, 4, 5, 2.75, 1$
19. **(a)** 5.67 kW; **(b)** left to the student **21.** 9 words **23.** 6
25. 15 **27.** $2, -3x, x^2$ **29.** $-2x^4, \frac{1}{3}x^3, -x, 3$ **31.** Trinomial
33. None of these **35.** Binomial **37.** Monomial **39.** $-3, 6$
41. $5, \frac{3}{4}, 3$ **43.** $-5, 6, -2.7, 1, -2$ **45.** $1, 0; 1$ **47.** $2, 1, 0; 2$
49. $3, 2, 1, 0; 3$ **51.** $2, 1, 6, 4; 6$
53.

Term	Coefficient	Degree of the Term	Degree of the Polynomial
$-7x^4$	-7	4	
$6x^3$	6	3	
$-x^2$	-1	2	4
$8x$	8	1	
-2	-2	0	

55. $6x^2$ and $-3x^2$ **57.** $2x^4$ and $-3x^4$; $5x$ and $-7x$ **59.** $3x^5$ and
$14x^5$; $-7x$ and $-2x$; 8 and -9 **61.** $-3x$ **63.** $-8x$
65. $11x^3 + 4$ **67.** $x^3 - x$ **69.** $4b^5$ **71.** $\frac{3}{4}x^5 - 2x - 42$
73. x^4 **75.** $\frac{15}{16}x^3 - \frac{7}{6}x^2$ **77.** $x^5 + 6x^3 + 2x^2 + x + 1$
79. $15y^9 + 7y^8 + 5y^3 - y^2 + y$ **81.** $x^6 + x^4$
83. $13x^3 - 9x + 8$ **85.** $-5x^2 + 9x$ **87.** $12x^4 - 2x + \frac{1}{4}$
89. x^2, x **91.** x^3, x^2, x^0 **93.** None missing
95. $x^3 + 0x^2 + 0x - 27; x^3 \qquad - 27$
97. $x^4 + 0x^3 + 0x^2 - x + 0x^0; x^4 \qquad - x$
99. None missing **101.** -19 **102.** -1 **103.** -2.25
104. $-\frac{3}{8}$ **105.** -19 **106.** $-\frac{17}{24}$ **107.** $\frac{5}{8}$ **108.** -2.6
109. 18 **110.** $-\frac{1}{3}$ **111.** -0.6 **112.** -24 **113.** 20
114. $-\frac{8}{5}$ **115.** 0 **116.** Not defined **117.** $10x^6 + 52x^5$
119. $4x^5 - 3x^3 + x^2 - 7x$; answers may vary
121. $x^3 - 2x^2 - 6x + 3$ **123.** $-4, 4, 5, 2.75, 1$ **125.** 9

101. 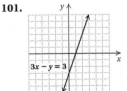 **102.**
103. 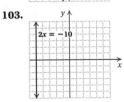 **104.**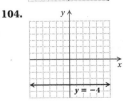
105. 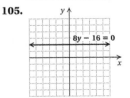 **106.**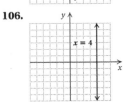
107. 2.478125×10^{-1} **109.** $\frac{1}{5}$ **111.** 3^{11} **113.** 7
115. $\frac{1}{0.4}$, or 2.5 **117.** False **119.** False **121.** True

Calculator Corner, p. 808

1. 3; 2.25; -27 **2.** 44; 0; 9.28

Exercise Set 13.4, p. 822

RC1. False **RC2.** True **RC3.** False **RC4.** False
1. $-x + 5$ **3.** $x^2 - \frac{11}{2}x - 1$ **5.** $2x^2$ **7.** $5x^2 + 3x - 30$
9. $-2.2x^3 - 0.2x^2 - 3.8x + 23$ **11.** $6 + 12x^2$
13. $-\frac{1}{2}x^4 + \frac{2}{3}x^3 + x^2$ **15.** $0.01x^5 + x^4 - 0.2x^3 + 0.2x + 0.06$
17. $9x^8 + 8x^7 - 6x^4 + 8x^2 + 4$
19. $1.05x^4 + 0.36x^3 + 14.22x^2 + x + 0.97$
21. $5x$ **23.** $x^2 - \frac{3}{2}x + 2$ **25.** $-12x^4 + 3x^3 - 3$ **27.** $-3x + 7$
29. $-4x^2 + 3x - 2$ **31.** $4x^4 - 6x^2 - \frac{3}{4}x + 8$ **33.** $7x - 1$
35. $-x^2 - 7x + 5$ **37.** -18 **39.** $6x^4 + 3x^3 - 4x^2 + 3x - 4$
41. $4.6x^3 + 9.2x^2 - 3.8x - 23$ **43.** $\frac{3}{4}x^3 - \frac{1}{2}x$
45. $0.06x^3 - 0.05x^2 + 0.01x + 1$ **47.** $3x + 6$
49. $11x^4 + 12x^3 - 9x^2 - 8x - 9$ **51.** $x^4 - x^3 + x^2 - x$
53. $\frac{23}{2}a + 12$ **55.** $5x^2 + 4x$
57. $(r + 11)(r + 9); 9r + 99 + 11r + r^2, \text{or } r^2 + 20r + 99$
59. $(x + 3)(x + 3), \text{or } (x + 3)^2; x^2 + 3x + 9 + 3x, \text{or } x^2 + 6x + 9$
61. $\pi r^2 - 25\pi$ **63.** $18z - 54$ **65.** 6 **66.** -19 **67.** $-\frac{7}{22}$
68. 5 **69.** 5 **70.** 1 **71.** $\frac{39}{2}$ **72.** $\frac{37}{2}$ **73.** $\{x \mid x \geq -10\}$
74. $\{x \mid x < 0\}$ **75.** $20w + 42$ **77.** $2x^2 + 20x$
79. $y^2 - 4y + 4$ **81.** $12y^2 - 23y + 21$
83. $-3y^4 - y^3 + 5y - 2$

Mid-Chapter Review: Chapter 13, p. 827

1. True **2.** False **3.** False **4.** True
5. $4w^3 + 6w - 8w^3 - 3w = (4 - 8)w^3 + (6 - 3)w = -4w^3 + 3w$ **6.** $(3y^4 - y^2 + 11) - (y^4 - 4y^2 + 5) = 3y^4 - y^2 + 11 - y^4 + 4y^2 - 5 = 2y^4 + 3y^2 + 6$
7. z **8.** 1 **9.** -32 **10.** 1 **11.** 5^7 **12.** $(3a)^9$
13. $\frac{1}{x^3}$ **14.** 1 **15.** 7^4 **16.** $\frac{1}{x^2}$ **17.** w^8 **18.** $\frac{1}{y^4}$ **19.** 3^{15}
20. $\frac{x^{18}}{y^{12}}$ **21.** $\frac{a^{24}}{5^6}$ **22.** $\frac{x^2 z^4}{4y^6}$ **23.** 2.543×10^7 **24.** 1.2×10^{-4}
25. 0.000036 **26.** $144{,}000{,}000$ **27.** 6×10^3 **28.** 5×10^{-7}
29. $16; 1$ **30.** $-16; 9$ **31.** $-2x^5 - 5x^2 + 4x + 2$
32. $8x^6 + 2x^3 - 8x^2$ **33.** $3, 1, 0; 3$ **34.** $1, 4, 6; 6$
35. Binomial **36.** Trinomial **37.** $8x^2 + 5$
38. $5x^3 - 2x^2 + 2x - 11$ **39.** $-4x - 10$
40. $-0.4x^2 - 3.4x + 9$ **41.** $3y + 3y^2$ **42.** The area of the smaller square is x^2, and the area of the larger square is $(3x)^2$, or $9x^2$, so the area of the larger square is nine times the area of the smaller square. **43.** The volume of the smaller cube is x^3, and the volume of the larger cube is $(2x)^3$, or $8x^3$, so the volume of the larger cube is eight times the volume of the smaller cube. **44.** Exponents are added when powers with like bases are multiplied. Exponents are multiplied when a power is raised to a power.
45. $3^{-29} = \frac{1}{3^{29}}$ and $2^{-29} = \frac{1}{2^{29}}$. Since $3^{29} > 2^{29}$, we have $\frac{1}{3^{29}} < \frac{1}{2^{29}}$.
46. It is better to evaluate a polynomial after like terms have been collected, because there are fewer terms to evaluate. **47.** Yes; consider the following: $(x^2 + 4) + (4x - 7) = x^2 + 4x - 3$.

Calculator Corner, p. 832

1. Correct **2.** Correct **3.** Not correct **4.** Not correct

Exercise Set 13.5, p. 833

RC1. (a) **RC2.** (a) **RC3.** (d) **RC4.** (e)
1. $40x^2$ **3.** x^3 **5.** $32x^8$ **7.** $0.03x^{11}$ **9.** $\frac{1}{15}x^4$ **11.** 0
13. $-24x^{11}$ **15.** $-2x^2 + 10x$ **17.** $-5x^2 + 5x$ **19.** $x^5 + x^2$
21. $6x^3 - 18x^2 + 3x$ **23.** $-6x^4 - 6x^3$ **25.** $18y^6 + 24y^5$
27. $x^2 + 9x + 18$ **29.** $x^2 + 3x - 10$ **31.** $x^2 + 3x - 4$
33. $x^2 - 7x + 12$ **35.** $x^2 - 9$ **37.** $x^2 - 16$
39. $3x^2 + 11x + 10$ **41.** $25 - 15x + 2x^2$ **43.** $4x^2 + 20x + 25$
45. $x^2 - 6x + 9$ **47.** $x^2 - \frac{21}{10}x - 1$ **49.** $x^2 + 2.4x - 10.81$
51. $(x + 2)(x + 6)$, or $x^2 + 8x + 12$ **53.** $(x + 1)(x + 6)$, or $x^2 + 7x + 6$

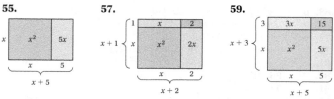

55. **57.** **59.**

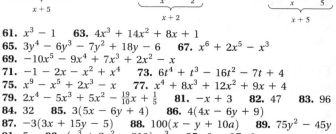

61. $x^3 - 1$ **63.** $4x^3 + 14x^2 + 8x + 1$
65. $3y^4 - 6y^3 - 7y^2 + 18y - 6$ **67.** $x^6 + 2x^5 - x^3$
69. $-10x^5 - 9x^4 + 7x^3 + 2x^2 - x$
71. $-1 - 2x - x^2 + x^4$ **73.** $6t^4 + t^3 - 16t^2 - 7t + 4$
75. $x^9 - x^5 + 2x^3 - x$ **77.** $x^4 + 8x^3 + 12x^2 + 9x + 4$
79. $2x^4 - 5x^3 + 5x^2 - \frac{19}{10}x + \frac{1}{5}$ **81.** $-x + 3$ **82.** 47 **83.** 96
84. 32 **85.** $3(5x - 6y + 4)$ **86.** $4(4x - 6y + 9)$
87. $-3(3x + 15y - 5)$ **88.** $100(x - y + 10a)$ **89.** $75y^2 - 45y$
91. 5 **93.** $(x^3 + 2x^2 - 210) \text{ m}^3$ **95.** 0 **97.** 0

Visualizing for Success, p. 842

1. E, F **2.** B, O **3.** K, S **4.** G, R **5.** D, M **6.** J, P
7. C, L **8.** N, Q **9.** A, H **10.** I, T

Exercise Set 13.6, p. 843

RC1. Outside; last **RC2.** Descending **RC3.** Difference
RC4. Square; binomial **RC5.** Binomials **RC6.** Difference
1. $x^3 + x^2 + 3x + 3$ **3.** $x^4 + x^3 + 2x + 2$ **5.** $y^2 - y - 6$
7. $9x^2 + 12x + 4$ **9.** $5x^2 + 4x - 12$ **11.** $9t^2 - 1$
13. $4x^2 - 6x + 2$ **15.** $p^2 - \frac{1}{16}$ **17.** $x^2 - 0.01$
19. $2x^3 + 2x^2 + 6x + 6$ **21.** $-2x^2 - 11x + 6$
23. $a^2 + 14a + 49$ **25.** $1 - x - 6x^2$ **27.** $\frac{9}{64}y^2 - \frac{5}{8}y + \frac{25}{36}$
29. $x^5 + 3x^3 - x^2 - 3$ **31.** $3x^6 - 2x^4 - 6x^2 + 4$
33. $13.16x^2 + 18.99x - 13.95$ **35.** $6x^7 + 18x^5 + 4x^2 + 12$
37. $4x^3 - 12x^2 + 3x - 9$ **39.** $4y^6 + 4y^5 + y^4 + y^3$
41. $x^2 - 16$ **43.** $4x^2 - 1$ **45.** $25m^2 - 4$ **47.** $4x^4 - 9$
49. $9x^8 - 16$ **51.** $x^{12} - x^4$ **53.** $x^8 - 9x^2$ **55.** $x^{24} - 9$
57. $4y^{16} - 9$ **59.** $\frac{25}{64}x^2 - 18.49$ **61.** $x^2 + 4x + 4$
63. $9x^4 + 6x^2 + 1$ **65.** $a^2 - a + \frac{1}{4}$ **67.** $9 + 6x + x^2$
69. $x^4 + 2x^2 + 1$ **71.** $4 - 12x^4 + 9x^8$ **73.** $25 + 60t^2 + 36t^4$
75. $x^2 - \frac{5}{4}x + \frac{25}{64}$ **77.** $9 - 12x^3 + 4x^6$ **79.** $4x^3 + 24x^2 - 12x$
81. $4x^4 - 2x^2 + \frac{1}{4}$ **83.** $9p^2 - 1$ **85.** $15t^5 - 3t^4 + 3t^3$
87. $36x^8 + 48x^4 + 16$ **89.** $12x^3 + 8x^2 + 15x + 10$
91. $64 - 96x^4 + 36x^8$ **93.** $t^3 - 1$ **95.** $25; 49$ **97.** $56; 16$
99. $a^2 + 2a + 1$ **101.** $t^2 + 10t + 24$ **103.** $\frac{28}{27}$ **104.** $-\frac{41}{7}$
105. $\frac{27}{4}$ **106.** $y = \frac{3x - 12}{2}$, or $y = \frac{3}{2}x - 6$ **107.** $b = \frac{C + r}{a}$
108. $a = \frac{5d + 4}{3}$, or $a = \frac{5}{3}d + \frac{4}{3}$ **109.** $30x^3 + 35x^2 - 15x$
111. $a^4 - 50a^2 + 625$ **113.** $81t^{16} - 72t^8 + 16$ **115.** -7
117. First row: $90, -432, -63$; second row: $7, -18, -36, -14, 12, -6, -21, -11$; third row: $9, -2, -2, 10, -8, -8, -8, -10, 21$; fourth row: $-19, -6$ **119.** Yes **121.** No

Exercise Set 13.7, p. 851

RC1. True **RC2.** False **RC3.** False **RC4.** False
1. -1 **3.** -15 **5.** 240 **7.** -145 **9.** 3.715 L
11. 2322 calories **13.** 44.46 in^2 **15.** 73.005 in^2
17. Coefficients: $1, -2, 3, -5$; degrees: $4, 2, 2, 0; 4$
19. Coefficients: $17, -3, -7$; degrees: $5, 5, 0; 5$ **21.** $-a - 2b$
23. $3x^2y - 2xy^2 + x^2$ **25.** $20au + 10av$ **27.** $8u^2v - 5uv^2$
29. $x^2 - 4xy + 3y^2$ **31.** $3r + 7$
33. $-b^2a^3 - 3b^3a^2 + 5ba + 3$ **35.** $ab^2 - a^2b$ **37.** $2ab - 2$
39. $-2a + 10b - 5c + 8d$ **41.** $6z^2 + 7zu - 3u^2$
43. $a^4b^2 - 7a^2b + 10$ **45.** $a^6 - b^2c^2$
47. $y^6x + y^4x + y^4 + 2y^2 + 1$ **49.** $12x^2y^2 + 2xy - 2$
51. $12 - c^2d^2 - c^4d^4$ **53.** $m^3 + m^2n - mn^2 - n^3$
55. $x^9y^9 - x^6y^6 + x^5y^5 - x^2y^2$ **57.** $x^2 + 2xh + h^2$
59. $9a^2 + 12ab + 4b^2$ **61.** $r^6t^4 - 8r^3t^2 + 16$

63. $p^8 + 2m^2n^2p^4 + m^4n^4$ **65.** $3a^3 - 12a^2b + 12ab^2$
67. $m^2 + 2mn + n^2 - 6m - 6n + 9$ **69.** $a^2 - b^2$
71. $4a^2 - b^2$ **73.** $c^4 - d^2$ **75.** $a^2b^2 - c^2d^4$
77. $x^2 + 2xy + y^2 - 9$ **79.** $x^2 - y^2 - 2yz - z^2$
81. $a^2 + 2ab + b^2 - c^2$ **83.** $3x^4 - 7x^2y + 3x^2 - 20y^2 + 22y - 6$
85. IV **86.** III **87.** I **88.** II **89.** 39 **90.** 1.125
91. $<$ **92.** -3 **93.** $4xy - 4y^2$ **95.** $2xy + \pi x^2$
97. $2\pi nh + 2\pi mh + 2\pi n^2 - 2\pi m^2$ **99.** 16 gal
101. \$12,351.94

Exercise Set 13.8, p. 860

RC1. Subtract; divide **RC2.** Divide **RC3.** Multiply; subtract
RC4. Multiply; add
1. $3x^4$ **3.** $5x$ **5.** $18x^3$ **7.** $4a^3b$ **9.** $3x^4 - \frac{1}{2}x^3 + \frac{1}{8}x^2 - 2$
11. $1 - 2u - u^4$ **13.** $5t^2 + 8t - 2$ **15.** $-4x^4 + 4x^2 + 1$
17. $6x^2 - 10x + \frac{3}{2}$ **19.** $9x^2 - \frac{5}{2}x + 1$ **21.** $6x^2 + 13x + 4$
23. $3rs + r - 2s$ **25.** $x + 2$ **27.** $x - 5 + \dfrac{-50}{x - 5}$
29. $x - 2 + \dfrac{-2}{x + 6}$ **31.** $x - 3$ **33.** $x^4 - x^3 + x^2 - x + 1$
35. $2x^2 - 7x + 4$ **37.** $x^3 - 6$ **39.** $t^2 + 1$
41. $y^2 - 3y + 1 + \dfrac{-5}{y + 2}$ **43.** $3x^2 + x + 2 + \dfrac{10}{5x + 1}$
45. $6y^2 - 5 + \dfrac{-6}{2y + 7}$ **47.** -1 **48.** $\frac{10}{3}$ **49.** $\frac{23}{19}$
50. $\{r \mid r \le -15\}$ **51.** 140% **52.** $\{x \mid x \ge 95\}$
53. 25,543.75 ft² **54.** 228, 229 **55.** $x^2 + 5$
57. $a + 3 + \dfrac{5}{5a^2 - 7a - 2}$ **59.** $2x^2 + x - 3$
61. $a^5 + a^4b + a^3b^2 + a^2b^3 + ab^4 + b^5$ **63.** -5 **65.** 1

Summary and Review: Chapter 13, p. 863

Vocabulary Reinforcement
1. Exponent **2.** Product **3.** Monomial **4.** Trinomial
5. Quotient **6.** Descending **7.** Degree **8.** Scientific

Concept Reinforcement
1. True **2.** False **3.** False **4.** True

Study Guide
1. z^8 **2.** a^2b^6 **3.** $\dfrac{y^6}{27x^{12}z^9}$ **4.** 7.63×10^5 **5.** 0.0003
6. 6×10^4 **7.** $2x^4 - 4x^2 - 3$ **8.** $3x^4 + x^3 - 2x^2 + 2$
9. $x^6 - 6x^4 + 11x^2 - 6$ **10.** $2y^2 + 11y + 12$ **11.** $x^2 - 25$
12. $9w^2 + 24w + 16$ **13.** $-2a^3b^2 - 5a^2b + ab^2 - 2ab$
14. $y^2 - 4y + \frac{8}{5}$ **15.** $x - 9 + \dfrac{48}{x + 5}$

Review Exercises
1. $\dfrac{1}{7^2}$ **2.** y^{11} **3.** $(3x)^{14}$ **4.** t^8 **5.** 4^3 **6.** $\dfrac{1}{a^3}$ **7.** 1
8. $9t^8$ **9.** $36x^8$ **10.** $\dfrac{y^3}{8x^3}$ **11.** t^{-5} **12.** $\dfrac{1}{y^4}$ **13.** 3.28×10^{-5}
14. 8,300,000 **15.** 2.09×10^4 **16.** 5.12×10^{-5}
17. 1.564×10^{10} slices **18.** 10 **19.** $-4y^5, 7y^2, -3y, -2$
20. x^2, x^0 **21.** 3, 2, 1, 0; 3 **22.** Binomial **23.** None of these
24. Monomial **25.** $-2x^2 - 3x + 2$ **26.** $10x^4 - 7x^2 - x - \frac{1}{2}$
27. $x^5 - 2x^4 + 6x^3 + 3x^2 - 9$ **28.** $-2x^5 - 6x^4 - 2x^3 - 2x^2 + 2$
29. $2x^2 - 4x$ **30.** $x^5 - 3x^3 - x^2 + 8$ **31.** Perimeter: $4w + 6$;
area: $w^2 + 3w$ **32.** $(t + 3)(t + 4), t^2 + 7t + 12$
33. $x^2 + \frac{7}{6}x + \frac{1}{3}$ **34.** $49x^2 + 14x + 1$ **35.** $12x^3 - 23x^2 + 13x - 2$
36. $9x^4 - 16$ **37.** $15x^7 - 40x^6 + 50x^5 + 10x^4$ **38.** $x^2 - 3x - 28$
39. $9y^4 - 12y^3 + 4y^2$ **40.** $2t^4 - 11t^2 - 21$ **41.** 49
42. Coefficients: 1, -7, 9, -8; degrees: 6, 2, 2, 0; 6

43. $-y + 9w - 5$ **44.** $m^6 - 2m^2n + 2m^2n^2 + 8n^2m - 6m^3$
45. $-9xy - 2y^2$ **46.** $11x^3y^2 - 8x^2y - 6x^2 - 6x + 6$
47. $p^3 - q^3$ **48.** $9a^8 - 2a^4b^3 + \frac{1}{9}b^6$ **49.** $5x^2 - \frac{1}{2}x + 3$
50. $3x^2 - 7x + 4 + \dfrac{1}{2x + 3}$ **51.** 0, 3.75, -3.75, 0 **52.** B
53. D **54.** $\frac{1}{2}x^2 - \frac{1}{2}y^2$ **55.** $400 - 4a^2$ **56.** $-28x^8$ **57.** $\frac{94}{13}$
58. $x^4 + x^3 + x^2 + x + 1$ **59.** 80 ft by 40 ft

Understanding Through Discussion and Writing
1. 578.6×10^{-7} is not in scientific notation because 578.6 is not a number greater than or equal to 1 and less than 10.
2. When evaluating polynomials, it is essential to know the order in which the operations are to be performed.
3. We label the figure as shown.

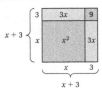

Then we see that the area of the figure is $(x + 3)^2$, or $x^2 + 3x + 3x + 9 \ne x^2 + 9$. **4.** Emma did not divide *each* term of the polynomial by the divisor. The first term was divided by $3x$, but the second was not. Multiplying Emma's "quotient" by the divisor $3x$, we get $12x^3 - 18x^2 \ne 12x^3 - 6x$. This should convince her that a mistake has been made. **5.** Yes; for example, $(x^2 + xy + 1) + (3x - xy + 2) = x^2 + 3x + 3$. **6.** Yes; consider $a + b + c + d$. This is a polynomial in 4 variables but it has degree 1.

Test: Chapter 13, p. 869

1. [13.1d, f] $\dfrac{1}{6^5}$ **2.** [13.1d] x^9 **3.** [13.1d] $(4a)^{11}$ **4.** [13.1e] 3^3
5. [13.1e, f] $\dfrac{1}{x^5}$ **6.** [13.1b, e] 1 **7.** [13.2a] x^6 **8.** [13.2a, b] $-27y^6$
9. [13.2a, b] $16a^{12}b^4$ **10.** [13.2b] $\dfrac{a^3b^3}{c^3}$ **11.** [13.1d], [13.2a, b] $-216x^{21}$
12. [13.1d], [13.2a, b] $-24x^{21}$ **13.** [13.1d], [13.2a, b] $162x^{10}$
14. [13.1d], [13.2a, b] $324x^{10}$ **15.** [13.1f] $\dfrac{1}{5^3}$ **16.** [13.1f] y^{-8}
17. [13.2c] 3.9×10^9 **18.** [13.2c] 0.00000005
19. [13.2d] 1.75×10^{17} **20.** [13.2d] 1.296×10^{22}
21. [13.2e] The mass of Saturn is 9.5×10 times the mass of Earth.
22. [13.3a] -43 **23.** [13.3c] $\frac{1}{3}, -1, 7$ **24.** [13.3c] 3, 0, 1, 6; 6
25. [13.3b] Binomial **26.** [13.3d] $5a^2 - 6$ **27.** [13.3d] $\frac{7}{4}y^2 - 4y$
28. [13.3e] $x^5 + 2x^3 + 4x^2 - 8x + 3$
29. [13.4a] $4x^5 + x^4 + 2x^3 - 8x^2 + 2x - 7$
30. [13.4a] $5x^4 + 5x^2 + x + 5$ **31.** [13.4c] $-4x^4 + x^3 - 8x - 3$
32. [13.4c] $-x^5 + 0.7x^3 - 0.8x^2 - 21$
33. [13.5b] $-12x^4 + 9x^3 + 15x^2$ **34.** [13.6c] $x^2 - \frac{2}{3}x + \frac{1}{9}$
35. [13.6b] $9x^2 - 100$ **36.** [13.6a] $3b^2 - 4b - 15$
37. [13.6a] $x^{14} - 4x^8 + 4x^6 - 16$ **38.** [13.6a] $48 + 34y - 5y^2$
39. [13.5d] $6x^3 - 7x^2 - 11x - 3$ **40.** [13.6c] $25t^2 + 20t + 4$
41. [13.7c] $-5x^3y - y^3 + xy^3 - x^2y^2 + 19$
42. [13.7e] $8a^2b^2 + 6ab - 4b^3 + 6ab^2 + ab^3$
43. [13.7f] $9x^{10} - 16y^{10}$ **44.** [13.8a] $4x^2 + 3x - 5$
45. [13.8b] $2x^2 - 4x - 2 + \dfrac{17}{3x + 2}$
46. [13.3a] 3, 1.5, -3.5, -5, -5.25
47. [13.4d] $(t + 2)(t + 2), t^2 + 4t + 4$ **48.** [13.4d] B
49. [13.5b], [13.6a] $V = l^3 - 3l^2 + 2l$
50. [11.3b], [13.6b, c] $-\frac{61}{12}$

CHAPTER 16

Calculator Corner, p. 1047
1. 17.3　**2.** 34

Calculator Corner, p. 1049

1. $y = x - 4$
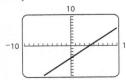

2. $y = -2x - 3$
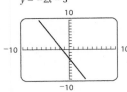

3. $y = 1 - x^2$
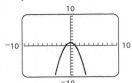

4. $y = 3x^2 - 4x + 1$
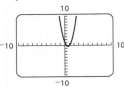

5. $y = x^3$
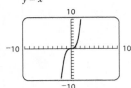

6. $y = |x + 3|$
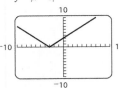

Exercise Set 16.1, p. 1051

RC1. $f(2) = 3$　**RC2.** $f(0) = 3$　**RC3.** $f(-2) = -5$
RC4. $f(3) = 0$
1. Yes　**3.** Yes　**5.** No　**7.** No　**9.** No　**11.** Yes
13. (a) 9; **(b)** 12; **(c)** 2; **(d)** 5; **(e)** 7.4; **(f)** $5\frac{2}{3}$　**15. (a)** -21;
(b) 15; **(c)** 2; **(d)** 0; **(e)** $18a$; **(f)** $3a + 3$　**17. (a)** 7;
(b) -17; **(c)** 6; **(d)** 4; **(e)** $3a - 2$; **(f)** $3a + 3h + 4$
19. (a) 0; **(b)** 5; **(c)** 2; **(d)** 170; **(e)** 65; **(f)** $32a^2 - 12a$
21. (a) 1; **(b)** 3; **(c)** 3; **(d)** 11; **(e)** $|a - 1| + 1$;
(f) $|a + h| + 1$　**23. (a)** 0; **(b)** -1; **(c)** 8; **(d)** 1000; **(e)** -125;
(f) $-27a^3$　**25.** 1980: about 60.5 years; 2013: about 62.0 years
27. $1\frac{20}{33}$ atm; $1\frac{10}{11}$ atm; $4\frac{1}{33}$ atm　**29.** 1.792 cm; 2.8 cm; 11.2 cm

31.

33.
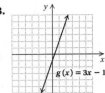

35.
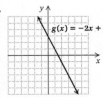

37.
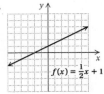

39.

41.
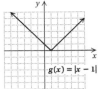

43.

45.
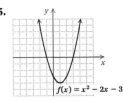

47.
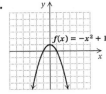

49.
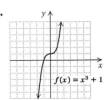

51. Yes　**53.** No　**55.** No　**57.** Yes　**59.** About 1.8 million
children　**61.** About 230,000 pharmacists　**63.** $\frac{3}{4}$　**64.** All real
numbers　**65.** $\{y | y < 4\}$　**66.** $-10, 8$　**67.** $(11t + 5)(11t - 5)$
68. $(w - 10)(w^2 + 10w + 100)$　**69.** x^5　**70.** $s^7 t^9$
71. $y^2 - 3y - 11$　**72.** $-5x^4 - x^2 + 5x - 2$　**73.** $g(-2) = 39$
75. 26; 99　**77.** $g(x) = \frac{15}{4}x - \frac{13}{4}$

Exercise Set 16.2, p. 1061

RC1. (a)　**RC2.** (b)　**RC3.** (a)　**RC4.** (b)　**RC5.** (a)
RC6. (d)
1. (a) 3; **(b)** $\{-4, -3, -2, -1, 0, 1, 2\}$; **(c)** $-2, 0$; **(d)** $\{1, 2, 3, 4\}$
3. (a) $2\frac{1}{2}$; **(b)** $\{x | -3 \le x \le 5\}$; **(c)** $2\frac{1}{4}$; **(d)** $\{y | 1 \le y \le 4\}$
5. (a) 1; **(b)** all real numbers; **(c)** 3; **(d)** all real numbers
7. (a) 1; **(b)** all real numbers; **(c)** $-2, 2$; **(d)** $\{y | y \ge 0\}$
9. $\{x | x \text{ is a real number } and\ x \ne -3\}$　**11.** All real numbers
13. All real numbers　**15.** $\left\{x | x \text{ is a real number } and\ x \ne \frac{14}{5}\right\}$
17. All real numbers　**19.** $\left\{x | x \text{ is a real number } and\ x \ne \frac{7}{4}\right\}$
21. $\{x | x \text{ is a real number } and\ x \ne 1\}$　**23.** All real numbers
25. All real numbers　**27.** $\left\{x | x \text{ is a real number } and\ x \ne \frac{5}{2}\right\}$
29. All real numbers　**31.** $\left\{x | x \text{ is a real number } and\ x \ne -\frac{5}{4}\right\}$
33. $-8; 0; -2$　**35.** $a - 1$　**36.** $\dfrac{2(y + 2)}{7(y + 7)}$　**37.** $\dfrac{5}{x + 2}$
38. $t - 4$　**39.** $w + 1$ with R 2, or $w + 1 + \dfrac{2}{w + 3}$
41. $\{y | y \text{ is a real number and } y \ne 0\}$; $\{y | y \ge 2\}$;
$\{y | y \ge -4\}$; $\{y | y \ge 0\}$　**43.** All real numbers

Mid-Chapter Review: Chapter 16, p. 1063
1. True　**2.** False　**3.** True　**4.** True　**5.** False

6.

x	$f(x)$
0	1
2	-2
-2	4
4	-5

7.

x	$f(x)$
-2	0
-2 and 3	0
0	-6
2	-4
-1	-4
1	-6

8. Yes   **9.** No   **10.** Domain: $\{x \mid -3 \le x \le 3\}$; range: $\{y \mid -2 \le y \le 1\}$ **11.** -3 **12.** -7 **13.** 8   **14.** 9   **15.** 9000   **16.** 0   **17.** Yes   **18.** No   **19.** Yes   **20.** $\{x \mid x \text{ is a real number } and \ x \ne 4\}$ **21.** All real numbers   **22.** $\{x \mid x \text{ is a real number } and \ x \ne -2\}$ **23.** All real numbers

24. 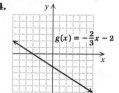 $g(x) = -\frac{2}{3}x - 2$

25. 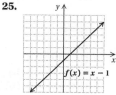 $f(x) = x - 1$

26. 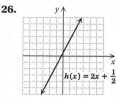 $h(x) = 2x + \frac{1}{2}$

27. 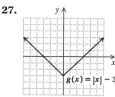 $g(x) = |x| - 3$

28. 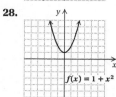 $f(x) = 1 + x^2$

29. 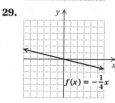 $f(x) = -\frac{1}{4}x$

30. No; since each input has exactly one output, the number of outputs cannot exceed the number of inputs.   **31.** When $x < 0$, then $y < 0$, and the graph contains points in quadrant III. When $0 < x < 30$, then $y < 0$, and the graph contains points in quadrant IV. When $x > 30$, then $y > 0$, and the graph contains points in quadrant I. Thus the graph passes through three quadrants.   **32.** The output -3 corresponds to the input 2. The number -3 in the range is paired with the number 2 in the domain. The point $(2, -3)$ is on the graph of the function.   **33.** The domain of a function is the set of all inputs, and the range is the set of all outputs.

Calculator Corner, p. 1066

1. The graph of $y_2 = x + 4$ is the graph of $y_1 = x$ moved 4 units up.   **2.** The graph of $y_3 = x - 3$ is the graph of $y_1 = x$ moved 3 units down.

Calculator Corner, p. 1069

1. The graph of $y = 10x$ will slant up from left to right. It will be steeper than the other graphs.   **2.** The graph of $y = 0.005x$ will slant up from left to right. It will be less steep than the other graphs.   **3.** The graph of $y = -10x$ will slant down from left to right. It will be steeper than the other graphs.   **4.** The graph of $y = -0.005x$ will slant down from left to right. It will be less steep than the other graphs.

Exercise Set 16.3, p. 1073

RC1. (f)   **RC2.** (b)   **RC3.** (d)   **RC4.** (c)   **RC5.** (e)   **RC6.** (a)
1. $m = 4$; y-intercept: $(0, 5)$ **3.** $m = -2$; y-intercept: $(0, -6)$ **5.** $m = -\frac{3}{8}$; y-intercept: $(0, -\frac{1}{5})$ **7.** $m = 0.5$; y-intercept: $(0, -9)$ **9.** $m = \frac{2}{3}$; y-intercept: $(0, -\frac{8}{3})$ **11.** $m = 3$; y-intercept: $(0, -2)$ **13.** $m = -8$; y-intercept: $(0, 12)$ **15.** $m = 0$; y-intercept: $(0, \frac{4}{17})$ **17.** $m = -\frac{1}{2}$ **19.** $m = \frac{1}{3}$ **21.** $m = 2$ **23.** $m = \frac{2}{3}$ **25.** $m = -\frac{1}{3}$ **27.** $\frac{2}{25}$, or 8%   **29.** $\frac{13}{41}$, or about 31.7%   **31.** The rate of change is about $2.74 billion per year.   **33.** The rate of change is $-$900 per year.   **35.** The rate of change is about $116.14 per year.   **37.** -1323 **38.** $45x + 54$ **39.** $350x - 60y + 120$ **40.** 25   **41.** Square: 15 yd; triangle: 20 yd   **42.** $(2 - 5x)(4 + 10x + 25x^2)$

43. $(c - d)(c^2 + cd + d^2)(c + d)(c^2 - cd + d^2)$ **44.** $7(2x - 1)(4x^2 + 2x + 1)$ **45.** $a - 10$, R -4, or $a - 10 + \dfrac{-4}{a - 1}$

Calculator Corner, p. 1077

1. $y = -3.2x - 16$
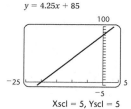
Xscl = 1, Yscl = 2

2. $y = 4.25x + 85$
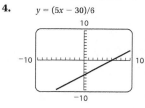
Xscl = 5, Yscl = 5

3. $y = (-6x + 90)/5$
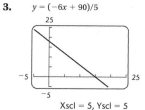
Xscl = 5, Yscl = 5

4. $y = (5x - 30)/6$
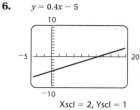

5. $y = (-8x + 9)/3$
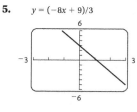

6. $y = 0.4x - 5$
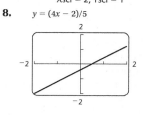
Xscl = 2, Yscl = 1

7. $y = 1.2x - 12$
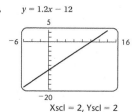

8. $y = (4x - 2)/5$
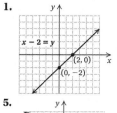
Xscl = 2, Yscl = 2

Visualizing for Success, p. 1083

1. D   **2.** I   **3.** H   **4.** C   **5.** F   **6.** A   **7.** G   **8.** B   **9.** E   **10.** J

Exercise Set 16.4, p. 1084

RC1. True   **RC2.** False   **RC3.** False   **RC4.** True   **RC5.** True   **RC6.** False

1. 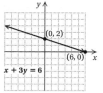 $x - 2 = y$; $(2, 0)$; $(0, -2)$

3. 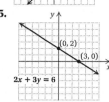 $(0, 2)$; $(6, 0)$; $x + 3y = 6$

5. $(0, 2)$; $(3, 0)$; $2x + 3y = 6$

7. 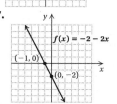 $f(x) = -2 - 2x$; $(-1, 0)$; $(0, -2)$

9.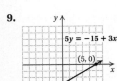

11.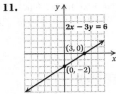

13.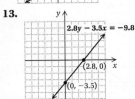

15.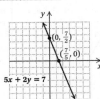

17.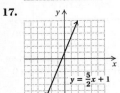

19.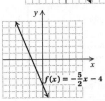

21.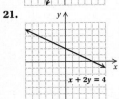

23.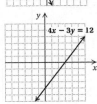

25.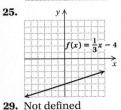

27.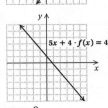

29. Not defined   **31.** $m = 0$

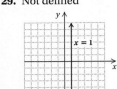

33. $m = 0$ **35.** $m = 0$

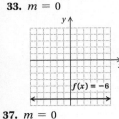

37. $m = 0$ **39.** Not defined

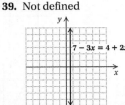

41. Yes   **43.** No   **45.** Yes   **47.** Yes   **49.** Yes   **51.** No
53. No   **55.** Yes   **57.** 5.3×10^{10} **58.** 4.7×10^{-5} **59.** 1.8×10^{-2}
60. 9.9902×10^{7} **61.** 0.0000213 **62.** $901,000,000$ **63.** $20,000$
64. 0.085677 **65.** $3(3x - 5y)$ **66.** $3a(4 + 7b)$
67. $7p(3 - q + 2)$ **68.** $64(x - 2y + 4)$ **69.** $2009\,\text{lb}$

70.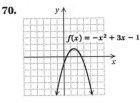

71. $a = 2$ **73.** $y = \frac{2}{15}x + \frac{2}{5}$
75. $y = 0$; yes   **77.** $m = -\frac{3}{4}$

Exercise Set 16.5, p. 1095

RC1. (a) $\frac{4}{11}$; **(b)** $-\frac{11}{4}$ **RC2. (a)** 0; **(b)** not defined
RC3. (a) 2; **(b)** $-\frac{1}{2}$ **RC4. (a)** $-\frac{5}{6}$; **(b)** $\frac{6}{5}$
RC5. (a) Not defined; **(b)** 0 **RC6. (a)** -2; **(b)** $\frac{1}{2}$
1. $y = -8x + 4$ **3.** $y = 2.3x - 1$ **5.** $f(x) = -\frac{7}{3}x - 5$
7. $f(x) = \frac{2}{3}x + \frac{5}{8}$ **9.** $y = 5x - 17$ **11.** $y = -3x + 33$
13. $y = x - 6$ **15.** $y = -2x + 16$ **17.** $y = -7$ **19.** $y = \frac{2}{3}x - \frac{8}{3}$
21. $y = \frac{1}{2}x + \frac{7}{2}$ **23.** $y = x$ **25.** $y = \frac{7}{4}x + 7$ **27.** $y = \frac{3}{2}x$
29. $y = \frac{1}{6}x$ **31.** $y = 13x - \frac{15}{4}$ **33.** $y = -\frac{1}{2}x + \frac{17}{2}$
35. $y = \frac{5}{7}x - \frac{17}{7}$ **37.** $y = \frac{1}{3}x + 4$ **39.** $y = \frac{1}{2}x + 4$
41. $y = \frac{4}{3}x - 6$ **43.** $y = \frac{5}{2}x + 9$
45. (a) $C(x) = 5x + 10$;   **(b)**   ; **(c)** $30

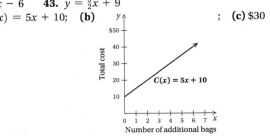

47. (a) $V(t) = 9400 - 85t$;
(b)   ; **(c)** $7870

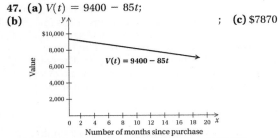

49. (a) $S(x) = 2x + 11$; **(b)** \$19 billion; \$37 billion
51. (a) $D(x) = -309.41x + 22,800$; **(b)** 21,253 dealerships;
(c) about 24 years after 1995, or in 2019
53. (a) $E(t) = 0.346t + 46.56$; **(b)** about 51.06 years   **55.** -1
56. $b + 1$ **57.** $\dfrac{x - 3}{2(x - 5)}$ **58.** $\dfrac{4}{y - 9}$ **59.** $\frac{2}{7}$ **60.** 0
61. Not defined   **62.** -2 **63.** -7.75

Summary and Review: Chapter 16, p. 1099

Vocabulary Reinforcement

1. Vertical   **2.** Point-slope   **3.** Function, domain, range,
domain, exactly one, range   **4.** Slope   **5.** Perpendicular
6. Slope-intercept   **7.** Parallel

Concept Reinforcement

1. False   **2.** True   **3.** False

Study Guide

1. No   **2.** $g(0) = -2; g(-2) = -3; g(6) = 1$
3.

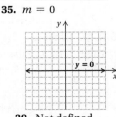

4. Yes   **5.** Domain: $\{x | -4 \le x \le 5\}$;
range: $\{y | -2 \le y \le 4\}$ **6.** $\{x | x$ is a
real number $and\, x \ne -3\}$ **7.** -2
8. Slope: $-\frac{1}{2}$; y-intercept: $(0, 2)$

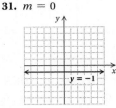

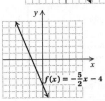

9.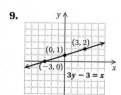

10.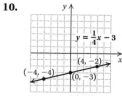

11.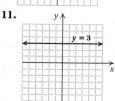

12.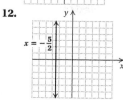

13. Parallel **14.** Perpendicular **15.** $y = -8x + 0.3$
16. $y = -4x - 1$ **17.** $y = -\frac{5}{3}x + \frac{11}{3}$ **18.** $y = \frac{4}{3}x - \frac{23}{3}$
19. $y = -\frac{3}{4}x - \frac{7}{2}$

Review Exercises

1. No **2.** Yes **3.** $g(0) = 5; g(-1) = 7$
4. $f(0) = 7; f(-1) = 12$ **5.** About $6810
6.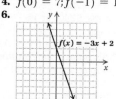

7.

8.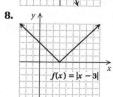

9.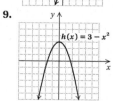

10. Yes **11.** No **12.** (a) $f(2) = 3$; (b) $\{x | -2 \leq x \leq 4\}$;
(c) -1; (d) $\{y | 1 \leq y \leq 5\}$ **13.** $\{x | x$ is a real number *and*
$x \neq 4\}$ **14.** All real numbers **15.** Slope: -3; y-intercept: $(0, 2)$
16. Slope: $-\frac{1}{2}$; y-intercept: $(0, 2)$ **17.** $\frac{11}{3}$
18.

19.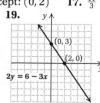

20.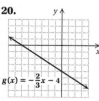

21.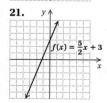

22.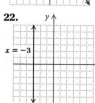

23.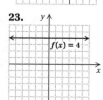

24. Perpendicular **25.** Parallel **26.** Parallel
27. Perpendicular **28.** $f(x) = 4.7x - 23$ **29.** $y = -3x + 4$
30. $y = -\frac{3}{2}x$ **31.** $y = -\frac{5}{7}x + 9$ **32.** $y = \frac{1}{3}x + \frac{1}{3}$
33. (a) $R(x) = -0.018x + 44.66$; (b) about 44.16 sec; about
43.98 sec **34.** C **35.** A **36.** $f(x) = 3.09x + 3.75$

Understanding Through Discussion and Writing

1. A line's x- and y-intercepts are the same only when the line
passes through the origin. The equation for such a line is of the form
$y = mx$. **2.** The concept of slope is useful in describing how a line
slants. A line with positive slope slants up from left to right. A line with

negative slope slants down from left to right. The larger the absolute
value of the slope, the steeper the slant. **3.** Find the slope-intercept
form of the equation:

$$4x + 5y = 12$$
$$5y = -4x + 12$$
$$y = -\tfrac{4}{5}x + \tfrac{12}{5}.$$

This form of the equation indicates that the line has a negative slope
and thus should slant down from left to right. The student may have
graphed $y = \frac{4}{5}x + \frac{12}{5}$. **4.** For $R(t) = 50t + 35$, $m = 50$ and
$b = 35$; 50 signifies that the cost per hour of a repair is $50;
35 signifies that the minimum cost of a repair job is $35.
5. $m = \dfrac{\text{change in } y}{\text{change in } x}$
As we move from one point to another on a vertical line, the
y-coordinate changes but the x-coordinate does not. Thus the change
in y is a nonzero number whereas the change in x is 0. Since division
by 0 is not defined, the slope of a vertical line is not defined. As we
move from one point to another on a horizontal line, the y-coordinate
does not change but the x-coordinate does. Thus the change in y is 0
whereas the change in x is a nonzero number, so the slope
is 0. **6.** Using algebra, we find that the slope-intercept form of the
equation is $y = \frac{5}{2}x - \frac{3}{2}$. This indicates that the y-intercept is $\left(0, -\frac{3}{2}\right)$,
so a mistake has been made. It appears that the student graphed
$y = \frac{5}{2}x + \frac{3}{2}$.

Test: Chapter 16, p. 1108

1. [16.1a] Yes **2.** [16.1a] No **3.** [16.1b] $-4; 2$ **4.** [16.1b] $7; 8$
5. [16.1b] $-6; -6$ **6.** [16.1b] $3; 0$
7. [16.1c] **8.** [16.1c] **9.** [16.1c]

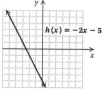

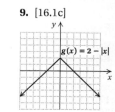

10. [16.1c] **11.** [16.4c] **12.** [16.4c]

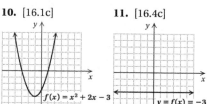

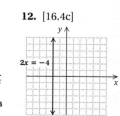

13. [16.1e] (a) About 9.4 years; (b) 1998 **14.** [16.1d] Yes
15. [16.1d] No **16.** [16.2a] $\{x | x$ is a real number *and* $x \neq -\frac{3}{2}\}$
17. [16.2a] All real numbers **18.** [16.2a] (a) 1;
(b) $\{x | -3 \leq x \leq 4\}$; (c) -3; (d) $\{y | -1 \leq y \leq 2\}$
19. [16.3b] Slope: $-\frac{3}{5}$; y-intercept: $(0, 12)$ **20.** [16.3b] Slope: $-\frac{2}{5}$;
y-intercept: $\left(0, -\frac{7}{5}\right)$ **21.** [16.3b] $\frac{5}{8}$ **22.** [16.3b] 0
23. [16.3c] $\frac{4}{5}$ km/min
24. [16.4a] **25.** [16.4b]

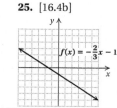

26. [16.4d] Parallel **27.** [16.4d] Perpendicular
28. [16.5a] $y = -3x + 4.8$ **29.** [16.5a] $f(x) = 5.2x - \frac{5}{8}$
30. [16.5b] $y = -4x + 2$ **31.** [16.5c] $y = -\frac{3}{2}x$
32. [16.5d] $y = \frac{1}{2}x - 3$ **33.** [16.5d] $y = 3x - 1$
34. [16.5e] (a) $A(x) = 0.125x + 23.2$; (b) 27.95 years; 28.825 years
35. [16.5b] B **36.** [16.4d] $\frac{24}{5}$ **37.** [16.1b] $f(x) = 3$; answers may vary

CHAPTER 17

Calculator Corner, p. 1117

1. $(2,3)$ **2.** $(-4,-1)$ **3.** $(-1,5)$ **4.** $(3,-1)$

Exercise Set 17.1, p. 1117

RC1. False   **RC2.** True   **RC3.** True   **RC4.** True
1. $(3,1)$; consistent; independent   **3.** $(1,-2)$; consistent; independent   **5.** $(4,-2)$; consistent; independent
7. $(2,1)$; consistent; independent   **9.** $\left(\frac{5}{2},-2\right)$; consistent; independent   **11.** $(3,-2)$; consistent; independent
13. No solution; inconsistent; independent   **15.** Infinitely many solutions; consistent; dependent   **17.** $(4,-5)$; consistent; independent   **19.** $(2,-3)$; consistent; independent
21. Consistent; independent; F   **23.** Consistent; dependent; B
25. Inconsistent; independent; D   **27.** $y=\frac{3}{5}x+\frac{22}{5}$
28. $y=\frac{3}{8}x+\frac{9}{4}$ **29.** $y=-\frac{3}{2}x$ **30.** $y=\frac{4}{3}x-14$
31. $(2.23,1.14)$ **33.** $(3,3),(-5,5)$

Exercise Set 17.2, p. 1124

RC1. True   **RC2.** True   **RC3.** False   **RC4.** False
1. $(2,-3)$ **3.** $\left(\frac{21}{5},\frac{12}{5}\right)$ **5.** $(2,-2)$ **7.** $(-2,-6)$ **9.** $(-2,1)$
11. No solution   **13.** $\left(\frac{19}{8},\frac{1}{8}\right)$ **15.** Infinitely many solutions
17. $\left(\frac{1}{2},\frac{1}{2}\right)$ **19.** Length: 25 m; width: 5 m   **21.** $48°$ and $132°$
23. Wins: 23; ties: 14   **25.** 1.3   **26.** $-15y-39$ **27.** $p=\dfrac{7A}{q}$
28. $\frac{7}{3}$ **29.** -23 **30.** $\frac{29}{22}$ **31.** $m=-\frac{1}{2};b=\frac{5}{2}$ **33.** Length: 57.6 in.; width: 20.4 in.

Exercise Set 17.3, p. 1133

RC1. Consistent   **RC2.** Inconsistent   **RC3.** Consistent
RC4. Dependent   **RC5.** Inconsistent   **RC6.** Independent
1. $(1,2)$ **3.** $(-1,3)$ **5.** $(-1,-2)$ **7.** $(5,2)$ **9.** Infinitely many solutions   **11.** $\left(\frac{1}{2},-\frac{1}{2}\right)$ **13.** $(4,6)$ **15.** No solution
17. $(10,-8)$ **19.** $(12,15)$ **21.** $(10,8)$ **23.** $(-4,6)$
25. $(10,-5)$ **27.** $(140,60)$ **29.** 36 and 27   **31.** 18 and -15
33. $48°$ and $42°$ **35.** Two-point shots: 21; three-point shots: 6
37. 3-credit courses: 25; 4-credit courses: 8   **39.** 1   **40.** 5
41. 15   **42.** $12a^2-2a+1$ **43.** $\{x\,|\,x$ is a real number $and\ x\neq-7\}$ **44.** Domain: all real numbers;
range: $\{y\,|\,y\leq5\}$ **45.** $y=-\frac{3}{5}x-7$ **46.** $y=x+12$
47. $(23.12,-12.04)$ **49.** $A=2,B=4$ **51.** $p=2,q=-\frac{1}{3}$

Translating for Success, p. 1144

1. G   **2.** E   **3.** D   **4.** A   **5.** J   **6.** B   **7.** C   **8.** I
9. F   **10.** H

Exercise Set 17.4, p. 1145

RC1. 10   **RC2.** 15   **RC3.** $0.15y$ **RC4.** 2   **1.** Books: 45; games: 23   **3.** Foil balloons: 2; latex balloons: 7
5. Olive oil: $22\frac{1}{2}$ oz; vinegar: $7\frac{1}{2}$ oz   **7.** 5 lb of each
9. 25%-acid: 4 L; 50%-acid: 6 L   **11.** Sweet-pepper packets: 11; hot-pepper packets: 5   **13.** \$7500 at 6%; \$4500 at 3%
15. Whole milk: $169\frac{3}{13}$ lb; cream: $30\frac{10}{13}$ lb   **17.** \$1800 at 5.5%; \$1400 at 4%   **19.** \$5 bills: 7; \$1 bills: 15   **21.** 375 mi
23. 14 km/h   **25.** Headwind: 30 mph; plane: 120 mph
27. $1\frac{1}{3}$ hr   **29.** About 1489 mi   **31.** All real numbers
32. $\{x\,|\,x$ is a real number $and\ x\neq14\}$ **33.** -4 **34.** -17
35. $-12h-7$ **36.** 3993   **37.** $4\frac{4}{7}$ L   **39.** City: 261 mi; highway: 204 mi

Mid-Chapter Review: Chapter 17, p. 1149

1. False   **2.** False   **3.** True   **4.** True

5.
$$x+2(x-6)=3$$
$$x+2x-12=3$$
$$3x-12=3$$
$$3x=15$$
$$x=5$$
$$y=5-6$$
$$y=-1$$
The solution is $(5,-1)$.

6.
$$6x-4y=10$$
$$\underline{2x+4y=14}$$
$$8x=24$$
$$x=3$$
$$2\cdot3+4y=14$$
$$6+4y=14$$
$$4y=8$$
$$y=2$$
The solution is $(3,2)$.

7. $(5,-1)$; consistent; independent   **8.** $(0,3)$; consistent; independent   **9.** Infinitely many solutions; consistent; dependent
10. No solution; inconsistent; independent   **11.** $(8,6)$
12. $(2,-3)$ **13.** $(-3,5)$ **14.** $(-1,-2)$ **15.** $(2,-2)$
16. $(5,-4)$ **17.** $(-1,-2)$ **18.** $(3,1)$ **19.** No solution
20. Infinitely many solutions   **21.** $(10,-12)$ **22.** $(-9,8)$
23. Length: 12 ft; width: 10 ft   **24.** \$2100 at 2%; \$2900 at 3%
25. 20% acid: 56 L; 50% acid: 28 L   **26.** 26 mph
27. *Graphically*: **1.** Graph $y=\frac{3}{4}x+2$ and $y=\frac{2}{5}x-5$ and find the point of intersection. The first coordinate of this point is the solution of the original equation. **2.** Rewrite the equation as $\frac{7}{20}x+7=0$. Then graph $y=\frac{7}{20}x+7$ and find the x-intercept. The first coordinate of this point is the solution of the original equation.
Algebraically: **1.** Use the addition and multiplication principles for equations. **2.** Multiply by 20 to clear the fractions and then use the addition and multiplication principles for equations.
28. (a) Answers may vary.
$$x+y=1,$$
$$x-y=7$$
(b) Answers may vary.
$$x+2y=5,$$
$$3x+6y=10$$
(c) Answers may vary.
$$x-2y=3,$$
$$3x-6y=9$$
29. Answers may vary. Form a linear expression in two variables and set it equal to two different constants. See Exercises 10 and 19 in this review for examples.   **30.** Answers may vary. Let any linear equation be one equation in the system. Multiply by a constant on both sides of that equation to get the second equation in the system. See Exercises 9 and 20 in this review for examples.

Exercise Set 17.5, p. 1155

RC1. (b)   **RC2.** (c)   **RC3.** (a)   **RC4.** (a)
1. $(1,2,-1)$ **3.** $(2,0,1)$ **5.** $(3,1,2)$ **7.** $(-3,-4,2)$
9. $(2,4,1)$ **11.** $(-3,0,4)$ **13.** $(2,2,4)$ **15.** $\left(\frac{1}{2},4,-6\right)$
17. $(-2,3,-1)$ **19.** $\left(\frac{1}{2},\frac{1}{3},\frac{1}{6}\right)$ **21.** $(3,-5,8)$ **23.** $(15,33,9)$
25. $(4,1,-2)$ **27.** $(17,9,79)$ **28.** $a=\dfrac{F}{3b}$ **29.** $a=\dfrac{Q-4b}{4}$, or $\dfrac{Q}{4}-b$ **30.** $d=\dfrac{tc-2F}{t}$, or $c-\dfrac{2F}{t}$ **31.** $c=\dfrac{2F+td}{t}$, or $\dfrac{2F}{t}+d$ **32.** $y=\dfrac{c-Ax}{B}$ **33.** $y=\dfrac{Ax-c}{B}$ **34.** Slope: $-\frac{2}{3}$; y-intercept: $\left(0,-\frac{5}{4}\right)$ **35.** Slope: -4; y-intercept: $(0,5)$
36. Slope: $\frac{2}{5}$; y-intercept: $(0,-2)$ **37.** Slope: 1.09375; y-intercept: $(0,-3.125)$ **39.** $(1,-2,4,-1)$

Exercise Set 17.6, p. 1161

RC1. (c)   **RC2.** (d)   **RC3.** (a)   **RC4.** (b)
1. Reading: 496; math: 514; writing: 488   **3.** $32°,96°,52°$

5. $-7, 20, 42$ **7.** Sixteen: 10; Original: 16; Power: 8
9. Egg: 274 mg; cupcake: 19 mg; pizza: 9 mg **11.** Automatic transmission: \$865; power door locks: \$520; air conditioning: \$375
13. Dog: \$200; cat: \$81; bird: \$9 **15.** Roast beef: 2; baked potato: 1; broccoli: 2 **17.** First fund: \$45,000; second fund: \$10,000; third fund: \$25,000 **19.** Par-3: 6 holes; par-4: 8 holes; par-5: 4 holes **21.** A: 1500 lenses; B: 1900 lenses; C: 2300 lenses

23. 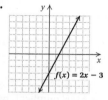 **24.**

25.

26. No **27.** Yes **28.** Yes **29.** $180°$ **31.** 464

Summary and Review: Chapter 17, p. 1165

Vocabulary Reinforcement
1. Pair **2.** Consistent **3.** Algebraic **4.** Triple
5. Independent

Concept Reinforcement
1. False **2.** True **3.** True **4.** False

Important Concepts
1. $(4, -1)$; consistent; independent **2.** $(-1, 4)$ **3.** $(-2, 3)$
4. \$8700 at 6%; \$14,300 at 5% **5.** $(3, -5, 1)$

Review Exercises
1. $(-2, 1)$; consistent; independent **2.** Infinitely many solutions; consistent; dependent **3.** No solution; inconsistent; independent **4.** $(1, -1)$ **5.** No solution **6.** $\left(\frac{2}{5}, -\frac{4}{5}\right)$
7. $(6, -3)$ **8.** $(2, 2)$ **9.** $(5, -3)$ **10.** Infinitely many solutions **11.** 150 mph **12.** 32 brushes at \$8.50; 13 brushes at \$9.75 **13.** 5 L of each **14.** $5\frac{1}{2}$ hr **15.** $(10, 4, -8)$
16. $(-1, 3, -2)$ **17.** $(2, 0, 4)$ **18.** $\left(2, \frac{1}{3}, -\frac{2}{3}\right)$
19. $90°, 67\frac{1}{2}°, 22\frac{1}{2}°$ **20.** Caramel nut crunch: \$30; plain: \$5; mocha choco latte: \$14 **21.** C **22.** C **23.** A
24. $(0, 2)$ and $(1, 3)$

Understanding Through Discussion and Writing
1. Answers may vary. One day, a florist sold a total of 23 hanging baskets and flats of petunias. Hanging baskets cost \$10.95 each and flats of petunias cost \$12.95 each. The sales totaled \$269.85. How many of each were sold? **2.** We know that machines A, B, and C can polish 5700 lenses in one week when working together. We also know that A and B together can polish 3400 lenses in one week, so C can polish $5700 - 3400$, or 2300, lenses in one week alone. We also know that B and C together can polish 4200 lenses in one week, so A can polish $5700 - 4200$, or 1500, lenses in one week alone. Also, B can polish $4200 - 2300$, or 1900, lenses in one week alone.
3. Let $x =$ the number of adults in the audience, $y =$ the number of senior citizens, and $z =$ the number of children. The total attendance is 100, so we have equation (1), $x + y + z = 100$. The amount taken in was \$100, so equation (2) is $10x + 3y + 0.5z = 100$. There is no other information that can

be translated to an equation. Clearing decimals in equation (2) and then eliminating z gives us equation (3), $95x + 25y = 500$. Dividing by 5 on both sides, we have equation (4), $19x + 5y = 100$. Since we have only two equations, it is not possible to eliminate z from another pair of equations. However, in $19x + 5y = 100$, note that 5 is a factor of both $5y$ and 100. Therefore, 5 must also be a factor of $19x$, and hence of x, since 5 is not a factor of 19. Then for some positive integer n, $x = 5n$. (We require n to be positive, since the number of adults clearly cannot be negative and must also be nonzero since the exercise states that the audience consists of adults, senior citizens, and children.) We have

$$19 \cdot 5n + 5y = 100$$
$$19n + y = 20. \quad \text{Dividing by 5}$$

Since n and y must both be positive, $n = 1$. (If $n > 1$, then $19n + y > 20$.) Then $x = 5 \cdot 1$, or 5.

$$19 \cdot 5 + 5y = 100 \quad \text{Substituting in (4)}$$
$$y = 1$$
$$5 + 1 + z = 100 \quad \text{Substituting in (1)}$$
$$z = 94$$

There were 5 adults, 1 senior citizen, and 94 children in the audience.

Test: Chapter 17, p. 1171

1. [17.1a] $(-2, 1)$; consistent; independent **2.** [17.1a] No solution; inconsistent; independent **3.** [17.1a] Infinitely many solutions; consistent; dependent **4.** [17.2a] $(2, -3)$
5. [17.2a] Infinitely many solutions **6.** [17.2a] $(-4, 5)$
7. [17.2a] $(8, -6)$ **8.** [17.3a] $(-1, 1)$ **9.** [17.3a] $\left(-\frac{3}{2}, -\frac{1}{2}\right)$
10. [17.3a] No solution **11.** [17.3a] $(10, -20)$
12. [17.2b] Length: 93 ft; width: 51 ft **13.** [17.4b] 120 km/h
14. [17.3b], [8.4a] Buckets: 17; dinners: 11
15. [17.4a] 20% solution: 12 L; 45% solution: 8 L
16. [17.5a] $\left(2, -\frac{1}{2}, -1\right)$ **17.** [17.6a] 3.5 hr **18.** [17.6a] B
19. [17.3a] $m = 7$; $b = 10$

CHAPTER 18

Translating for Success, p. 1183
1. F **2.** I **3.** C **4.** E **5.** D **6.** J **7.** O **8.** M
9. B **10.** L

Exercise Set 18.1, p. 1184
RC1. (b) **RC2.** (h) **RC3.** (c) **RC4.** (a) **RC5.** (g)
RC6. (d) **1.** No, no, no, yes **3.** No, yes, yes, no, no
5. $(-\infty, 5)$ **7.** $[-3, 3]$ **9.** $(-8, -4)$ **11.** $(-2, 5)$
13. $(-\sqrt{2}, \infty)$
15. $\{x | x > -1\}$, or $(-1, \infty)$ **17.** $\{y | y < 6\}$, or $(-\infty, 6)$

19. $\{a | a \leq -22\}$, or $(-\infty, -22]$

21. $\{t | t \geq -4\}$, or $[-4, \infty)$ **23.** $\{y | y > -6\}$, or $(-6, \infty)$

25. $\{x | x \leq 9\}$, or $(-\infty, 9]$ **27.** $\{x | x \geq 3\}$, or $[3, \infty)$

29. $\{x | x < -60\}$, or $(-\infty, -60)$ **31.** $\{x | x > 3\}$, or $(3, \infty)$

33. $\{x | x \leq 0.9\}$, or $(-\infty, 0.9]$ **35.** $\left\{x | x \leq \frac{5}{6}\right\}$, or $\left(-\infty, \frac{5}{6}\right]$
37. $\{x | x < 6\}$, or $(-\infty, 6)$ **39.** $\{y | y \leq -3\}$, or $(-\infty, -3]$

41. $\left\{y\,\middle|\,y > \frac{2}{3}\right\}$, or $\left(\frac{2}{3}, \infty\right)$ **43.** $\{x\,|\,x \geq 11.25\}$, or $[11.25, \infty)$
45. $\left\{x\,\middle|\,x \leq \frac{1}{2}\right\}$, or $\left(-\infty, \frac{1}{2}\right]$ **47.** $\left\{y\,\middle|\,y \leq -\frac{75}{2}\right\}$, or $\left(-\infty, -\frac{75}{2}\right]$
49. $\left\{x\,\middle|\,x > -\frac{2}{17}\right\}$, or $\left(-\frac{2}{17}, \infty\right)$ **51.** $\left\{m\,\middle|\,m > \frac{7}{3}\right\}$, or $\left(\frac{7}{3}, \infty\right)$
53. $\{r\,|\,r < -3\}$, or $(-\infty, -3)$ **55.** $\{x\,|\,x \geq 2\}$, or $[2, \infty)$
57. $\{y\,|\,y < 5\}$, or $(-\infty, 5)$ **59.** $\left\{x\,\middle|\,x \leq \frac{4}{7}\right\}$, or $\left(-\infty, \frac{4}{7}\right]$
61. $\{x\,|\,x < 8\}$, or $(-\infty, 8)$ **63.** $\left\{x\,\middle|\,x \geq \frac{13}{2}\right\}$, or $\left[\frac{13}{2}, \infty\right)$
65. $\left\{x\,\middle|\,x < \frac{11}{18}\right\}$, or $\left(-\infty, \frac{11}{18}\right)$ **67.** $\left\{x\,\middle|\,x \geq -\frac{51}{31}\right\}$, or $\left[-\frac{51}{31}, \infty\right)$
69. $\{a\,|\,a \leq 2\}$, or $(-\infty, 2]$
71. $\{W\,|\,W < \text{(approximately)}\ 136.7\,\text{lb}\}$ **73.** $\{S\,|\,S \geq 84\}$
75. $\{B\,|\,B \geq \$11{,}500\}$ **77.** $\{S\,|\,S > \$7000\}$ **79.** $\{c\,|\,c > \$735\}$
81. $\{p\,|\,p > 80\}$ **83.** $\{s\,|\,s < 980\,\text{ft}^2\}$ **85. (a)** 2010: \$2333;
2014: \$2661; **(b)** more than 13.13 years since 2005, or $\{t\,|\,t > 13.13\}$
87. $3x^2 + 20x - 32$ **88.** $6r^2 - 23rs - 4s^2$ **89.** $6a^2 + 7a - 55$
90. $t^2 - 7st - 18s^2$ **91.** $(2x - 9)^2$ **92.** $16(5y + 1)(5y - 1)$
93. $(3w - 2)(9w^2 + 6w + 4)$ **94.** $2(8 - 3x)(5 + x)$
95. $\{x\,|\,x\ \text{is a real number}\ and\ x \neq -8\}$ **96.** All real numbers
97. All real numbers **98.** $\left\{x\,\middle|\,x\ \text{is a real number}\ and\ x \neq \frac{2}{3}\right\}$
99. (a) $\{p\,|\,p > 10\}$; **(b)** $\{p\,|\,p < 10\}$ **101.** True
103. All real numbers **105.** All real numbers

Exercise Set 18.2, p. 1197

RC1. True **RC2.** False **RC3.** True **RC4.** True **1.** $\{9, 11\}$
3. $\{b\}$ **5.** $\{9, 10, 11, 13\}$ **7.** $\{a, b, c, d, f, g\}$ **9.** $\varnothing$
11. $\{3, 5, 7\}$ **13.** ; $(-4, 1]$
15. ; $(1, 6)$
17. $\{x\,|\,-4 \leq x < 5\}$, or $[-4, 5)$;
19. $\{x\,|\,x \geq 2\}$, or $[2, \infty)$; **21.** $\varnothing$
23. $\{x\,|\,-8 < x < 6\}$, or $(-8, 6)$
25. $\{x\,|\,-6 < x \leq 2\}$, or $(-6, 2]$
27. $\{x\,|\,-1 < x \leq 6\}$, or $(-1, 6]$
29. $\{y\,|\,-1 < y \leq 5\}$, or $(-1, 5]$
31. $\left\{x\,\middle|\,-\frac{5}{3} \leq x \leq \frac{4}{3}\right\}$, or $\left[-\frac{5}{3}, \frac{4}{3}\right]$
33. $\left\{x\,\middle|\,-\frac{7}{2} < x \leq \frac{11}{2}\right\}$, or $\left(-\frac{7}{2}, \frac{11}{2}\right]$
35. $\{x\,|\,10 < x \leq 14\}$, or $(10, 14]$
37. $\left\{x\,\middle|\,-\frac{13}{3} \leq x \leq 9\right\}$, or $\left[-\frac{13}{3}, 9\right]$
39. ; $(-\infty, -2) \cup (1, \infty)$
41. ; $(-\infty, -3] \cup (1, \infty)$
43. $\{x\,|\,x < -5\ or\ x > -1\}$, or $(-\infty, -5) \cup (-1, \infty)$;
45. $\left\{x\,\middle|\,x \leq \frac{5}{2}\ or\ x \geq 4\right\}$, or $\left(-\infty, \frac{5}{2}\right] \cup [4, \infty)$;
47. $\{x\,|\,x \geq -3\}$, or $[-3, \infty)$;
49. $\left\{x\,\middle|\,x \leq -\frac{5}{4}\ or\ x > -\frac{1}{2}\right\}$, or $\left(-\infty, -\frac{5}{4}\right] \cup \left(-\frac{1}{2}, \infty\right)$
51. All real numbers, or $(-\infty, \infty)$
53. $\{x\,|\,x < -4\ or\ x > 2\}$, or $(-\infty, -4) \cup (2, \infty)$
55. $\left\{x\,\middle|\,x < \frac{79}{4}\ or\ x > \frac{89}{4}\right\}$, or $\left(-\infty, \frac{79}{4}\right) \cup \left(\frac{89}{4}, \infty\right)$
57. $\left\{x\,\middle|\,x \leq -\frac{13}{2}\ or\ x \geq \frac{29}{2}\right\}$, or $\left(-\infty, -\frac{13}{2}\right] \cup \left[\frac{29}{2}, \infty\right)$
59. $\{d\,|\,0\,\text{ft} \leq d \leq 198\,\text{ft}\}$ **61.** Between 23 beats and 27 beats
63. $\{W\,|\,101.2\,\text{lb} \leq W \leq 136.2\,\text{lb}\}$
65. $\{d\,|\,250\,\text{mg} < d < 500\,\text{mg}\}$ **67.** $(-1, 2)$ **68.** $(-3, -5)$
69. $(4, -4)$ **70.** $y = -11x + 29$ **71.** $y = -4x + 7$
72. $y = -x + 2$ **73.** $6a^2 + 7ab - 5b^2$ **74.** $25y^2 + 35y + 6$
75. $21x^2 - 59x + 40$ **76.** $13x^2 + 37xy - 6y^2$
77. $\{x\,|\,-4 < x \leq 1\}$, or $(-4, 1]$ **79.** $\left\{x\,\middle|\,\frac{2}{5} \leq x \leq \frac{4}{5}\right\}$, or $\left[\frac{2}{5}, \frac{4}{5}\right]$
81. $\{x\,|\,10 < x \leq 18\}$, or $(10, 18]$ **83.** True **85.** False

Mid-Chapter Review: Chapter 18, p. 1202

1. True **2.** False **3.** True **4.** True
5. $8 - 5x \leq x + 20$ **6.** $-17 < 3 - x < 36$
${-5x} \leq x + 12$ $\phantom{-17 <}{-20} < -x < 33$
${-6x} \leq 12$ $\phantom{-17 <}{20} > x > -33$
${x} \geq -2$
7. G **8.** B **9.** H **10.** E **11.** F **12.** A **13.** $\{0, 10\}$
14. $\{b, d, e, f, g, h\}$ **15.** $\left\{\frac{1}{4}, \frac{3}{8}\right\}$ **16.** $\varnothing$
17. $\{y\,|\,y \leq -2\}$, or $(-\infty, -2]$ **18.** $\{x\,|\,x \leq 4\}$, or $(-\infty, 4]$
19. $\{x\,|\,x < -9\ or\ x > 1\}$, or $(-\infty, -9) \cup (1, \infty)$
20. $\{x\,|\,3 \leq x < 24\}$, or $[3, 24)$ **21.** $\{x\,|\,x < -10\ or\ x > 1\}$, or
$(-\infty, -10) \cup (1, \infty)$ **22.** $\{t\,|\,t > 2\}$, or $(2, \infty)$
23. $\{x\,|\,x \geq -3\}$, or $[-3, \infty)$ **24.** $\left\{y\,\middle|\,y > \frac{10}{11}\right\}$, or $\left(\frac{10}{11}, \infty\right)$
25. $\left\{x\,\middle|\,-\frac{17}{2} < x \leq \frac{25}{2}\right\}$, or $\left(-\frac{17}{2}, \frac{25}{2}\right)$ **26.** $\left\{x\,\middle|\,x \geq \frac{35}{6}\right\}$, or $\left[\frac{35}{6}, \infty\right)$
27. $\{x\,|\,4 < x < 17\}$, or $(4, 17)$ **28.** $\{x\,|\,x\ \text{is a real number}\}$, or
$(-\infty, \infty)$ **29.** $\{S\,|\,S \geq 86\}$ **30.** \$3000
31. When the signs of the quantities on either side of the
inequality symbol are changed, their relative positions on the
number line are reversed. **32. (1)** $-9(x + 2) = -9x - 18$, not
$-9x + 2$. **(2)** The left side should be $7 - 3x$. The inequality symbol
should not have been reversed. Using the incorrect right side in
step (1), this should now be $2 + x$. **(3)** If (2) were correct, the right-
hand side would be -5, not 8. **(4)** The inequality symbol should be
reversed. The correct solution is

$$7 - 9x + 6x < -9(x + 2) + 10x$$
$$7 - 9x + 6x < -9x - 18 + 10x$$
$$7 - 3x < x - 18$$
$$-4x < -25$$
$$x > \tfrac{25}{4}.$$

33. By definition, the notation $3 < x < 5$ indicates that $3 < x\ and$
$x < 5$. A solution of the disjunction $3 < x\ or\ x < 5$ must be in
at least one of these sets but not necessarily in both, so the
disjunction cannot be written as $3 < x < 5$.

Exercise Set 18.3, p. 1211

RC1. (f) **RC2.** (b) **RC3.** (e) **RC4.** (c) **RC5.** (a)
RC6. (d) **1.** $9|x|$ **3.** $2x^2$ **5.** $2x^2$ **7.** $6|y|$ **9.** $\dfrac{2}{|x|}$
11. $\dfrac{x^2}{|y|}$ **13.** $4|x|$ **15.** $\dfrac{y^2}{3}$ **17.** 38 **19.** 19 **21.** 6.3
23. 5 **25.** $\{-3, 3\}$ **27.** $\varnothing$ **29.** $\{0\}$ **31.** $\{-9, 15\}$
33. $\left\{-\frac{1}{2}, \frac{7}{2}\right\}$ **35.** $\left\{-\frac{5}{4}, \frac{23}{4}\right\}$ **37.** $\{-11, 11\}$ **39.** $\{-291, 291\}$
41. $\{-8, 8\}$ **43.** $\{-7, 7\}$ **45.** $\{-2, 2\}$ **47.** $\{-7, 8\}$
49. $\{-12, 2\}$ **51.** $\left\{-\frac{5}{2}, \frac{7}{2}\right\}$ **53.** $\varnothing$ **55.** $\left\{-\frac{13}{54}, -\frac{7}{54}\right\}$
57. $\left\{-\frac{11}{2}, \frac{3}{4}\right\}$ **59.** $\left\{\frac{3}{2}\right\}$ **61.** $\left\{5, -\frac{3}{5}\right\}$ **63.** All real numbers
65. $\left\{-\frac{3}{2}\right\}$ **67.** $\left\{\frac{24}{23}, 0\right\}$ **69.** $\left\{32, \frac{8}{3}\right\}$ **71.** $\{x\,|\,-3 < x < 3\}$,
or $(-3, 3)$ **73.** $\{x\,|\,x \leq -2\ or\ x \geq 2\}$, or $(-\infty, -2] \cup [2, \infty)$
75. $\{x\,|\,0 < x < 2\}$, or $(0, 2)$ **77.** $\{x\,|\,-6 \leq x \leq -2\}$, or
$[-6, -2]$ **79.** $\left\{x\,\middle|\,-\frac{1}{2} \leq x \leq \frac{7}{2}\right\}$, or $\left[-\frac{1}{2}, \frac{7}{2}\right]$
81. $\left\{y\,\middle|\,y < -\frac{3}{2}\ or\ y > \frac{17}{2}\right\}$, or $\left(-\infty, -\frac{3}{2}\right) \cup \left(\frac{17}{2}, \infty\right)$
83. $\left\{x\,\middle|\,x \leq -\frac{5}{4}\ or\ x \geq \frac{23}{4}\right\}$, or $\left(-\infty, -\frac{5}{4}\right] \cup \left[\frac{23}{4}, \infty\right)$
85. $\{y\,|\,-9 < y < 15\}$, or $(-9, 15)$ **87.** $\left\{x\,\middle|\,-\frac{7}{2} \leq x \leq \frac{1}{2}\right\}$, or
$\left[-\frac{7}{2}, \frac{1}{2}\right]$ **89.** $\left\{y\,\middle|\,y < -\frac{4}{3}\ or\ y > 4\right\}$, or $\left(-\infty, -\frac{4}{3}\right) \cup (4, \infty)$
91. $\left\{x\,\middle|\,x \leq -\frac{5}{4}\ or\ x \geq \frac{23}{4}\right\}$, or $\left(-\infty, -\frac{5}{4}\right] \cup \left[\frac{23}{4}, \infty\right)$
93. $\left\{x\,\middle|\,-\frac{9}{2} < x < 6\right\}$, or $\left(-\frac{9}{2}, 6\right)$ **95.** $\left\{x\,\middle|\,x \leq -\frac{25}{6}\ or\ x \geq \frac{23}{6}\right\}$,
or $\left(-\infty, -\frac{25}{6}\right] \cup \left[\frac{23}{6}, \infty\right)$ **97.** $\{x\,|\,-5 < x < 19\}$, or $(-5, 19)$
99. $\left\{x\,\middle|\,x \leq -\frac{2}{15}\ or\ x \geq \frac{14}{15}\right\}$, or $\left(-\infty, -\frac{2}{15}\right] \cup \left[\frac{14}{15}, \infty\right)$
101. $\{m\,|\,-12 \leq m \leq 2\}$, or $[-12, 2]$ **103.** $\left\{x\,\middle|\,\frac{1}{2} \leq x \leq \frac{5}{2}\right\}$,
or $\left[\frac{1}{2}, \frac{5}{2}\right]$ **105.** $\{x\,|\,-1 \leq x \leq 2\}$, or $[-1, 2]$ **107.** $\frac{1}{11}$
108. Not defined **109.** 2 **110.** 0
111. $(3w - 10)(9w^2 + 30w + 100)$

112. $(2 + 5t)(4 - 10t + 25t^2)$

113. $\dfrac{3w^2}{w + z}$ **114.** $\dfrac{5}{3}$ **115.** $\left\{ d \mid 5\frac{1}{2}\,\text{ft} \le d \le 6\frac{1}{2}\,\text{ft} \right\}$

117. All real numbers **119.** $\left\{ 1, -\frac{1}{4} \right\}$ **121.** $\varnothing$

123. $|x| < 3$ **125.** $|x| \ge 6$ **127.** $|x + 3| > 5$

Visualizing for Success, p. 1224

1. D **2.** B **3.** E **4.** C **5.** I **6.** G **7.** F **8.** H
9. A **10.** J

Exercise Set 18.4, p. 1225

RC1. (d) **RC2.** (e) **RC3.** (f) **RC4.** (c) **RC5.** (a)
RC6. (b) **1.** Yes **3.** Yes

5. 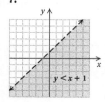 **7.** 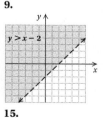 **9.**

11. 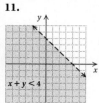 **13.** 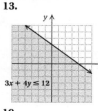 **15.**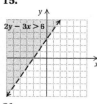

17. 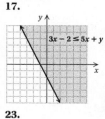 **19.** 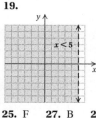 **21.**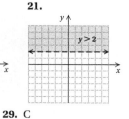

23.   **25.** F **27.** B **29.** C

31. 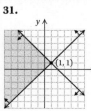 **33.** 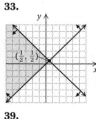 **35.**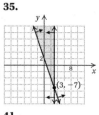

37. 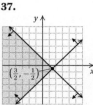 **39.** 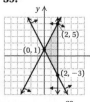 **41.**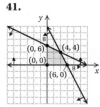

43. $\dfrac{10}{17}$ **44.** $-\dfrac{14}{13}$ **45.** -2 **46.** $\dfrac{29}{11}$ **47.** -12 **48.** $\dfrac{333}{245}$
49. 2 **50.** 3 **51.** 1 **52.** 8 **53.** 4 **54.** $|2 - 2a|$, or $2|1 - a|$ **55.** 6 **56.** 0.2

57. $w > 0$,
$h > 0$,
$w + h + 30 \le 62$, or
$w + h \le 32$,
$2w + 2h + 30 \le 130$, or
$w + h \le 50$

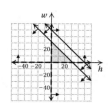

Summary and Review: Chapter 18, p. 1229

Vocabulary Reinforcement

1. Inequality **2.** Set-builder **3.** Interval **4.** Intersection
5. Conjunction **6.** Empty set **7.** Disjoint sets **8.** Union
9. Disjunction **10.** At most **11.** At least **12.** Distance

Concept Reinforcement

1. True **2.** False **3.** False **4.** False **5.** True **6.** False
7. True

Study Guide

1. -2 is not a solution; 5 is a solution. **2. (a)** $(-\infty, -8)$;
(b) $[-7, 10)$; **(c)** $[3, \infty)$ **3.** $\{y \mid y < -2\}$, or $(-\infty, -2)$;

4. $\{z \mid -2 \le z < 1\}$, or $[-2, 1)$;

5. $\{z \mid z < -1 \text{ or } z \ge 1\}$, or $(-\infty, -1) \cup [1, \infty)$;

6. $\left\{ -\frac{8}{5}, 2 \right\}$ **7.** $\left\{ 3, -\frac{1}{2} \right\}$ **8. (a)** $\{x \mid -4 < x < 1\}$, or $(-4, 1)$;
(b) $\left\{ x \mid x \le -\frac{10}{3} \text{ or } x \ge 2 \right\}$, or $\left(-\infty, -\frac{10}{3} \right] \cup [2, \infty)$

9. 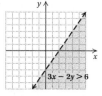 **10.**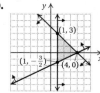

Review Exercises

1. -3 is not a solution; 7 is a solution. **2.** $[-8, 9)$ **3.** $(-\infty, 40]$

4. $\longleftrightarrow$; $(-\infty, -2]$

5. $\longleftrightarrow$; $(1, \infty)$

6. $\{a \mid a \le -21\}$, or $(-\infty, -21]$ **7.** $\{y \mid y \ge -7\}$, or $[-7, \infty)$
8. $\{y \mid y > -4\}$, or $(-4, \infty)$ **9.** $\{y \mid y > -30\}$, or $(-30, \infty)$
10. $\{x \mid x > -3\}$, or $(-3, \infty)$ **11.** $\left\{ y \mid y \le -\frac{6}{5} \right\}$, or $\left(-\infty, -\frac{6}{5} \right]$
12. $\{x \mid x < -3\}$, or $(-\infty, -3)$ **13.** $\{y \mid y > -10\}$, or $(-10, \infty)$
14. $\left\{ x \mid x \le -\frac{5}{2} \right\}$, or $\left(-\infty, -\frac{5}{2} \right]$ **15.** $\left\{ t \mid t > 4\frac{1}{4}\,\text{hr} \right\}$
16. \$10,000 **17.** $\longleftrightarrow$; $[-2, 5)$
18. $\longleftrightarrow$; $(-\infty, -2] \cup (5, \infty)$
19. $\{1, 5, 9\}$ **20.** $\{1, 2, 3, 5, 6, 9\}$ **21.** $\varnothing$
22. $\{x \mid -7 < x \le 2\}$, or $(-7, 2]$ **23.** $\left\{ x \mid -\frac{5}{4} < x < \frac{5}{2} \right\}$, or
$\left(-\frac{5}{4}, \frac{5}{2} \right)$ **24.** $\{x \mid x < -3 \text{ or } x > 1\}$, or $(-\infty, -3) \cup (1, \infty)$
25. $\{x \mid x < -11 \text{ or } x \ge -6\}$, or $(-\infty, -11) \cup [-6, \infty)$
26. $\{x \mid x \le -6 \text{ or } x \ge 8\}$, or $(-\infty, -6] \cup [8, \infty)$
27. $\dfrac{3}{|x|}$ **28.** $\dfrac{2|x|}{y^2}$ **29.** $\dfrac{4}{|y|}$ **30.** 62 **31.** $\{-6, 6\}$
32. $\{-5, 9\}$ **33.** $\left\{ -14, \frac{4}{3} \right\}$ **34.** $\varnothing$ **35.** $\left\{ x \mid -\frac{17}{2} < x < \frac{7}{2} \right\}$,
or $\left(-\frac{17}{2}, \frac{7}{2} \right)$ **36.** $\{x \mid x \le -3.5 \text{ or } x \ge 3.5\}$, or
$(-\infty, -3.5] \cup [3.5, \infty)$ **37.** $\left\{ x \mid x \le -\frac{11}{3} \text{ or } x \ge \frac{19}{3} \right\}$, or
$\left(-\infty, -\frac{11}{3} \right] \cup \left[\frac{19}{3}, \infty \right)$ **38.** $\varnothing$

39. 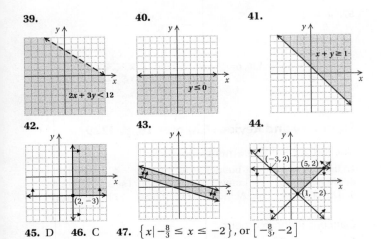 **40.** **41.**

42. 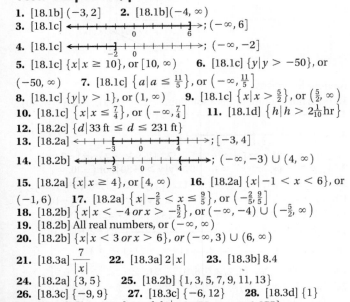 **43.** **44.**

34. [18.4b]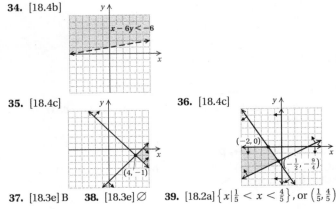

35. [18.4c] **36.** [18.4c]

37. [18.3e] B **38.** [18.3e] ∅ **39.** [18.2a] $\left\{x\mid\frac{1}{5}<x<\frac{4}{5}\right\}$, or $\left(\frac{1}{5},\frac{4}{5}\right)$

45. D **46.** C **47.** $\left\{x\mid -\frac{8}{3}\le x\le -2\right\}$, or $\left[-\frac{8}{3},-2\right]$

Understanding Through Discussion and Writing

1. When $b\ge c$, then the intervals overlap and $[a,b]\cup[c,d]=[a,d]$. **2.** The distance between x and -5 is $|x-(-5)|$, or $|x+5|$. Then the solutions of the inequality $|x+5|\le 2$ can be interpreted as "all those numbers x whose distance from -5 is at most 2 units." **3.** No; the symbol $\ge$ does not always yield a graph in which the half-plane above the line is shaded. For the inequality $-y\ge 3$, for example, the half-plane below the line $y=-3$ is shaded. **4.** The solutions of $|x|\ge 6$ are those numbers whose distance from 0 is greater than or equal to 6. In addition to the numbers in $[6,\infty)$, the distance from 0 of the numbers in $(-\infty,-6]$ is also greater than or equal to 6. Thus, $[6,\infty)$ is only part of the solution of the inequality.

Test: Chapter 18, p. 1235

1. [18.1b] $(-3,2]$ **2.** [18.1b] $(-4,\infty)$
3. [18.1c] ⟵┼┼┼┼┼┼┼┼┼┼┤⟶; $(-\infty,6]$
        0   6
4. [18.1c] ⟵┼┼┼┤┼┼┼┼┼┼┼┼⟶; $(-\infty,-2]$
        -2 0
5. [18.1c] $\{x\mid x\ge 10\}$, or $[10,\infty)$ **6.** [18.1c] $\{y\mid y>-50\}$, or $(-50,\infty)$ **7.** [18.1c] $\left\{a\mid a\le \frac{11}{5}\right\}$, or $\left(-\infty,\frac{11}{5}\right]$
8. [18.1c] $\{y\mid y>1\}$, or $(1,\infty)$ **9.** [18.1c] $\left\{x\mid x>\frac{5}{2}\right\}$, or $\left(\frac{5}{2},\infty\right)$
10. [18.1c] $\left\{x\mid x\le \frac{7}{4}\right\}$, or $\left(-\infty,\frac{7}{4}\right]$ **11.** [18.1d] $\left\{h\mid h>2\frac{1}{10}\,\text{hr}\right\}$
12. [18.2c] $\{d\mid 33\,\text{ft}\le d\le 231\,\text{ft}\}$
13. [18.2a] ⟵┼┼┼[┼┼┼┼┼]┼┼⟶; $[-3,4]$
        -3   0   4
14. [18.2b] ⟵┼┼┼)┼┼┼┼┼(┼┼⟶; $(-\infty,-3)\cup(4,\infty)$
        -3   0   4
15. [18.2a] $\{x\mid x\ge 4\}$, or $[4,\infty)$ **16.** [18.2a] $\{x\mid -1<x<6\}$, or $(-1,6)$ **17.** [18.2a] $\left\{x\mid -\frac{2}{5}<x\le \frac{9}{5}\right\}$, or $\left(-\frac{2}{5},\frac{9}{5}\right]$
18. [18.2b] $\left\{x\mid x<-4\;or\;x>-\frac{5}{2}\right\}$, or $(-\infty,-4)\cup\left(-\frac{5}{2},\infty\right)$
19. [18.2b] All real numbers, or $(-\infty,\infty)$
20. [18.2b] $\{x\mid x<3\;or\;x>6\}$, or $(-\infty,3)\cup(6,\infty)$
21. [18.3a] $\dfrac{7}{|x|}$ **22.** [18.3a] $2|x|$ **23.** [18.3b] 8.4
24. [18.2a] $\{3,5\}$ **25.** [18.2b] $\{1,3,5,7,9,11,13\}$
26. [18.3c] $\{-9,9\}$ **27.** [18.3c] $\{-6,12\}$ **28.** [18.3d] $\{1\}$
29. [18.3c] ∅ **30.** [18.3e] $\{x\mid -0.875<x<1.375\}$, or $(-0.875,1.375)$ **31.** [18.3e] $\{x\mid x<-3\;or\;x>3\}$, or $(-\infty,-3)\cup(3,\infty)$ **32.** [18.3e] $\{x\mid -99\le x\le 111\}$, or $[-99,111]$ **33.** [18.3e] $\left\{x\mid x\le -\frac{13}{5}\;or\;x\ge \frac{7}{5}\right\}$, or $\left(-\infty,-\frac{13}{5}\right]\cup\left[\frac{7}{5},\infty\right)$